Innovative Statistics in Regulatory Science

Chapman & Hall/CRC Biostatistics Series

Shein-Chung Chow, Duke University School of Medicine
Byron Jones, Novartis Pharma AG
Jen-pei Liu, National Taiwan University
Karl E. Peace, Georgia Southern University
Bruce W. Turnbull, Cornell University

Recently Published Titles

Cancer Clinical Trials: Current and Controversial Issues in Design and Analysis
Stephen L. George, Xiaofei Wang, Herbert Pang

Data and Safety Monitoring Committees in Clinical Trials 2nd Edition
Jay Herson

Clinical Trial Optimization Using R
Alex Dmitrienko, Erik Pulkstenis

Mixture Modelling for Medical and Health Sciences
Shu-Kay Ng, Liming Xiang, Kelvin Kai Wing Yau

Economic Evaluation of Cancer Drugs: Using Clinical Trial and Real-World Data
Iftekhar Khan, Ralph Crott, Zahid Bashir

Bayesian Analysis with R for Biopharmaceuticals: Concepts, Algorithms, and Case Studies
Harry Yang and Steven J. Novick

Mathematical and Statistical Skills in the Biopharmaceutical Industry: A Pragmatic Approach
Arkadiy Pitman, Oleksandr Sverdlov, L. Bruce Pearce

Bayesian Applications in Pharmaceutical Development
Mani Lakshminarayanan, Fanni Natanegara

Innovative Statistics in Regulatory Science
Shein-Chung Chow

For more information about this series, please visit:
https://www.crcpress.com/go/biostats

Innovative Statistics in Regulatory Science

Shein-Chung Chow
Duke University School of Medicine
Durham, North Carolina

CRC Press
Taylor & Francis Group
Boca Raton London New York

CRC Press is an imprint of the
Taylor & Francis Group, an informa business
A CHAPMAN & HALL BOOK

CRC Press
Taylor & Francis Group
6000 Broken Sound Parkway NW, Suite 300
Boca Raton, FL 33487-2742

First issued in paperback 2021

© 2020 by Taylor & Francis Group, LLC
CRC Press is an imprint of Taylor & Francis Group, an Informa business

No claim to original U.S. Government works

ISBN-13: 978-0-367-22476-9 (hbk)
ISBN-13: 978-1-03-208653-8 (pbk)

Visit the Taylor & Francis Web site at
http://www.taylorandfrancis.com

and the CRC Press Web site at
http://www.crcpress.com

Contents

Preface xvii
Author xxi

1. **Introduction** 1
   1.1 Introduction 1
   1.2 Key Statistical Concepts 2
        1.2.1 Confounding and Interaction 2
               1.2.1.1 Confounding 3
               1.2.1.2 Interaction 6
        1.2.2 Hypotheses Testing and p-values 8
               1.2.2.1 Hypotheses Testing 8
               1.2.2.2 p-value 10
        1.2.3 One-Sided versus Two-Sided Hypotheses 12
        1.2.4 Clinical Significance and Clinical Equivalence 14
        1.2.5 Reproducibility and Generalizability 17
               1.2.5.1 Reproducibility 17
               1.2.5.2 Generalizability 19
   1.3 Complex Innovative Designs 22
        1.3.1 Adaptive Trial Design 22
               1.3.1.1 Adaptations 23
               1.3.1.2 Types of Adaptive Design 24
        1.3.2 The n-of-1 Trial Design 24
               1.3.2.1 Complete n-of-1 Trial Design 24
               1.3.2.2 Merits and Limitations 25
        1.3.3 The Concept of Master Protocols 26
        1.3.4 Bayesian Approach 26
   1.4 Practical, Challenging, and Controversial Issues 27
        1.4.1 Totality-of-the-Evidence 27
        1.4.2 $(1-\alpha)$ CI for New Drugs versus $(1-2\alpha)$ CI for
               Generics/Biosimilars 28
        1.4.3 Endpoint Selection 29
        1.4.4 Criteria for Decision-Making at Interim 30
        1.4.5 Non-inferiority or Equivalence/Similarity Margin
               Selection 31
        1.4.6 Treatment of Missing Data 32
        1.4.7 Sample Size Requirement 36
        1.4.8 Consistency Test 37

1.4.9   Extrapolation...38
1.4.10   Drug Products with Multiple Components..........................39
1.4.11   Advisory Committee..40
1.4.12   Recent FDA Critical Clinical Initiatives...........................41
1.5   Aim and Scope of the Book...43

2. Totality-of-the-Evidence..47
2.1   Introduction...47
2.2   Substantial Evidence 48
2.3   Totality-of-the-Evidence..49
       2.3.1   Stepwise Approach...49
       2.3.2   Fundamental Biosimilarity Assumptions.......................51
       2.3.3   Examples—Recent Biosimilar Regulatory Submissions 52
       2.3.4   Remarks..53
2.4   Practical Issues and Challenges..54
       2.4.1   Link among Analytical Similarity, PK/PD Similarity,
                and Clinical Similarity 54
       2.4.2   Totality-of-the-Evidence versus Substantial Evidence.......57
       2.4.3   Same Regulatory Standards 58
2.5   Development of Totality-of-the-Evidence.................................59
2.6   Concluding Remarks...63

3. Hypotheses Testing versus Confidence Interval 65
3.1   Introduction...65
3.2   Hypotheses Testing 66
       3.2.1   Point Hypotheses Testing 67
       3.2.2   Interval Hypotheses Testing...68
       3.2.3   Probability of Inconclusiveness.....................................69
3.3   Confidence Interval Approach...69
       3.3.1   Confidence Interval Approach with Single Reference 69
       3.3.2   Confidence Interval Approach with Multiple
                References 70
                3.3.2.1   Pairwise Comparisons 70
                3.3.2.2   Simultaneous Confidence Interval 70
                3.3.2.3   Example 1 (False Negative)...........................71
                3.3.2.4   Example 2 (False Positive)............................72
3.4   Two One-Sided Tests versus Confidence Interval Approach.........76
       3.4.1   Two One-Sided Tests (TOST) Procedure...........................76
       3.4.2   Confidence Interval Approach..78
                3.4.2.1   Level $1 - \alpha$ versus Level $1 - 2\alpha$..................78
                3.4.2.2   Significance Level versus Size 79
                3.4.2.3   Sizes of Tests Related to Different Confidence
                          Intervals...79
       3.4.3   Remarks..80

3.5   A Comparison 81
    3.5.1   Performance Characteristics 81
    3.5.2   Simulation Studies 82
    3.5.3   An Example—Binary Responses 86
3.6   Sample Size Requirement 89
3.7   Concluding Remarks 90
Appendix 91

4. Endpoint Selection 93
4.1   Introduction 93
4.2   Clinical Strategy for Endpoint Selection 95
4.3   Translations among Clinical Endpoints 96
4.4   Comparison of Different Clinical Strategies 99
    4.4.1   Test Statistics, Power and Sample Size Determination 99
    4.4.2   Determination of the Non-inferiority Margin 102
    4.4.3   A Numerical Study 102
        4.4.3.1   Absolute Difference versus Relative Difference 103
        4.4.3.2   Responders' Rate Based on Absolute Difference 103
        4.4.3.3   Responders' Rate Based on Relative Difference 103
4.5   Development of Therapeutic Index Function 110
    4.5.1   Introduction 110
    4.5.2   Therapeutic Index Function 114
        4.5.2.1   Selection of ω_i 114
        4.5.2.2   Determination of $f_i(\cdot)$ and the Distribution of e 115
        4.5.2.3   Derivation of $\Pr(I_i \,|\, e_j)$ and $\Pr(e_j \,|\, I_i)$ 115
4.6   Concluding Remarks 121

5. Non-inferiority/Equivalence Margin 123
5.1   Introduction 123
5.2   Non-inferiority versus Equivalence 124
    5.2.1   Relationship among Non-inferiority, Equivalence,
        and Superiority 125
    5.2.2   Impact on Sample Size Requirement 126
5.3   Non-inferiority Hypothesis 127
    5.3.1   Regulatory Requirements 127
    5.3.2   Hypothesis Setting and Clinically Meaningful Margin 128
    5.3.3   Retention of Treatment Effect in the Absence of Placebo 129
5.4   Methods for Selection of Non-inferiority Margin 130
    5.4.1   Classical Method 130
    5.4.2   FDA's Recommendations 130
    5.4.3   Chow and Shao's Method 131
    5.4.4   Alternative Methods 132
    5.4.5   An Example 133
    5.4.6   Remarks 135

5.5 Strategy for Margin Selection.. 135
    5.5.1 Criteria for Risk Assessment 136
    5.5.2 Risk Assessment with Continuous Endpoints.................... 138
    5.5.3 Numerical Studies.. 143
    5.5.4 An Example.. 149
5.6 Concluding Remarks 151

6. Missing Data.. 153
6.1 Introduction 153
6.2 Missing Data Imputation 155
    6.2.1 Last Observation Carried Forward 155
        6.2.1.1 Bias-variance Trade-off................................ 156
        6.2.1.2 Hypothesis Testing...................................... 157
    6.2.2 Mean/Median Imputation... 158
    6.2.3 Regression Imputation 159
6.3 Marginal/Conditional Imputation for Contingency 159
    6.3.1 Simple Random Sampling 160
    6.3.2 Goodness-of-Fit Test 161
6.4 Test for Independence 162
    6.4.1 Results Under Stratified Simple Random Sampling.......... 162
    6.4.2 When Number of Strata Is Large................................. 163
6.5 Recent Development... 164
    6.5.1 Other Methods for Missing Data................................. 164
    6.5.2 The Use of Estimand in Missing Data 165
    6.5.3 Statistical Methods Under Incomplete Data Structure...... 166
        6.5.3.1 Introduction.. 166
        6.5.3.2 Statistical Methods for 2×3 Crossover Designs
            with Incomplete Data................................ 168
        6.5.3.3 A Special Case... 172
        6.5.3.4 An Example 174
6.6 Concluding Remarks.. 176

7. Multiplicity 179
7.1 General Concepts 179
7.2 Regulatory Perspective and Controversial Issues...................... 180
    7.2.1 Regulatory Perspectives... 180
    7.2.2 Controversial Issues... 181
7.3 Statistical Method for Adjustment of Multiplicity 182
    7.3.1 Bonferroni Method 183
    7.3.2 Tukey's Multiple Range Testing Procedure.................... 184
    7.3.3 Dunnett's Test.. 184
    7.3.4 Closed Testing Procedure 185
    7.3.5 Other Tests 186

7.4   Gate-Keeping Procedures 187
        7.4.1   Multiple Endpoints 187
        7.4.2   Gate-Keeping Testing Procedures 188
7.5   Concluding Remarks 192

8. Sample Size 195
8.1   Introduction 195
8.2   Traditional Sample Size Calculation 196
8.3   Selection of Study Endpoints 200
        8.3.1   Translations among Clinical Endpoints 200
        8.3.2   Comparison of Different Clinical Strategies 202
8.4   Multiple-stage Adaptive Designs 205
8.5   Sample Size Adjustment with Protocol Amendments 208
8.6   Multi-regional Clinical Trials 211
8.7   Current Issues 214
        8.7.1   Is Power Calculation the Only Way? 214
        8.7.2   Instability of Sample Size 215
        8.7.3   Sample Size Adjustment for Protocol Amendment 216
        8.7.4   Sample Size Based on Confidence Interval Approach 216
8.8   Concluding Remarks 217

9. Reproducible Research 219
9.1   Introduction 219
9.2   The Concept of Reproducibility Probability 220
9.3   The Estimated Power Approach 222
        9.3.1   Two Samples with Equal Variances 222
        9.3.2   Two Samples with Unequal Variances 225
        9.3.3   Parallel-Group Designs 227
9.4   Alternative Methods for Evaluation of Reproducibility
        Probability 228
        9.4.1   The Confidence Bound Approach 228
        9.4.2   The Bayesian Approach 230
9.5   Applications 235
        9.5.1   Substantial Evidence with a Single Trial 235
        9.5.2   Sample Size 236
        9.5.3   Generalizability between Patient Populations 236
9.6   Future Perspectives 240

10. Extrapolation 241
10.1   Introduction 241
10.2   Shift in Target Patient Population 242

10.3   Assessment of Sensitivity Index 244
        10.3.1   The Case Where ε Is Random and C Is Fixed 244
        10.3.2   The Case Where ε Is Fixed and C Is Random 247
        10.3.3   The Case Where Both ε and C Are Random 250
10.4   Statistical Inference 253
        10.4.1   The Case Where ε Is Random and C Is Fixed 254
        10.4.2   The Case Where ε Is Fixed and C Is Random 255
        10.4.3   The Case Where ε and C Are Random 256
10.5   An Example 258
        10.5.1   Case 1: ε Is Random and C Is Fixed 258
        10.5.2   Case 2: ε Is Fixed and C Is Random 259
        10.5.3   Case 3: ε and C Are Both Random 259
10.6   Concluding Remarks 259
Appendix 260

11. **Consistency Evaluation** 263
11.1   Introduction 263
11.2   Issues in Multi-regional Clinical Trials 264
        11.2.1   Multi-center Trials 264
        11.2.2   Multi-regional, Multi-center Trials 265
11.3   Statistical Methods 266
        11.3.1   Test for Consistency 266
        11.3.2   Assessment of Consistency Index 267
        11.3.3   Evaluation of Sensitivity Index 269
        11.3.4   Achieving Reproducibility and/or Generalizability 270
                 11.3.4.1   Specificity Reproducibility Probability for
                            Inequality Test 270
                 11.3.4.2   Superiority Reproducibility Probability 271
                 11.3.4.3   Reproducibility Probability Ratio for
                            Inequality Test 272
                 11.3.4.4   Reproducibility Probability Ratio for
                            Superiority Test 273
        11.3.5   Bayesian Approach 273
        11.3.6   Japanese Approach 275
        11.3.7   The Applicability of Those Approaches 275
11.4   Simulation Study 276
        11.4.1   The Case of the Matched-Pair Parallel Design with
                 Normal Data and Superiority Test 276
        11.4.2   The Case of the Two-Group Parallel Design with
                 Normal Data and Superiority Test 281
        11.4.3   Remarks 285
11.5   An Example 286
11.6   Other Considerations/Discussions 290
11.7   Concluding Remarks 291

12. Drug Products with Multiple Components—Development of TCM 293
  12.1   Introduction 293
  12.2   Fundamental Differences 295
       12.2.1 Medical Theory/Mechanism and Practice 295
              12.2.1.1 Medical Practice 296
       12.2.2 Techniques of Diagnosis 297
              12.2.2.1 Objective versus Subjective Criteria for Evaluability 297
       12.2.3 Treatment 298
              12.2.3.1 Single Active Ingredient versus Multiple Components 298
              12.2.3.2 Fixed Dose versus Flexible Dose 298
       12.2.4 Remarks 299
  12.3   Basic Considerations 300
       12.3.1 Study Design 300
       12.3.2 Validation of Quantitative Instrument 301
       12.3.3 Clinical Endpoint 302
       12.3.4 Matching Placebo 303
       12.3.5 Sample Size Calculation 303
  12.4   TCM Drug Development 304
       12.4.1 Statistical Quality Control Method for Assessing Consistency 304
              12.4.1.1 Acceptance Criteria 308
              12.4.1.2 Sampling Plan 308
              12.4.1.3 Testing Procedure 311
              12.4.1.4 Strategy for Statistical Quality Control 311
              12.4.1.5 Remarks 315
       12.4.2 Stability Analysis 317
              12.4.2.1 Models and Assumptions 319
              12.4.2.2 Shelf-Life Determination 320
              12.4.2.3 An Example 321
              12.4.2.4 Discussion 323
       12.4.3 Calibration of Study Endpoints in Clinical Development 323
              12.4.3.1 Chinese Diagnostic Procedure 324
              12.4.3.2 Calibration 325
              12.4.3.3 Validity 326
              12.4.3.4 Reliability 328
              12.4.3.5 Ruggedness 329
  12.5   Challenging Issues 331
       12.5.1 Regulatory Requirements 331
       12.5.2 Test for Consistency 332
       12.5.3 Animal Studies 333

          12.5.4   Shelf-Life Estimation 333
          12.5.5   Indication and Label..334
    12.6   Recent Development...335
          12.6.1   Introduction 335
          12.6.2   Health Index and Efficacy Measure 336
          12.6.3   Assessment of Efficacy 336
          12.6.4   Remarks..339
    12.7   Concluding Remarks..339

13. **Adaptive Trial Design** 341
    13.1   Introduction...341
    13.2   What Is Adaptive Design? 343
          13.2.1   Adaptations..344
          13.2.2   Types of Adaptive Designs...344
                    13.2.2.1   Adaptive Randomization Design 344
                    13.2.2.2   Group Sequential Design...............................345
                    13.2.2.3   Flexible Sample Size Re-estimation
                                (SSRE) Design..346
                    13.2.2.4   Drop-the-Losers Design...................................346
                    13.2.2.5   Adaptive Dose Finding Design......................347
                    13.2.2.6   Biomarker-Adaptive Design 348
                    13.2.2.7   Adaptive Treatment-Switching Design.........349
                    13.2.2.8   Adaptive-Hypotheses Design 349
                    13.2.2.9   Seamless Adaptive Trial Design 350
                    13.2.2.10   Multiple Adaptive Design...............................350
    13.3   Regulatory/Statistical Perspectives...351
    13.4   Impact, Challenges, and Obstacles..352
          13.4.1   Impact of Protocol Amendments.....................................352
          13.4.2   Challenges in By Design Adaptations 352
          13.4.3   Obstacles of Retrospective Adaptations.........................354
    13.5   Some Examples 354
    13.6   Strategies for Clinical Development 363
    13.7   Concluding Remarks..364

14. **Criteria for Dose Selection** 367
    14.1   Introduction...367
    14.2   Dose Selection Criteria..368
          14.2.1   Conditional Power 369
          14.2.2   Precision Analysis Based on Confidence Interval..........370
          14.2.3   Predictive Probability of Success.....................................370
          14.2.4   Probability of Being the Best Dose 370
    14.3   Implementation and Example..371
          14.3.1   Single Primary Endpoint 371
          14.3.2   Co-primary Endpoints 372
          14.3.3   A Numeric Example 376

14.4 Clinical Trial Simulation..377
    14.4.1 Single Primary Endpoint...377
    14.4.2 Co-primary Endpoints...377
14.5 Concluding Remarks...386

15. Generics and Biosimilars...387
15.1 Introduction...387
15.2 Fundamental Differences..388
15.3 Quantitative Evaluation of Generic Drugs...............................389
    15.3.1 Study Design...390
    15.3.2 Statistical Methods...391
    15.3.3 Other Criteria for Bioequivalence Assessment.............392
        15.3.3.1 Population Bioequivalence and Individual
                    Bioequivalence (PBE/IBE).............................392
        15.3.3.2 Scaled Average Bioequivalence (SABE)..........392
        15.3.3.3 Scaled Criterion for Drug
                    Interchangeability (SCDI)..............................393
        15.3.3.4 Remarks..394
15.4 Quantitative Evaluation of Biosimilars....................................395
    15.4.1 Regulatory Requirement..395
    15.4.2 Biosimilarity...396
        15.4.2.1 Basic Principles...396
        15.4.2.2 Criteria for Biosimilarity................................397
        15.4.2.3 Study Design...397
        15.4.2.4 Statistical Methods..398
    15.4.3 Interchangeability...398
        15.4.3.1 Definition and Basic Concepts.......................398
        15.4.3.2 Switching and Alternating.............................399
        15.4.3.3 Study Design...399
    15.4.4 Remarks..400
15.5 General Approach for Assessment of Bioequivalence/
    Biosimilarity..400
    15.5.1 Development of Bioequivalence/Biosimilarity Index.....400
    15.5.2 Remarks..403
15.6 Scientific Factors and Practical Issues for Biosimilars.................404
    15.6.1 Fundamental Biosimilarity Assumption.........................404
    15.6.2 Endpoint Selection...405
    15.6.3 How Similar Is Similar?..405
    15.6.4 Guidance on Analytical Similarity Assessment..............405
    15.6.5 Practical Issues..406
        15.6.5.1 Criteria for Biosimilarity (in Terms of
                    Average, Variability, or Distribution)..............406
        15.6.5.2 Criteria for Interchangeability........................407
        15.6.5.3 Reference Product Changes.............................407
        15.6.5.4 Extrapolation..407

              15.6.5.5   Non-medical Switch..408

              15.6.5.6   Bridging Studies for Assessing
                          Biosimilarity...408

   15.7   Concluding Remarks..408

16. Precision Medicine.. **411**

   16.1   Introduction... 411

   16.2   The Concept of Precision Medicine.................................... 412

           16.2.1   Definition of Precision Medicine......................... 412

           16.2.2   Biomarker-Driven Clinical Trials........................ 412

           16.2.3   Precision Medicine versus Personalized Medicine 414

   16.3   Design and Analysis of Precision Medicine...................... 415

           16.3.1   Study Designs... 415

           16.3.2   Statistical Methods... 417

           16.3.3   Simulation Results 422

   16.4   Alternative Enrichment Designs 423

           16.4.1   Alternative Designs with/without Molecular Targets 423

           16.4.2   Statistical Methods...425

           16.4.3   Remarks...428

   16.5   Concluding Remarks..430

17. Big Data Analytics...433

   17.1   Introduction...433

   17.2   Basic Considerations..435

           17.2.1   Representativeness of Big Data 435

           17.2.2   Selection Bias 435

           17.2.3   Heterogeneity 435

           17.2.4   Reproducibility and Generalizability 436

           17.2.5   Data Quality, Integrity, and Validity..................436

           17.2.6   FDA Part 11 Compliance 437

           17.2.7   Missing Data...437

   17.3   Types of Big Data Analytics 438

           17.3.1   Case-Control Studies..438

                   17.3.1.1   Propensity Score Matching...............439

                   17.3.1.2   Model Building..................................439

                   17.3.1.3   Model Diagnosis and Validation......441

                   17.3.1.4   Model Generalizability.....................441

           17.3.2   Meta-analysis..442

                   17.3.2.1   Issues in Meta-analysis.....................443

   17.4   Bias of Big Data Analytics 444

   17.5   Statistical Methods for Estimation of Δ and $\mu_P - \mu_N$ 446

           17.5.1   Estimation of Δ 446

            17.5.2   Estimation of $\mu_P - \mu_N$ 448

            17.5.3   Assumptions and Application 448

17.6 Simulation Study..449
17.7 Concluding Remarks...454

18. Rare Diseases Drug Development..457
18.1 Introduction...457
18.2 Basic Considerations..458
    18.2.1 Historical Data...458
    18.2.2 Ethical Consideration 459
    18.2.3 The Use of Biomarkers 459
    18.2.4 Generalizability..460
    18.2.5 Sample Size 460
18.3 Innovative Trial Designs...461
    18.3.1 n-of-1 Trial Design...461
        18.3.1.1 Complete n-of-1 Trial Design.....................462
        18.3.1.2 Merits and Limitations 463
    18.3.2 Adaptive Trial Design 463
    18.3.3 Other Designs..464
        18.3.3.1 Master Protocol......................................464
        18.3.3.2 Bayesian Approach 466
18.4 Statistical Methods for Data Analysis....................................466
    18.4.1 Analysis under a Complete n-of-1 Trial Design 466
        18.4.1.1 Statistical Model 466
        18.4.1.2 Statistical Analysis 467
        18.4.1.3 Sample Size Requirement 468
    18.4.2 Analysis under an Adaptive Trial Design....................470
        18.4.2.1 Two-Stage Adaptive Design.......................473
        18.4.2.2 Remarks 476
18.5 Evaluation of Rare Disease Clinical Trials...............................476
    18.5.1 Predictive Confidence Interval (PCI) 477
    18.5.2 Probability of Reproducibility....................................477
18.6 Some Proposals for Regulatory Consideration 478
    18.6.1 Demonstrating *Effectiveness* or Demonstrating
        Not Ineffectiveness..478
    18.6.2 Two-Stage Adaptive Trial Design for Rare Disease
        Product Development..480
    18.6.3 Probability Monitoring Procedure for Sample Size.......481
18.7 Concluding Remarks...482

Bibliography...483

Index...517

Author

Shein-Chung Chow, PhD, is a professor of Biostatistics and Bioinformatics at Duke University School of Medicine, Durham, NC. Dr. Chow is also a special government employee (SGE) appointed by the US Food and Drug Administration (FDA) as an advisory committee voting member and statistical advisor to the FDA. Between 2017 and 2019, Dr. Chow was on leave for the FDA as an associate director at Office of Biostatistics (OB), Center for Drug Research and Evaluation (CDER), FDA. Dr. Chow is an editor-in-chief of the *Journal of Biopharmaceutical Statistics* and editor-in-chief of the *Biostatistics Book Series* at Chapman and Hall and CRC Press, Taylor & Francis Group. Dr. Chow is a fellow of the American Statistical Association, who is the author or co-author of over 300 methodology papers and 30 books including *Innovative Statistics in Regulatory Science* (Chapman and Hall/CRC Press).

1

Introduction

1.1 Introduction

For approval of pharmaceutical products (including drugs, biological products, and medical devices) in the United States the Food and Drug Administration (FDA) requires that *substantial evidence* regarding the safety and effectiveness of the test treatment under investigation be provided in the regulatory submission process Section 314 of 21 Codes of Federal Regulation (CFR). The substantial evidence regarding safety and effectiveness of the test treatment under investigation will then be evaluated by the reviewers (including statistical reviewers, medical reviewers, and reviewers from other relevant disciplines) in the review and approval process of the test treatment under investigation. Statistics plays an important role to ensure the accuracy, reliability, and reproducibility of the substantial evidence obtained from the studies conducted in the process of product development. Statistical methods and/or tools that are commonly used in the review and approval process of regulatory submissions are usually referred to as statistics in regulatory science or *regulatory statistics*. Thus, in a broader sense, regulatory statistics can be defined as valid statistics that may be used in the review and approval of regulatory submissions of pharmaceutical products. The purpose of regulatory statistics is to provide an objective, unbiased and reliable assessment of the test treatment under investigation.

Regulatory statistics generally follow several principles to ensure the validity of the statistics used in the review and approval process of regulatory submissions. The first principle is to provide unbiased and reliable assessment of the substantial evidence regarding the safety and effectiveness of the test treatment under investigation. The second principle is to ensure quality, validity and integrity of the data collected for supporting the substantial evidence required for regulatory approval. The third principle is to make sure that the observed substantial evidence is not by chance alone and it is reproducible if the same studies were conducted under similar experimental conditions. To ensure the validity of regulatory statistics, it is suggested that statistical principles for Good Statistics Practice (GSP) that outlined in the ICH (International Conference Harmonization) E9 guideline should be followed (ICH, 2018).

The general statistical principles (or key statistical concepts) are the foundation of GSP in regulatory science, which not only ensure the quality, validity and integrity of the intended clinical research during the process of pharmaceutical development, but also provide unbiased and reliable assessment of the test treatment under investigation. Key statistical concepts include, but are not limited to, confounding and interaction; hypotheses testing and *p*-values; one-sided hypotheses versus two-sided hypotheses; clinical significance/ equivalence; and reproducibility and generalizability. In practice, some challenging and controversial issues in the review and approval process of regulatory submissions may arise. These issues include totality-of-the-evidence versus substantial evidence; confusion between the use of $(1-\alpha)\times 100\%$ confidence interval (CI) approach for evaluation of new drugs versus the use of $(1-2\alpha)\times 100\%$ CI approach for assessment of generics/biosimilars; endpoint selection; selecting the proper criteria for decision-making at interim; non-inferiority or equivalence/similarity margin selection; treatment of missing data, the issue of multiplicity; sample size requirement; consistency test in multi-regional trials; extrapolation; drug products with multiple components; and the role of Advisory Committees (e.g., Oncologic Drug Advisory Committee). In addition, there are several critical clinical initiatives recently established by the FDA. These critical clinical initiatives concern precision and/or personalized (individualized) medicine; biomarker-driven clinical research; complex innovative design (CID); model-informed drug development (MIDD); rare diseases drug development; big data analytics; real-world data and real-world evidence, and machine learning for mobile individualized medicine (MIM) and imaging medicine (IM).

In Section 1.2, some key statistical concepts are briefly introduced. Section 1.3 describes some complex innovative designs and corresponding statistical methods. These complex innovative designs include adaptive trial designs, complete *n*-of-1 trial design, master protocols, and Bayesian approach. Challenging and controversial issues that are commonly encountered in the review and approval process of regulatory submissions are outlined in Section 1.4. Also included in this section are introduction of FDA recent critical clinical initiatives. Section 1.5 provides the aim and scope of the book.

1.2 Key Statistical Concepts

1.2.1 Confounding and Interaction

In pharmaceutical/clinical research and development, confounding and interaction effects are probably the most common distortions in the evaluation of a test treatment under investigation. *Confounding effects* are contributed by various factors such as race and gender that cannot be separated by the design under study, while an *interaction effect* between factors is a joint effect with one or more contributing factors (Chow and Liu, 2013). Confounding and interaction effects are important considerations in pharmaceutical/ clinical development. For example, when confounding effects are observed, we cannot assess the treatment effect because it has been contaminated. On the other hand, when interactions among factors are observed, the treatment must be carefully evaluated to isolate those effects.

1.2.1.1 Confounding

In clinical trials, there are many sources of variation that have an impact on the primary clinical endpoints for evaluation relating to a certain new regimen or intervention. If some of these variations are not identified and properly controlled, they can become mixed in with the treatment effect that the trial is designed to demonstrate. Then the treatment effect is said to be confounded by effects due to these variations. To gain a better understanding, consider the following example. Suppose that last winter Dr. Smith noticed that the temperature in the emergency room of a hospital was relatively low and caused some discomfort among medical personnel and patients. Dr. Smith suspected that the heating system might not be functioning properly and called on a technician to improve it. As a result, the temperature of the emergency room was at a comfortable level this winter. However, this winter is not as cold as last winter. Therefore, it is not clear whether the improvement (temperature control) in the emergency room was due to the improvement in the heating system or the effect of a warmer winter. In fact, the effect due to the improvement of the heating system and that due to a warmer winter are confounded and cannot be separated from each other. In clinical trials, there are many subtle, unrecognizable, and seemingly innocent confounding factors that can cause ruinous results of clinical trials.

Moses (1985) discussed an example of the devastating result with the confounder being the personal choice of a patient. The example concerns a polio-vaccine trial that was conducted on two million children worldwide to investigate the effect of Salk poliomyelitis vaccine. This trial reported that the incidence rate of polio was lower in the children whose parents refused injection than whose who received placebo after their parent gave permission (Meier, 1989). After an exhaustive examination of the data, it was found that susceptibility to poliomyelitis was related to the differences between the families who gave the permission and those who did not.

In many cases, confounding factors are inherent in the design of the studies. For example, dose-titration studies in escalating levels are often used to investigate the dose-response relationship of the antihypertensive agents during phase II stage of clinical development. For a typical dose titration study, after a washout period during which previous medication stops and the placebo is prescribed, N subjects start at the lowest dose for a prespecified time interval. At the end of the interval, each patient is evaluated as a responder to the treatment or a non-responder according to some criteria prespecified in the protocol. In a titration study, a subject will continue to receive the next higher dose if he or she fails, at the current level, to meet some objective physiological criteria such as reduction of diastolic blood pressure by a specific amount and has not experienced any unacceptable adverse experience. Figure 1.1 provides a graphical presentation of a typical titration study (Shih et al., 1989). Dose titration studies are quite popular among clinicians because they mimic real clinical practice in the care of patients. The major problem with this typical design for a dose-titration study is that the dose-response relationship is often confounded with time course and the unavoidable carryover effects from the previous dose levels which cannot be estimated and eliminated. One can always argue that the relationship found in a dose titration study is not due to the dose but to the time. Statistical methods for binary data from dose-titration studies have been suggested under some assumptions (e.g., see Chuang, 1987; Shih et al., 1989). Because the dose level is confounded with time, estimation of the dose-response relationship based on continuous data has not yet been resolved in general.

Another type of design that can induce confounding problems when it is conducted inappropriately is the crossover design. For a standard 2×2 crossover design, each subject is randomly assigned to one of the two sequences.

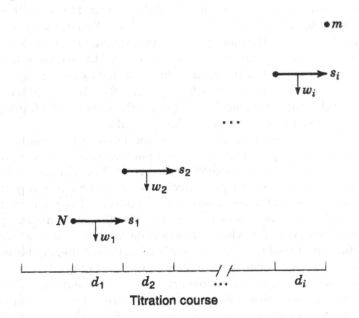

FIGURE 1.1
Graphical display of a titration trial. d_i, the ith dose level; s_i, the number of subjects who responded at the ith dose; w_i, the number of subjects who withdrew at the ith dose; and m, the number of subjects who completed the study without a response. (From Shih, W.J. et al., *Stat. Med.*, 8, 583–591, 1989.)

In sequence 1, subjects receive the reference (or control) treatment at the first dosing period and the test treatment at the second dosing period after a washout period of sufficient length. The order of treatments is reversed for the subjects in sequence 2. The issues in analysis of the data from a 2×2 crossover design are twofold. First, unbiased estimates of treatment effect cannot be obtained from the data of both periods in the presence of a non-zero carryover effect. The second problem is that the carryover effect is confounded with sequence effect and treatment-by-period interaction. In the absence of a significant sequence effect, however, an unbiased estimate of the treatment effect can be estimated from the data of both periods. In practice, it is not clear whether an observed statistically significant sequence effect (or carryover effect) is a true sequence effect (or carryover effect). As a result, this remains a major drawback of the standard 2×2 crossover design, since the primary interest is to estimate a treatment effect that is still an issue in the presence of a significant nuisance parameter. The sequence and carryover effects, however, are not confounded to each other in higher-order crossover designs that compare two treatments and can provide unbiased estimation of treatment effect in the presence of a significant carryover effect (Chow and Liu, 1992a, 2000, 2008).

Bailar (1992) provided another example of subtle and unrecognizable confounding factors. Wilson et al. (1985) and Stampfer et al. (1985) both reported the results on the incidence of cardiovascular diseases in postmenopausal women who had been taking hormones compared to those who had not. Their conclusions, however, were quite different. One reported that the incidence rate of cardiovascular disease among the women taking hormones was twice that in the control group, while the other reported a totally opposite result in which the incidence of the experimental group was only half that of women who were not taking hormones. Although these trials were not randomized studies, both studies were well planned and conducted. Both studies had carefully considered the differences in known risk factors between the two groups in each study. As a result, the puzzling difference in the two studies may be due to some subtle confounding factors such as the dose of hormones, study populations, research methods, or other related causes. This example indicates that it is imperative to identify and take into account all confounding factors for the two adequate, well-controlled studies that are required for demonstration of effectiveness and safety of the study medication under review.

In clinical trials, it is not uncommon for some subjects not to follow instructions concerning taking the prescribed dose at the scheduled time as specified in the protocol. If the treatment effect is related to (or confounded with) patients' compliance, any estimates of the treatment effect are biased unless there is a placebo group in which the differences in treatment effects between subjects with good compliance and poor compliance can be estimated. As a result, interpretation and extrapolation of the findings are inappropriate. In practice, it is very difficult to identify compliers and noncompliers and to quantify the relationship between treatment and compliance. On the other hand, subject withdrawals or dropouts from clinical trials are the ultimate examples of noncompliance. There are several possible reasons for dropouts. For example, a subject with severe disease did not improve and hence dropped out from the study. This will cause the estimate of treatment effect to be biased in favor of a false positive efficacy, if the subjects with mild disease remain and improve. On the other hand, if the subjects whose conditions improve withdraw from a study, and those who did not improve remain until the scheduled, the estimation of efficacy will then be biased and hence indicate a false negative efficacy. Noncompliance and subject dropouts are only two of the many confounding factors that can occur in many aspects of clinical trials. If there is an unequal proportion of the subjects who withdraw from the study or comply to the dosing regimen among different treatment groups, it is very important to perform an analysis on these two groups of subjects to determine whether confounded factors exist and the direction of possible bias. In addition, every effort must be made to continue subsequent evaluation of withdrawals in primary clinical endpoints such as survival or any serious adverse events. For analyses of data with noncompliance or withdrawals, it is suggested that an intention-to-treat analysis be performed. An *intention-to-treat* analysis includes all available data based on all randomized subjects with the degree of compliance or reasons for withdrawal as possible covariates.

1.2.1.2 Interaction

The objective of a statistical interaction investigation is to conclude whether the joint contribution of two or more factors is the same as the sum of the contributions from each factor when considered alone. The factors may be different drugs, different doses of two drugs, or some stratification variables such as severity of underlying disease, gender, or other important covariates. To illustrate the concept of statistical interaction, we consider the *Second International Study of Infarct Survival* (ISIS-2 Group, 1988). This study employed a 2×2 factorial design (two factors with two levels at each factor) to study the effect of streptokinase and aspirin in the reduction of vascular mortality in patients with suspected acute myocardial infarction. The two factors are one-hour intravenous infusion of 1.5 MU of streptokinase and one month of 150 mg per day enteric-coated aspirin. The two levels for each factor are either active treatment and their respective placebo infusion or tablets. A total of 17,187 patients were enrolled in this study. The numbers of the patients randomized to each arm is illustrated in Table 1.1.

The key efficacy endpoint is the cumulative vascular mortality within 35 days after randomization. Table 1.2 provides the cumulative vascular mortality for each of the four arms as well as those for streptokinase and aspirin alone.

TABLE 1.1

Treatment of ISIS-2 with Number of Patients Randomized

Aspirin	IV Infusion of Streptokinase		
	Active	Placebo	Total
Active	4292	4295	8,587
Placebo	4300	4300	8,600
Total	8592	8595	17,187

Source: ISIS-2 Group, *Lancet*, 13, 349–360, 1988.

TABLE 1.2

Cumulative Vascular Mortality in Days 0-35 of ISIS-2

Aspirin	IV Infusion of Streptokinase		
	Active	Placebo	Total
Active	8.0%	10.7%	9.4%
Placebo	10.4%	13.2%	11.8%
Total	9.2%	12.0%	

Source: ISIS-2 Group, *Lancet*, 13, 349–360, 1988.

From Table 1.2 the mortality of streptokinase group is about 9.2%, with the corresponding placebo mortality being 12.0%. The improvement in mortality rate attributed to streptokinase is 2.8% (12.0%–9.2%). This is referred to as the main effect of streptokinase. Similarly, the main effect of aspirin tablets can also be estimated from Table 1.2 as 2.4% (11.8%–9.4%). From Table 1.2, the joint contribution of both streptokinase and aspirin in improvement in mortality is 5.2% (13.2%–8.0%), which is exactly equal to the contribution in mortality by streptokinase (2.8%) plus that by aspirin (2.4%). This is a typical example that no interaction exists between streptokinase and aspirin because the reduction in mortality by joint administration of both streptokinase and aspirin can be expected as the sum of reduction in mortality attributed to each anti-thrombolytic agent when administrated alone. In other words, the difference between the two levels in one factor does not depend on the level of the other factor. For example, the difference in vascular mortality between streptokinase and placebo for the patients taking aspirin tablets is 2.7% (10.7%–8.0%). A similar difference of 2.8% is observed between streptokinase (10.4%) and placebo (13.2%) for the patients taking placebo tablets. Therefore, the reduction in mortality attributed to streptokinase is homogeneous for the two levels of aspirin tablets. As a result, there is no interaction between streptokinase infusion and aspirin tablets.

The ISIS-2 trial provides an example of an investigation of interaction between two treatments. However, in the clinical trial it is common to check for interaction between treatment and other important prognostic and stratification factors. For example, almost all adequate well-controlled studies for the establishment of effectiveness and safety for approval of pharmaceutical agents are multicenter studies. For multicenter trials, the FDA requires that the treatment-by-center interaction be examined to evaluate whether the treatment effect is consistent across all centers.

1.2.2 Hypotheses Testing and *p*-values

1.2.2.1 Hypotheses Testing

In clinical trials a hypothesis is a postulation, assumption, or statement that is made about the population regarding the efficacy, safety, or other pharmacoeconomic outcomes (e.g., quality of life) of a test treatment under study. This statement or hypothesis is usually a scientific question that needs to be investigated. A clinical trial is often designed to address the question by translating it into specific study objective(s). Once the study objective(s) has been carefully selected and defined, a random sample can be drawn through an appropriate study design to evaluate the hypothesis about the drug product. For example, a scientific question regarding a test treatment, say treatment A, of interest could be either (i) "Is the mortality reduced by treatment A?" or (ii) "Is treatment A superior to treatment B in treating hypertension?" For the questions regarding treatment A described above, the null hypotheses are that (i) there is no difference between treatment A and the placebo in the reduction of mortality and (ii) there is no difference between treatment A and treatment B in treating hypertension, respectively. The alternative hypotheses are that (i) treatment A reduces the mortality and (ii) treatment A is superior to treatment B in treating hypertension, respectively. These scientific questions or hypotheses to be tested can then be translated into specific study objectives. Chow and Liu (2000) recommended the following steps be taken to perform a hypothesis testing:

Step 1. Choose the null hypothesis that is to be questioned.

Step 2. Choose an alternative hypothesis that is of particular interest to the investigators.

Step 3. Select a test statistic, and define the rejection region (or a rule) for decision making about when to reject the null hypothesis and when not to reject it.

Step 4. Draw a random sample by conducting a clinical trial.

Step 5. Calculate the test statistic and its corresponding *p*-value.

Step 6. Make conclusion according to the predetermined rule specified in step 3.

When performing a hypotheses testing, basically two kinds of errors (i.e., type I error and type III error) occur. Table 1.3 summarizes the relationship between type I and type II errors when testing hypotheses.

TABLE 1.3

Relationship Between Type I and Type II Errors

	If H_0 is	
	True	False
When		
Fail to reject	No error	Type II error
Reject	Type I error	No error

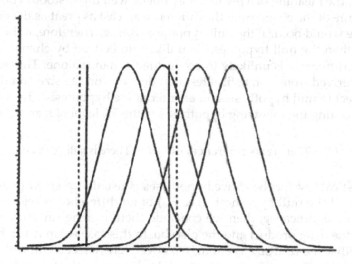

FIGURE 1.2
Relationship between probabilities of type I and type II errors.

A graph based on the null hypothesis of no difference is presented in Figure 1.2 to illustrate the relationship between α and β (or power) for various βs under H_0 for various alternatives at $\alpha = 5\%$ and 10%. It can be seen that α decreases as β increases or α increases as β decreases. The only way of decreasing both α and β is to increase the sample size. In clinical trials a typical approach is to first choose a significant level α and then select a sample size to achieve a desired test power. In other words, a sample size is chosen to reduce type II error such that β is within an acceptable range at a prespecified significant level of α. From Table 1.3 and Figure 1.2 it can be seen that α and β depend on the selected null and alternative hypotheses. As indicated earlier, the hypothesis to be questioned is usually chosen as the null hypothesis. The alternative hypothesis is usually of particular interest to the investigators. In practice, the choice of the null hypothesis and the alternative hypothesis has an impact on the parameter to be tested. Chow and Liu (2000) indicate that the null hypothesis may be selected based on the importance of the type I error. In either case, however, it should be noted that we would never be able to prove that H_0 is true even though the data fail to reject it.

1.2.2.2 p-value

In medical literature p-values are often used to summarize results of clinical trials in a probabilistic way. The probability statement indicated that a difference at least as great as the observed would occur in less than 1 in 100 trials if a 1% level of significance were chosen or in less than 1 in 20 trials if a 5% level of significance were selected provided that the null hypothesis of no difference between treatments is true and the assumed statistical model is correct. In practice, the smaller the p-value shows, the stronger the result is. However, the meaning of a p-value may not be well understood. The p-value is a measure of the chance that the difference at least as great as the observed difference would occur if the null hypothesis is true. Therefore, if the p-value is small, then the null hypothesis is unlikely to be true by chance, and the observed difference is unlikely to occur due to chance alone. The p-value is usually derived from a statistical test that depends on the size and direction of the effect (a null hypothesis and an alternative hypothesis). To show this, consider testing the following hypotheses at the 5% level of significance:

$$H_0 : \text{There is no difference vs. } H_a : \text{There is a difference.} \qquad (1.1)$$

The statistical test for the above hypotheses is usually referred to as a two-sided test. If the null hypothesis (i.e., H_0) of no difference is rejected at the 5% level of significance, then we conclude there is a significant difference between the drug product and the placebo. In this case we may further evaluate whether the trial size is enough to effectively detect a clinically important difference (i.e., a difference that will lead the investigators to believe the drug is of clinical benefit and hence of effectiveness) when such difference exists. Typically, the FDA requires at least 80% power for detecting such difference. In other words, the FDA requires there be at least 80% chance of correctly detecting such difference when the difference indeed exists.

Figure 1.3 displays the sampling distribution of a two-sided test under the null hypothesis in (1.1). It can be seen from Figure 1.3 that a two-sided test has (an) equal chance to show that the drug is either effective—one side—or ineffective—the other side. In Figure 1.3, C and $-C$ are critical values. The area under the probability curve between $-C$ and C constitutes the acceptance region for the null hypothesis. In other words, any observed difference in means in this region is a piece of supportive information of the null hypothesis. The area under the probability curve below $-C$ and beyond C is the rejection region. An observed difference in means in this region is a doubt of the null hypothesis. Based on this concept, we can statistically evaluate whether the null hypothesis is a true statement. Let μ_D and μ_P be the population means of the primary efficacy variable of the drug product and the placebo, respectively. Under the null hypothesis of no difference (i.e., $\mu_D = \mu_P$), a statistical test, say T, can be derived. Suppose that t, the observed difference in means of the drug product and the placebo, is a realization of T. Under the null hypothesis

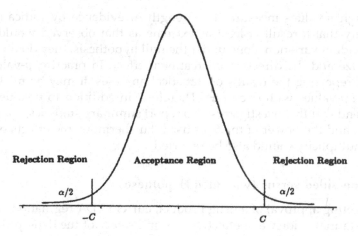

FIGURE 1.3
Sampling distribution of two-sided test.

we can expect that the majority of t will fall around the center, $\mu_D - \mu_P = 0$. There is a 2.5% chance that we would see t will fall in each tail. That is, there is a 2.5% chance that t will be either below the critical value $-C$ or beyond the critical value C. If t falls below $-C$, then the drug is worse than the placebo. On the other hand, if t falls beyond C, then the drug is superior to the placebo. In both cases we would suspect the validity of the statement under the null hypothesis. Therefore we would reject the null hypothesis of no difference if

$$t > C \ \text{ or } t < -C.$$

Furthermore we may want to evaluate how strong the evidence is. In this case, we calculate the area under the probability curve beyond the point t. This area is known as the observed p-value. Therefore the p-value is the probability that a result at least as extreme as that observed would occur by chance if the null hypothesis is true. It can be seen from Figure 1.3 that

$$p\text{-value} < 0.05 \text{ if and only if } t > C \ \text{ or } t < -C.$$

A smaller p-value indicates that t is further away from the center (i.e., $\mu_D - \mu_P = 0$) and consequently provides stronger evidence that supports the alternative hypothesis of a difference. In practice, we can construct a confidence interval for $\mu_D - \mu_P = 0$. If the constructed confidence interval does not contain 0, then we reject the null hypothesis of no difference at the 5% level of significance. It should be noted that the above evaluations for the null hypothesis reach the same conclusion regarding the rejection of the null hypothesis. However, a typical approach is to present the observed p-value. If the observed p-value is less than the level of significance, then the investigators would reject the null hypothesis in favor of the alternative hypothesis.

Although *p*-values measure the strength of evidence by indicating the probability that a result at least as extreme as that observed would occur due to random variation alone under the null hypothesis, they do not reflect sample size and the direction of treatment effect. In practice, *p*-values are a way of reporting the results of statistical analyses. It may be misleading to equate *p*-values with decisions. Therefore, in addition to *p*-values, they recommend that the investigators also report summary statistics, confidence intervals, and the power of the tests used. Furthermore, the effects of selection or multiplicity should also be reported.

1.2.3 One-Sided versus Two-Sided Hypotheses

For marketing approval of a drug product, current FDA regulations require that substantial evidence of effectiveness and safety of the drug product be provided. Substantial evidence can be obtained through conducting two adequate well-controlled clinical trials. The evidence is considered substantial if the results from the two adequate well-controlled studies are consistent in the positive direction. In other words, both trials show that the drug product is significantly different from the placebo in the positive direction. If the primary objective of a clinical trial is to establish that the test drug under investigation is superior to an active control agent, it is referred to as a superiority trial (ICH, 1998). However, the hypotheses given in (1.1) do not specify the direction once the null hypothesis is rejected. As an alternative, the following hypotheses are proposed:

$$H_0 : \text{There is no difference vs.} \quad H_a : \text{The drug is better than placebo.} \quad (1.2)$$

The statistical test for the above hypotheses is known as a one-sided test. If the null hypothesis of no difference is rejected at the 5% level of significance, then we conclude that the drug product is better than the placebo and hence is effective. Figure 1.4 gives the rejection region of a one-sided test. To further compare a one-sided and a two-sided test, let's consider the level of proof required for marketing approval of a drug product at the 5% level of significance. For a given clinical trial, if a two-sided test is employed, the level of proof required is one out of 40. In other words, at the 5% level of significance, there is 2.5% chance (or one out of 40) that we may reject the null hypothesis of no difference in the positive direction and conclude the drug is effective at one side. On the other hand, if a one-sided test is used, the level of proof required is one out of 20. It turns out that the one-sided test allows more ineffective drugs to be approved by chance as compared to the two-sided test. As indicated earlier, to demonstrate the effectiveness and safety of a drug product, the FDA requires two adequate well-controlled clinical trials be conducted. Then the level of proof required should be squared regardless of which test is used. Table 1.4 summarizes the levels of proof required for the marketing approval of a drug product. As Table 1.4 indicates, the levels of proof required

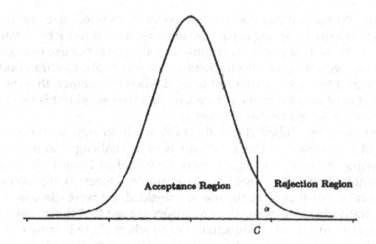

FIGURE 1.4
Sampling distribution of one-sided test.

TABLE 1.4

Level of Proof Required for Clinical Investigation

	Type of Tests	
Number of Trials	One-Sided	Two-Sided
One trial	1/20	1/40
Two trials	1/400	1/1600

for one-sided and two-sided tests are one out of 400 and one out of 1600, respectively. Fisher (1991) argues that the level of proof of one out of 400 is a strong proof and is sufficient to be considered as substantial evidence for marketing approval, so the one-sided test is appropriate. However, there is no universal agreement among the regulatory agencies (e.g., FDA), academia, and the pharmaceutical industry as to whether a one-sided test or a two-sided test should be used. The concern raised is based on the following two reasons:

1. Investigators would not run a trial if they thought the drug would be worse than the placebo. They would study the drug only if they believe that it might be of benefit.

2. When testing at the 0.05 significance level with 80% power, the sample size required is increased by 27% for the two-sided test as opposed to the one-sided test. As a result there is a substantial impact on cost when a one-sided test is used.

It should be noted that although investigators may believe that a drug is better than the placebo, it is likely that the placebo turns out to be superior to the drug (Fleiss, 1987). Ellenberg (1990) indicates that the use of a one-sided test is usually a signal that the trial has too small a sample size and that the investigators are attempting to squeeze out a significant result by a statistical maneuver. These observations certainly argue against the use of one-sided test for the evaluation of effectiveness in clinical trials. Cochran and Cox (1957) suggest that a one-sided test is used when it is known that the drug must be at least as good as the placebo, while a two-sided test is used when it is not known which treatment is better.

As indicated by Dubey (1991), the FDA tends to oppose the use of a one-sided test. However, this position has been challenged at administrative hearings by several drug sponsors on behalf of Drug Efficacy Study Implementation (DESI) drugs. As an example, Dubey (1991) points out that several views that favor the use of one-sided test were discussed in an administrative hearing. Some drug sponsors argued that the one-sided test is appropriate in the following situations: (i) where there is truly only concern with outcomes in one tail and (ii) where it is completely inconceivable that the results can go in the opposite direction. In this hearing the sponsors inferred that the prophylactic value of the combination drug is greater than that posted by the null hypothesis of equal incidence, and therefore the risk of finding an effect when none in fact exists is located only in the upper tail. As a result, a one-sided test is called for. However, the FDA feels that a two-sided test should be applied to account for not only the possibility that the combination drugs are better than the single agent alone at preventing candidiasis but also the possibility that they are worse at doing so.

Dubey's opinion is that one-sided tests may be justified in some situations such as toxicity studies, safety evaluation, analysis of occurrences of adverse drug reactions data, risk evaluation, and laboratory research data. Fisher (1991) argues that one-sided tests are appropriate for drugs that are tested against placebos at the 0.05 level of significance for two well-controlled trials. If, on the other hand, only one clinical trial rather than two is conducted, a one-sided test should be applied at the 0.025 level of significance. However, Fisher agrees that two-sided tests are more appropriate for active control trials.

It is critical to specify hypotheses to be tested in the protocol. A one-sided test or two-sided test can then be justified based on the hypotheses. It should be noted that the FDA is against a post hoc decision to create significance or near significance on any parameters when significance did not previously exist. This critical switch cannot be adequately explained and hence is considered an invalid practice by the FDA. More discussion regarding the use of one-sided test versus two-sided test from the perspectives of the pharmaceutical industry, academe, an FDA Advisory Committee member, and the FDA can be found in Peace (1991), Koch (1991), Fisher (1991), and Dubey (1991), respectively.

1.2.4 Clinical Significance and Clinical Equivalence

As indicated in the hypotheses of (1.1), the objective of most clinical trials is to detect the existence of predefined clinical difference using a statistical testing procedure such as the unpaired two-sample t-test. If this predefined difference is clinically meaningful, then it is of clinical significance. If the null hypothesis in (1.1) is rejected at the α level of significance, then we conclude that a statistically significant difference exists between treatments. In other words, an observed difference that is unlikely to occur by chance alone is considered a statistically significant difference. However, a statistically significant difference depends on the sample size of the trial. A trial with a small sample size usually provides little information regarding the efficacy and safety of the test drug under investigation. On the other hand, a trial with a large sample size provides substantial evidence of the efficacy of the safety of the test drug product. An observed statistically significant difference, which is of little or no clinical significance, will not be able to address the scientific/clinical questions that a clinical trial was intended to answer in the first place.

The magnitude of a clinically significant difference varies. In practice, no precise definition exists for the clinically significant difference, which depends on the disease, indication, therapeutic area, class of drugs, and primary efficacy and safety endpoints. For example, for antidepressant agents (e.g., Serzone), a change from a baseline of 8 in the Hamilton depression (Ham-D) scale or a 50% reduction from baseline in the Hamilton depression (Ham-D) scale with a baseline score over 20 may be considered of clinical importance. For antimicrobial agents (e.g., Cefil), a 15% reduction in bacteriologic eradication rate could be considered a significant improvement. Similarly, we could also consider a reduction of 10 mm Hg in sitting diastolic blood pressure as clinically significant for ACE inhibitor agents in treating hypertensive patients.

The examples of clinical significance of antidepressant or antihypertensive agents are those of individual clinical significance, which can be applied to evaluation of the treatment for individual patients in usual clinical practice. Because individual clinical significance only reflects the clinical change after the therapy, it cannot be employed to compare the clinical change of a therapy to that of no therapy or of a different therapy. Temple (1982) pointed out that in evaluation of one of phase II clinical trials for an ACE inhibitor, although the ACE inhibitor at 150 mg t.i.d. can produce a mean reduction from baseline in diastolic blood pressure of 16 mm Hg, the corresponding mean reduction from baseline for the placebo is also 9 mm Hg. It is easy to see that a sizable proportion of the patients in the placebo group reached the level of individual clinical significance of 10 mm Hg. Therefore, this example illustrates a fact that individual clinical significance alone cannot be used to establish the effectiveness of a new treatment.

For assessment of efficacy/safety of a new treatment modality, it is, within the same trial, compared with either a placebo or another treatment, usually the standard therapy. If the concurrent competitor in the same study is a placebo, the effectiveness of the new modality can then be established, based on some primary endpoints, by providing the evidence of an average difference between the new modality and placebo that is larger than some prespecified difference of clinical importance to investigators or to the medical/scientific community. This observed average difference is said to be of the comparative clinical significance. The ability of a placebo-controlled clinical trial to provide such observed difference of both comparative clinical significance and statistical significance is referred to as assay sensitivity. A similar definition of assay sensitivity is also given in the ICH E10 guideline entitled, *Choice of Control Group in Clinical Trials* (ICH, 1999).

On the other hand, when the concurrent competitor in the trial is the standard treatment or other active treatment, then efficacy of the new treatment can be established by showing that the test treatment is as good as or at least no worse than standard treatment. However, under this situation, the proof of efficacy for the new treatment is based on a crucial assumption that the standard treatment or active competitor has established its own efficacy by demonstrating a difference of comparative clinical significance with respect to placebo in adequate placebo-controlled studies. This assumption is referred to as the sensitivity-to-drug-effects (ICH E10, 1999).

Table 1.5 presents the results first reported in Leber (1989), which was again used by Temple (1983) and Temple and Ellenberg (2000) to illustrate the issues and difficulties in evaluating and interpreting active-controlled trials. All six trials compare nomifensine (a test antidepressant) to imipramine (a standard tricyclic antidepressant) concurrently with placebo. The common baseline means and 4-week adjusted group means based on the Hamilton depression scale are given in Table 1.5. Except for trial V311(2), based on the Hamilton depression scale, both nomifensine and imipramine showed more than 50% mean reduction. However, magnitudes of average reduction on the Hamilton depression scale at 4 weeks for the placebo are almost the same as the other two active treatments for all five trials. Therefore, these five trials do not have assay sensitivity. It should be noted that trial V311(2) is the smallest trial, with a total sample size only of 22 patients. However, it was the only

TABLE 1.5

Summary of Means of Hamilton Depression Scales of Six Trials Comparing Nomifensine, Imipramine, and Placebo

Study	Common Baseline Mean	Four-Week Adjusted Mean (Number of Subjects)		
		Nomifensine	Imipramine	Placebo
R301	23.9	13.4(33)	12.8(33)	14.8(36)
G305	26.0	13.0(39)	13.4(30)	13.9(36)
C311(1)	28.1	19.4(11)	20.3(11)	18.9(13)
V311(2)	29.6	7.3(7)	9.5(8)	23.5(7)
F313	37.6	21.9(7)	21.9(8)	22.0(8)
K317	26.1	11.2(37)	10.8(32)	10.5(36)

Source: Temple, R. and Ellenberg, S.S., *Ann Intern. Med.*, 133, 455–463, 2000.

trial in Table 1.5 that demonstrates that both nomifensine and imipramine are better than placebo in the sense of both comparative clinical significance and statistical significance.

1.2.5 Reproducibility and Generalizability

As indicated in the previous chapter, for marketing approval of a new drug product, the FDA requires that substantial evidence of the effectiveness and safety of the drug product be provided through the conduct of at least two adequate and well-controlled clinical trials. The purpose of requiring at least two pivotal clinical trials is not only to assure the reproducibility, but also to provide valuable information regarding generalizability. Shao and Chow (2002) define reproducibility as (i) whether the clinical results in the same target patient population are reproducible from one location (e.g., study site) to another within the same region (e.g., the United States of America, European Union, or Asian Pacific region) or (ii) whether the clinical results are reproducible from one region to another region in the same target patient population. Generalizability is referred to as (i) whether the clinical results can be generalized from the target patient population (e.g., adult) to another similar but slightly different patient population (e.g., elderly) within the same region or (ii) whether the clinical results can be generalized from the target patient population (e.g., white) in one region to a similar but slightly different patient population (e.g., Asian) in another region. In what follows, we will provide the concept of reproducibility and generalizability for providing substantial evidence in clinical research and development.

1.2.5.1 Reproducibility

In clinical research, two questions are commonly asked. First, what is the chance that we will observe a negative result in a future clinical study under the same study protocol given that positive results have been observed in the two pivotal trials? In practice, two positive results observed from the two pivotal trials, which have fulfilled the regulatory requirement for providing substantial evidence, may not guarantee that the clinical results are reproducible in a future clinical trial with the same study protocol with a high probability. This is very likely, especially when the positive results observed from the two pivotal trials are marginal (i.e., their p-values are close to but less than the level of significance). Second, it is often of interest to determine whether a large clinical trial that produced positive clinical results can be used to replace two pivotal trials for providing substantial evidence for regulatory approval. Although the FDA requires at least two pivotal trials be conducted for providing substantial evidence regarding the effectiveness and safety of the drug product under investigation for regulatory review, under the circumstances, the FDA Modernization Act (FDAMA) of 1997 includes a provision (Section 115 of FDAMA) to allow data from one adequate and well-controlled clinical trial investigation and confirmatory evidence to establish effectiveness for risk/benefit assessment of drug and biological candidates for approval. To address the above two questions, Shao and Chow (2002) suggested evaluating the probability of observing a positive result in a future clinical study with the same study protocol, given that a positive clinical result has been observed.

Let H_0 and H_a be the null hypothesis and the alternative hypothesis of (1.1). Thus, the null hypothesis is that there is no difference in mean response between a test drug and a control (e.g., placebo). Suppose that the null hypothesis is rejected if and only if $|T| > C$, where C is a positive known constant and T is a test statistic, which is usually related to a two-sided alternative hypothesis. In statistical theory, the probability of observing a significant clinical result when H_a is indeed true is referred to as the power of the test procedure. If the statistical model under H_a is a parametric model, then the power can be evaluated at θ, where θ is an unknown parameter or vector of parameters. Suppose now that one clinical trial has been conducted and the result is significant. Then, what is the probability that the second trial will produce a significant result, i.e., the significant result from the first trial is reproducible? Statistically, if the two trials are independent, the probability of observing a significant result in the second trial when H_a is true is the same as that of the first trial regardless of whether the result from the first trial is significant. However, it is suggested that information from the first clinical trial should be used in the evaluation of the probability of observing a significant result in the second trial. This leads to the concept of reproducibility probability (Shao and Chow, 2002).

In general, the reproducibility probability is a person's subjective probability of observing a significant clinical result from a future trial, when he/she observes significant results from one or several previous trials. Goodman (1992) considered the reproducibility probability as the power of the trial (evaluated at θ) by simply replacing θ with its estimate based on the data from previous trials. In other words, the reproducibility probability can be defined as an *estimated power* of the future trial using the data from previous studies. Shao and Chow (2002) studied how to evaluate the reproducibility probability using this approach under several study designs for comparing means with both equal and unequal variances. When the reproducibility probability is used to provide substantial evidence of the effectiveness of a drug product, the estimated power approach may produce an optimistic result. Alternatively, Shao and Chow (2002) suggested that the reproducibility probability be defined as a lower confidence bound of the power of the second trial. In addition, they also suggested a more sensible definition of reproducibility probability using the Bayesian approach. Under the Bayesian approach, the unknown parameter θ is a random vector with a prior distribution, say, $\pi(\theta)$, which is assumed known. Thus, the reproducibility probability can be defined as the conditional probability of $|T| > C$ in the future trial, given the data set.

TABLE 1.6

Reproducibility Probability $\hat{P}$

$|T(x)|$	Known σ^2		Unknown σ^2 ($n = 30$)	
	p-value	$\hat{P}$	*p*-value	$\hat{P}$
1.96	0.050	0.500	0.060	0.473
2.05	0.040	0.536	0.050	0.508
2.17	0.030	0.583	0.039	0.554
2.33	0.020	0.644	0.027	0.614
2.58	0.010	0.732	0.015	0.702
2.81	0.005	0.802	0.009	0.774
3.30	0.001	0.910	0.003	0.890

Source: Shao and Chow (2002).

In practice, the reproducibility probability is useful when the clinical trials are conducted sequentially. It provides important information for regulatory agencies in determining whether it is necessary to require the second clinical trial when the result from the first clinical trial is strongly significant. To illustrate the concept of reproducibility probability, reproducibility probabilities for various values of $|T(x)|$ with $n = 30$ are given in Table 1.6. Table 1.6 suggests that it is not necessary to conduct the second trial if the observed *p*-value of the first trial is less than or equal to 0.001 because the reproducibility probability is about 0.91. On the other hand, even when the observed *p*-value is less than the 5% level of significance, say, the observed *p*-value is less than or equal to 0.01, a second trial is recommended because the reproducibility probability may not reach the level of confidence for the regulatory agency to support the substantial evidence of effectiveness of the drug product under investigation. When the second trial is necessary, the reproducibility probability can be used for sample size adjustment of the second trial. More details regarding sample size calculation based on reproducibility can be found in Shao and Chow (2002) and Chow et al. (2003).

1.2.5.2 Generalizability

As discussed in Section 1.2.5.1, the concept of reproducibility involves whether clinical results observed from the *same* targeted patient population are reproducible from study site to study site within the same region or from region to region. In clinical development, after the drug product has been shown to be effective and safe with respect to the targeted patient population, it is often of interest to determine how likely the clinical results can be reproducible to a *different but similar* patient population with the same disease. We will refer to the reproducibility of clinical results in a different but similar patient population as the generalizability of the clinical results. For example, if the approved drug product is intended for the adult patient population, it is often of interest to study the effect of the drug product on a different but similar patient population, such as the elderly or pediatric patient population with the same disease. In addition, it is also of interest to determine whether the clinical results can be generalized to patient populations with ethnic differences. Similarly, Shao and Chow (2002) proposed to consider the so-called generalizability probability, which is the reproducibility probability with the population of a future trial slightly deviated from the targeted patient population of previous trials, to determine whether the clinical results can be generalized from the targeted patient population to a different but similar patient population with the same disease.

In practice, the response of a patient to a drug product under investigation is expected to vary from patient to patient, especially from patients from the target patient population to patients from a different but similar patient population. The responses of patients from a different but similar patient population could be different from those from the target patient population. As an example, consider a clinical trial, which was conducted to compare the efficacy and safety of a test drug with an active control agent for treatment of schizophrenia patients and patients with schizoaffective disorder. The primary study endpoint is the positive and negative symptom score (PANSS). The treatment duration of the clinical trial was 1 year with a 6-month follow-up. Table 1.7 provides summary statistics of PANSS by race. As it can be seen from Table 1.7, the means and standard deviations of PANSS are different across different races. Oriental patients tend to have higher PANSS with less variability as compared to those in white patients. Black patients seem to have lower PANSS with less variability at both baseline and endpoint. Thus, it is of interest to determine that the observed clinical results can be generalized to a different but similar patient population such as black or Oriental.

Chow (2001) indicated that the responses of patients from a different but similar patient population could be described by the changes in mean and variance of the responses of patients from the target patient population. Consider a parallel-group clinical trial comparing two treatments with population means μ_1 and μ_2 and an equal variance σ^2. Suppose that in the future trial, the population mean difference is changed to $\mu_1 - \mu_2 + \varepsilon$ and the population variance is changed to $C^2\sigma^2$, where $C > 0$. The signal-to-noise ratio for the population difference in the previous trial is $|\mu_1 - \mu_2|/\sigma$ whereas the signal-to-noise ratio for the population difference in the future trial is

$$\frac{|\mu_1 - \mu_2 + \varepsilon|}{C\sigma} = \frac{|\Delta(\mu_1 - \mu_2)|}{\sigma},$$

where

$$\Delta = \frac{1 + \varepsilon / (\mu_1 - \mu_2)}{C}.$$

TABLE 1.7

Summary Statistics of PANSS

Race		Baseline			Endpoint		
		All Subjects	Test	Active Control	All Subjects	Test	Active Control
All Subjects	N	364	177	187	359	172	187
	Mean	66.3	65.1	67.5	65.6	61.8	69.1
	S.D.	16.85	16.05	17.54	20.41	19.28	20.83
	Median	65.0	63.0	66.0	64.0	59.0	67.0
	Range	(30–131)	(30–115)	(33–131)	(31–146)	(31–145)	(33–146)
White							
	N	174	81	93	169	77	92
	Mean	68.6	67.6	69.5	69.0	64.6	72.7
	S.D.	17.98	17.88	18.11	21.31	21.40	20.64
	Median	65.5	64.0	66.0	66.0	61.0	70.5
	Range	(30–131)	(30–115)	(33–131)	(31–146)	(31–145)	(39–146)
Black							
	N	129	67	62	129	66	63
	Mean	63.8	63.3	64.4	61.7	58.3	65.2
	S.D.	13.97	12.83	15.19	18.43	16.64	19.64
	Median	64.0	63.0	65.5	61.0	56.5	66.0
	Range	(34–109)	(38–95)	(34–109)	(31–129)	(31–98)	(33–129)
Oriental							
	N	5	2	3	5	2	3
	Mean	71.8	72.5	71.3	73.2	91.5	61.0
	S.D.	4.38	4.95	5.03	24.57	20.51	20.95
	Median	72.0	72.5	72.0	77.0	91.5	66.0
	Range	(66–76)	(69–76)	(66–76)	(38–106)	(77–106)	(38–79)
Hispanic							
	N	51	24	27	51	24	27
	Mean	64.5	61.4	67.3	64.6	61.9	67.1
	S.D.	18.71	16.78	20.17	20.60	16.71	23.58
	Median	63.0	60.0	68.0	66.0	59.5	67.0
	Range	(33–104)	(35–102)	(33–104)	(33–121)	(33–90)	(33–121)

is a measure of change in the signal-to-noise ratio for the population difference. Note that the above can be expressed by $|\Delta|$ multiplying the *effect size* of the first trial. As a result, Shao and Chow (2002) refer to Δ as the sensitivity index, which is useful when assessing similarity in bridging studies. For most practical problems, $|\varepsilon| < |\mu_1 - \mu_2|$ and thus $\Delta > 0$.

If the power for the previous trial is $p(\theta)$, then the power for the future trial is $p(\Delta\theta)$. Suppose that Δ is known. As discussed earlier, the generalizability probability is given by $\hat{P}_{\Delta}$, which can be obtained by simply replacing $T(x)$ with $\Delta T(x)$. Under the Bayesian approach, the generalizability probability can be obtained by replacing $p(\delta/u)$ with $p(\Delta\delta/u)$.

In practice, the generalizability probability is useful when assessing similarity between clinical trials conducted in different regions (e.g., Europe and the United States of America or the United States of America and the Asian Pacific region). It provides important information for local regulatory health authorities in determining whether it is necessary to require a bridging clinical study based on the analysis of the sensitivity index for assessment of possible difference in ethnic factors (Chow et al., 2002). When a bridging study is deemed necessary, the assessment of the generalizability probability based on the sensitivity index can be used for sample size adjustment of the bridging clinical study.

1.3 Complex Innovative Designs

1.3.1 Adaptive Trial Design

In the past several decades, it is recognized that increasing spending of biomedical research does not reflect an increase of the success rate of pharmaceutical (clinical) development. Woodcock (2005) indicated that the low success rate of pharmaceutical development could be due to: (i) a diminished margin for improvement that escalates the level of difficulty in proving drug benefits; (ii) genomics and other new science have not yet reached their full potential; (iii) mergers and other business arrangements have decreased candidates; (iv) easy targets are the focus of effort, as chronic diseases are harder to study; (v) failure rates have not improved; and (vi) rapidly escalating costs and complexity has decreased willingness/ability to bring many candidates forward for clinical trials. As a result, the FDA established a Critical Path Initiative to assist sponsors in identifying the scientific challenges underlying the medical product pipeline problems. In 2006, the FDA released a Critical Path Opportunities List that calls for advancing innovative trial designs, especially for the use of prior experience or accumulated information in trial design. Many researchers have interpreted the FDA's action as an encouragement for the use of innovative adaptive design methods in clinical trials, while some researchers believe it is an encouragement for the use of Bayesian approach.

The purpose of adaptive design methods in clinical trials is to give the investigator the flexibility for identifying any signals or trends (preferably best or optimal clinical benefit) of the test treatment under investigation without undermining the validity and integrity of the intended study. The concept of adaptive design can be traced back to 1970s when the adaptive randomization and a class of designs for sequential clinical trials were introduced.

As a result, most adaptive design methods in clinical research are referred to as adaptive randomization, group sequential designs with the flexibility for stopping a trial early due to safety, futility and/or efficacy, and sample size re-estimation at interim for achieving the desired statistical power. The use of adaptive design methods for modifying the trial and/or statistical procedures of on-going clinical trials based on accrued data has been practiced for years in clinical research. Adaptive design methods in clinical research are very attractive to clinical scientists due to the following reasons. First, it reflects medical practice in real-world. Second, it is ethical with respect to both e efficacy and safety (toxicity) of the test treatment under investigation. Third, it is not only flexible, but also efficient in the early phase of clinical development. However, it is a concern whether the *p*-value or confidence interval regarding the treatment effect obtained after the modification is reliable or correct. In addition, it is also a concern that the use of adaptive design methods in a clinical trial may lead to a totally different trial that is unable to address scientific/medical questions that the trial is intended to answer.

In its recent draft guidance, *Adaptive Design Clinical Trials for Drugs and Biologics*, the FDA defines an adaptive design clinical study as a study that includes a prospectively planned opportunity for modification of one or more specified aspects of the study design and hypotheses based on analysis of data (usually interim data) from subjects in the study. The FDA emphasizes that one of the major characteristics of an adaptive design is the prospectively planned opportunity. Changes should be made based on analysis of data, usually interim data (FDA, 2010b, 2018). Note that the FDA's definition excludes changes made through protocol amendments. Thus, it does not reflect real practice in clinical trials. In many cases, an adaptive design is also known as a flexible design (EMEA, 2002, 2006).

1.3.1.1 Adaptations

An adaptation is referred to as a modification or a change made to trial procedures and/or statistical methods during the conduct of a clinical trial. By definition, adaptations that are commonly employed in clinical trials can be classified into the categories of prospective adaptation, concurrent (or ad hoc) adaptation, and retrospective adaptation. Prospective adaptations include, but are not limited to, adaptive randomization; stopping a trial early due to safety, futility or efficacy at interim analysis; dropping the losers (or inferior treatment groups), sample size re-estimation, etc. Thus, prospective adaptations are usually referred to as by design adaptations as described in the PhRMA white paper (Gallo et al., 2006). Concurrent adaptations are usually referred to as any ad hoc modifications or changes made as the trial continues. Concurrent adaptations include, but are not limited to, modifications in inclusion/exclusion criteria, evaluability criteria, dose/regimen and treatment duration, changes in hypotheses and/or study endpoints, and etc. Retrospective adaptations are usually referred to as modifications and/or changes made to statistical analysis plan prior to database lock or unblinding of treatment codes. In practice, prospective, ad hoc, and retrospective adaptations are implemented by study protocol, protocol amendments, and statistical analysis plan with regulatory reviewer's consensus, respectively.

1.3.1.2 Types of Adaptive Design

Based on the adaptations employed, commonly considered adaptive designs in clinical trials include, but are not limited to: (i) an adaptive randomization design; (ii) a group sequential design; (iii) a sample size re-estimation (SSRE) design or an N-adjustable design; (iv) a drop-the-loser (or pick-the-winner) design; (v) an adaptive dose finding design; (iv) a biomarker-adaptive design; (vii) an adaptive treatment-switching design; (viii) an adaptive-hypothesis design; (ix) an adaptive seamless (e.g., a two-stage phase I/II or phase II/III) trial design; and (x) a multiple adaptive design. More detailed information regarding these designs can be found in Chapter 13 (see also Chow and Chang, 2011).

1.3.2 The *n*-of-1 Trial Design

One of the major dilemmas for rare diseases clinical trials is the in-availability of patients with the rare diseases under study. In addition, it is unethical to consider a placebo control in the intended clinical trial. Thus, it is suggested that an n-of-1 crossover design be considered. An n-of-1 trial design is to apply n treatments (including placebo) to an individual at different dosing periods with sufficient washout in between dosing periods. A complete n-of-1 trial design is a crossover design consisting of all possible combinations of treatment assignment at different dosing periods.

1.3.2.1 Complete n-of-1 Trial Design

Suppose there are p dosing periods and two test treatments, e.g., a test (T) treatment and a reference (R) product, to be compared. A complete n-of-1 trial design for comparing two treatments consists of $\Pi_{i=1}^{p} 2$, where $p \geq 2$, sequences of p treatments (either T or R at each dosing period). In this case, $n = p$. If $p = 2$, then the n-of 1 trial design is a 4 × 2 crossover design, i.e., (RR, RT, TT, TR), which is a typical Balaam's design. When $p = 3$, the n-of-1 trial design becomes an 8 × 3 crossover design, while the complete n-of-1 trial design with $p = 4$ is a 16 × 4 crossover design, which is illustrated in Table 1.8.

As indicated in a recent FDA draft guidance, a two-sequence dual design, i.e., (RTR, TRT) and a 4 × 2 crossover designs, i.e., (RTRT, RRRR)

TABLE 1.8

Examples of Complete n-of-1 Designs with $p = 4$

Group	Period 1	Period 2	Period 3	Period 4
1	R	R	R	R
2	R	T	R	R
3	T	T	R	R
4	T	R	R	R
5	R	R	T	R
6	R	T	T	T
7	T	R	T	R
8	T	T	T	T
9	R	R	R	T
10	R	R	T	T
11	R	T	R	T
12	R	T	T	R
13	T	R	R	T
14	T	R	T	T
15	T	T	R	T
16	T	T	T	R

Note: The first block (a 4 × 2 crossover design) is a complete n-of-1 design with 2 periods, while the second block is a complete n-of-1 design with 3 periods.

recommended by the FDA are commonly considered switching designs for assessing interchangeability in biosimilar product development (FDA, 2017). These two switching designs, however, are limited for fully characterizing relative risk (i.e., reduction in efficacy or increase incidence rate of adverse event rate). On the other hand, these two trial designs are special case of the complete n-of-1 trial design with 3 or 4 dosing periods, respectively. Under the complete n-of-1 crossover design with 4 dosing periods, all possible switching and alternations can be assessed and the results can be compared within the same group of patients and between different groups of patients.

1.3.2.2 Merits and Limitations

A complete n-of-1 trial design has the following advantages: (i) each subject is at his/her own control; (ii) it allows a comparison between the test product and the placebo if the intended trial is a placebo-controlled study (this has lifted the ethical issue of using placebo on the patients with critical conditions); (iii) it allows estimates of intra-subject variability; (iv) it provides estimates for treatment effect in the presence of possible carry-over effect, and most importantly; (v) it requires less subjects for achieving the study objectives of the intended trial design. However, the n-of-1 trial design suffers from the drawbacks of (i) possible dropouts or missing data and (ii) patients' disease status may change at each dosing period prior to dosing.

1.3.3 The Concept of Master Protocols

Woodcock and LaVange (2017) introduced the concept of master protocol for studying multiple therapies, multiple diseases, or both in order to answer more questions in a more efficient and timely fashion (see also, Redman and Allegra, 2015). Master protocols include the following types of trials: umbrella, basket, and platform. The umbrella trial is to study multiple targeted therapies in the context of a single disease, while the basket trial is to study a single therapy in the context of multiple diseases or disease subtypes. The platform trial is used to study multiple targeted therapies in the context of a single disease in a perpetual manner, with therapies allowed to enter or leave the platform on the basis of decision algorithm. As indicated by Woodcock and LaVange (2017), if designed correctly, master protocols offer a number of benefits include streamlined logistics; improved data quality, collection and sharing; as well as the potential to use innovative statistical approaches to study design and analysis. Master protocols may be a collection of sub-studies or a complex statistical design or platform for rapid learning and decision-making.

In practice, master protocol is intended for the addition or removal of drugs, arms, and study hypotheses. Thus, in practice, master protocols may or may not be adaptive, umbrella, or basket studies. Since master protocol has the ability to combine a variety of logistical, innovative, and correlative elements, it allows learning more from smaller patient populations. Thus, the concept of master protocols in conjunction with adaptive trial design described in the previous section may be useful for rare diseases clinical investigation although it has been most frequently implemented in oncology research.

1.3.4 Bayesian Approach

Under the assumption that historical data (e.g., previous studies or experience) are available, Bayesian methods for borrowing information from different data sources may be useful. These data sources could include, but are not limited to, natural history studies and expert's opinion regarding prior distribution about the relationship between endpoints and clinical outcomes. The impact of borrowing on results can be assessed through the conduct of sensitivity analysis. One of the key questions of particular interest to the investigator and regulatory reviewer is that how much to borrow in order to (i) achieve desired statistical assurance for substantial evidence; and (ii) maintain the quality, validity, and integrity of the study.

Although the Bayesian approach provides a formal framework for borrowing historical information, which is useful in rare disease clinical trials, borrowing can only be done under the assumption that there is a well-established relationship between patient populations (e.g., from previous studies to current study). In practice, it is suggested to not borrow any data from previous studies whenever possible. The primary analysis should rely on the data collected from the current study. When borrowing, the associated risk should be carefully evaluated for scientific/statistical validity of the final conclusion. It should be noted that Bayesian approach may not be feasible if there are no prior experience or study available. The determination of prior in Bayesian is always debatable because the primary assumption of the selected prior is often difficult, if not impossible, to be verified.

1.4 Practical, Challenging, and Controversial Issues

1.4.1 Totality-of-the-Evidence

For regulatory approval of new drugs, Section 314.126 of 21 CFR states that substantial evidence needs to be provided to support the claims of new drugs. For regulatory approval of a proposed biosimilar product, the FDA requires totality-of-the-evidence be provided to support a demonstration of biosimilarity between the proposed biosimilar product and the US-licensed drug product. In practice, it should be noted that there is no clear distinction between the substantial evidence in new drug development and the totality-of-the-evidence in biosimilar drug product development.

For approval of a proposed biosimilar product, the FDA requires that totality-of-the-evidence be provided to support a demonstration that the proposed biosimilar product is highly similar to the US-licensed product, notwithstanding minor differences in clinically inactive components, and that there are no clinically meaningful differences between the proposed biosimilar product and the US-licensed product in terms of the safety, purity and potency of the product.

To assist the sponsor in biosimilar product development, the FDA recommends a stepwise approach for obtaining the totality-of-the-evidence for demonstrating biosimilarity between the proposed biosimilar product and its corresponding innovative drug product in terms of safety, purity, and efficacy (Chow, 2013; FDA, 2015a, 2017a; Endrenyi et al., 2017). The stepwise approach starts with a similarity assessment in critical quality attributes (CQAs) in analytical studies, followed by a similarity assessment in pharmacological activities in pharmacokinetic and pharmacodynamic (PK/PD) studies and a similarity assessment in safety and efficacy in clinical studies. For analytical similarity assessment in CQAs, the FDA further recommends a tiered approach that classifies CQAs into three tiers depending upon their criticality or risk ranking relevant to clinical outcomes. For determination of criticality or risk ranking, FDA suggests establishing a predictive (statistical) model based on either mechanism of action (MOA) or PK relevant to clinical outcome. Thus, the following assumptions are made for the stepwise approach for obtaining the totality-of-the-evidence.

1. Analytical similarity is predictive of PK/PD similarity;
2. Analytical similarity is predictive of clinical outcomes;
3. PK/PD similarity is predictive of clinical outcomes.

These assumptions, however, are difficult (if not impossible) to verify in practice. For assumptions (1) and (2), although many in vitro and in vivo correlations (IVIVC) have been studied in the literature, the correlations between specific CQAs and PK/PD parameters or clinical endpoints are not fully studied and understood. In other words, most predictive models are not well established or are established but not validated. Thus, it is not clear how a (notable) change in a specific CQA can be translated to a change in drug absorption or clinical outcome. For (3), unlike bioequivalence assessment for generic drug products, there does not exist Fundamental Biosimilarity Assumption indicating that PK/PD similarity implies clinical similarity in terms of safety and efficacy. In other words, PK/PD similarity or dis-similarity may or may not lead to clinical similarity. Note that the assumptions (1) and (3) together do not imply (2) automatically.

The validity of assumptions (1)–(3) is critical for the success of obtaining totality-of-the-evidence for assessing biosimilarity between the proposed biosimilar and the innovative biological product. This is because the validity of these assumptions ensures the relationships among analytical, PK/PD, and clinical similarity assessment and consequently the validity of the overall biosimilarity assessment. Table 1.1 illustrates relationships among analytical, PK/PD, and clinical assessments in the stepwise approach for obtaining the totality-of-the-evidence in biosimilar product development.

1.4.2 $(1-\alpha)$ CI for New Drugs versus $(1-2\alpha)$ CI for Generics/Biosimilars

Recall that, for review and approval of regulatory submissions of drug products, a $(1-\alpha) \times 100\%$ confidence interval (CI) approach is commonly used for evaluation of safety and efficacy of new drugs, while a $(1-2\alpha) \times 100\%$ CI is often considered for assessment of bioequivalence for generic drug products and biosimilarity assessment for biosimilar products. If, α is chosen to be 5%, this leads to a 95% confidence interval approach for evaluation of new drugs and a 90% confidence interval approach for assessment of generics and biosimilars. In the past decade, the FDA has been challenged for adopting different standards (i.e., 5% type I error rate for new drugs and 10% for generics/biosimilars) in the review and approval process of regulatory submissions of drugs and biologics. The issue of using a 95% CI for new drugs and a 90% CI for generics/biosimilars has received much attention lately.

This controversy surrounding this issue is probably due to the mixed-up use of the concepts of hypotheses testing and the confidence interval approach for evaluation of drug products. The statistical methods for evaluation of new drugs and for assessment of generics/biosimilars recommended by the FDA are a two-sided test (TST) for testing point hypotheses of equality and a two one-sided tests (TOST) for testing interval hypotheses of bioequivalence or biosimilarity, respectively. Under the hypotheses testing framework, test results are interpreted using confidence interval approach regardless that their corresponding statistical inferences may not be (operationally) equivalent.

In practice, there are fundamental differences (i) between the concepts of point hypotheses and interval hypotheses, and (ii) between the concepts of hypotheses testing and confidence interval approach. For (i), a TST is often performed for testing point hypotheses of equality at the 5% level of significance, while a TOST is commonly used for testing interval hypotheses of equivalence or similarity at the 5% level of significance. For (ii), hypotheses testing focuses on power analysis for achieving a desired power (i.e., a type II error), while confidence interval focuses on precision analysis (i.e., a type I error) for controlling the maximum error allowed. The confusion between the use of a 95% CI and the use of a 90% CI for evaluation of drug products will inevitably occur if we use a confidence interval mindset to interpret the results obtained under the hypotheses testing framework.

1.4.3 Endpoint Selection

Utility function for endpoint selection, Chow and Lin (2015) developed valid statistical tests for analysis of two-stage adaptive designs with different study objectives and study endpoints at different stages under the assumption that there is well-established relationship between different but similar study endpoints at different stages. To apply the methodology developed by Chow and Lin (2015), we propose the development of a utility function to link all clinical outcomes relevant to NASH for endpoint selection at different stages as follows.

Let $y = \{y_1, y_2, \ldots, y_m\}$ be the clinical outcomes of interest, which could be efficacy or safety/toxicity of a test treatment under investigation. Clinical outcomes could be NAFLD Activity Score (NAS) including steatosis, laula inflammation, and ballooning, fibrosis progression at different stages, and abnormalities on liver biopsy. Each of these clinical outcomes, y_i is a function of criteria $y_i(x)$, $x \in X$, where X is a space of criteria. We can then define a utility function (e.g., a composite index such as $NAS \geq 4$ and/or F2/F3 (fibrosis stage 2 ad fibrosis stage 3) or co-primary endpoints (NAS score and fibrosis stage) for endpoint selection as follows.

$$U_s = \sum_{j=1}^{m} w_{sj} = \sum_{j=1}^{m} w(y_{sj}),$$

where U_s denotes the endpoint derived from the utility at the sth stage of the multiple-stage adaptive design and $w_j, j = 1, \ldots, m$ are pre-specified weights.

The above single utility function, which takes different clinical outcomes with pre-specified criteria into consideration, is based on the single utility index rather than individual clinical outcomes. The single utility index model allows the investigator to accurately and reliably assess the treatment effect in a more efficient way as follows:

$$p = P\{U_s \geq \tau_s, \ s = 1,2,\ldots,k\}, \qquad (2)$$

where τ_s may be suggested by the regulatory agencies such as the FDA. It should be noted that U_s, $s = 1,2,\ldots,k$ (stages) are similar but different. In practice, if $k = 2$, we can apply statistical methods for two-stage adaptive design with different study endpoints and objectives at different stages (Chow and Lin, 2015).

1.4.4 Criteria for Decision-Making at Interim

For a multiple-stage adaptive design, sample size is often selected to empower the study to detect a clinically meaningful treatment effect at the end of the last stage. As a result, the chosen sample size may not provide adequate or sufficient power for making critical decision (e.g., detecting clinically meaningful difference for dose selection or dropping the inferior treatment groups) at earlier stages in the study. In this case, a precision analysis is recommended to assure that (1) the selected dose has achieved statistical significance (i.e., the observed difference is not by chance alone) and (2) with the sample size available at interim, desired statistical inference for (critical) decision-making. Let (L_i, U_i) be the 95% confidence interval for the difference between the ith dose group and the control group, where $i = 1,\ldots,k$. If $L_i > 0$, we claim that the observed difference between the ith dose group and the control group has achieved statistical significance. In other words, the observed difference is not by chance alone and hence is reproducible. In this case, the dose level with $L_* = \max\{L_i, i = 1,\ldots,k\}$ will be selected moving forward to the next stage. On the other hand, if (L_i, U_i) covers 0, we conclude that there is no difference between the ith dose group and the control group. In this case, the confidence level for achieving statistical significance is defined as the probability that the true mean difference is within $(0, U_i)$. In the case where all (L_i, U_i), $i = 1,\ldots,k$ cover 0, the following criteria are often considered for dose selection:

1. The dose with highest confidence level for achieving statistical significance will be selected; or
2. The doses with confidence levels for achieving statistical significance less than 75% will be dropped.

Note that sample size increases, the corresponding confidence level for achieving statistical difference increases. It should also be noted that precision assessment in terms of confidence interval approach is operationally equivalent to hypotheses testing for comparing means.

1.4.5 Non-inferiority or Equivalence/Similarity Margin Selection

In clinical trials, it is unethical to treat patients with critical/severe and/or life-threatening diseases such as cancer when approved and effective therapies such as standard of care or active control agents are available. In this case, an active control trial is often conducted for investigation of a new test treatment. The goal of an active control trial is to demonstrate that the test treatment is not inferior to or equivalent to the active control agent in the sense that the effect of the test treatment is not below some non-inferiority margin or equivalence limit when compared with the efficacy of the active control agent. In practice, there may be a need to develop a new treatment or therapy that is non-inferior (but not necessarily superior) to an established efficacious treatment due to the following reasons: (i) the test treatment is less toxic; (ii) the test treatment has a better safety profile; (iii) the test treatment is easy to administer; (iv) the test treatment is less expensive; (v) the test treatment provides better quality of life; (vi) the test treatment provides an alternative treatment with some additional clinical benefits, e.g., generics or biosimilars. Clinical trials of this kind are referred to as non-inferiority trials. A comprehensive overview of design concepts and important issues that are commonly encountered in active control or non-inferiority trials can be found in D'Agostino et al. (2003).

For a non-inferiority trial, the idea is to reject the null hypothesis that a test treatment is inferior to a standard therapy or an active control agent and conclude that the difference between the test treatment and the active control agent is less than a clinically meaningful difference (non-inferiority margin) and hence the test treatment is at least as effective as (or not worsen than) the active control agent. The test treatment can then serve as an alternative to the active control agent. In practice, however, it should be noted that unlike equivalence testing, non-inferiority testing is a one-sided equivalence testing which consists of the concepts of equivalence and superiority. In other words, superiority may be tested after the non-inferiority has been established. We conclude equivalence if we fail to reject the null hypothesis of non-superiority. On the other hand, superiority may be concluded if the null hypothesis of non-superiority is rejected.

One of the major considerations in a non-inferiority trial is the selection of the non-inferiority margin. A different choice of non-inferiority margin will have an impact on sample size requirement for achieving a desired power for establishment of non-inferiority. It should be noted that non-inferiority margin could be selected based on either absolute change or relative change of the primary study endpoint, which will affect the method for data analysis of the collected clinical data, and consequently may alter the conclusion of the clinical study. In practice, despite the existence of some studies (e.g., Tsong et al., 1999; Hung et al., 2003; Laster and Johnson, 2003; Phillips, 2003), there is no established rule or gold standard for determination of non-inferiority margins in active control trials until early 2000. In 2000, the

International Conference on Harmonization (ICH) published a guideline to assist the sponsors for selection of an appropriate non-inferiority margin (ICH E10, 2000). ICH E10 guideline suggests that a non-inferiority margin may be selected based on past experience in placebo control trials with valid design under conditions similar to those planned for the new trial and the determination of a non-inferiority margin should not only reflect uncertainties in the evidence on which the choice is based, but also be suitably conservative. Along this line, in 2010, the FDA also published draft guidance on non-inferiority clinical trials and recommends a couple of approaches for selection of non-inferiority margin (FDA, 2010).

1.4.6 Treatment of Missing Data

Missing values or incomplete data are commonly encountered in clinical trials. One of the primary causes of missing data is the dropout. Reasons for dropout include, but are limited to, refusal to continue in the study (e.g., withdrawal of informed consent), perceived lack of efficacy, relocation, adverse events, unpleasant study procedures, worsening of disease, unrelated disease, non-compliance with the study, need to use prohibited medication, and death (DeSouza et al., 2009). Following the idea of Little and Rubin (1987), DeSouza et al. (2009) provided an overview of three types of missingness mechanisms for dropouts. These three types of missingness mechanisms include (1) missing completely at random (MCAR), (2) missing at random (MAR), and (3) missing not at random (MNAR). Missing completely at random is referred to the dropout process that is independent of the observed data and the missing data. Missing at random indicates that the dropout process is dependent on the observed data but is independent of the missing data. For missing not at random, the dropout process is dependent on the missing data and possibly the observed data. Depending upon the missingness mechanisms, appropriate missing data analysis strategies can then be considered based on existing analysis methods in the literature. For example, commonly considered methods under MAR include (1) discard incomplete cases and analyze complete-case only, (2) impute or fill-in missing values and then analyze the filled-in data, (3) analyze the incomplete data by a method such as likelihood-based method (e.g., maximum likelihood, restricted maximum likelihood, and Bayesian approach), moment-based method (e.g., generalized estimating equations and their variants), and survival analysis method (e.g., Cox proportional hazards model) that does not require a complete data set. On the other hand, under MNAR, commonly considered methods are derived under pattern mixture models (Little, 1994) which can be divided into two types, parametric (see, e.g., Diggle and Kenward, 1994) and semi-parametric (e.g., Rotnitzky et al., 1998).

In practice, the possible causes of missing values in a study can generally be classified into two categories. The first category includes the reasons that are not directly related to the study. For example, a patient may be lost to follow-up because he/she moves out of the area. This category of missing values can be considered as missing completely at random. The second category includes the reasons that are related to the study. For example, a patient may withdraw from the study due to treatment-emergent adverse events. In clinical research, it is not uncommon to have multiple assessments from each subject. Subjects with all observations missing are called unit non-respondents. Because unit non-respondents do not provide any useful information, these subjects are usually excluded from the analysis. On the other hand, the subjects with some, but not all, observations missing are referred to as item non-respondents. In practice, excluding item non-respondents from the analysis is considered against the intent-to-treat (ITT) principle and, hence is not acceptable. In clinical research, the primary analysis is usually conducted based on ITT population, which includes all randomized subjects with at least post-treatment evaluation. As a result, most item non-respondents may be included in the ITT population. Excluding item non-respondents may seriously decrease power/efficiency of the study. Statistical methods for missing values imputation have been studied by many authors (see, e.g., Kalton and Kasprzyk, 1986; Little and Rubin, 1987; Schafer, 1997).

To account for item non-respondents, two methods are commonly considered. The first method is the so-called likelihood-based method. Under a parametric model, the marginal likelihood function for the observed responses is obtained by integrating out the missing responses. The parameter of interest can then be estimated by the maximum likelihood estimator (MLE). Consequently, a corresponding test (e.g., likelihood ratio test) can be constructed. The merit of this method is that the resulting statistical procedures are usually efficient. The drawback is that the calculation of the marginal likelihood could be difficult. As a result, some special statistical or numerical algorithms are commonly applied for obtaining the MLE. For example, the expectation–maximization (EM) algorithm is one of the most popular methods for obtaining the MLE when there are missing data. The other method for item non-respondents is imputation. Compared with the likelihood-based method, the method of imputation is relatively simple and easy to apply. The idea of imputation is to treat the imputed values as the observed values and then apply the standard statistical software for obtaining consistent estimators. However, it should be noted that the variability of the estimator obtained by imputation is usually different from the estimator obtained from the complete data. In this case, the formulas designed to estimate the variance of the complete data set cannot be used to estimate the variance of estimator produced by the imputed data. As an alternative, two methods are considered for estimation of its variability. One is based on Taylor's expansion. This method is referred to as the linearization method. The merit of the linearization method is that it requires less computation. However, the drawback is that its formula could be very complicated and/or not trackable. The other approach is based on re-sampling method (e.g., bootstrap and jackknife). The drawback of the re-sampling method is that it requires an intensive computation. The merit is that it is very easy to apply. With the help of a fast-speed computer, the re-sampling method has become much more attractive in practice.

Note that imputation is very popular in clinical research. The simple imputation method of last observation carried forward (LOCF) at endpoint is probably the most commonly used imputation method in clinical research. Although the LOCF is simple and easy for implementation in clinical trials, many researchers have challenged its validity. As a result, the search for alternative valid statistical methods for missing values imputation has received much attention in the past decade. In practice, the imputation methods in clinical research are more diversified due to the complexity of the study design relative to sample survey. As a result, statistical properties of many commonly used imputation methods in clinical research are still unknown, while most imputation methods used in sample survey are well studied. Hence, the imputation methods in clinical research provide a unique challenge and also an opportunity for the statisticians in the area of clinical research.

The issue of multiplicity—Lepor et al. (1996) report the results of a double-blind, randomized multicenter clinical trial that evaluated the efficacy and safety of terazosin (10 mg daily), and αl-adrenergicantagonist, finasteride, a 5 α-reductase inhibitor (5 mg daily) or both with a placebo control in equal allocation in 1229 men with benign prostatic hyperplasia. The primary efficacy endpoints of this trial are the American Urological Association (AUA) symptom score (Barry et al., 1992) and the maximum uroflow rate. These endpoints were evaluated twice during the four-week placebo run-in period and at 2, 4, 13, 26, 39, and 52 weeks of therapy. The primary comparisons of interest included pairwise comparisons among the active drugs and combination therapy, while the secondary comparisons consisted of a pairwise comparison of the active drugs and combination therapy with the placebo. The results for the primary efficacy endpoints presented in Lepor et al. (1996) were obtained by performing analyses of covariance with repeated measurements based on the intention-to-treat population.

One of the objectives of the trial is to determine the time when the treatments reach therapeutic effects. Therefore, comparisons among treatment groups were performed at each scheduled post-randomization visits at 2, 4, 13, 26, 39, and 52 weeks. In addition to the original observations of the primary endpoints, the change from baseline can also be employed to characterize the change after treatment for each patient. It may be of interest to see whether the treatment effects are homogeneous across race, age, and baseline disease severity. Therefore, some subgroup analyses can be performed such as for Caucasians and for non-Caucasians patients, for patients below or at least 65 years of age, for patients with the baseline AUA symptom score below 16 or at least 16, or for patients with the maximum uroflow rate below 10 mL/s. The number of the total comparisons for the primary efficacy endpoints can be as large as 1,344. If there is no difference among the four treatment groups and each of the 1,344 comparisons are performed at the 5% level of significance, we can expect 67 statistically significant comparisons with reported p-values smaller than 0.05. The probability of observing at least one statistically significant difference among 1,344 comparisons could be as large as 1 under the assumption that all 1,344 comparisons are statistically independent. The number of p-values does not include those from the center-specific treatment comparisons and from other types of comparisons such as treatment-by-center interaction.

Although the above example is a bit exaggerated, it does point out that the multiplicity in multicenter clinical trials is an important issue that has an impact on statistical inference of the overall treatment effect. In practice, however, it is almost impossible to characterize a particular disease by a single efficacy measure due to: (i) the multifaceted nature of the disease, (ii) lack of understanding of the disease, and (iii) lack of consensus on the characterization of the disease. Therefore, multiple efficacy endpoints are often considered to evaluate the effectiveness of test drugs in treatment of most diseases such as AIDS, asthma, benign prostatic hyperplasia, arthritis, postmenopausal osteoporosis, and ventricular tachycardia. Some of these endpoints are objective histological or physiological measurements such as the maximum uroflow rate for benign prostatic hyperplasia or pulmonary function FEV_1 (forced expiratory volume in one second) for asthma. Other may include the symptoms or subjective judgment of the well-being of the patients improved by the treatments such as the AUA symptom scores for benign prostatic hyperplasia, asthma-specific symptom score for asthma, or the Greene climacteric scale for postmenopausal osteoporosis (Greene and Hart, 1987). Hence one type of multiplicity in statistical inference for clinical trials results from the source of multiple endpoints.

On the other hand, a clinical trial may be conducted to compare several drugs of different classes for the same indication. For example, the study by Lepor et al. (1996) compares two monotherapies of terazosin, finasteride with the combination therapy, and a placebo control for treatment of patients with benign prostatic hyperplasia. Some other trials might be intended for the investigation of a dose-response relationship of the test drug. For example, Gormley et al. (1992) evaluate the efficacy and safety of 1 and 5 mg of finasteride with a placebo control. This type of multiplicity is inherited from the fact that the number of treatment groups evaluated in a clinical trial is greater than 2. Other types of multiplicity are caused by subgroup analyses. Examples include a trial reported by the National Institute for Neurological Disorders and Stroke rt-PA Study Group (1995) in which stratified analyses were performed according to the time from the onset of stroke to the start of treatment (0–90 or 91–180 minutes). In addition, the BHAT (1982) and CAST (1989) studies were terminated early because of overwhelming evidence of either beneficial efficacy or serious safety concern before the scheduled conclusion of the trials by the technique of repeated interim analyses. In summary, multiplicity in clinical trials can be classified as repeated interim analyses, multiple comparisons, multiple endpoints, and subgroup analyses. Since the causes of these multiplicities are different, special attention should be paid to (i) the formulation of statistical hypotheses based on the objectives of the trial, (ii) the proper control of experiment-wise false positive rates in subsequent analyses of the data, and (iii) the interpretation of the results.

1.4.7 Sample Size Requirement

In clinical trials, a pre-study power analysis for sample size calculation (estimation or determination) is often performed based on either (i) information obtained from small scale pilot studies with limited number of subjects or (ii) purely guess based on the best knowledge of the investigator (with or without scientific justification). The observed data and/or the investigator's best guess could be far away from the truth. Such deviation may bias the sample size calculation for reaching the desired power for achieving the study objectives at a pre-specified level of significance. Sample size calculation is a key to the success of pharmaceutical/clinical research and development. Thus, how to select the minimum sample size for achieving the desired power at a pre-specified significance level has become an important question to clinical scientists (Chow et al., 2008; Chow and Liu, 1998b). A study without sufficient number of subjects cannot guarantee the desired power (i.e., the probability of correctly detecting a clinically meaningful difference). On the other hand, an unnecessarily large sample size could be quite a waste of limited resources.

Sample size calculation plays an important role in pharmaceutical/clinical research and development. In order to determine the minimum sample size required for achieving a desired power, one needs to have some information regarding study parameters such as variability associated with the observations and the difference (e.g., treatment effect) that the study is designed to detect. In practice, it is well recognized that sample size calculation depends upon the assumed variability associated with the observation, which is often unknown. Thus, the classical pre-study power analysis for sample size calculation based on information obtained from a small pilot study (with large variability) could vary widely and hence instable depending upon the sampling variability. As a result, one of the controversial issues regarding sample size calculation is the stability (sensitivity or robustness) of the obtained sample size. To overcome the instability of sample size calculation, alternatively, Chow (2011) suggested that a bootstrap-median approach be considered to select a stable (required minimum) sample size. Such an improved stable sample size can be derived theoretically by the method of an Edgeworth-type expansion. Chow (2011) showed that the bootstrap-median approach performs quite well for providing a stable sample size in clinical trial through an extensive simulation study.

It should be noted that procedures used for sample size calculation could be very different from one another depending on different study objectives and hypotheses (e.g., testing for equality, testing for superiority, or testing for non-inferiority/equivalence) and different data types (e.g., continuous, binary, and time-to-event). For example, see Lachin and Foulkes (1986), Lakatos (1986), Wang and Chow (2002a), Wang et al. (2002) and Chow and Liu (2008). For a good introduction and summary, one can refer to Chow et al. (2008). In this chapter, for simplicity, we will focus on the most commonly seen situation where the primary response is continuous and the hypotheses of interest are about the mean under the normality assumption. Most of our discussions thereafter focus on the one sample problem for the purpose of simplicity. However, the extension to two-sample problem is straightforward.

1.4.8 Consistency Test

In recent years, multi-regional (or multi-national) multi-center clinical trials have become very popular in global pharmaceutical/clinical development. The main purpose of a multi-regional clinical trial is not only to assess the efficacy of the test treatment over all regions in the trial, but also to bridge the overall effect of the test treatment to each of the region in the trial. Most importantly, a multi-regional clinical trial is to shorten the time for drug development and regulatory development, submission, and approval around the world (Tsong ad Tsou, 2013). Although multi-regional clinical trials provide the opportunity to fully utilize clinical data from all regions to support regional (local) registration, some critical issues such as regional differences (e.g., culture and medical practice/perception) and possible treatment-by-region interaction may occur, which may have an impact on the validity of the multi-regional trials.

In multi-regional trials for global pharmaceutical/clinical development, one of the commonly encountered critical issues is that clinical results observed from some regions (e.g., Asian Pacific region) are inconsistent with clinical results from other regions (e.g., European Community) or global results (i.e., all regions combined). The inconsistency in clinical results between different regions (e.g., Asian Pacific region and European Community and/or United States) could be due to difference in ethnic factors. In this case, the dose or dose regimen may require adjustment or a bridging study may be required before the data can be pooled for an overall combined assessment of the treatment effect. As a result, evaluation of consistency between specific regions (sub-population) and all regions combined (global population) is necessarily conducted before regional registration (e.g., Japan and China). It should be noted that different regions may have different requirements regarding sample sizes at specific regions in order to comply with regulatory requirement for registration (see, e.g., MHLW, 2007).

In practice, consistency between clinical results observed from a subpopulation (specific region such as Japan or China) and the entire population (all regions combined) is often interpreted as similarity and/or equivalence between the two populations in terms of treatment effect (i.e., safety and/or efficacy). Along this line, several statistical methods including test for consistency (Shih, 2001b), assessment of consistency index (Tse et al., 2006), evaluation of sensitivity index (Chow, Shao, and Hu, 2002), achieving reproducibility and/or generalizability (Shao and Chow, 2002), Bayesian approach (Hsiao et al., 2007; Chow and Hsiao, 2010), and the Japanese approach for evaluation of assurance probability (MHLW, 2007) have been proposed in the literature (see also, Liu, Chow, and Hsiao, 2013). The purpose of this section is not only to provide an overview of these methods, but also to compare the relative performances of these methods through extensive clinical trial simulation.

1.4.9 Extrapolation

For marketing approval of a new drug product, the FDA requires that at least two adequate and well-controlled clinical trials be conducted to provide substantial evidence regarding the effectiveness and safety of the drug product under investigation. The purpose of requiring at least two clinical studies is not only to assure reproducibility, but also to provide valuable information regarding generalizability. Generalizability can have several distinct meanings. First, it can refer to whether the clinical results of the original target patient population under study (e.g., adults) can be generalized to other similar but different patient populations (e.g., pediatrics or elderly). Second, it can be a measure of whether a newly developed or approved drug product in one region (e.g., United States or European Union) can be approved at another region (e.g., countries in Asian-Pacific Region) particularly if there exists a concern that differences in ethnic factors could alter the efficacy and safety of the drug product in the new region. Third, for case-control studies, it is often of interest to determine whether the developed or established predictive model based on a database at a medical center can be applied to a different medical center that has similar but different database for patients with similar diseases under study. In practice, since it is of interest to determine whether the observed clinical results from the original target patient population can be generalized to a similar but different patient population, we will focus on the first scenario. Statistical methods for assessment of generalizability of clinical results, however, can also be applied to other scenarios.

Although the ICH E5 guideline establishes the framework for the acceptability of foreign clinical data, it does not clearly define the desired similarity in terms of dose response, safety, and efficacy between the original region and a new region. Shih (2001) interpreted similarity as *consistency* among study centers by treating the new region as a new center of multicenter clinical trials. Under this definition, Shih proposed a method for assessment of consistency to determine whether the study is capable of bridging the foreign data to the new region. Alternatively, Shao and Chow (2002) proposed the concepts of reproducibility and *generalizability* probabilities for assessing bridging studies. In addition, Chow et al. (2002) proposed to assess similarity by analysis using a *sensitivity index*, which is a measure of population shift between the original region and the new region. For assessment of generalizability of clinical results from one population to another, Chow (2010) proposed to evaluate a so-called generalizability probability of a positive clinical result observed from the original patient population by studying the impact of shift in target patient through a model that link the population means with some covariates (see also, Chow and Shao, 2005; Chow and Chang, 2006). However, in many cases, such covariates may not exist or exist but not observable. In this case, it is suggested that the degree of shift in location and scale of patient population be studied based on a mixture distribution by assuming the location or scale parameter is random variable (Shao and Chow, 2002). The purpose of this section is to assess the generalizability of clinical results by evaluating the sensitivity index under different models given the different situations: (1) shift in location parameter is random, (2) shift in scale parameter is random, and (3) shifts in both location and scale parameters are random.

1.4.10 Drug Products with Multiple Components

In recent years, as more and more innovative drug products are going off patent protection, the search for new medicines that treat critical and/or life-threatening diseases such as cardiovascular diseases and cancer has become the center of attention of many pharmaceutical companies and research organizations such as National Institute of Health (NIH). This leads to the study of the potential use of promising traditional Chinese medicines (TCM), especially for critical and/or life-threatening diseases. Bensoussan et al. (1998) used randomized clinical trial (RCT) to assess the effect of Chinese herb medicine in treating the Irritable Bowel Syndrome. However, RCT is not in common use when studying TCM. There are fundamental differences between Western medicines and TCM in terms of diagnostic procedures, therapeutic indices, medical mechanism, medical theory and practice (Chow, Pong and Chang, 2006; Chow, 2015). Besides, TCM often consists of multiple components with flexible dose.

Chinese doctors believe that all of the organs within a healthy subject should reach the so-called *global dynamic balance and harmony* among organs. Once the global balance is broken at certain sites such as heart, liver, or kidney, some signs and symptoms will appear to reflect the imbalance at these sites. The collective signs and symptoms are then used to determine what disease the individual patient have. An experienced

Chinese doctor usually assesses the causes of global imbalance before a TCM with flexible doses is prescribed to fix the problem. This approach is sometimes referred to as a personalized (or individualized) medicine approach. In practice, TCM consider inspection, auscultation and olfaction, interrogation, and pulse taking and palpation as the primary diagnostic procedures. The scientific validity of these subjective and experience-based diagnostic procedures has been criticized due to lack of reference standards and anticipated large evaluator-to-evaluator (i.e., Chinese doctor-to-Chinese doctor) variability. For a systematic discussion of the statistical issues of TCM, see Chow (2015).

In this book, we attempt to propose a unified approach to developing a composite illness index based on a number of indices collected from a given subject under the concept of global dynamic balance among organs. Dynamic balance among organs can be defined as follows. Following the concept of testing bioequivalence or biosimilarity, if the 95% confidence upper bound is less than some health limit, we conclude that the treatment achieves dynamic balance among the organs of the subject hence is considered as efficacious. If we fail to reject the null hypothesis, we conclude that the treatment is not efficacious since there is still a signal of illness (e.g., some of signs and symptoms are still out of the health limit). In practice, these signals of illness can be grouped to diagnose specific diseases based on some pre-specified reference standards for diseases status of specific diseases which are developed based on indices related to specific organs (or diseases).

1.4.11 Advisory Committee

The FDA has established advisory committees each consisting of clinical, pharmacological, and statistical experts and one consumer advocate (not employed by the FDA) in designated drug classes and sub-specialties. The responsibilities of the committees are to review data presented in NDA's and to advise FDA as to whether there exists substantial evidence of safety and effectiveness based on adequate and well-controlled clinical studies. In addition the committee may also be asked at times to review certain INDs, protocols, or important issues relating to marketed drugs and biologics. The advisory committees not only supplement the FDA's expertise but also allow an independent peer review during the regulatory process. Note that the FDA usually prepares a set of questions for the advisory committee to address at the meeting. The following is a list of some typical questions:

1. Are there two or more adequate and well-controlled trials?
2. Have the patient populations been well enough characterized?
3. Has the dose-response relationship been sufficiently characterized?

4. Do you recommend the use of the drug for the indication sought by the sponsor for the intended patient population?

The FDA usually will follow the recommendations made by the Advisory Committee for marketing approval, though they do not have to legally.

1.4.12 Recent FDA Critical Clinical Initiatives

In addition to the above-mentioned practical, challenging, and controversial issues, the FDA also kicked off several critical clinical initiatives to assist sponsors in pharmaceutical product development. These critical clinical initiatives include, but are limited to, (i) statistical methodology development for generic drugs and biosimilar product (see Chapter 15); (ii) precision/personalized medicine (see Chapter 16); (iii) biomarker-driven clinical trials (see Chapter 16); (iv) big data analytics (see Chapter 17); (v) rare diseases drug development (see Chapter 18); (vi) real-world data and real-world evidence; (vii) model-informed drug development (MIDD); and (viii) machine learning for mobile individualized medicine (MIM) and imaging medicine.

Among these critical clinical initiatives, more details can be found in Chapters 13–18. For example, more details regarding an overview of adaptive trial designs such as phase II/III seamless adaptive trial design can be found in Chapters 13 and 14. Statistical methods for assessment of bioequivalence for generic drugs and for demonstration of biosimilarity for biosimilar products can be found in Chapter 15. The difference between precision medicine and personalized medicine is discussed in Chapter 16. The concept of big data analytics can be found in Chapter 17. Innovative thinking for rare disease drug development including innovative trial designs, statistical methods for data analysis, and sample size requirement can be found in Chapter 18.

Regarding real-world data and real-world evidence, the following are commonly asked questions:

1. Does the information include clinical data collected from randomized clinical trials (RCT)?
2. Does real-world evidence constitute substantial evidence for evaluation of safety and efficacy of a drug product?

To provide a better understanding of the issue associated with real-world data and real-world evidence, Table 1.9 provides a comparison between real-world evidence and substantial evidence obtained from adequate well-controlled clinical studies. As it can be seen from Table 1.9, analysis of real-world data for obtaining real-world evidence could lead to incorrect and unreliable conclusion regarding safety and efficacy of the test treatment under investigation

TABLE 1.9

A Comparison between Real Word Evidence (RWE) and Substantial Evidence Obtained from Randomized Clinical Trial (RCT)

Real-World Evidence	Randomized Clinical Trials
• General population	• Specific population
• Selection bias	• Bias is minimized
• Variability—expected and not controllable	• Variability – expected and Controllable
• Real-world evidence from multiple/diverse sources	• Substantial evidence form RCTs
• Reflect real clinical practice	• Reflect controlled clinical practice
• Statistical methods are not fully established	• Statistical methods are well-established
• Could generate incorrect and unreliable conclusion	• Accurate and reliable conclusion

due to (i) selection bias; (ii) uncontrollable variability; and (iii) the real-world data are from multiple/diverse sources. Thus, it is suggested that real-world evidence should be used for safety assessment but not for efficacy evaluation in regulatory approval process.

Regarding model-informed drug development (MIDD), as indicated by PDUFA VI, MIDD can be classified into six categories: (i) PK/PD; (ii) PK, population PK (POPPK); and physiologically based PK (PBPK) modeling; (iii) disease models including clinical trial model; (iv) system biology: quantitative system pharmacology (QSP) and congenital insensitivity to pain with anhidrosis (CIPA); (v) quantitative structure activity relationship (QSAR) and quantitative structure property relationship (QSPR); and (vi) clinical trial simulation (see Figure 1.5). Statistically speaking, MIDD is to study response-exposure relationship, which can be performed in the following steps: (i) model building, (ii) model validation, and (iii) model

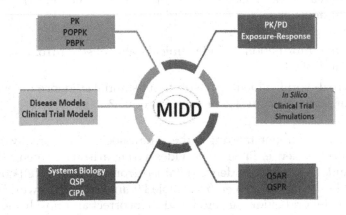

FIGURE 1.5
The scope of model-informed drug development.

generalizability. Model building may involve the identification of risk factors (predictors), test for collinearity, and goodness-of-fit. For model validation, a typical approach is to randomly split the data into two sets: one for model building and one for model validation. At this stage of model validation, it is considered internal validation. For external validation, it is usually referred to as model generalizability, i.e., the predictive model can be generalized from one patient population to another or from one medical center to another.

Machine learning is the scientific study of algorithms and statistical models that computer systems use to effectively perform a specific task without using explicit instructions, relying on patterns and inference instead. The application of machine learning in drug research and development includes, but are not limited to, mobile individual medicine (MIM) and imaging medicine (IM). For mobile individualized medicine, machine learning can be applied for (i) safety monitoring, e.g., the detection of the death of overdose of opioid, and (ii) capture of real time data in the conduct of clinical trials. For imaging medicine, machine learning is useful in analysis and interpretation of imaging data collected from clinical research.

1.5 Aim and Scope of the Book

This is intended to be the first book entirely devoted to statistics in regulatory science that are relevant to the review and approval process of regulatory submissions of pharmaceutical product development. It covers general principles for GSP and key statistical concepts that are commonly employed in the review and approval process of regulatory submissions. Some practical challenging and controversial issues that may occur during the review and approval process of regulatory submissions are discussed. In addition to complex innovative designs, the FDA recently kicked off several critical clinical initiatives regarding precision medicine, big data analytics, rare disease product development, biomarker-driven clinical research, model-informed drug development (MIDD), Patient-focused risk assessment drug development, and real-world data and real-world evidence. These initiatives are to assist the sponsors in expedite pharmaceutical research and development process in a more efficient way. The purpose of this book is to outline these challenging and controversial issues and recent development regarding FDA critical clinical initiatives that are commonly encountered in the review and approval process of regulatory submissions.

It is my goal to provide a useful desk reference and state-of-the-art examination of statistics in regulatory science in drug development, to those in government regulatory agencies who have to make critical decisions in regulatory submissions, and to biostatisticians who provide the statistical support to studies conducted for regulatory submissions of pharmaceutical product development. More importantly we would like to provide graduate students in pharmacokinetics, clinical pharmacology, biopharmaceutics, clinical investigation, and biostatistics an advanced textbook in biosimilar studies. We hope that this book can serve as a bridge among government regulatory agencies, the pharmaceutical industry and academia.

The scope of this book is restricted to practical issues that are commonly seen in regulatory science of pharmaceutical research and development. This book consists of 20 chapters concerning topics related to research activities, review of regulatory submissions, policy/guidance development, and FDA critical clinical initiatives in regulatory science. Chapter 1 provides key statistical concepts, and innovative design methods in clinical trials that are commonly considered in regulatory science. Also included in this chapter are some practical, challenging, and controversial issues that are commonly seen in the review and approval process of regulatory submissions. Chapter 2 provides interpretation of substantial evidence required for demonstration of the safety and effectiveness of drug products under investigation. The related concepts of totality-of-the-evidence and real-world evidence are also described in this chapter. Chapter 3 distinguishes the concepts of hypotheses testing and confidence interval approach for evaluation of the safety and effectiveness of drug products under investigation. Also included in this chapter is the comparison between the use of a 90% confidence interval approach for evaluation of generics/biosimilars and the use of a 95% confidence interval approach for assessment of new drugs. Chapter 4 deals with endpoint selection in clinical research and development. Also included in this chapter is the development of therapeutic index for endpoint selection in complex innovative designs such as multiple-stage adaptive designs. Chapter 5 focuses on non-inferiority margin selection. Also included in this chapter is a proposed clinical strategy for margin selection based on risk assessment of false positive rate. Chapters 6 and 7 provide discussions on statistical methods for missing data imputation and multiplicity adjustment for multiple comparisons in clinical trials, respectively. Sample size requirements under various designs are summarized in Chapter 8. Chapter 9 introduces that concept of reproducible research. Chapter 10 discusses the concept and statistical methods for assessment of extrapolation across patient populations and/or indications. Chapter 11 compares statistical methods for evaluation of consistency in multi-regional clinical trials. Chapter 19 provides an overview of drug products with multiple components such as botanical drug products and traditional Chinese medicine. Chapter 13 provides an overview of adaptive trial designs that are commonly used in pharmaceutical/clinical research and development. Chapter 14 evaluates several selection criteria that are commonly considered in adaptive dose finding studies. Chapter 15 compares bioequivalence assessment for generic drug products and biosimilarity assessment for biosimilar products. Also included in this chapter is a proposed general approach for assessment of bioequivalence for generics and biosimilarity for biosimilars. Chapter 16 discusses the difference between precision medicine and personalized medicine. Chapter 17 introduces the concept of big data analytics. Also included in this chapter is types of big data analytics and potential bias of big data analytics. Chapter 18 focuses rare disease clinical development including innovative trial designs, statistical methods for data analysis, and some commonly seen challenging issues.

2

Totality-of-the-Evidence

2.1 Introduction

As indicated in the previous Chapter 1, for approval of pharmaceutical products, the United States (US) Food and Drug Administration (FDA) requires that substantial evidence regarding the safety and effectiveness of the test treatment under investigation be provided for review and regulatory approval. The substantial evidence, however, can only be obtained through the conduct of *adequate and well-controlled* studies (Section 314 of 21 CFR). The FDA requires that reports of adequate and well-controlled investigations provide the primary basis for determining whether there is substantial evidence to support the claims of drugs, biologics, and medical devices.

Recently, the FDA proposed a new concept of totality-of-the-evidence for review and approval of regulatory submissions of biosimilar (follow-on biological) products due to structural and functional complexity of large molecular biological drug products (FDA, 2015a). Following similar concepts of substantial evidence, the totality-of-the-evidence can be obtained through an FDA recommended stepwise approach that starts with analytical studies for functional and structural characterization of critical quality attributes (CQAs) that are identified at various stages of manufacturing process and considered relevant to clinical outcomes. The stepwise approach continues with animal studies for toxicity assessment, pharmacological studies such as pharmacokinetic (PK) and pharmacodynamic (PD) studies, and clinical studies such as immunogenicity, safety, and efficacy studies. The FDA indicated that stepwise approach is an approved approach for obtaining totality-of-the-evidence for demonstration of highly similar between a proposed biosimilar product and an innovative biological product. Thus, the totality-of-the-evidence comprises analytical similarity evaluation, PK/PD similarity assessment, and clinical similarity demonstration. Note that there is no one-size-fits-all assessment in the stepwise approach.

In practice, it is very likely that the proposed biosimilar product does not meet requirements for all similarity tests (i.e., analytical similarity, PK/PD similarity, and clinical similarity). For example, the proposed biosimilar product may fail analytical similarity evaluation but pass both PK/PD and clinical

similarity assessment. In this case, the sponsor is often asked to provide justification that the observed difference has little impact on the clinical outcomes.
On the other hand, if the proposed biosimilar product passes tests for analytical similarity and PK/PD similarity but fails to pass clinical similarity
test, most likely the proposed biosimilar product will be viewed as not highly
similar to the innovative product. Thus, a few of questions have been raised.
First, *are there any links among analytical similarity, PK/PD similarity, and clinical similarity*? In other words, are analytical similarities predictive of PK/PD
similarity and/or clinical similarity? Second, *does analytical similarity evaluation
carry the same weight as PK/PD similarity assessment and clinical similarity demonstration in the review and approval process of the proposed biosimilar product*? If they
carry the same weight (i.e., they are equally important), then one would expect
all analytical similarity, PK/PD similarity, and clinical similarity should be
demonstrated for obtaining the totality-of-the-evidence before the proposed
biosimilar product can be approved. Third, *does totality-of-the-evidence constitute
substantial evidence for approval of the proposed biosimilar product*? This chapter
will attempt to address these three questions.

In the next section, an introduction of substantial evidence that is required
for evaluation of pharmaceutical products is briefly described. Section 2.3
will examine the concept of the totality-of-the-evidence for demonstration
of biosimilarity between a proposed biosimilar product and an innovative
biological product. Also included in this section is an example concerning a
recent regulatory submission utilizing the concept of totality-of-the-evidence
for regulatory approval. Section 2.4 posts some practical and challenging
issues regarding the use of totality-of-the-evidence in the regulatory approval
process for biosimilar products. Section 2.5 outlines the development of
an index for quantitation of the totality-of-the-evidence. Some concluding
remarks are given in Section 2.6.

2.2 Substantial Evidence

For approval of a test treatment under investigation, Section 314.126 of
21 CFR requires that substantial evidence regarding the safety and effectiveness of the test treatment under investigation be provided in the review and
approval process of the test treatment under investigation. Section 314.126 of
21 CFR also indicates that substantial evidence, however, can only be obtained
through the conduct of *adequate and well-controlled* studies. Section 314.126 of
21 CFR provides the definition of an adequate and well-controlled study,
which is summarized in Table 2.1.

As it can be seen from Table 2.1, an adequate and well-controlled study is
judged by eight criteria specified in the CFR. These criteria include: (i) study objectives; (ii) method of analysis; (iii) design of studies; (iv) selection of subjects; (v)

TABLE 2.1

Characteristics of an Adequate and Well-Controlled Study

Criteria	Characteristics
Objectives	Clear statement of investigation's purpose
Methods of analysis	Summary of proposed or actual methods of analysis
Design	Valid comparison with a control to provide a quantitative assessment of drug effect
Selection of subjects	Adequate assurance of the disease or conditions under study
Assignment of subjects	Minimization of bias and assurance of comparability of groups
Participants of studies	Minimization of bias on the part of subjects, observers, and analysts
Assessment of responses	Well-defined and reliable
Assessment of the effect	Requirements of appropriate statistical methods

Source: Section 314.126 of 21 CFR.

assignment of subjects; (vi) participants of studies; (vii) assessment of responses; and (viii) the assessment of effect, all of which are objective and are closely related to statistics used for assessment of test treatment under investigation.

In summary, an adequate and well-controlled study starts with clearly stated study objectives. Under the clearly stated study objective(s), valid study designs and methods will then be applied for collecting quality data from a random sample (a well representative sample) drawn from the target population. For this purpose, statistical methods such as randomization and blinding are used to minimize potential biases and variations for an accurate and reliable assessment of the treatment responses and effects of the test treatment under investigation.

2.3 Totality-of-the-Evidence

For approval of a proposed biosimilar product, the FDA requires that totality-of-the-evidence be provided to support a demonstration that the proposed biosimilar product is highly similar to the original US-licensed product, notwithstanding minor differences in clinically inactive components, and that there are no clinically meaningful differences between the proposed biosimilar product and the US-licensed product in terms of the safety, purity and potency of the product.

2.3.1 Stepwise Approach

As defined in the Biologics Price Competition and Innovation Act of 2009 (BPCI Act), a biosimilar product is a product that is *highly similar* to the reference product, notwithstanding minor differences in clinically inactive components and for which there are no clinically meaningful differences in terms of safety, purity, and potency. Based on the definition of the BPCI Act, biosimilarity requires that there are no *clinically meaningful differences* in terms of *safety, purity* and *potency*. Safety could include PK/PD, safety and tolerability, and immunogenicity studies. Purity includes all critical quality attributes during manufacturing process. Potency is referred to as efficacy studies. As indicated earlier, in the 2015 FDA guidance on scientific considerations, the FDA recommends that a stepwise approach be considered for providing the totality-of-the-evidence to demonstrating biosimilarity of a proposed biosimilar product as compared to a reference product (FDA, 2015a).

To assist the sponsor in biosimilar product development, the FDA recommends a stepwise approach for obtaining the totality-of-the-evidence for demonstrating biosimilarity between the proposed biosimilar product and its innovative drug product in terms of safety, purity, and efficacy (Chow, 2013; FDA, 2015a, 2017; Endrenyi et al., 2017). The stepwise approach starts with similarity assessment in CQAs, in analytical studies, followed by the similarity assessment in pharmacological activities in PK/PD studies and similarity assessment in safety and efficacy in clinical studies. The pyramid illustrated in Figure 2.1 briefly summarizes the stepwise approach.

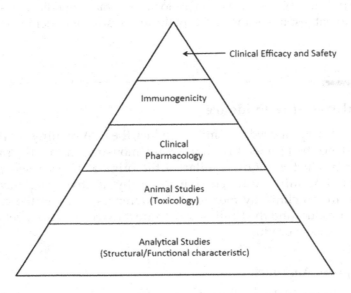

FIGURE 2.1
Stepwise approach for biosimilar product development.

2.3.2 Fundamental Biosimilarity Assumptions

As indicated in Section 2.3.1, the assessment of analytical data constitutes the first step in the FDA recommended totality-of-the-evidence as stepwise approach. In practice, the assessment of analytical data is performed to achieve the following primary objectives (see, e.g., BLA 781028 and BLA 761074).

1. The assessment of analytical data is to demonstration that the proposed biosimilar product can be manufactured in a well-controlled and consistent manner that meets appropriate quality standards;
2. The assessment of analytical data is to support a demonstration that the proposed biosimilar product and the reference product are highly similar;
3. The assessment of analytical data may be used to serve as a bridging for PK/PD similarity and/or clinical similarity;
4. The assessment of analytical data may be used to provide scientific justification for extrapolation of data to support biosimilarity in each of the additional indications for which the sponsor is seeking licensure.

Note that, regarding (4) for analytical similarity assessment in CQAs, the FDA further recommends a tiered approach that classifies CQAs into three tiers depending upon their criticality or risk ranking relevant to clinical outcomes. For determination of criticality or risk ranking, the FDA suggests establishing a predictive (statistical) model based on either mechanism of action (MOA) or PK relevant to clinical outcome.

To achieve these objectives, the study of the relationship among analytical, PK/PD, and clinical data is essential. For this purpose, the following assumptions are made for the stepwise approach for obtaining the totality-of-the-evidence.

1. Analytical similarity is predictive of PK/PD similarity;
2. Analytical similarity is predictive of clinical outcomes;
3. PK/PD similarity is predictive of clinical outcomes.

These assumptions, however, are difficult (if not impossible) to verify in practice. For assumptions (1) and (2), although many *in vitro* and *in vivo* correlations (IVIVC) have been studied in the literature, the correlations between specific CQAs and PK/PD parameters or clinical endpoints are not fully studied and understood. In other words, most predictive models are not well established or are established but not validated. Thus, it is not clear how a (notable) change in a specific CQA can be translated to a change in drug absorption or clinical outcome. For (3), unlike bioequivalence assessment for generic drug products, there does not exist a *Fundamental Biosimilarity Assumption* indicating that PK/PD similarity implies clinical similarity in terms of safety and efficacy. In other words, PK/PD similarity or dissimilarity may or may not lead to clinical similarity. Note that assumptions (1) and (3) being met simultaneously does not lead to the validity of assumption (2) automatically.

The validity of assumptions (1)–(3) is critical for the success of obtaining totality-of-the-evidence for assessing biosimilarity between the proposed biosimilar and the innovative biological product. This is because the validity of these assumptions ensures the relationships among analytical, PK/PD, and clinical similarity assessment and consequently the validity of the overall biosimilarity assessment. Figure 2.2 illustrates relationships among analytical, PK/PD, and clinical assessments in the stepwise approach for obtaining the totality-of-the-evidence in biosimilar product development.

2.3.3 Examples—Recent Biosimilar Regulatory Submissions

For illustration purpose, consider two recent FDA biosimilar regulatory submissions, i.e., Avastin biosimilar (ABP215 sponsored by Amgen) and Herceptin biosimilar (MYL-1401O sponsored by Mylan). These two regulatory submissions were reviewed and discussed at an Oncologic Drug Advisory Committee (ODAC) meeting held on July 13th, 2017 in Silver Spring, Maryland. Table 2.2 briefly summarizes the results of the review based on the concept of totality-of-the-evidence.

As for ABP215, a proposed biosimilar to Genetech's Avastin, although ABP215 passed both PK/PD similarity and clinical similarity tests, several quality attribute differences were noted. These notable differences include glycosylation content, Fc γ RIIIa binding and product related species (aggregates, fragments, and charge variants). The glycosylation and Fc γ RIIIa binding differences were addressed by means of *in vitro* cell based on ADCC and CDC activity, which were not detected for all products (ABP215,

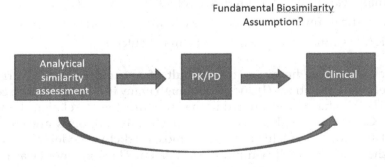

FIGURE 2.2
Relationships among analytical, PK/PD, and clinical assessment.

TABLE 2.2

Examples of Assessment of Totality-of-the-Evidence

			Totality-of-the-Evidence		
Regulatory Submission	Innovative Product	Proposed Biosimilar	Analytical Similarity	PK/PD Similarity	Clinical Similarity
BLA 761028 (Amgen)	Avastin	ABP215	Notable differences observed in glycosylation content and FcgRIIIa binding	Pass	Pass
BLA 761074 (Mylan)	Herceptin	MYL-1401O	Subtle shifts in glycosylation (sialic acid, high mannose, and NG-HC)	Pass	Pass

US-licensed Avastin, and EU-approved Avastin). In considering the totality-of-the-evidence, the ODAC panel considered the data submitted by the sponsor sufficient to support a demonstration that ABP215 is highly similar to the US-licensed Avastin, notwithstanding minor differences in clinically inactive components, and supports that there are no clinically meaningful differences between ABP215 and the US-licensed Avastin in terms of the safety, purity and potency of the product.

For MYL-1401O, a proposed biosimilar to Genetech's Herceptin, although MYL-1401O passed both PK/PD similarity and clinical similarity tests, there are subtle shifts in glycosylation (sialic acid, high mannose, and NG-HC). However, the residual uncertainties related to increase in total mannose forms and sialic acid and decrease in NG-HC were addressed by ADCC similarity and by the PK similarity. Thus, the ODAC panel considered the data submitted by the sponsor was sufficient to support a demonstration that MYL-1401O is highly similar to the US-licensed Herceptin, notwithstanding minor differences in clinically inactive components, and to support that there are no clinically meaningful differences between MYL-1401O and the US-licensed Herceptin in terms of the safety, purity and potency of the product.

2.3.4 Remarks

As discussed in Section 2.3.3, it is not clear whether totality-of-the-evidence of highly similarity can only be achieved if the proposed biosimilar product has passed all similarity tests across different domains of analytical, PK/PD, and clinical assessment. When notable differences in some CQAs in Tier 1 are observed, these notable differences may be ignored if the sponsors can provide scientific rationales/justification to rule out that the observed difference have an impact on clinical outcomes. This, however, is somewhat controversial because Tier 1 CQAs are considered most relevant to clinical outcomes depending upon their criticalities or risk rankings that impact the clinical outcomes. The criticalities and/or risk rankings may be determined using model (3). If a notable difference is considered having little or no impact on the clinical outcome, then the CQA should not be classified into Tier 1 in the first place. This controversy could be due to the classification of CQAs based on subjective judgment rather than via objectively statistical modeling.

In the two examples concerning biosimilar regulatory submissions of ABP215 (Avastin biosimilar) and MYL-1401O (Herceptin biosimilar), the sponsors also sought approval across different indications. There has been a tremendous amount of discussion regarding whether totality-of-the-evidence observed from one indication or a couple of indications can be used to extrapolate to other indications even different indications have similar mechanism of actions. The ODAC panel expressed their concern over extrapolation in the absence of any collection of clinical data and encouraged further research on the scientific validity of extrapolation and/or generalizability of the proposed biosimilar product.

2.4 Practical Issues and Challenges

The use of the concept of totality-of-the-evidence for demonstrating that a proposed biosimilar product is highly similar to an innovative biological product has been challenged through the three questions described in Section 2.1. To address the first question whether analytical similarity is predictive of PK/PD similarity and/or clinical similarity, the relationship among analytical similarity, PK/PD similarity and clinical similarity must be studied.

2.4.1 Link among Analytical Similarity, PK/PD Similarity, and Clinical Similarity

Relationships among CQAs, PK/PD responses, and clinical outcomes can be described in Figure 2.2. In practice, for simplicity, CQAs, PK/PD responses, and clinical outcomes are usually assumed linearly correlated. For example, let x, y, and z be the test result of a CQA, PK/PD response, and clinical outcome, respectively. Under assumptions (1)–(3), we have

(1) $y = a_1 + b_1 x + e_1$;

(2) $z = a_2 + b_2 y + e_2$

(3) $z = a_3 + b_3 x + e_3$;

where e_1, e_2, and e_3 follow a normal distribution with mean 0 and variances σ_1^2, σ_2^2, and σ_3^2, respectively. In practice, each of the above models is often difficult, if it is not impossible, to validate due to insufficient data being collected during the biosimilar product development. Under each of the above models, we may consider the criterion for examination of the closeness between an observed response and its predictive value to determine whether the respective model is a good predictive model. As an example, under model (1), we may consider the following two measures of closeness, which are based on either the *absolute* difference or the *relative* difference, respectively, between an observed value y and its predictive value $\hat{y}$

$$\text{Criterion I.} \quad p_1 = P\left\{\left|y - \hat{y}\right| < \delta\right\},$$

$$\text{Criterion II.} \quad p_2 = P\left\{\left|\frac{y - \hat{y}}{y}\right| < \delta\right\}.$$

It is desirable to have a high probability that the difference or the relative difference between y and $\hat{y}$, given by p_1 and p_2, respectively, is less than a clinically meaningful difference δ.

Suppose there is a well-established relationship between x (e.g., test result of a given CQA) and y (e.g., PK/PD response). Model (1) indicates that a change in CQA, say Δ_x corresponds to a change of $a_1 + b_1 \Delta_x$ in PK/PD response. Similarly, model (2) indicates that a change in PK/PD response, say Δ_y corresponds to a change of $a_2 + b_2 \Delta_y$ in clinical outcomes. Models (2) and (3) allows us to evaluate the impact of the change in CQA (i.e., x) on PK/PD (i.e., y) and consequently clinical outcome (i.e., z). Under models (2) and (3), we have

$$a_2 + b_2 y + e_2 = a_3 + b_3 x + e_3.$$

This leads to

$$a_1 = \frac{a_3 - a_2}{b_2}, \quad b_1 = \frac{b_3}{b_2}, \quad \text{and} \quad e_1 = \frac{e_3 - e_2}{b_2}.$$

with

$$b_2^2 \sigma_1^2 = \sigma_3^2 + \sigma_2^2$$

or

$$\sigma_1 = \frac{1}{b_2}\sqrt{\sigma_2^2 + \sigma_3^2}.$$

In practice, the above relationships can be used to verify primary assumptions as described in the previous section provided that models (1)–(3) have been validated. Suppose models (1)–(3) are well-established, validated, and fully understood. A commonly asked question is whether PK/PD studies and/or clinical studies can be waived if analytical similarity and/or PK/PD similarity have been demonstrated. Note that the above relationships hold only under our linearity assumption. When there is a departure from linearity in any one of models (1)–(3), the above relationships are necessarily altered.

Considering multiple CQAs and several endpoints in PK/PD and clinical outcomes, the model (1)–(3) can be easily extended to a general linear model of the form

$$(4)\, Y = B_1 X + E_1;$$

$$(5)\, Z = B_2 Y + E_2;$$

$$(6)\, Z = B_3 X + E_3;$$

where, each of $E_1, E_2,$ and E_3 follow a multivariate normal distribution, $N(0, \sigma_1^2 I), N(0, \sigma_2^2 I),$ and $N(0, \sigma_3^2 I),$ respectively. Thus, we have

$$B_1 = B_2^{-1} B_3,$$

with

$$\sigma_1^2 I = \left(\sigma_2^2 + \sigma_3^2\right) B_2^{-1},$$

where

$$B_1 = \left(X'X\right)^{-1} X'Y,$$

$$B_2 = \left(Y'Y\right)^{-1} Y'Z,$$

and

$$B_3 = \left(X'X\right)^{-1} X'Z.$$

The existence of unique solution depends on the rank of matrices, X and Y. One way to obtain those solutions is to use numerical computations. In this case, no clinical meaningful difference might be obtained if the minimum of the

$$P\left\{ norm\left(Z - Z\left(X'X\right)^{-1} X'Z\right) < \delta \right\}$$

and

$$P\left\{norm\left(Z - Z(X'X)^{-1}Y'Z\right) < \delta\right\}$$

is sufficiently large.

2.4.2 Totality-of-the-Evidence versus Substantial Evidence

The second and third questions described in Section 2.1 are basically to challenge the FDA to explain (i) what constitute the totality-of-the-evidence and (ii) whether the totality-of-the-evidence is equivalent to regulatory standard of substantial evidence for approval of drug products. As indicated earlier, the FDA's recommended stepwise approach focuses on three major domains, namely, analytical, PK/PD, and clinical similarity, which may be highly correlated under models (1)–(3). Some pharmaceutical scientists interpret the stepwise approach as a scoring system (perhaps, with appropriate weights) that includes the domains of analytical, PK/PD, and clinical similarity assessment. In this case, the totality-of-the-evidence can be assessed based on information regarding biosimilarity obtained from each domain. In practice, for each domain, we may consider either the FDA's recommended binary response (i.e., similar or dis-similar) or resort to the use of the concept of biosimilarity index (Chow et al., 2011) to assess similarity information and consequently the totality-of-the-evidence across domains.

For the FDA's recommended approach, Table 2.3 provides possible scenarios that may be encountered when performing analytical similarity assessment, PK/PD similarity test, and clinical similarity assessment. As it can be seen from Table 2.1, if the proposed biosimilar product passes similarity test in all domains, the FDA considers the sponsor has provided totality-of-the-evidence for demonstration of highly similarity between the proposed biosimilar and the innovative biological product. On the other hand, if the proposed

TABLE 2.3

Assessment of Totality-of-the-Evidence

No. of Dis-similarities	Analytical Similarity Assessment	PK/PD Similarity Assessment	Clinical Similarity	Overall Assessment
0	Yes	Yes	Yes	Yes
1	Yes	Yes	No	No
1	Yes	No	Yes	*
1	No	Yes	Yes	*
2	Yes	No	No	No
2	No	Yes	No	No
2	No	No	Yes	No
3	No	No	No	No

* Scientific rationale are necessarily provided.

biosimilar product fails to pass any of the suggested similarity assessments (i.e., analytical similarity, PK/PD similarity, and clinical similarity), then regulatory agency will reject the proposed biosimilar product.

In practice, it is uncommon to see that the proposed biosimilar may fail in one of the three suggested similarity assessments, namely analytical similarity, PK/PD similarity, and clinical similarity assessments. In this case, the regulatory agency may be reluctant to grant approval of the proposed biosimilar product. A typical example of this sort of failure is that notable differences in some CQAs between the proposed biosimilar product and the innovative biological product may be observed in analytical similarity assessment. In this case, the sponsors often provide scientific rationales/justifications to indicate that the notable differences have little or no impact on clinical outcomes. This move by the sponsors may cause a contentious debate between the FDA and the Advisory Committee during the review/approval process of the proposed biosimilar product because it is not clearly stated in the FDA guidance whether a proposed biosimilar product is required to passes all similarity tests, regardless of whether they are Tier 1 CQAs or Tier 2/Tier 3 CQAs, before the regulatory agency can grant approval of the proposed biosimilar product. In this situation, if the FDA and the ODAC panel accept the sponsors' scientific rationales and justifications that the notable differences have little or no impact on the clinical outcomes, the proposed biosimilar is likely to be granted for approval.

Such occurrences, however, have raised the interesting question of whether the proposed biosimilar product is required to pass all similarity tests (i.e., analytical similarity, PK/PD similarity, and clinical similarity) for regulatory approval.

2.4.3 Same Regulatory Standards

Recently, the FDA has been challenged by several sponsors who claim that the FDA employs inconsistent standards in the drug review and approval process because the FDA adopts a 95% confidence interval for evaluation of new drugs but uses a 90% confidence interval approach for assessment of generic drugs and biosimilar products. The FDA has tried to clarify the issue by pointing out the difference between a two-sided test (TST) for point hypotheses for the evaluation of new drugs and a two one-sided test (TOST) procedure for testing interval hypotheses in the case of bioequivalence and biosimilarity for generic drugs and biosimilar products. Both TST (for point hypotheses) and TSOT (for interval hypotheses) are size-α tests. Thus, the FDA upholds the same regulatory standards for both new drugs and generic drugs and biosimilar products. The confusion about FDA standards arose because of the mixed use of hypotheses testing and confidence interval approaches for the evaluation of drug products. More details can be found in Chapter 3.

Analytical similarity evaluation usually involves a large number of CQAs relevant to clinical outcomes with high-risk ranking. In practice, it is not clear whether α-adjustment for multiple comparisons (i.e., analytical similarity, PK/PD similarity, and clinical similarity) should be done for obtaining the totality-of-the-evidence for demonstration of biosimilarity.

2.5 Development of Totality-of-the-Evidence

Chow (2009) proposed the development of a composite index for assessing the biosimilarity of follow-on biologics based on the facts that (1) the concept of biosimilarity for biologic products (made of living cells) is very different from that of bioequivalence for drug products, and (2) biologic products are very sensitive to small changes in the variation during the manufacturing process (i.e., it might have a drastic change in clinical outcome). Some research on the comparison of moment-based criteria and probability-based criteria for the assessment of (1) average biosimilarity, and (2) the variability of biosimilarity for some given study endpoints by applying the criteria for bioequivalence are available in the literature (see, e.g., Chow et al., 2010 and Hsieh et al., 2010). Yet, universally acceptable criteria for biosimilarity are not available in the regulatory guidelines/guidances. Thus, Chow (2009) and Chow et al. (2011) proposed a biosimilarity index based on the concept of the probability of reproducibility as follows:

Step 1. Assess the average biosimilarity between the test product and the reference product based on a given biosimilarity criterion. For the purpose of an illustration, consider a bioequivalence criterion as a biosimilarity criterion. That is, biosimilarity is claimed if the 90% confidence interval of the ratio of means of a given study endpoint falls within the biosimilarity limits of (80%, 125%) or (−0.2231, 0.2231) based on log-transformed data or based on raw (original) data.

Step 2. Once the product passes the test for biosimilarity in Step 1, calculate the reproducibility probability based on the observed ratio (or observed mean difference) and variability. Thus, the calculated reproducibility probability will take the variability and the sensitivity of heterogeneity in variances into consideration for the assessment of biosimilarity.

Step 3. We then claim biosimilarity if the calculated 95% confidence lower bound of the reproducibility probability is larger than a prespecified number p_0, which can be obtained based on an estimated of reproducibility probability for a study comparing a "reference product" to itself (the "reference product"). We will refer to such a study as an R-R study. Alternatively, we can then claim (local) biosimilarity if the 95% confidence lower bound of the biosimilarity index is larger than p_0.

In an R-R study, define

$$P_{TR} = P \left(\begin{array}{l} \text{concluding average biosimiliarity between the test and the} \\ \text{reference products in a future trial given that the average} \\ \text{biosimiliarity based on average bioequivalence (ABE)} \\ \text{criterion has been established in first trial} \end{array} \right) \quad (2.1)$$

Alternatively, a reproducibility probability for evaluating the biosimilarity of the same two reference products based on the ABE criterion is defined as:

$$P_{RR} = P \left(\begin{array}{l} \text{concluding average biosimiliarity of the two same reference} \\ \text{products in a future trial given that the average biosimilarity} \\ \text{based on ABE criterion have been established in first trial} \end{array} \right) \quad (2.2)$$

Since the idea of the biosimilarity index is to show that the reproducibility probability is higher in a study for comparing "a reference product" with "the reference product" than the study for comparing a follow-on biologic with the innovative (reference) product, the criterion of an acceptable reproducibility probability (i.e., p_0) for the assessment of biosimilarity can be obtained based on the R-R study. For example, if the R-R study suggests the reproducibility probability of 90%, i.e., $P_{RR} = 90\%$, the criterion of the reproducibility probability for the biosimilarity study could be chosen as 80% of the 90% which is $p_0 = 80\% \times P_{RR} = 72\%$.

The biosimilarity index described above has the advantages that (i) it is robust with respect to the selected study endpoint, biosimilarity criteria, and study design, and (ii) the probability of reproducibility will reflect the sensitivity of heterogeneity in variance.

Note that the proposed biosimilarity index can be applied to different functional areas (domains) of biological products such as pharmacokinetics (PK), biological activities, biomarkers (e.g., pharmacodynamics), immunogenicity, manufacturing process, efficacy, etc. An overall biosimilarity index or totality biosimilarity index across domains can be similarly obtained as follows:

Step 1. Obtain $\hat{p}_i$, the probability of reproducibility for the i-th domain, $i = 1,.., K$.

Step 2. Define the biosimilarity index $\hat{p} = \sum_{i=1}^{K} w_i \hat{p}_i$, where w_i is the weight for the i-th domain.

Step 3. Claim global biosimilarity if we reject the null hypothesis that $p \leq p_0$, where p_0 is a pre-specified acceptable reproducibility probability. Alternatively, we can claim (global) biosimilarity if the 95% confidence lower bound of p is larger than p_0,

Let T and R be the parameters of interest (e.g., a pharmacokinetic response) with means of μ_T and μ_R, for a test product and a reference product, respectively. Thus, the interval hypotheses for testing the ABE of two products can be expressed as

$$H_0 : \theta'_L \geq \frac{\mu'_T}{\mu'_R} \quad \text{or} \quad \theta'_U \leq \frac{\mu'_T}{\mu'_R} \quad \text{vs.} \quad H_a : \theta'_L < \frac{\mu'_T}{\mu'_R} < \theta'_U$$

where (θ'_L, θ'_U) is the ABE limit. For *in vivo* bioequivalence testing, (θ'_L, θ'_U) is chosen to be (80%, 125%). The above hypotheses can be re-expressed as

$$H_0 : \theta_L \geq \mu_T - \mu_R \quad \text{or} \quad \theta_U \leq \mu_T - \mu_R \quad \text{vs.} \quad H_a : \theta_L < \mu_T - \mu_R < \theta_U$$

where μ_T and μ_R are the means of log-transformed data which are equal to the log-transformed values of μ'_T and μ'_R. (θ_L, θ_U) is (−0.2231, 0.2231), which is equal to the log-transformed values of (80%, 125%). To calculate the reproducibility probability under the above interval hypotheses, the probability of P_{TR} can be expressed when considering a parallel design (since it is a common design for biologic products) as follows:

$$P(\delta_L, \delta_U)$$

$$= P\left(T_L\left(\overline{Y}_T, \overline{Y}_R, s_T, s_R\right) > t_{\alpha, dfp} \text{ and } T_U\left(\overline{Y}_T, \overline{Y}_R, s_T, s_R\right) < -t_{\alpha, dfp} \mid \delta_L, \delta_U \right) \tag{2.3}$$

where $s_T, s_R, n_T,$ and n_R are the sample standard deviations and sample sizes for the test and reference formulations, respectively. The value of dfp can be calculated by

$$dfp = \frac{\left(\dfrac{s_T^2}{n_1} + \dfrac{s_R^2}{n_2}\right)^2}{\dfrac{\left(\dfrac{s_T^2}{n_T}\right)^2}{n_T - 1} + \dfrac{\left(\dfrac{s_R^2}{n_T}\right)^2}{n_R - 1}},$$

$$T_L\left(\overline{Y}_T, \overline{Y}_R, \sigma_T, \sigma_R\right) = \frac{\left(\overline{Y}_T - \overline{Y}_R\right) - \theta_L}{\sqrt{\dfrac{s_T^2}{n_T} + \dfrac{s_R^2}{n_R}}}, \quad T_U\left(\overline{Y}_T, \overline{Y}_R, \sigma_T, \sigma_R\right) = \frac{\left(\overline{Y}_T - \overline{Y}_R\right) - \theta_U}{\sqrt{\dfrac{s_T^2}{n_T} + \dfrac{s_R^2}{n_R}}},$$

$$\delta_L = \frac{\mu_T - \mu_R - \theta_L}{\sqrt{\dfrac{\sigma_T^2}{n_T} + \dfrac{\sigma_R^2}{n_R}}} \quad \text{and} \quad \delta_U = \frac{\mu_T - \mu_R - \theta_U}{\sigma_d\sqrt{\dfrac{\sigma_T^2}{n_T} + \dfrac{\sigma_R^2}{n_R}}} \tag{2.4}$$

σ_T^2 and σ_R^2 are the variances for test and reference formulations, respectively.

The vectors (T_L, T_U) can be shown to follow a bivariate noncentral t-distribution with $n_1 + n_2 - 2$ and *dfp* degrees of freedom, correlation of 1, and non-centrality parameters δ_L and δ_U (Phillips, 1990; Owen, 1965). Owen (1965) showed that the integral of the above bivariate noncentral t distribution can be expressed as the difference of the integrals between two univariate noncentral t-distributions. Therefore, the power function in (2.3) can be obtained by:

$$P(\delta_L, \delta_U) = Q_f(t_U, \delta_U; 0, R) - Q_f(t_L, \delta_L; 0, R) \tag{2.5}$$

where

$$Q_f(t, \delta; 0, R) = \frac{\sqrt{2\pi}}{\Gamma(f/2)2^{(f-2)/2}} \int_0^R G\left(tx/\sqrt{f} - \delta\right) x^{f-1} G'(x)\,dx$$

$$R = (\delta_L - \delta_U)\sqrt{f}/(t_L - t_U) \,,\; G'(x) = \frac{1}{\sqrt{2\pi}}e^{-x^2/2} \,,\; G(x) = \int_{-\infty}^x G'(t)\,dt$$

and

$$t_L = t_{\alpha, dfp} \,,\; t_U = -t_{\alpha, dfp} \quad \text{and} \quad f = dfp \;\text{ for parallel design}$$

Note that when $0 < \theta_U = -\theta_L, P(\delta_L, \delta_U) = P(-\delta_U, -\delta_L)$.

The reproducibility probabilities increase when the sample size increases and the means ratio is close to 1, while it decreases when the variability increases for the same setting of sample size and means ratio which show the impact of variability on reproducibility probabilities.

Since the true values of δ_L and δ_U are unknown, we proceed by using the idea of replacing δ_L and δ_U in (2.4) with their estimates based on the sample from the first study. The estimated reproducibility probability can be obtained as:

$$\hat{P}(\hat{\delta}_L, \hat{\delta}_U) = Q_f(t_L, \hat{\delta}_U; 0, \hat{R}) - Q_f(t_U, \hat{\delta}_L; 0, \hat{R}) \tag{2.6}$$

where

$$\hat{\delta}_L = \frac{\overline{Y}_T - \overline{Y}_L - \theta_L'}{\sqrt{\dfrac{s_T^2}{n_T} + \dfrac{s_R^2}{n_R}}} \,,\; \hat{\delta}_U = \frac{\overline{Y}_T - \overline{Y}_L - \theta_U'}{\sqrt{\dfrac{s_T^2}{n_T} + \dfrac{s_R^2}{n_R}}} \,,\; \hat{R} = (\hat{\delta}_L - \hat{\delta}_U)\sqrt{f}/(t_L - t_U)$$

2.6 Concluding Remarks

For regulatory approval of new drugs, Section 314.126 of 21 CFR states that substantial evidence needs to be provided to support the claims of new drugs. For regulatory approval of a proposed biosimilar product as compared to an innovative biological product (usually US-licensed drug product), the FDA requires totality-of-the-evidence be provided to support a demonstration of biosimilarity between the proposed biosimilar product and the US-licensed drug product. In practice, it should be noted that there is no clear distinction between the substantial evidence in new drug development and the totality-of-the-evidence in biosimilar drug product development.

In addition, it is not clear whether totality-of-the-evidence provide the same degree of substantial evidence for the assessment of the safety and effectiveness of the test treatment under investigation. Changes in quality attributes or responses in different domains may not be translated to other domains in terms of their criticality or risk ranking relevant to clinical outcomes. It is then suggested that the proposed index for totality-of-the-evidence be used for a more accurate and reliable assessment of the test treatment under investigation.

3

Hypotheses Testing versus Confidence Interval

3.1 Introduction

For review and approval of regulatory submissions of drug products, a $(1-\alpha)\times100\%$ confidence interval (CI) approach is commonly used for evaluation of safety and efficacy of new drugs, while a $(1-2\alpha)\times100\%$ CI is often considered for assessment of bioequivalence for generic drugs and for assessment of biosimilarity for biosimilar products. If α is chosen to be 5%, this leads to a 95% CI approach for evaluation of new drugs and a 90% CI approach for assessment of generics and biosimilars. In the past decade, the FDA has been challenged for adopting different standards (i.e., 5% type I error rate with respect to 95% CI for new drugs and 10% with respect to 90% CI for generics/biosimilars) in the review and approval process of regulatory submissions of drugs and biologics. The issue of using a 95% CI for new drugs and a 90% CI for generics/biosimilars has received much attention lately.

This controversial issue is probably due to the mixed-up use of the concepts of hypotheses testing and confidence interval approach for evaluation of drugs and biologics. The statistical methods for evaluation of new drugs and for assessment of generics/biosimilars recommended by the FDA are a two-sided test (TST) for testing *point hypotheses* of equality and a two one-sided tests (TOST) procedure for testing *interval hypotheses* (i.e., testing bioequivalence or biosimilarity), respectively. However, even under the hypotheses testing framework, test results are interpreted using a confidence interval approach regardless that their corresponding statistical inferences may not be (operationally) equivalent. As a result, many sponsors and/or reviewers are confused when to use a 90% CI approach or a 95% CI approach for pharmaceutical/clinical research and development.

In practice, there are fundamental differences (i) between the concepts of point hypotheses and interval hypotheses, and (ii) between the concepts of hypotheses testing and confidence interval approach. For (i), a TST is often performed for testing point hypotheses of equality at the 5% level of significance, while a TOST is a valid testing procedure for testing interval hypotheses of equivalence or similarity at the 5% level of significance for each one-sided test. For (ii), hypotheses testing focuses on power analysis for achieving a desired power (i.e., a type II error), while confidence interval focuses on precision analysis (i.e., a type I error) for controlling the maximum error allowed. The confusion between the use of a 90% CI and the use of a 95% CI for evaluation of drugs and biologics inevitably occur if we use confidence interval approach to interpret the results obtained under the hypotheses testing framework.

The purpose of this chapter is to clarify the confusion between (i) the use of hypotheses testing (both point hypotheses and interval hypotheses) and the use of confidence interval approach and (ii) the use of a 90% CI versus a 95% CI for evaluation of drugs and biologics. In the next couple of sections, the use of the method of hypotheses testing and the confidence interval approach for evaluation of safety and effectiveness of a test treatment under investigation are briefly described, respectively. Section 3.4 explores the relationship between a TOST procedure and its corresponding confidence interval approach for assessment of the safety and effectiveness of the test treatment under investigation. A comparison between the method of hypotheses testing and confidence interval approach is given in Section 3.5. Sample size requirement based on hypotheses testing and confidence interval is discussed in Section 3.6. Some concluding remarks are given in Section 3.7.

3.2 Hypotheses Testing

In pharmaceutical/clinical development, hypotheses testing and CI approach are often used interchangeably for evaluation of the safety and efficacy of a test treatment under investigation. It, however, should be noted that the method of hypotheses testing is not generally equivalent to the confidence interval approach. In the next couple of sections, we will explore the distinction between the two approaches although in many cases they are operationally equivalent under certain conditions.

For evaluation the safety and efficacy of a new drug or a test treatment under investigation, a typical approach is to test the hypotheses of equality, i.e., the null hypothesis (H_0) of *equality* versus an alternative hypothesis (H_a) of inequality. We would then reject the null hypothesis of equality (i.e., there is no treatment effect or efficacy) in favor of the alternative hypothesis of inequality (i.e., there is treatment effect or efficacy). In practice, we often select an appropriate sample size for achieving a desired power (i.e., the probability of correctly concluding efficacy when the test treatment is efficacious) at a pre-specified level of significance (to rule out that the observed efficacy is not by chance alone).

For assessment of generics or biosimilar products, on the other hand, the goal is to demonstrate that a proposed generic or biosimilar (test) product is bioequivalent or highly similar to an innovative or brand-name (reference) product. In this case, a test for interval hypotheses of equivalence (for generic drugs) or similarity (for biosimilar products) rather than a test for point hypotheses of equality (for new drugs) is often employed. In practice, however, we consider the two drug products are equivalent or highly similar if their difference falls within a pre-specified equivalent or similar range. In other words, we often start with hypotheses testing but draw conclusion based on confidence interval approach.

In practice, there is a distinction between equality and equivalence (similarity). In what follows, the concepts for testing point hypotheses and interval hypotheses are briefly described, respectively.

3.2.1 Point Hypotheses Testing

Let μ_T and μ_R be the population means of the test product and the reference product, respectively. The following point hypotheses are often considered for testing equality between means of a test (T) product (e.g., a new drug) and a reference (R) product (e.g., a placebo control or an active control):

$$H_0: \mu_T = \mu_R \text{ vs. } H_a: \mu_T \neq \mu_R. \tag{3.1}$$

For testing equality hypotheses, a two-sided test (TST) at the $\alpha = 5\%$ level of significance is often performed. Rejection of the null hypothesis of equality leads to the conclusion that there is a statistical significant difference between μ_T and μ_R. In clinical evaluation of a new drug, an appropriate sample size is often selected for achieving a desired power for detecting a clinically meaningful difference (or treatment effect) if such a difference truly exists at a pre-specified level of significance.

It can be verified that TST at the α level of significance is equivalent to $(1-\alpha) \times 100\%$ CI approach for evaluation of the treatment effect under investigation. Thus, in practice, point hypotheses testing for equality is often mixed up with the CI approach for assessment of the treatment effect under investigation. It should be noted that sample size calculations for point hypotheses testing and the CI approach are different. Sample size calculation for point hypotheses testing is typically performed based on power analysis (which focuses on type II error), while sample size calculation for the CI approach is performed based on precision analysis (which focuses on type I error). Thus, the required sample size for point hypotheses testing and the CI approach could be very different.

3.2.2 Interval Hypotheses Testing

On the other hand, the following interval hypotheses are usually used for testing bioequivalence (for generic drugs) or biosimilarity (for biosimilar products) between the test product and the reference product:

$$H_0 : \text{Bio}inequivalence \text{ or } dissimilarity$$

$$\text{vs. } H_a : \text{Bioequivalence or similarity} \tag{3.2}$$

Thus, we would reject the null hypothesis of bio*in*equivalence or dissimilarity and in favor the alternative hypothesis of bioequivalence or similarity. Interval hypotheses (3.2) are usually written as follows:

$$H_0 : \mu_T - \mu_R \leq -\delta \text{ or } \mu_T - \mu_R \geq \delta \quad \text{vs. } H_a : -\delta < \mu_T - \mu_R < \delta, \tag{3.3}$$

where δ is the so-called bioequivalence limit or similarity margin. Interval hypotheses (3.3) can be re-written as the following two one-sided hypotheses:

$$H_{01} : \mu_T - \mu_R \leq -\delta \quad \text{vs. } H_{a1} : -\delta < \mu_T - \mu_R ;$$

$$H_{02} : \mu_T - \mu_R \geq \delta \quad \text{vs. } H_{a2} : \mu_T - \mu_R < \delta. \tag{3.4}$$

For testing interval hypotheses (3.4), a two one-sided tests (TOST) procedure is recommended for testing bioequivalence for generic drugs or biosimilarity for biosimilar products (Schuirmann, 1987; FDA, 2003b). For interval hypotheses (3.4), the idea is to test one-side to determine whether the test product is inferior to the reference product at the α level of significance and then test the other side to determine whether the test product is not superior to the reference product at the α level of significance after the non-inferiority of the test product has been established. Note that the FDA recommended Schuirmann's TOST procedure should be used for testing the above interval hypotheses for bioequivalence or biosimilarity (FDA, 1992, 2003b).

In its guidance on bioequivalence assessment, the FDA suggested a log-transformation of data be performed before data analysis and recommended the bioequivalence limit be selected as $\delta = 0.8$ for PK responses (FDA, 2003b). Thus, interval hypotheses (3.3) can be rewritten as

$$H_0 : \mu_T / \mu_R \leq -0.8 \text{ or } \mu_T / \mu_R \geq 1.25 \quad \text{vs. } H_a : 0.8 < \mu_T / \mu_R < 1.25, \tag{3.5}$$

where 0.8 (80%) and 1.25 (125%) are lower and upper equivalence or similarity limits. Under hypotheses (3.5), a TOST procedure is to test one side (say non-inferiority) at the $\alpha = 5\%$ level of significance and then test the other side (say non-superiority) at the $\alpha = 5\%$ level of significance once the non-inferiority has been established.

Chow and Shao (2002b) indicated that the two one-sided tests procedure is a size-α test. This indicates that the FDA uses the same standard (i.e., at the α level of significance) for evaluation of new drug products based on TST under hypotheses (3.1) and for assessment of generic and biosimilar drug products based on TOST under hypotheses (3.3) or (3.4).

3.2.3 Probability of Inconclusiveness

In practice, one thing regarding point hypotheses testing that is worth mentioning is that the *probability of inconclusiveness*. In practice, we usually reject the null hypothesis at the $\alpha = 5\%$. Some investigators, however, prefer $\alpha = 1\%$ and consider p-values between 1% and 5% as inconclusive. If we let $z_{95\%}$ and $z_{99\%}$ be the critical values corresponding to $\alpha = 5\%$ and $\alpha = 1\%$, respectively, the area under the probability density between $z_{95\%}$ and $z_{99\%}$ is referred to as the probability of inconclusiveness.

The concept of interval hypotheses is very different from that of point hypotheses. Thus, interval hypotheses testing is generally not equivalent to the $(1-\alpha) \times 100\%$ CI approach for evaluation of the treatment effect under investigation. Interval hypotheses testing, somehow, overcomes the issue of inconclusiveness. For example, if we consider one-side of interval hypotheses (3.4) for testing non-inferiority, testing this one-side of interval hypotheses (3.4) is actually testing point hypotheses of inferiority. The rejection of the null hypothesis is in favor of non-inferiority. The concept of non-inferiority incorporates the concepts of equivalence and superiority. Thus, testing for non-inferiority does not imply testing for equivalence or similarity. Thus, for testing equivalence or similarity, the TOST is recommended.

3.3 Confidence Interval Approach

3.3.1 Confidence Interval Approach with Single Reference

For clinical investigation of new drugs, a typical approach is to testing the following point hypotheses for equality:

$$H_0 : \mu_T / \mu_R = 1 \text{ vs. } H_a : \mu_T / \mu_R \neq 1. \tag{3.6}$$

We then reject the null hypothesis of no treatment difference and conclude that there is statistically significant treatment difference. The statistical difference is then evaluated to determine whether such a difference is of clinically meaningful difference. In practice, a power calculation is usually performed for determination of sample size for achieving a desired power (e.g., 80%) for detecting a clinically meaningful difference (or treatment effect) under the alternative hypothesis that such a difference truly exists. For point hypotheses testing for equality, a two-sided test (TST) at the α level of significance is usually performed. The TST at the α level of significance is equivalent to the $(1-\alpha)\times 100\%$ confidence interval. In practice, the $(1-\alpha)\times 100\%$ confidence interval approach is often used to assess treatment effect instead of hypotheses testing for equality.

3.3.2 Confidence Interval Approach with Multiple References

3.3.2.1 Pairwise Comparisons

In comparative clinical trials, it is not uncommon to have multiple controls (or references). For example, for assessment of biosimilarity between a proposed biosimilar product (test product) and an innovative biological product (reference product), there may be multiple references, e.g., a US-licensed reference product and an EU-approved reference version of the same product. In this case, the method of pairwise comparisons is often applied. When two reference products (e.g., a US-licensed reference and an EU-approved reference) are considered, the method of pairwise comparisons includes three comparisons (i.e., the proposed biosimilar product versus the US-licensed reference product, the proposed biosimilar product versus the EU-approved reference product, and the US-licensed reference product versus the EU-approved reference product).

The method of pairwise comparisons sounds reasonable. However, at the Oncologic Drugs Advisory Committee (ODAC) meeting on July 13, 2017 for review of biosimilar products of Avastin and Herceptin, the method of pairwise comparisons was criticized by the ODAC panel. The first criticism was regarding the lack of accuracy and reliability of each pairwise comparison, since each comparison does not fully utilize all data collected from the three groups. In addition, since the equivalence criterion for analytical similarity is based on the variability of reference product, the pairwise comparisons method uses different equivalence criteria in the three comparisons, which may lead to inconsistent conclusions regarding the assessment of biosimilarity.

3.3.2.2 Simultaneous Confidence Interval

Alternatively, the ODAC suggested the potential use of simultaneous confidence approach, which has the advantages of utilizing all data collected from the study and using consistent equivalence criterion. As a result, Zheng et al. (2019) proposed three different types (namely, the original version, integrated version, and least favorable version) of simultaneous confidence intervals based on the fiducial inference theory under a parallel-group design (see also Chow, 2018).

Zheng et al. (2019) conducted several simulations for evaluation of the performance of the proposed simultaneous confidence intervals approach as compared to the method of pairwise comparisons. Simulation results indicated that current pairwise comparison methods lacks the accuracy and reliability of each pairwise comparison since each comparison does not fully utilize all data collected from the three groups, and suffers from the inconsistent use of different equivalence criteria in the three comparisons. The simulation results also showed that the methods using the original version and integrated version of simultaneous confidence interval have significantly larger power compared to the pairwise comparisons method and meanwhile can well control the type I error rate. While the method using the least favorable version of simultaneous confidence interval demonstrated the smallest power among the four methods, it was better able to control type I error rate—thus it is a conservative approach which is preferred for avoiding false positive conclusions.

To provide a better understanding and to illustrate the inappropriateness of pairwise comparisons, Zheng et al. (2019) provided a couple of examples: one regarding the case where pairwise comparisons failed to conclude similarity when similarity truly exists (i.e., false negative), the other is concerning the case where pairwise comparisons wrongly concludes similarity while similarity does not hold (i.e., false positive). These two examples are briefly described below.

3.3.2.3 Example 1 (False Negative)

Suppose we have two reference products US reference and EU reference, denoted by US and EU, and one test product, denoted by T. Assume US, EU, and T follow normal distributions and share equal variance. The true means of the three products were set to be 99, 101, 100, and the true standard deviations were assumed to be equal among the three products were set to be 6. Three groups of samples of 10 were randomly generated from US, EU, and T population, respectively. Another two groups of samples of size 10 were randomly taken from the US and EU population to obtain the 'true' standard deviations. The type I error allowed was set to be 0.1. Three pairwise comparisons, US versus EU, US versus T, EU versus T, were analyzed using the FDA recommended approach, with US, US, and EU as the references, respectively. The data are displayed in Table 3.1 and corresponding scatter plot is given in Figure 3.1.

Under this setting, an effective test should be able to reject the null hypothesis and conclude the similarity. However, from Table 3.2, the pairwise comparisons approach failed to reject one of the null hypotheses that the two reference drugs are not similar enough (EU vs. US, 90% CI: 0.42–5.33, exceeds the equivalence acceptance criterion (EAC) margin = 5.01). While two out of the three simultaneous confidence interval methods had fiducial probabilities calculated higher than 0.9 (0.92 for both original version and integrated version), and the corresponding two versions of confidence intervals lie within

TABLE 3.1

Random Samples Generated from the Three Population (Example 1)

Group	\multicolumn{10}{c}{Lot}

Group	1	2	3	4	5	6	7	8	9	10
US	102.13	102.07	92.69	92.09	96.99	101.83	95.15	102.72	95.02	103.45
EU	102.93	95.29	100.21	105.77	100.87	100.72	98.33	108.55	97.74	102.46
T	96.70	101.63	110.70	89.96	98.25	105.39	101.13	103.91	90.92	102.99
US (ref)[a]	104.38	99.39	102.71	103.81	95.35	102.41	97.56	101.56	95.10	99.96
EU (ref)[a]	101.39	104.90	98.09	98.32	101.82	107.23	83.62	100.30	106.98	98.52

[a] Samples randomly taken from the US and EU population to obtain the "true" standard deviations.

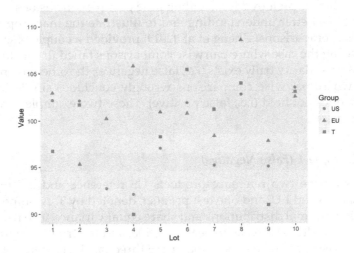

FIGURE 3.1
Scatter plot of the random samples generated for each group (Example 1).

the simultaneous margin, thus could successfully reject all three hypotheses, i.e., conclude similarity among US, EU, and T. However, the least favorable version failed to conclude the similarity (fiducial probability = 0.79). This example illustrates the case that pairwise method failed to conclude similarity when true similarity holds (i.e., false negative towards the hypothesis tests), and compared to this, the new proposed simultaneous interval approach was able to reject the null hypothesis and thus more powerful in this case.

3.3.2.4 Example 2 (False Positive)

Suppose the true means of the three products was set to be 95, 105, 100, and the true standard deviations were assumed to be equal among the three products and were set to be 6. Similarly, three groups of samples of size 10 were

TABLE 3.2

The Results of Pairwise Comparisons Method vs. Simultaneous Confidence Interval Approach (Example 1)

	Pairwise Comparisons Approach				Simultaneous Confidence Interval Approach				
Comparison	Mean Difference	90% CI	EAC Margin[a]	Equivalence Test	Method	Fiducial Probability	Type I 90% CI	Type II 90% CI	Simultaneous Similarity
EU vs. US	2.87	(0.42, 5.33)	5.01	Fail	Original	0.92	(−4.51, 4.51)	(−4.79, 4.79)	Pass
T vs. US	1.74	(−0.72, 4.20)	5.01	Pass	Integrated	0.92	(−4.51, 4.51)	(−4.84, 4.84)	Pass
T vs. EU	−1.13	(−6.08, 3.82)	10.09	Pass	Least Favorable	0.79	NA	(−4.15, 4.15)	Fail

[a] Similarity margin = $1.5 * \sigma_R$.

randomly generated from US, EU, and T population, respectively. Another two groups of samples with size 10 were randomly taken from the US and EU population to obtain the 'true' standard deviations. The type I error allowed was set to be 0.1. Three pairwise comparisons, US versus EU, US versus T, EU versus T, were analyzed using the FDA recommended approach, with US, US, and EU as the references, respectively. The data were displayed in Table 3.3 and corresponding scatter plot was showed in Figure 3.2.

Under this setting, an effective test should be able to accept the null hypothesis and not conclude the similarity. However, Table 3.4 indicates that follow the pairwise comparisons approach, the data passes all the three equivalence tests and incorrectly concluded the similarity between three drug products (all the three 90% CIs lie within the EAC margins). Considering the simultaneous confidence interval approach, although the original and integrated versions of simultaneous confidence interval approaches also incorrectly

TABLE 3.3

Random Samples Generated from the Three Population (Example 2)

Group	1	2	3	4	5	6	7	8	9	10
					Lot					
US	96.41	101.81	100.58	90.98	88.06	108.19	95.49	105.62	99.98	100.66
EU	94.13	119.26	106.72	99.86	101.54	101.33	105.83	112.27	94.25	93.75
T	100.46	95.53	107.85	106.19	112.90	95.75	99.85	101.27	97.50	97.32
US (ref)[a]	89.81	105.22	95.79	93.76	95.01	96.70	104.06	98.06	90.52	88.88
EU (ref)[a]	94.19	100.15	104.23	116.74	103.17	109.69	106.76	118.46	104.73	106.25

[a] Samples randomly taken from the US and EU population to obtain the "true" standard deviations.

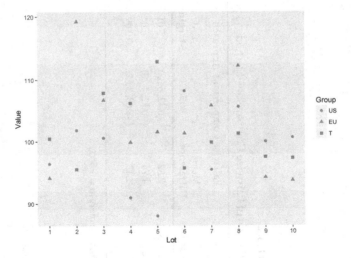

FIGURE 3.2
Scatter plot of the random samples generated for each group (Example 2).

TABLE 3.4

The Results of Pairwise Comparisons Method vs. Simultaneous Confidence Interval Approach (Example 2)

| | Pairwise Comparisons Approach | | | Simultaneous Confidence Interval Approach | | | | |
Comparison	Mean Difference	90% CI	EAC Margin[a]	Equivalence Test	Method	Fiducial Probability	Type I 90% CI	Type II 90% CI	Simultaneous Similarity
EU vs. US	4.12	(0.02, 8.21)	8.36	Pass	Original	0.94	(−6.48, 6.48)	(−7.63, 7.63)	Pass
T vs. US	2.68	(−1.41, 6.78)	8.36	Pass	Integrated	0.95	(−6.38, 6.38)	(−7.72, 7.72)	Pass
T vs. EU	−1.43	(−6.74, 3.88)	10.83	Pass	Least Favorable	0.84	NA	(−6.60, 6.60)	Fail

[a] Similarity margin = $1.5 * \sigma_R$.

concluded the similarity (had fiducial probabilities calculated higher than 0.9:0.94 for original version and 0.95 for integrated version, and the corresponding two versions of confidence intervals lie within the simultaneous margin), the least favorable version successfully detected the difference and did not conclude the similarity (fiducial probability = 0.79). This example illustrates the case that pairwise method incorrectly conclude similarity when significant difference between the three groups truly exists (i.e., false positive towards the hypothesis tests), and compared to this, the new proposed least favorable version of simultaneous interval approach was more conservative and avoided the type I error in this case. Further discussion of the new methods' performance under different parameter settings can be found in the simulation studies of the following sections.

3.4 Two One-Sided Tests versus Confidence Interval Approach

For testing equivalence (either bioequivalence, biosimilarity, or therapeutic equivalence), the two one-sided tests procedure and the confidence interval approach are often considered (Schuirmann, 1987; Chow and Liu, 2003). However, some confusion arises. For example, what is the difference between the two approaches, given the fact that in some cases the two approaches produce the same test? For a confidence interval approach, current practice considers $1 - \alpha$ for establishing therapeutic equivalence and $1 - 2\alpha$ for demonstration of bioequivalence. Thus, should we use level $1 - \alpha$ or $1 - 2\alpha$ when applying the confidence interval approach for establishing equivalence? When different confidence intervals are available, which confidence interval should be used? These questions have an impact on sample size calculation for establishing equivalence in clinical trials.

3.4.1 Two One-Sided Tests (TOST) Procedure

Chow and Shao (2002b) indicated that the approach of using $1 - \alpha$ confidence intervals produces level α-tests, but the sizes of these tests may be smaller than α, and that the use of $1 - 2\alpha$ confidence intervals generally does not ensure that the corresponding test be of level α, although there are exceptional cases. In this section, we will re-visit these questions. Let μ_T and μ_S denote, respectively, the mean responses of a study primary endpoint for a test drug and a standard therapy (or active control agent), and let $\delta > 0$ be the magnitude of difference of clinical importance. If the concern is whether the test drug is non-inferior to the standard therapy (or active control agent), then the following hypotheses are tested:

$$H_0: \mu_T - \mu_S \leq -\delta \text{ versus } H_a: \mu_T - \mu_S > -\delta.$$

The hypotheses related to whether the test drug is superior to the standard therapy (or active control agent) are

$$H_0: \mu_T - \mu_S \leq \delta \text{ versus } H_a: \mu_T - \mu_S > \delta.$$

(see, for example, Hwang and Morikawa, 1999). If it is of interest to show whether the test drug and the standard therapy (or active control agent) are therapeutically equivalent, then we consider the following hypotheses:

$$H_0: |\mu_T - \mu_S| \geq \delta \text{ versus } H_a: |\mu_T - \mu_S| < \delta. \tag{3.7}$$

Note that non-inferiority, superiority, or therapeutic equivalence is concluded if the null hypothesis H_0 is rejected at a given significance level α. In this section we focus on the two-sided hypotheses given in (3.7), which is also of interest in assessing bioequivalence between two drug products. There are two commonly employed statistical tests for hypotheses in (3.7). One is the TOST procedure and the other is the CI approach. Berger and Hsu (1996) studied statistical properties of the tests based on the TOST and CI approaches. However, the fact that the TOST approach with level α is operationally the same as the CI approach with a particular $1 - 2\alpha$ confidence interval (e.g., Blair and Cole, 2002; Chow and Liu, 2008) has caused some confusion in the pharmaceutical industry. For example, should we use level $1 - \alpha$ or $1 - 2\alpha$ when applying the CI approach? If the TOST and CI approaches are operationally the same, then why are they considered to be different approaches? Furthermore, there are several tests corresponding to different confidence intervals, which one should be recommended? Chow and Shao (2002b) clarified this confusion by comparing the two one-sided tests (TOST) approach with the confidence interval (CI) approaches as follows.

The TOST approach is based on the fact that the null hypothesis H_0 in (3.7) is the union of the following two one-sided hypotheses:

$$H_{01}: \mu_T - \mu_S \geq \delta \text{ and } H_{02}: \mu_T - \mu_S \leq -\delta. \tag{3.8}$$

Hence, we reject the null hypothesis H_0 when both H_{01} and H_{02} are rejected. For example, when observed responses are normally distributed with a constant variance, the TOST procedure rejects H_0 in (3.7) if and only if

$$(\bar{y}_T - \bar{y}_S + \delta) / se > t_\alpha \text{ and } (\bar{y}_T - \bar{y}_S - \delta) / se < -t_\alpha, \tag{3.9}$$

where $\bar{y}_T$ and $\bar{y}_S$ are the sample means of the test product and the standard therapy (or active control agent), respectively, se is an estimated standard deviation and t_α is the upper αth percentile of a central t-distribution with appropriate degrees of freedom (e.g., Berger and Hsu, 1996; Blair and Cole, 2002). Note that each of the two statements in (3.9) defines the rejection of a level α test for one of the null hypotheses in (3.8).

3.4.2 Confidence Interval Approach

For the confidence interval approach, the following confidence intervals are commonly considered:

$$\text{CI}_W = \left[\bar{y}_T - \bar{y}_S - t_{\alpha_1} se, \ \bar{y}_T - \bar{y}_S + t_{\alpha_2} se \right]$$

$$\text{CI}_L = \left[-\mid \bar{y}_T - \bar{y}_S \mid - t_\alpha se, \ \mid \bar{y}_T - \bar{y}_S \mid + t_\alpha se \right]$$

$$\text{CI}_E = \left[\min\left(0, \bar{y}_T - \bar{y}_S - t_\alpha se\right), \ \max\left(0, \bar{y}_T - \bar{y}_S + t_\alpha se\right) \right]$$

Here, CI_W is the so-called Westlaske symmetric confidence interval with appropriate choices of $\alpha_1 > 0$, $\alpha_2 > 0$, $\alpha_1 + \alpha_2 = \alpha$ (Westlake, 1976), CI_L is derived from the confidence interval of $\mid \mu_T - \mu_S \mid$ and CI_E is the expanded confidence interval derived by Hsu (1984) and Bofinger (1985, 1992). If we define

$$\text{CI}_{2\alpha} = \left[\bar{y}_T - \bar{y}_S - t_\alpha se, \ \bar{y}_T - \bar{y}_S + t_\alpha se \right],$$

then the result in (3.9) is equivalent to the result that $\text{CI}_{2\alpha}$ falls within the interval $(-\delta, \delta)$. Therefore, the TOST approach is operationally the same as the CI approach with the confidence interval $\text{CI}_{2\alpha}$.

3.4.2.1 Level 1 – α versus Level 1 – 2α

Note that CI_W, CI_L, and CI_E are $1 - \alpha$ confidence intervals, whereas $\text{CI}_{2\alpha}$ is a $1 - 2\alpha$ confidence interval. Thus, there is confusion whether level $1 - \alpha$ or level $1 - 2\alpha$ should be used when applying the CI approach. This directly affects statistical analysis, starting from sample size calculation. When applying the CI approach to the problem of assessing average bioequivalence, Berger and Hsu (1996) indicated that the misconception that size-α bioequivalence tests generally correspond to $(1 - 2\alpha) \times 100\%$ confidence sets will be shown to lead to incorrect statistical practices, and should be abandoned. This is because the use of a $1 - \alpha$ confidence interval guarantees that the corresponding test is of level α, whereas the use of a $1 - 2\alpha$ confidence interval can only ensure that the corresponding test is of level 2α. However, a further problem arises. Why does the use of $1 - 2\alpha$ confidence interval $\text{CI}_{2\alpha}$ produce a level α test? To clarify this, we need to first understand the difference between the significance level and the size of a statistical test.

3.4.2.2 Significance Level versus Size

Let T be a test procedure. The size of T is defined to be

$$\alpha_T = \sup_{P \text{ under } H_0} P(T \text{ rejects } H_0).$$

On the other hand, any $\alpha_1 \geq \alpha_T$ is called a significance level of T. Thus, a test of level α is also of level α_2 for any $\alpha_2 > \alpha_1$ and it is possible that the test is of level α_0, which is smaller than α_1. The size of a test is its smallest possible level and any number larger than the size (and smaller than 1) is a significance level of the test. Thus, if α (e.g., 1% or 5%) is a desired level of significance and T is a given test procedure, then we must first ensure that $\alpha_T \leq \alpha$ (i.e., T is of level α) and then try to have $\alpha_T = \alpha$; otherwise T is too conservative.

Chow and Shao (2002b) discussed the difference between the TOST approach and the CI approach, although they may be operationally the same. It also reveals a disadvantage of using the CI approach. That is, the test obtained using a $1 - \alpha$ confidence interval may be too conservative in the sense that the size of the test may be much smaller than α. An immediate question is: What are the sizes of the tests obtained by using level $1 - \alpha$ confidence intervals CI_W, CI_L, and CI_E?

3.4.2.3 Sizes of Tests Related to Different Confidence Intervals

If can be verified that $CI_W \supset CI_L \supset CI_E \supset CI_{2\alpha}$. If the tests obtained by using these confidence intervals are T_W, T_L, T_E, and $T_{2\alpha}$, respectively, then their sizes satisfy $\alpha_{Tw} \leq \alpha_{TL} \leq \alpha_{TE} \leq \alpha_{T2\alpha}$. In the previous section, we conclude that the size of $T_{2\alpha}$ is α. Berger and Hsu (1996) showed that the size of T_E is α, although CI_E is always wider than $CI_{2\alpha}$. The size of T_L is also α, although CI_L is even wider than CI_E. This is because

$$\alpha_{T_L} = \sup_{|\mu_T - \mu_S| \geq \delta} P(|\bar{y}_T - \bar{y}_S| + t_\alpha se < \delta)$$

$$= \max \left\{ \sup_{\mu_T - \mu_S = \delta} P(|\bar{y}_T - \bar{y}_S| + t_\alpha se < \delta, \sup_{\mu_S - \mu_T = \delta} P(|\bar{y}_T - \bar{y}_S| + t_\alpha se < \delta \right\}$$

$$= \alpha.$$

However, the size of T_W is $\max(\alpha_1, \alpha_2)$, which is usually smaller than α since $\alpha_1 + \alpha_2 = \alpha$. Thus, T_W (or CI_W) is not recommended if the desired size is α. It should be noted that the size of a test is not the only measure of its successfulness. The TOST procedure is of size α yet biased because the probability of rejection of the null hypothesis when it is false (power) may be lower than α. Similarly, tests T_E and T_L are biased. Berger and Hsu (1996) proposed a nearly unbiased and uniformly more powerful test than the TOST procedure. Brown et al. (1997) provided an improved test that is unbiased and uniformly more powerful than the TOST procedure. These improved tests are, however, more complicated than the TOST procedure.

To avoid confusion, it is strongly suggested that the approaches of confidence interval and two one-sided tests should be applied separately, although sometimes they are operationally equivalent.

3.4.3 Remarks

Chow and Liu (1992a) indicated that Schuirmann's two one-sided tests procedure is *operationally* (algebraically) equivalent to confidence interval approach in many cases under certain assumptions. In other words, we claim bioequivalence or biosimilarity if the constructed $(1-2\alpha)\times 100\%$ confidence interval falls completely within the bioequivalence or biosimilarity limits. As a result, interval hypotheses testing (i.e., two one-sided tests procedure) for bioequivalence or biosimilarity has been mixed up with the use of $(1-2\alpha)\times 100\%$ confidence interval approach since then.

To provide a better understanding, Table 3.5 summarizes the fundamental differences between hypotheses testing for equality (new drugs) and hypotheses testing for equivalence (generic drugs and biosimilar products).

As it can be seen from Table 3.5 that TST and TOST are official test procedures recommended by the FDA for evaluation of new drugs and generic/biosimilar drugs, respectively. Both TST and TOST are size-α test procedures. In other words, the overall type I error rates are well controlled. In practice, it is suggested that the concepts of hypotheses testing (based on type II error) and confidence interval approach (based on type I error) should not be mixed up to avoid possible confusion in evaluation of new drugs and generic/biosimilar products.

TABLE 3.5

Comparison of Statistical Methods for Assessment of Generic/Biosimilar Drugs and New Drugs

Characteristics	Generic/Biosimilar Drugs	New Drugs
Hypotheses testing	Interval hypotheses	Point hypotheses
FDA recommended approach	TOST	TST
Control of α	Yes	Yes
Confidence interval approach	Operationally equivalent $(1-2\alpha)\times 100\%$ CI	Equivalent $(1-\alpha)\times 100\%$ CI
	90% CI if $\alpha=5\%$	95% CI if $\alpha=5\%$
Sample size requirement	Based on TOST	Based on TST

Note: TOST = two one-sided tests; TST = two-sided test.

3.5 A Comparison

In this section, we will compare the method of hypotheses testing (i.e., TOST) and the confidence interval approaches (i.e., 90% CI and 95% CI) for assessment of bioequivalence for generic drugs and biosimilarity for biosimilar drug products. The comparison will be made under the framework of hypotheses testing in terms of the probability of correctly concluding bioequivalence or biosimilarity.

3.5.1 Performance Characteristics

Under interval hypotheses (3.7), we would reject the null hypothesis of bioinequivalence (or dis-similarity) and conclude bioequivalence (biosimilarity) at the 5% level of significance. The probability of correctly concluding bioequivalence or biosimilarity, when in fact the test product is bioequivalent or biosimilar to the reference product, is the power of the TOST. In practice, power of TOST can be obtained by calculating the probability of rejecting H_0 or accepting H_a under H_a (i.e., assuming H_a is true, which is given below.

$$\text{power} = \{\text{accept}\, H_a \,\text{under}\, H_a \,\text{of hyotheses} (3.7)\,\text{or}(3.8)\}.$$

Since TOST is operationally equivalent to the 90% confidence interval in many cases, we will consider the 90% CI as an alternative method for assessment of bioequivalence for generic drug products and biosimilarity for biosimilar drug products. We would conclude bioequivalence or biosimilarity if the constructed 90% CI falls within (80%, 125%) entirely. Thus, the probability of concluding bioequivalence for generic drug products or biosimilarity for biosimilar drug products is given below

$$p_{90\%CI} = P\{90\%\,CI \subset (0.8, 1.25)\}.$$

The performance of the 95% CI approach for assessment of bioequivalence for generic drug products and biosimilarity for biosimilar drug products can be similarly evaluated. For the 95% CI approach, if the constructed 95% CI falls within (80%, 125%) entirely. Thus, the probability of concluding bioequivalence for generic drug products or biosimilarity for biosimilar drug products is given below

$$p_{95\%\,CI} = P\{95\%\,CI \subset (0.8, 1.25)\}.$$

Denote $\hat{F}$ the point estimate of $\log(\mu_T/\mu_R)$, se an estimate of the standard deviation of $\hat{F}$, and t_α the upper αth percentile of a central t-distribution with appropriate degrees of freedom. The following $(1-\alpha)\times 100\%$ confidence intervals are commonly considered in the CI approach:

$$\text{CI}_W = \left[\hat{F} - t_{\alpha_1}se, \hat{F} + t_{\alpha_2}se\right];$$

$$\text{CI}_L = \left[-\left|\hat{F}\right| - t_\alpha se, \left|\hat{F}\right| + t_\alpha se\right];$$

$$\text{CI}_E = \left[\min\left(0, \hat{F} - t_{\alpha_1}se\right), \max\left(0, \hat{F} + t_{\alpha_2}se\right)\right].$$

Here CI_W is the so-called Westlake symmetric confidence interval with appropriate choices of $\alpha_1 > 0$, $\alpha_2 > 0$, $\alpha_1 + \alpha_2 = \alpha$ (Westlake, 1976), CI_L is derived from the confidence interval of $\left|\log\left(\frac{\mu_T}{\mu_R}\right)\right|$ (Liu, 1990), and CI_E is an expended confidence interval derived by Hsu (1984) and Bofinger (1985, 1992). Denote the classic $(1-2\alpha)\times 100\%$ confidence interval as

$$\text{CI}_{2\alpha} = \left[\hat{F} - t_\alpha se, \hat{F} + t_\alpha se\right],$$

the results of the TOST approach at the α level of significance is equivalent to the result that $\text{CI}_{2\alpha}$ falls within the interval $(-\delta, \delta)$, where $\delta = \log(1.25)$ for the data after log-transformation. In addition, it can be derived that $\text{CI}_{2\alpha} \subset (-\delta, \delta)$ implies $\text{CI}_L \subset (-\delta, \delta)$ and $\text{CI}_E \subset (-\delta, \delta)$. Thus, the $(1-\alpha)\times 100\%$ CI approaches (CI_L and CI_E), the classical $(1-2\alpha)\times 100\%$ CI approach ($\text{CI}_{2\alpha}$), and the TOST approach at the α level of significance are operationally equivalent for testing hypotheses of (3.8).

With $\text{CI}_{2\alpha} \subset \text{CI}_W$, we have the following power relations between the TOST approach and the four CI approaches with the confidence levels of 90% and 95%:

$$p_{90\%\text{CI}_L} = p_{90\%\text{CI}_E} > p_{90\%\text{CI}_W} > p_{90\%\text{CI}_{2\alpha}}$$

$$= p_{\text{TOST}} = p_{95\%\text{CI}_L} = p_{95\%\text{CI}_E} > p_{95\%\text{CI}_W} > p_{95\%\text{CI}_{2\alpha}}$$

(3.10)

3.5.2 Simulation Studies

To compare the power (or type I error) of those approaches, we also conducted simulation studies to evaluate small sample performances. We consider the classic 2 × 2 crossover design without carryover effects (a brief introduction of the statistical model for this design is given in Appendix 1). A variety of scenarios were considered with parameter specifications in Table 3.6.

The power (when the alternative hypothesis holds) and the type I error (when the null hypothesis holds) were compared between the nine approaches: 90%CI_L, 90%CI_E, 90%CI_W, 90%$\text{CI}_{2\alpha}$, 95%CI_L, 95%CI_E, 95%CI_W,

TABLE 3.6

Parameter Specification for Simulation Studies

μ_T	80	85	95	100	110	120	125
μ_R	100	100	100	100	100	100	100
σ_T	10	20	30				
σ_R	10	20	30				
$\dfrac{\sigma_{BT}^2}{\sigma_{TT}^2} = \dfrac{\sigma_{BR}^2}{\sigma_{TR}^2}$	0.75	0.25					
ρ	0	0.3	0.6				
n_1	12	24					
n_2	12	24					

Note: σ_T and σ_R are the total standard deviations of drug T and drug R before log-transformation, respectively. The meanings of σ_{BT}^2, σ_{TT}^2, σ_{BR}^2, σ_{TR}^2 and ρ can be found in the Appendix 1. n_1 and n_2 are numbers of subjects with the sequences of RT and TR, respectively. Without loss of generality, the fixed effects of periods and sequences in all bioequivalence trials are set to be 0 for all scenarios. There are 7 combinations of (μ_T, μ_R), 3 combinations of (σ_T, σ_R), 2 combinations of $\left(\dfrac{\sigma_{BT}^2}{\sigma_{TT}^2}, \dfrac{\sigma_{BR}^2}{\sigma_{TR}^2} \right)$, 3 combinations of ρ, and 2 combinations of (n_1, n_2), resulting in $7 \times 3 \times 2 \times 3 \times 2$ scenarios in total.

$95\%\text{CI}_{2\alpha}$, and the FDA recommended TOST approach. Note that the $90\%\text{CI}_{2\alpha}$, the TOST approach, $95\%\text{CI}_L$, and $95\%\text{CI}_E$ are operationally equivalent. So are $90\%\text{CI}_L$ and $90\%\text{CI}_E$. For illustration, we only present the results for the scenarios of $(n_1, n_2) = (12, 12)$ and the last four combination of (μ_T, μ_R) only. The results for other scenarios are similar and we do not present them here for simplicity.

The type I error rates are presented in Table 3.7. We see the type I error rates of $90\%\text{CI}_L$, $90\%\text{CI}_E$, and $90\%\text{CI}_W$ are highly similar and range from 6% to 9% with all values greater than the 5% level of significance, while the rates for all the other methods are well controlled at the level of 5%. Among them, the rates of $90\%\text{CI}_{2\alpha}$, $95\%\text{CI}_L$, $95\%\text{CI}_E$, $95\%\text{CI}_W$, and the TOST approach are highly similar and range appropriately from 2.7% to 4.5%, while $95\%\text{CI}_{2\alpha}$ is too conservative with the type one errors no greater than 2.3%.

The power for those approaches is presented in Table 3.8. The results are consistent with the relationships in (3.10). The power of three approaches, $90\%\text{CI}_L$, $90\%\text{CI}_E$, and $90\%\text{CI}_W$, are very similar and higher than the other approaches. For $90\%\text{CI}_{2\alpha}$, $95\%\text{CI}_L$, $95\%\text{CI}_E$, $95\%\text{CI}_W$, and the TOST approach, their power are very similar and the power of $95\%\text{CI}_W$ is slightly lower than the other four. The $95\%\text{CI}_{2\alpha}$ approach has the lowest power.

Regarding controlling the 5% level of significance and power, the TOST approach, as well as the three CI approaches ($90\%\text{CI}_{2\alpha}$, $95\%\text{CI}_L$, and $95\%\text{CI}_E$), perform best among the nine approaches.

TABLE 3.7

Type I Error Rate When H_0 in (3.4) Is True for $n_1 = n_2 = 12$

GMR	μ_T	σ_R	$\dfrac{\sigma_{BT}^2}{\sigma_{TT}^2}$	ρ	90%CI$_L$	90%CI$_W$	TOST	95%CI$_W$	95%CI$_{2\alpha}$
1.25	125	10	0.75	0	0.09	0.09	0.045	0.044	0.021
1.25	125	10	0.75	0.3	0.084	0.084	0.043	0.043	0.023
1.25	125	10	0.75	0.6	0.09	0.09	0.045	0.045	0.023
1.25	125	10	0.25	0	0.089	0.089	0.044	0.043	0.021
1.25	125	10	0.25	0.3	0.09	0.09	0.042	0.042	0.02
1.25	125	10	0.25	0.6	0.088	0.088	0.043	0.043	0.022
1.26	125	20	0.75	0	0.081	0.081	0.039	0.039	0.019
1.26	125	20	0.75	0.3	0.078	0.078	0.037	0.037	0.018
1.26	125	20	0.75	0.6	0.075	0.075	0.034	0.034	0.016
1.26	125	20	0.25	0	0.077	0.077	0.038	0.038	0.017
1.26	125	20	0.25	0.3	0.078	0.078	0.037	0.037	0.018
1.26	125	20	0.25	0.6	0.076	0.076	0.036	0.036	0.014
1.27	125	30	0.75	0	0.07	0.069	0.035	0.033	0.016
1.27	125	30	0.75	0.3	0.065	0.064	0.031	0.03	0.013
1.27	125	30	0.75	0.6	0.06	0.06	0.027	0.027	0.013
1.27	125	30	0.25	0	0.071	0.069	0.031	0.03	0.016
1.27	125	30	0.25	0.3	0.069	0.068	0.034	0.033	0.016
1.27	125	30	0.25	0.6	0.069	0.068	0.033	0.032	0.015

Note: GMR: geometric mean ratio between T and R. For the type I error rate, deeper color indicates higher value.

TABLE 3.8

Power When H_a in (3.4) Is True for $n_1 = n_2 = 12$

GMR	μ_T	σ_R	$\dfrac{\sigma_{BT}^2}{\sigma_{TT}^2}$	ρ	90%CI$_L$	90%CI$_W$	TOST	95%CI$_W$	95%CI$_{2\alpha}$
1	100	10	0.75	0	1	1	1	1	1
1	100	10	0.75	0.3	1	1	1	1	1
1	100	10	0.75	0.6	1	1	1	1	1
1	100	10	0.25	0	1	1	1	1	1
1	100	10	0.25	0.3	1	1	1	1	1
1	100	10	0.25	0.6	1	1	1	1	1
1	100	20	0.75	0	0.989	0.989	0.968	0.968	0.925
1	100	20	0.75	0.3	0.998	0.998	0.991	0.991	0.977
1	100	20	0.75	0.6	1	1	1	1	0.997
1	100	20	0.25	0	0.989	0.989	0.965	0.965	0.925
1	100	20	0.25	0.3	0.993	0.993	0.978	0.978	0.946
1	100	20	0.25	0.6	0.996	0.996	0.985	0.985	0.963
1	100	30	0.75	0	0.804	0.798	0.632	0.606	0.421
1	100	30	0.75	0.3	0.902	0.899	0.789	0.78	0.627

(Continued)

TABLE 3.8 (*Continued*)

Power When H_a in (3.4) Is True for $n_1 = n_2 = 12$

GMR	μ_T	σ_R	$\dfrac{\sigma_{BT}^2}{\sigma_{TT}^2}$	ρ	90%CI$_L$	90%CI$_W$	TOST	95%CI$_W$	95%CI$_{2\alpha}$
1	100	30	0.75	0.6	0.971	0.97	0.926	0.924	0.851
1	100	30	0.25	0	0.8	0.794	0.625	0.6	0.422
1	100	30	0.25	0.3	0.837	0.832	0.683	0.667	0.492
1	100	30	0.25	0.6	0.878	0.874	0.745	0.73	0.562
1.1	110	10	0.75	0	0.999	0.999	0.998	0.998	0.994
1.1	110	10	0.75	0.3	1	1	1	1	0.999
1.1	110	10	0.75	0.6	1	1	1	1	1
1.1	110	10	0.25	0	1	1	0.998	0.998	0.994
1.1	110	10	0.25	0.3	1	1	0.999	0.999	0.996
1.1	110	10	0.25	0.6	1	1	0.999	0.999	0.998
1.1	110	20	0.75	0	0.832	0.831	0.718	0.717	0.591
1.1	110	20	0.75	0.3	0.891	0.891	0.807	0.807	0.694
1.1	110	20	0.75	0.6	0.961	0.961	0.909	0.908	0.833
1.1	110	20	0.25	0	0.834	0.834	0.719	0.718	0.593
1.1	110	20	0.25	0.3	0.849	0.849	0.737	0.736	0.61
1.1	110	20	0.25	0.6	0.874	0.874	0.773	0.773	0.659
1.11	110	30	0.75	0	0.567	0.562	0.41	0.398	0.269
1.11	110	30	0.75	0.3	0.637	0.632	0.486	0.48	0.353
1.11	110	30	0.75	0.6	0.759	0.759	0.622	0.619	0.484
1.11	110	30	0.25	0	0.559	0.554	0.403	0.391	0.258
1.11	110	30	0.25	0.3	0.592	0.587	0.431	0.423	0.302
1.11	110	30	0.25	0.6	0.617	0.612	0.467	0.46	0.329
1.2	120	10	0.75	0	0.572	0.572	0.422	0.422	0.294
1.2	120	10	0.75	0.3	0.632	0.631	0.483	0.483	0.351
1.2	120	10	0.75	0.6	0.736	0.736	0.608	0.608	0.477
1.2	120	10	0.25	0	0.562	0.561	0.41	0.409	0.288
1.2	120	10	0.25	0.3	0.592	0.591	0.44	0.439	0.318
1.2	120	10	0.25	0.6	0.615	0.615	0.462	0.462	0.338
1.21	120	20	0.75	0	0.263	0.263	0.163	0.162	0.096
1.21	120	20	0.75	0.3	0.29	0.29	0.174	0.174	0.103
1.21	120	20	0.75	0.6	0.349	0.349	0.223	0.223	0.139
1.21	120	20	0.25	0	0.269	0.269	0.162	0.161	0.097
1.21	120	20	0.25	0.3	0.264	0.264	0.158	0.157	0.091
1.21	120	20	0.25	0.6	0.277	0.277	0.168	0.167	0.101
1.22	120	30	0.75	0	0.176	0.173	0.096	0.093	0.049
1.22	120	30	0.75	0.3	0.192	0.191	0.107	0.105	0.06
1.22	120	30	0.75	0.6	0.209	0.209	0.117	0.116	0.065
1.22	120	30	0.25	0	0.177	0.174	0.094	0.092	0.05
1.22	120	30	0.25	0.3	0.181	0.179	0.099	0.096	0.051
1.22	120	30	0.25	0.6	0.183	0.182	0.101	0.1	0.056

Note: GMR: geometric mean ratio between T and R. For the power, deeper color indicates higher value.

In addition, we see from the simulation results that the TOST approach is not equivalent to some 90% approaches, $90\%\text{CI}_L$, $90\%\text{CI}_E$, and $90\%\text{CI}_W$. The later three approaches have higher type I error rates and higher power than the TOST approach. Therefore, if we start with hypotheses testing but use 90% CI approaches (say, $90\%\text{CI}_L$, $90\%\text{CI}_E$, and $90\%\text{CI}_W$) for bioequivalence assessment, we may have the risk of undesirable high type I error rate, which can be greater than 5%.

3.5.3 An Example—Binary Responses

For binary responses, the following interval hypotheses are usually used for testing bioequivalence or biosimilarity between the test product and the reference product:

$$H_0 : p_T - p_R \leq -0.2 \text{ or } p_T - p_R \geq 0.2 \quad \text{vs.} \quad H_a : -0.2 < p_T - p_R < 0.2, \quad (3.11)$$

where p_T and p_R are response rates of the test product and the reference product, respectively. To compare the power (or type I error rates) of the TOST approach and the CI approach, we conducted simulations to evaluate their performances. The TOST approach we use is based on the adjusted Wald test (Agresti and Min, 2005). The CI approach we use is based on Tango's score confidence interval (Tango, 1998). We consider the classic 2×2 crossover design without carryover effects for total sample sizes of $n = 24, 48, 100$, and 200, respectively. Unlike the crossover design for the continuous responses, here we do not consider period effect and sequence effect. Thus the samples can be seen as n independently and identically distributed matched pairs. For parameter specifications, we consider a variety of response rate combinations $(p_{00}, p_{01}, p_{10}, p_{11})$, where p_{jk}, $j = 0,1$, $k = 0,1$ is the rate of binary reference response being equal to j and binary test response being equal to k. 10,000 repetitions were implemented for each scenario.

The type I error rates are presented in Table 3.9. With sample size increasing, the type I error rates of both approaches converge to the level of 5%. When sample size is small, the type I error rate can depart far away from the level of 5%. In addition, the two approaches have different type I error rates.

The power is presented in Table 3.10. We see the two approaches have different power. The power of the TOST approach is higher than the other, especially when sample size is small. With limited sample size (say, $n = 24$ as in the conventional 2×2 crossover design for continuous responses), the power of both approaches is not high (sometimes very low).

Simulation results indicate that compared with continuous responses, binary responses require larger sample sizes to achieve satisfactory power and type I error rate. In addition, the TOST approach and the CI approach are not operationally equivalent under our simulation settings.

TABLE 3.9

Type I Error Rate for Binary Responses When H_o in (3.11) Is True

p_{00}	p_{01}	p_{10}	p_{11}	n	L_1	U_1	L_2	U_2	CR_1	CR_2
0.1	0	0.2	0.7	24	0.049	0.322	0.079	0.359	0.118	0.036
				48	0.096	0.288	0.123	0.308	0.067	0.067
				100	0.130	0.263	0.143	0.273	0.081	0.046
				200	0.152	0.245	0.158	0.250	0.062	0.041
0.1	0.1	0.3	0.5	24	−0.009	0.376	−0.006	0.394	0.034	0.012
				48	0.053	0.331	0.055	0.340	0.060	0.049
				100	0.098	0.293	0.100	0.297	0.055	0.050
				200	0.129	0.268	0.130	0.269	0.052	0.049
0.3	0	0.2	0.5	24	0.048	0.322	0.079	0.358	0.116	0.033
				48	0.096	0.289	0.123	0.309	0.060	0.060
				100	0.130	0.262	0.143	0.272	0.080	0.045
				200	0.151	0.245	0.158	0.250	0.063	0.043

Note: L_1 and U_1 are the means of lower and upper bounds of the 95% Adjusted Wald interval for a difference of proportions with matched pairs (Agresti and Min, 2005). L_2 and U_2 are the means of lower and upper bounds of the 95% Tango's score confidence interval for a difference of proportions with matched pairs (Tango, 1998). CR_1 and CR_2 are the rejection rates of the TOST approach based on Adjusted Wald interval and the CI approach based on Tango's score confidence interval, respectively.
The darker the color, the larger the value.

TABLE 3.10

Power for Binary Responses When H_a in (3.11) Is True

p_{00}	p_{01}	p_{10}	p_{11}	n	L_1	U_1	L_2	U_2	CR_1	CR_2
0	0.1	0.1	0.8	24	−0.148	0.147	−0.163	0.162	0.483	0.306
				48	−0.105	0.105	−0.111	0.111	0.853	0.817
				100	−0.074	0.073	−0.076	0.075	0.994	0.992
				200	−0.052	0.052	−0.053	0.053	1.000	1.000
0.1	0.05	0.05	0.8	24	−0.112	0.113	−0.133	0.135	0.820	0.645
				48	−0.076	0.079	−0.085	0.088	0.978	0.967
				100	−0.052	0.053	−0.056	0.056	1.000	1.000
				200	−0.037	0.037	−0.038	0.039	1.000	1.000
0.05	0.05	0.1	0.8	24	−0.084	0.177	−0.093	0.202	0.581	0.412
				48	−0.043	0.140	−0.045	0.152	0.834	0.793
				100	−0.015	0.112	−0.015	0.117	0.983	0.974
				200	0.005	0.094	0.005	0.097	1.000	1.000
0.1	0	0.1	0.8	24	−0.016	0.202	−0.011	0.241	0.561	0.290
				48	0.022	0.171	0.040	0.192	0.651	0.651
				100	0.047	0.148	0.061	0.159	0.932	0.879
				200	0.063	0.134	0.070	0.140	0.996	0.992
0	0.05	0.15	0.8	24	−0.054	0.236	−0.053	0.262	0.334	0.210

(Continued)

TABLE 3.10 (*Continued*)

Power for Binary Responses When H_a in (3.11) Is True

p_{00}	p_{01}	p_{10}	p_{11}	n	L_1	U_1	L_2	U_2	CR_1	CR_2
				48	−0.007	0.199	−0.004	0.212	0.499	0.450
				100	0.027	0.170	0.030	0.176	0.743	0.697
				200	0.049	0.150	0.050	0.153	0.939	0.927
0.1	0.1	0.1	0.7	24	−0.147	0.148	−0.162	0.163	0.488	0.310
				48	−0.105	0.106	−0.111	0.112	0.854	0.822
				100	−0.073	0.073	−0.076	0.075	0.996	0.993
				200	−0.052	0.052	−0.053	0.053	1.000	1.000
0.15	0.05	0.1	0.7	24	−0.085	0.175	−0.094	0.200	0.588	0.412
				48	−0.043	0.140	−0.044	0.153	0.836	0.793
				100	−0.014	0.113	−0.013	0.118	0.983	0.974
				200	0.005	0.095	0.006	0.097	1.000	1.000
0.05	0.1	0.15	0.7	24	−0.114	0.210	−0.122	0.228	0.321	0.177
				48	−0.068	0.164	−0.071	0.173	0.667	0.621
				100	−0.032	0.130	−0.033	0.134	0.922	0.908
				200	−0.008	0.107	−0.008	0.109	0.996	0.994
0.1	0.05	0.15	0.7	24	−0.053	0.237	−0.053	0.263	0.336	0.214
				48	−0.008	0.198	−0.005	0.211	0.501	0.451
				100	0.026	0.169	0.029	0.176	0.752	0.710
				200	0.048	0.150	0.050	0.153	0.944	0.932
0.1	0.2	0.2	0.5	24	−0.202	0.201	−0.211	0.211	0.079	0.023
				48	−0.144	0.147	−0.148	0.151	0.475	0.420
				100	−0.102	0.103	−0.104	0.104	0.883	0.874
				200	−0.074	0.073	−0.074	0.073	0.996	0.996
0.3	0.1	0.1	0.5	24	−0.148	0.147	−0.164	0.163	0.482	0.310
				48	−0.105	0.106	−0.111	0.112	0.855	0.821
				100	−0.073	0.073	−0.076	0.075	0.995	0.993
				200	−0.052	0.052	−0.052	0.053	1.000	1.000
0.3	0.05	0.15	0.5	24	−0.053	0.237	−0.053	0.264	0.331	0.209
				48	−0.007	0.199	−0.003	0.212	0.499	0.448
				100	0.026	0.169	0.029	0.176	0.749	0.704
				200	0.049	0.150	0.050	0.153	0.941	0.927
0.1	0.15	0.25	0.5	24	−0.107	0.290	−0.112	0.304	0.065	0.017
				48	−0.049	0.239	−0.050	0.246	0.294	0.256
				100	−0.004	0.199	−0.004	0.201	0.506	0.489
				200	0.027	0.171	0.027	0.172	0.744	0.733

Note: L_1 and U_1 are the means of lower and upper bounds of the 95% Adjusted Wald interval for a difference of proportions with matched pairs (Agresti and Min, 2005). L_2 and U_2 are the means of lower and upper bounds of the 95% Tango's score confidence interval for a difference of proportions with matched pairs (Tango, 1998). CR_1 and CR_2 are the rejection rates of the TOST approach based on Adjusted Wald interval and the CI approach based on Tango's score confidence interval, respectively.
The darker the color, the larger the value.

3.6 Sample Size Requirement

As the TOST approach at 5% level of significance is operationally equivalent to the approaches of 90%CI$_{2\alpha}$, 95%CI$_L$, and 95%CI$_E$, for testing (3), the required sample sizes to achieve a desirable power for those approaches are identical. Similarly, the TOST approach at 2.5% level of significance is operationally equivalent to the approach of 95%CI$_{2\alpha}$, and the TOST approach at 10% level of significance is operationally equivalent to the approaches of 90%CI$_L$, and 90%CI$_E$. Thus, the required sample sizes for those CI approaches except CI$_W$ can be derived equivalently as those derived for the corresponding TOST approach with appropriate level of significance. Thus, without loss of generality, we only need to focus on the TOST approach and the CI$_W$ approach for calculating the required sample size for the TOST approach and the CI approaches. The equations for sample size calculation of the two approaches are given below.

Denote θ_0 and σ as the true values that are estimated by $\hat{F}$ and se. The power of the TOST approach and the CI$_W$ approach can be easily expressed as follows.

$$p_{\text{TOST}} = P\left\{-\delta < \hat{F} - t_\alpha se \text{ and } \hat{F} + t_\alpha se < \delta\right\}$$

$$= P\left\{-\delta + t_\alpha se < \hat{F} < \delta - t_\alpha se\right\}$$

$$= E\left[P\left\{\frac{-\delta + t_\alpha se}{\sigma} - \frac{\theta_0}{\sigma} < \frac{\hat{F} - \theta_0}{\sigma} < \frac{\delta - t_\alpha se}{\sigma} - \frac{\theta_0}{\sigma}\bigg|se\right\}\right]$$

$$p_{\text{CI}_W} = P\left\{-\delta < \hat{F} - t_{\alpha_1} se \text{ and } \hat{F} + t_{\alpha_2} se < \delta\right\}$$

$$= P\left\{-\delta + t_{\alpha_1} se < \hat{F} < \delta - t_{\alpha_2} se\right\}$$

Note that in above expressions, t_α, se and σ are functions of the sample size, $\frac{\hat{F} - \theta_0}{\sigma}$ follows a certain distribution (say, a standard normal distribution) which does not depend on the sample size and design parameters, and se over which the expectation is taken follows a certain distribution (say, some variation of the chi-square distribution). Given the desirable power $1 - \beta$, the specified parameters for the alternative hypothesis, the required sample size for the TOST approach can be calculated by solving $p_{\text{TOST}} \geq 1 - \beta$ with respect to the sample size. For the CI$_W$ approach, p_{CI_W} does not have an explicit expression and the required sample size can be obtained by numeric simulations. The simulated power for some parameter specifications can be found in Table 3.4. For the TOST approach, an approximate formula to calculate the required sample size can be found in Chow and Liu (2008).

3.7 Concluding Remarks

For evaluation of the safety and efficacy of drug product, the method of hypotheses testing (point hypotheses testing for new drugs and interval hypotheses testing for generics/biosimilars) and confidence interval approach are commonly considered. The concepts and interpretations between the method of hypotheses testing and the confidence interval are different. For evaluation of drug products including new drugs and generics/biosimilars, the method of hypotheses testing including point hypotheses testing for *equality* and interval hypotheses testing for *equivalence* is the official test procedure recommended by the FDA. In practice, however, the use of the method of hypotheses testing and the confidence interval approach are often confused and misused.

The mixed-up use of the method of hypotheses testing and confidence interval approach has raised the controversial issues that (i) why a 95% confidence interval approach is used for evaluation of new drugs while a 90% confidence interval approach is considered for assessment of generics/biosimilars? and (ii) why the regulatory adopts different standards (i.e., 95% versus 90% confidence levels or 5% versus 10% significance levels) for drug products (i.e., new drugs versus generics/biosimilars).

This chapter clarifies the following controversial issues that are commonly encountered in drug product development and regulatory review/approval process. First, there is a distinction between point hypotheses testing for equality (for evaluation of new drugs) and interval hypotheses testing for equivalence (for assessment generic and biosimilar drug products). Second, for point hypotheses testing, a two-sided test (TST) is often used. TST at the α level of significance is equivalent to $(1-\alpha)\times100\%$ confidence interval approach. Third, for interval hypotheses testing, TOST procedure is recommended by the FDA. TOST consists of two one-sided test (each at the α level of significance) is a size-α test (Chow and Shao, 2002b) which is *operationally* equivalent to the $(1-2\alpha)\times100\%$ confidence interval approach in many cases under certain conditions. TOST (each at the α level of significance) is *not* equivalence to the $(1-2\alpha)\times100\%$ confidence interval approach in general. Thus, in practice, it is suggested that the concepts of hypotheses testing and confidence interval approach should not be mixed-up used to avoid possible confusion.

Sample size estimation should be performed under the framework used for the evaluation of the test treatment under investigation. In other words, under the framework of interval hypotheses testing, the power function should be used for selecting an appropriate sample size for achieving the desired power for establishment of equivalence.

Appendix

Consider the conventional 2×2 crossover designs without carryover effects. Let Y_{ijk} be the log-transformation of the pharmacokinetic response of interest, of the ith subject in the jth period, and of the kth sequence of the trial. As suggested by the FDA, the following statistical model is useful in describing Y_{ijk}:

$$Y_{ijk} = \mu + F_l + P_j + Q_k + S_{ikl} + e_{ijk},$$

Where μ is the overall mean; P_j is the fixed effect of the jth period, where $j = 1, 2$, and $\sum_{j=1}^{2} P_j = 0$; Q_k is the fixed effect of the kth sequence, where $k = 1, 2$, and $\sum_{k=1}^{2} Q_k = 0$; F_l is the direct fixed effect of the lth drug formulation ($F_T + F_R = 0$) when $j = k$, $l = T$, the test formulation; otherwise, $l = R$, the reference formulation; S_{ikl} is the random effect of the ith subject in the kth sequence under drug formulation l; and $S_{ik} = (S_{ikT}, S_{ikR})$, $i = 1, \ldots, n_k$, and $k = 1, 2$ are independently and identically distributed bivariate normal random vectors with mean $(0, 0)$ and an unknown variance-covariance matrix:

$$\begin{pmatrix} \sigma_{BT}^2 & \rho \sigma_{BT} \sigma_{BR} \\ \rho \sigma_{BT} \sigma_{BR} & \sigma_{BR}^2 \end{pmatrix};$$

e_{ijk}'s are the independent random errors distributed as $N\left(0, \sigma_{WI}^2\right)$, and S_{ik}'s and e_{ijk}'s are independent. Note that σ_{BT}^2 and σ_{BR}^2 are between-subject variances and σ_{WT}^2 and σ_{WR}^2 are within-subject variances and that $\sigma_{TT}^2 = \sigma_{BT}^2 + \sigma_{WT}^2$ and $\sigma_{TR}^2 = \sigma_{BR}^2 + \sigma_{WR}^2$ are called total variances for the test and reference formulations, respectively.

The average bioequivalence index, denoted as v, is $F_T - F_R = \mu_T - \mu_R$. Let $\bar{y}_{jk}$ be the sample average of the observations in the jth period and the kth sequence. Under the assumed statistical model,

$$\bar{y}_{11} - \bar{y}_{21} \sim N\left(v + P_1 - P_2, \tau^2 / n_1\right)$$

and

$$\bar{y}_{12} - \bar{y}_{22} \sim N\left(-v + P_1 - P_2, \tau^2 / n_2\right),$$

where

$$\tau^2 = \sigma_{BT}^2 + \sigma_{BR}^2 - 2\rho \sigma_{BT} \sigma_{BR} + \sigma_{WT}^2 + \sigma_{WR}^2 = \sigma_{TT}^2 + \sigma_{TR}^2 - 2\rho \sigma_{BT} \sigma_{BR}.$$

Consequently, we have the point estimators of v and τ^2 as

$$\hat{v} = \left(\bar{y}_{11} - \bar{y}_{21} - \bar{y}_{12} + \bar{y}_{22}\right)/2$$

and

$$\hat{\tau}^2 = \left\{(n_1 - 1)s_{D1}^2 + (n_2 - 1)s_{D2}^2\right\}/(n_1 + n_2 - 2),$$

where s_{Dk}^2 is the sample variance based on the differences $\{y_{i1k} - y_{i2k}, i = 1, \ldots, n_k\}$, $k = 1, 2$.

Denote $c = (1/n_1 + 1/n_2)/4$; the point estimators $\hat{v}$ and $\hat{\tau}^2$ satisfy

$$\hat{v} \sim N\left(v, c\tau^2\right)$$

and

$$\frac{(n_1 + n_2 - 2)\hat{\tau}^2}{\tau^2} \sim \chi^2(n_1 + n_2 - 2)$$

Then, the $(1 - 2\alpha) \times 100\%$ confidence lower limit and upper limit of $\mu_T - \mu_R$ are given as

$$L = \hat{v} - Z_\alpha \sqrt{c\hat{\tau}^2}$$

and

$$U = \hat{v} + Z_\alpha \sqrt{c\hat{\tau}^2},$$

where $Z_\alpha = t_\alpha(n_1 + n_2 - 2)$ is the αth quantile of t distribution with degree of freedom $n_1 + n_2 - 2$.

4

Endpoint Selection

4.1 Introduction

In clinical trials, it is important to determine the primary response variables for addressing the scientific and/or medical questions of interest. The response variables, which are also known as the clinical endpoints, are usually chosen to meet the study objectives. Once the response variables are chosen, the possible outcomes of treatment are defined and the corresponding information would be used to assess the safety and efficacy of a study drug. Typically, to assess the safety and efficacy of a study drug, the study drug is first shown to be statistically significant from a placebo control. If there is a statistically significant difference, the trial is demonstrated to have a high probability of correctly detecting a clinically meaningful difference, which is known as the (statistical) power of the trial. Therefore, in practice, a pre-study power analysis for sample size estimation is usually performed to ensure that the trial with the intended sample size has a desired power, say 80%, for addressing the scientific/medical question of interest. The purpose is to find an appropriate sample size based on the information (the desired power, variability and clinically meaningful differences, etc.) provided by clinical scientists.

In many clinical studies, it is not uncommon that the sample size of a study is determined based on expected absolute change from baseline of a primary study endpoint but the collected data are analyzed based on relative change from baseline (e.g., percent change from baseline) of the primary study endpoint, or based on the percentage of patients who show some improvement (i.e., responder analysis). The definition of a responder could be based on either absolute change from baseline or relative change from baseline of the primary study endpoint. It is very controversial in terms of the interpretation of the analysis results, especially when a significant result is observed based on a study endpoint (e.g., absolute change from baseline, relative change from baseline, or the responder analysis) but not on the other study endpoint (e.g., absolute change from baseline, relative change from baseline, or responder analysis). In practice, it is then of interest to explore how an observed significant difference of a study endpoint (e.g., absolute change from baseline, relative change from baseline, or responder's analysis) can be translated to that of the other study endpoint (e.g., absolute change from baseline, relative change from baseline, or responder's analysis). An immediate impact on the assessment of treatment effect based on different study endpoints is the power analysis for sample size calculation. For example, sample size required for achieving a desired power based on the absolute change could be very different from that obtained based on the percent change, or the percentage of patients who show an improvement based on the absolute change or relative change at α level of significance. As an example, consider a clinical trial for evaluation of possible weight reduction of a test treatment in female patients. Weight data from 10 subjects are given in Table 4.1.

As it can be seen from Table 4.1, mean absolute change and mean percent change from pre-treatment are 5.3 lbs and 5.1%, respectively. If a subject is considered a responder if there is weight reduction by more than 5 lbs (absolute change) or by more than 5% (relative change), the response rates based on absolute change and relative change are given by 40% and 30%, respectively. It should be noted that sample sizes required for achieving a desired power for detecting a clinically meaningful difference, say, by an absolute change of 5.5 lbs and a relative change of 5.5%, for the two study endpoints would not be the same. Similarly, the required sample sizes are also different using the response rates based on absolute change and relative change. Table 4.2 summarizes sample size calculation based on absolute change, relative change, and responders (defined based on either absolute change or relative change).

In clinical trials, one of the most controversial issues regarding clinical endpoint selection is determining which clinical endpoint is telling the truth. The other controversial issue is how to translate clinical results among

TABLE 4.1

Weight Data from Ten Female Subjects

Pre-treatment	Post-treatment	Absolute Change	Relative Change
110	106	4	3.6
90	80	10	11.1
105	100	5	4.8
95	93	2	2.2
170	163	7	4.1
90	84	8	8.9
150	145	5	3.3
135	131	4	3.0
160	159	1	0.6
100	91	9	9.0
120.5	115.2	5.3	5.1
(30.5)	(31.53)		

TABLE 4.2

Sample Size Calculation

Study Endpoint	Clinical Meaningful Difference	Sample Size Required
Absolute change	5 lb	262
Relative change	5%	146
Responder (based on absolute change)*	>5 lb	12
Responder (based on relative change)**	>5%	19

* Response rate based on absolute change greater than 5 lb is 60%.
** Response rate based on relative change greater than 5% is 30%.

the study endpoints. In practice, the sponsors always choose the clinical endpoints to their best interest. The regulatory agencies, however, require the primary clinical endpoint be specified in the study protocol. Positive results from other clinical endpoints will not be considered as the primary analysis results for regulatory approval. This, however, does not have any scientific or statistical justification for assessment of the treatment effect of the test drug under investigation.

In this chapter, we attempt to provide some insight to the above issues. In particular, the focus is to evaluate the effect on the power of the test when the sample size of the clinical study is determined by an alternative clinical strategy based on different study endpoint and non-inferiority margin. In the next section, model and assumptions for studying the relationship among these study endpoints are described. Under the model, translations among different study endpoints are studied. Section 4.4 provides a comparison of different clinical strategies for endpoint sections in terms of sample size and the corresponding power. A numerical study is given in Section 4.5 to provide some insight regarding the effect to the different clinical strategies for endpoint selection. Development of therapeutic index function for endpoint selection is given in Section 4.6. Brief concluding remarks are presented in the last section.

4.2 Clinical Strategy for Endpoint Selection

In clinical trials, for a given primary response variable, commonly considered study endpoints include: (i) measurements based on absolute change (e.g., endpoint change from baseline); (ii) measurements based on relative change; (iii) proportion of responders based on absolute change; and (iv) proportion of responders based on relative change. We will refer these study endpoints to as the *derived study endpoints* because they are derived from the original data collected from the same patient population. In practice, it will be more

TABLE 4.3

Clinical Strategy for Endpoint Selection in Non-inferiority Trials

	Non-inferiority Margin	
Study Endpoint	Absolute Difference (δ_1)	Relative Difference (δ_2)
Absolute change (E_1)	$I = E_1\delta_1$	$II = E_1\delta_2$
Relative change (E_2)	$III = E_2\delta_1$	$IV = E_2\delta_2$
Responder based on Absolute change (E_3)	$V = E_3\delta_1$	$VI = E_3\delta_2$
Responder based on Relative change (E_4)	$VII = E_4\delta_1$	$VII = E_4\delta_2$

complicated if the intended trial is to establish non-inferiority of a test treatment to an active control (reference) treatment. In this case, sample size calculation will also depend on the size of the non-inferiority margin, which may be based on either absolute change or relative change of the derived study endpoint. For example, based on responder's analysis, we may want to detect a 30% difference in response rate or to detect a 50% relative improvement in response rate. Thus, in addition to the four types of derived study endpoints, there are also two different ways to define a non-inferiority margin. Thus, there are many possible clinical strategies with different combinations of the derived study endpoint and the selection of non-inferiority margin for assessment of the treatment effect. These clinical strategies are summarized in Table 4.3.

To ensure the success of an intended clinical trial, the sponsor will usually carefully evaluate all possible clinical strategies for selecting the type of study endpoint, clinically meaningful difference, and non-inferiority margin during the stage of protocol development. In practice, some strategies may lead to the success of the intended clinical trial (i.e., achieve the study objectives with the desired power), while some strategies may not. A common practice for the sponsor is to choose a strategy to their best interest. However, regulatory agencies such as the FDA may challenge the sponsor as to the inconsistent results. This has raised the following questions. First, which study endpoint is telling the truth regarding the efficacy and safety of the test treatment under study? Second, how to translate the clinical information among different derived study endpoints since they are obtained based on the same data collected from the some patient population? These questions, however, remain unanswered.

4.3 Translations among Clinical Endpoints

Suppose that there are two test treatments, namely, a test treatment (T and a reference treatment (R). Denote the corresponding measurements of the *i*th subject in the *j*th treatment group before and after the treatment by W_{1ij} and

W_{2ij}, respectively, where $j = T$ or R corresponds to the test and the reference treatment, respectively. Assume that the measurement W_{1ij} is lognormal distributed with parameters μ_j and σ_{1j}^2, i.e.,

$$W_{1ij} \sim \text{lognormal}\,(\mu_j, \sigma_{1j}^2).$$

Let $W_{2ij} = W_{1ij}(1 + \Delta_{ij})$, where Δ_{ij} denotes the percentage change after receiving the treatment. In addition, assume that Δ_{ij} is lognormal distributed with parameters μ_{Δ_j} and $\sigma_{\Delta_j}^2$, i.e.,

$$\Delta_{ij} \sim \text{lognormal}\,(\mu_{\Delta_j}, \sigma_{\Delta_j}^2).$$

Thus, the difference and the relative difference between the measurements before and after the treatment are given by $W_{2ij} - W_{1ij}$ and $\left(W_{2ij} - W_{1ij}\right)/W_{1ij}$, respectively. In particular,

$$W_{2ij} - W_{1ij} = W_{1ij}\Delta_{ij} \sim \text{lognormal}\,(\mu_j + \mu_{\Delta_j}, \sigma_j^2 + \sigma_{\Delta_j}^2),$$

and

$$\frac{W_{2ij} - W_{1ij}}{W_{1ij}} \sim \text{lognormal}\,(\mu_{\Delta_j}, \sigma_{\Delta_j}^2).$$

To simplify the notations, define X_{ij} and Y_{ij} as $X_{ij} = \log(W_{2ij} - W_{1ij})$, $Y_{ij} = \log\left(\frac{W_{2ij} - W_{1ij}}{W_{1ij}}\right)$. Then, both X_{ij} and Y_{ij} are normally distributed with means $\mu_j + \mu_{\Delta_j}$ and μ_{Δ_j}, $i = 1, 2, \ldots, n_j$, $j = T, R$, respectively.

Thus, possible derived study endpoints based on the responses observed before and after the treatment as described earlier include X_{ij}, the absolute difference between "before treatment" and "after treatment" responses of the subjects, Y_{ij}, the relative difference between "before treatment" and "after treatment" responses of the subjects,

$$r_{A_j} = \#\{x_{ij} > c_1, i = 1, \ldots, n_j\}/n_j,$$

the proportion of responders, which is defined as a subject whose absolute difference between "before treatment" and "after treatment" responses is larger than a pre-specified value c_1,

$$r_{R_j} = \#\{y_{ij} > c_2, i = 1, \ldots, n_j\}/n_j,$$

the proportion of responders, which is defined as a subject whose relative difference between "before treatment" and "after treatment" responses is larger than a pre-specified value c_2.

To define notation, for $j = T, R$, let $p_{A_j} = E(r_{A_j})$ and $p_{R_j} = E(r_{R_j})$. Given the above possible types of derived study endpoints, we may consider the following hypotheses for testing non-inferiority with non-inferiority margins determined based on either absolute difference or relative difference:

1. The absolute difference of the responses

$$H_0 : (\mu_R - \mu_{\Delta_R}) - (\mu_T - \mu_{\Delta_T}) \geq \delta_1 \text{ vs. } H_a : (\mu_R - \mu_{\Delta_R}) - (\mu_T - \mu_{\Delta_T}) < \delta_1 \quad (4.1)$$

2. The relative difference of the responses

$$H_0 : (\mu_{\Delta_R} - \mu_{\Delta_T}) \geq \delta_2 \text{ vs. } H_a : (\mu_{\Delta_R} - \mu_{\Delta_T}) < \delta_2 \quad (4.2)$$

3. The difference of responders' rates based on the absolute difference of the responses

$$H_0 : p_{A_R} - p_{A_T} \geq \delta_3 \text{ vs. } H_a : p_{A_R} - p_{A_T} < \delta_3 \quad (4.3)$$

4. The relative difference of responders' rates based on the absolute difference of the responses

$$H_0 : \frac{p_{A_R} - p_{A_T}}{p_{A_R}} \geq \delta_4 \text{ vs. } H_a : \frac{p_{A_R} - p_{A_T}}{p_{A_R}} < \delta_4 \quad (4.4)$$

5. The absolute difference of responders' rates based on the relative difference of the responses

$$H_0 : p_{R_R} - p_{R_T} \geq \delta_5 \text{ vs. } H_a : p_{R_R} - p_{R_T} < \delta_5 \quad (4.5)$$

6. The relative difference of responders' rate based on the relative difference of the responses

$$H_0 : \frac{p_{R_R} - p_{R_T}}{p_{R_R}} \geq \delta_6 \text{ vs. } H_a : \frac{p_{R_R} - p_{R_T}}{p_{R_R}} < \delta_6 \quad (4.6)$$

For a given clinical study, the above are the possible clinical strategies for assessment of the treatment effect. Practitioners or sponsors of the study often choose the strategy to their best interest. It should be noted that current regulatory position is to require the sponsor to pre-specify which study endpoint will be used for assessment of the treatment effect in the study protocol without any scientific justification.

In practice, however, it is of particular to study the effect to power analysis for sample size calculation based on different clinical strategies. As pointed out earlier, the required sample size for achieving a desired power based on the absolute difference of a given primary study endpoint may be quite different from that obtained based on the relative difference of the given primary study endpoint. Thus, it is of interest to clinician or clinical scientist to investigate this issue under various scenarios. In particular, hypotheses (4.1) may be used for sample size determination but hypotheses (4.3) are used for testing treatment effect. However, the comparison of these two clinical strategies would be affected by the value of c_1, which is used to determine the proportion of responders. However, in the interest of a simple and easier comparison, the number of parameters is kept as small as possible.

4.4 Comparison of Different Clinical Strategies

4.4.1 Test Statistics, Power and Sample Size Determination

Note that X_{ij} denotes the absolute difference between "before treatment" and "after treatment" responses of the ith subjects under the jth treatment, and Y_{ij} denotes the relative difference between "before treatment" and "after treatment" responses of the ith subjects under the jth treatment. Let $\bar{x}_{\cdot j} = \frac{1}{n_j} = \sum_{i=1}^{n_j} x_{ij}$ and $\bar{y}_{\cdot j} = \frac{1}{n_j} = \sum_{i=1}^{n_j} y_{ij}$ be the sample means of X_{ij} and Y_{ij} for the jth treatment group, $j = T, R$, respectively.

Based on normal distribution, the null hypothesis in (4.1) is rejected at a level α of significance if

$$\frac{\bar{x}_{\cdot R} - \bar{x}_{\cdot T} + \delta_1}{\sqrt{\left(\frac{1}{n_T} + \frac{1}{n_R}\right)\left[\left(\sigma_T^2 + \sigma_{\Delta T}^2\right) + \left(\sigma_R^2 + \sigma_{\Delta R}^2\right)\right]}} > z_\alpha. \tag{4.7}$$

Thus, the power of the corresponding test is given as

$$\Phi\left(\frac{(\mu_T + \mu_{\Delta T}) - (\mu_R + \mu_{\Delta R}) + \delta_1}{\sqrt{\left(n_T^{-1} + n_R^{-1}\right)\left[\left(\sigma_T^2 + \sigma_{\Delta T}^2\right) + \left(\sigma_R^2 + \sigma_{\Delta R}^2\right)\right]}} - z_\alpha\right), \tag{4.8}$$

where $\Phi(.)$ is the cumulative distribution function of the standard normal distribution. Suppose that the sample sizes allocated to the reference and test treatments are in the ratio of r, where r is a known constant. Using these results, the required total sample size for the test the hypotheses (4.1) with a power level of $(1-\beta)$ is $N = n_T + n_R$, with

$$n_T = \frac{(z_\alpha + z_\beta)^2(\sigma_1^2 + \sigma_2^2)(1+1/\rho)}{\left[(\mu_R + \mu_{\Delta R}) - (\mu_T + \mu_{\Delta T}) - \delta_1\right]^2},\tag{4.9}$$

$n_R = \rho n_T$ and z_u is $1-u$ quantile of the standard normal distribution.

Note that y_{ij}s are normally distributed. The testing statistic based on $\bar{y}_{.j}$ would be similar to the above case. In particular, the null hypothesis in (4.2) is rejected at a significance level α if

$$\frac{\bar{y}_{T.} - \bar{y}_{R.} + \delta_2}{\sqrt{\left(\dfrac{1}{n_T} + \dfrac{1}{n_R}\right)(\sigma_{\Delta T}^2 + \sigma_{\Delta R}^2)}} > z_\alpha.\tag{4.10}$$

The power of the corresponding test is given as

$$\Phi\left(\frac{\mu_{\Delta T} - \mu_{\Delta R} + \delta_2}{\sqrt{\left(n_T^{-1} + n_R^{-1}\right)(\sigma_{\Delta T}^2 + \sigma_{\Delta R}^2)}} - z_\alpha\right).\tag{4.11}$$

Suppose that $n_R = \rho n_T$, where r is a known constant. Then the required total sample size to test hypotheses (4.2) with a power level of $(1-\beta)$ is $(1+\rho)n_T$, where

$$n_T = \frac{(z_\alpha + z_\beta)^2(\sigma_{\Delta T}^2 + \sigma_{\Delta R}^2)(1+1/\rho)}{\left[(\mu_R + \mu_{\Delta R}) - (\mu_T + \mu_{\Delta T}) - \delta_2\right]^2}.\tag{4.12}$$

For sufficiently large sample size n_j, r_{A_j} is asymptotically normal with mean p_{A_j} and variance $\frac{p_{A_j}(1-p_{A_j})}{n_j}$, $j = T, R$. Thus, based on Slutsky Theorem, the null hypothesis in (4.3) is rejected at an approximate α level of significance if

$$\frac{r_{A_T} - r_{A_R} + \delta_3}{\sqrt{\dfrac{1}{n_T}r_{A_T}(1-r_{A_T}) + \dfrac{1}{n_R}r_{A_R}(1-r_{A_R})}} > z_\alpha.\tag{4.13}$$

The power of the above test can be approximated by

$$\Phi\left(\frac{p_{A_T} - p_{A_R} + \delta_3}{\sqrt{n_T^{-1}p_{A_T}(1-p_{A_T}) + n_R^{-1}r_{A_R}(1-p_{A_R})}} - z_\alpha\right).\tag{4.14}$$

If $n_R = \rho n_T$, where r is a known constant. Then, the required sample size to test hypotheses (4.3) with a power level of $(1-\beta)$ is $(1+\rho)n_T$, where

$$n_T = \frac{(z_\alpha + z_\beta)^2 \left[p_{A_T}(1-p_{A_T}) + p_{A_R}(1-p_{A_R})/\rho \right]}{(p_{A_R} - p_{A_T} - \delta_3)^2} \tag{4.15}$$

Note that, by definition,

$$p_{A_j} = 1 - \Phi\left(\frac{c_1 - (\mu_j + \mu_{\Delta_j})}{\sqrt{\sigma_j^2 + \sigma_{\Delta_j}^2}} \right),$$

where $j = T, R$. Therefore, following similar arguments, the above results also apply to test hypotheses (4.5) with p_{A_j} replaced by $p_{R_j} = 1 - \Phi\left(\frac{c_2 - \mu_{\Delta_j}}{\sigma_{\Delta_j}} \right)$ and δ_3 replaced by δ_5.

The hypotheses in (4.4) are equivalent to

$$H_0 : (1-\delta_4)p_{A_R} - p_{A_T} \geq 0 \quad \text{vs.} \quad H_1 : (1-\delta_4)p_{A_R} - p_{A_T} < 0. \tag{4.16}$$

Therefore, the null hypothesis in (4.4) is rejected at an approximate a level of significance if

$$\frac{r_{A_T} - (1-\delta_4)r_{A_R}}{\sqrt{\dfrac{1}{n_T} r_{A_T}(1-r_{A_T}) + \dfrac{(1-\delta_4)^2}{n_R} r_{A_R}(1-r_{A_R})}} > z_\alpha. \tag{4.17}$$

Using normal approximation to the test statistic when both n_T and n_R are sufficiently large, the power of the above test can be approximated by

$$\Phi\left(\frac{p_{A_T} - (1-\delta_4)p_{A_R}}{\sqrt{n_T^{-1}p_{A_T}(1-p_{A_T}) + n_R^{-1}(1-\delta_4)^2 p_{A_R}(1-p_{A_R})}} - z_\alpha \right) \tag{4.18}$$

Suppose that $n_R = \rho n_T$, where r is a known constant. Then the required total sample size to test hypotheses (4.10), or equivalently (4.16), with a power level of $(1-\beta)$ is $(1+\rho)n_T$, where

$$n_T = \frac{(z_\alpha + z_\beta)^2 \left[p_{A_T}(1-p_{A_T}) + (1-\delta_4)^2 p_{A_R}(1-p_{A_R})/\rho \right]}{\left[p_{A_T} - (1-\delta_4)p_{A_R} \right]^2}. \tag{4.19}$$

Similarly, the results derived in (4.17) through (4.19) for the hypotheses (4.4) also apply to the hypotheses in (4.6) with p_{A_j} replaced by $p_{R_j} = 1 - \Phi\left(\frac{c_2 - \mu_{\Delta j}}{\sigma_{\Delta j}}\right)$ and δ_4 replaced by δ_6.

4.4.2 Determination of the Non-inferiority Margin

Based on the results derived in the previous section, the non-inferiority margins corresponding to the tests based on the absolute difference and the relative difference can be chosen in such a way that the two tests would have the same power. In particular, hypothesis (4.1) and (4.2) would give the power level if the power function given in (4.8) is the same as that given in (4.11). Consequently, the non-inferiority margins δ_1 and δ_2 would satisfy the following equation

$$\frac{(\sigma_T^2 + \sigma_{\Delta_T}^2) + (\sigma_R^2 + \sigma_{\Delta_R}^2)}{\left[(\mu_T + \mu_{\Delta_T}) - (\mu_R + \mu_{\Delta_R}) + \delta_1\right]^2} = \frac{(\sigma_{\Delta_T}^2 + \sigma_{\Delta_R}^2)}{\left[(\mu_{\Delta_T} - \mu_{\Delta_R}) + \delta_2\right]^2}. \tag{4.20}$$

Similarly, for hypotheses (4.3) and (4.4), the non-inferiority margins δ_3 and δ_4 would satisfy the following relationship

$$\frac{p_{A_T}(1 - p_{A_T}) + p_{A_R}(1 - p_{A_R})/\rho}{(p_{A_R} - p_{A_T} - \delta_3)^2} = \frac{p_{A_T}(1 - p_{A_T}) + (1 - \delta_4)^2 p_{A_R}(1 - p_{A_R})/\rho}{\left[p_{A_R} - (1 - \delta_4)p_{A_T}\right]^2} \tag{4.21}$$

For hypotheses (4.5) and (4.6), the non-inferiority margins δ_5 and δ_6 satisfy

$$\frac{p_{R_T}(1 - p_{R_T}) + p_{R_R}(1 - p_{R_R})/\rho}{(p_{R_R} - p_{R_T} - \delta_5)^2} = \frac{p_{R_T}(1 - p_{R_T}) + (1 - \delta_6)^2 p_{R_R}(1 - p_{R_R})/\rho}{\left[p_{R_R} - (1 - \delta_6)p_{R_T}\right]^2} \tag{4.22}$$

Results given in (4.20) through (4.22) provide a way of translating the non-inferiority margins between endpoints based on the difference and the relative difference. In the next section, we would present a numerical study to provide some insight how the power level of these tests would be affected by the choices of different study endpoints for various combinations of parameters values.

4.4.3 A Numerical Study

In this section, a numerical study was conducted to provide some insight about the effect to the different clinical strategies.

4.4.3.1 Absolute Difference versus Relative Difference

Table 4.4, provides the required sample sizes for the test of non-inferiority based on the absolute difference (X_{ij}) and relative difference (Y_{ij}). In particular, the nominal power level $(1 - \beta)$ is chosen to be 0.80 and α is 0.05. The corresponding sample sizes are calculated using the formulae in (4.9) and (4.12). It is difficult to conduct any comparison because the corresponding non-inferiority margins are based on different measurement scales. However, to provide some idea to assess the impact of switching from a clinical endpoint based on absolute difference to that based on relative difference, a numerical study on the power of the test was conducted. In particular, Table 4.5 presents the power of the test for non-inferiority based on relative difference (Y) with the sample sizes determined by the power based on absolute difference (X). The power was calculated using the result given in (4.11). The results demonstrate that the effect is, in general, very significant. In many cases, the power is much smaller than the nominal level 0.8.

4.4.3.2 Responders' Rate Based on Absolute Difference

Similar computation was conducted for the case when the hypotheses are defined in terms of the responders' rate based on the absolute difference, i.e., hypotheses defined (4.3) and (4.4). Table 4.6 gives the required sample sizes, with the derived results given in (4.15) and (4.19), for the corresponding hypotheses with non-inferiority margins given both in terms of absolute difference and relative difference of the responders' rates. Similarly, Table 4.7 presents the power of the test for non-inferiority based on relative difference of the responders' rate with the sample sizes determined by the power based on absolute difference of the responders' rate. The power was calculated using the result given in (4.14). Again, the results demonstrate that the effect is, in general, very significant. In many cases, the power is much smaller than the nominal level 0.8.

4.4.3.3 Responders' Rate Based on Relative Difference

Similar to the issues considered in the above paragraph with the exception that the responders' rate is defined based on the relative difference, the required sample sizes for the corresponding hypotheses with non-inferiority margins given both in terms of absolute difference and relative difference of the responders' rates defined based on the relative difference, i.e., hypotheses defined in (4.5) and (4.6). The results are showed in Table 4.8. Following the similar steps, Table 4.9 presents the power of the test for non-inferiority based on relative difference of the responders' rate with the sample sizes determined by the power based on absolute difference of the responders' rate. The similar pattern emerges and the results demonstrate that the power is usually much smaller than the nominal level 0.8.

TABLE 4.4

Sample Sizes for Non-inferiority Testing Based on Absolute Difference and Relative Difference ($\alpha = 0.05$, $\beta = 0.20$, $\rho = 1$)

$\sigma_T^2 + \sigma_R^2$	\multicolumn{9}{c}{$(\mu_R + \mu_{\Delta R}) - (\mu_T + \mu_{\Delta T}) = 0.20$}									\multicolumn{9}{c}{$(\mu_R + \mu_{\Delta R}) - (\mu_T + \mu_{\Delta T}) = 0.30$}								
	1.0			2.0			3.0			1.0			2.0			3.0		
$\sigma_{\Delta T}^2 + \sigma_{\Delta R}^2$	1.0	1.5	2.0	1.0	1.5	2.0	1.0	1.5	2.0	1.0	1.5	2.0	1.0	1.5	2.0	1.0	1.5	2.0
Absolute Difference																		
$\delta_1 = .50$	275	344	413	413	481	550	550	619	687	619	773	928	928	1082	1237	1237	1392	1546
$\delta_1 = .55$	202	253	303	303	354	404	404	455	505	396	495	594	594	693	792	792	891	990
$\delta_1 = .60$	155	194	232	232	271	310	310	348	387	275	344	413	413	481	550	550	619	687
$\delta_1 = .65$	123	153	184	184	214	245	245	275	306	202	253	303	303	354	404	404	455	505
$\delta_1 = .70$	99	124	149	149	174	198	198	223	248	155	194	232	232	271	310	310	348	387
Relative Difference																		
$\delta_2 = .40$	310	464	619	310	464	619	310	464	619	1237	1855	2474	1237	1855	2474	1237	1855	2474
$\delta_2 = .50$	138	207	275	138	207	275	138	207	275	310	464	619	310	464	619	310	464	619
$\delta_2 = .60$	78	116	155	78	116	155	78	116	155	138	207	275	138	207	275	138	207	275

TABLE 4.5

Power of the Test of Non-inferiority Based on Relative Difference

$\sigma_T^2+\sigma_R^2$	$\sigma_{\Delta T}^2+\sigma_{\Delta R}^2$	$(\mu_R+\mu_{\Delta R})-(\mu_T+\mu_{\Delta T})=0.20$									$(\mu_R+\mu_{\Delta R})-(\mu_T+\mu_{\Delta T})=0.30$								
		1.0			2.0			3.0			1.0			2.0			3.0		
δ_1	δ_2	1.0	1.5	2.0	1.0	1.5	2.0	1.0	1.5	2.0	1.0	1.5	2.0	1.0	1.5	2.0	1.0	1.5	2.0
$\delta_1=.50$	$\delta_2=.4$	75.8	69.0	65.1	89.0	81.3	75.8	95.3	89.0	83.6	54.6	48.4	45.2	69.5	60.0	54.5	80.0	69.5	62.6
	$\delta_2=.5$	96.9	94.2	92.0	99.6	98.4	96.9	100.0	99.6	98.9	97.0	94.1	91.9	99.6	98.4	96.9	100.0	99.6	98.9
	$\delta_2=.6$	99.9	99.6	99.2	100.0	100.0	99.9	100.0	100.0	100.0	100.0	99.9	99.8	100.0	100.0	100.0	100.0	100.0	100.0
$\delta_1=.55$	$\delta_2=.4$	64.2	57.6	53.8	79.3	70.1	64.2	88.4	79.3	72.7	40.6	35.9	33.5	53.1	45.0	40.6	63.5	53.1	47.1
	$\delta_2=.5$	91.5	86.7	83.3	98.0	94.7	91.5	99.6	98.0	95.8	87.9	82.2	78.6	96.4	91.8	87.9	99.0	96.4	93.3
	$\delta_2=.6$	99.1	97.9	96.7	99.9	99.7	99.1	100.0	99.9	99.8	99.5	98.6	97.8	100.0	99.8	99.5	100.0	100.0	99.9
$\delta_1=.60$	$\delta_2=.4$	54.6	48.5	45.2	69.5	60.1	54.6	80.1	69.5	62.6	31.8	28.3	26.5	41.8	35.2	31.8	50.5	41.7	36.9
	$\delta_2=.5$	84.0	77.9	73.9	94.4	88.6	84.0	98.2	94.4	90.4	75.8	69.0	65.1	89.0	81.3	75.8	95.3	89.0	83.6
	$\delta_2=.6$	97.0	94.2	91.9	99.6	98.4	97.0	100.0	99.6	98.9	96.9	94.2	92.0	99.6	98.4	96.9	100.0	99.6	98.9
$\delta_1=.65$	$\delta_2=.4$	47.0	41.4	38.7	60.8	51.8	46.8	71.5	60.6	54.2	26.1	23.4	21.9	33.9	28.8	26.1	41.2	34.0	30.1
	$\delta_2=.5$	76.0	69.1	65.2	89.1	81.3	75.9	95.3	89.0	83.6	64.2	57.6	53.8	79.3	70.1	64.2	88.4	79.3	72.7
	$\delta_2=.6$	93.2	88.7	85.7	98.6	95.8	93.1	99.7	98.6	96.8	91.5	86.7	83.3	98.0	94.7	91.5	99.6	98.0	95.8
$\delta_1=.70$	$\delta_2=.4$	40.6	36.0	33.6	53.2	45.2	40.6	63.5	53.2	47.2	22.2	20.0	18.9	28.5	24.4	22.2	34.5	28.5	25.4
	$\delta_2=.5$	67.9	61.2	57.4	82.8	73.9	67.9	91.0	82.7	76.3	54.6	48.5	45.2	69.5	60.1	54.6	80.1	69.5	62.6
	$\delta_2=.6$	87.9	82.3	78.7	96.5	91.9	87.9	99.0	96.4	93.4	84.0	77.9	73.9	94.4	88.6	84.0	98.2	94.4	90.4

TABLE 4.6

Sample Sizes for Non-inferiority Testing Based on Absolute Difference and Relative Difference of Response Rates Defined by the Absolute Difference (X_{ij}) ($\alpha = 0.05$, $\beta = 0.20$, $\rho = 1$, $c_1 - (\mu_T + \mu_{\Delta T}) = 0$)

		$c_1-(\mu_R+\mu_{\Delta R})=-0.60$									$c_1-(\mu_R+\mu_{\Delta R})=-0.80$								
		1.0			2.0			3.0			1.0			2.0			3.0		
$\sigma_T^2+\sigma_R^2$		1.0	1.5	2.0	1.0	1.5	2.0	1.0	1.5	2.0	1.0	1.5	2.0	1.0	1.5	2.0	1.0	1.5	2.0
$\sigma_{\Delta T}^2+\sigma_{\Delta R}^2$																			
Absolute Difference																			
$\delta_3 = .25$		399	284	228	228	195	173	173	157	146	2191	898	558	558	410	329	329	279	245
$\delta_3 = .30$		159	128	111	111	99	91	91	85	81	382	253	195	195	162	141	141	127	116
$\delta_3 = .35$		85	73	65	65	60	56	56	53	51	153	117	98	98	86	78	78	72	68
$\delta_3 = .40$		53	47	43	43	40	38	38	37	35	82	68	59	59	54	50	50	47	44
$\delta_3 = .45$		36	33	31	31	29	28	28	27	26	51	44	40	40	37	34	34	33	31
Relative Difference																			
$\delta_4 = .35$		458	344	285	285	249	224	224	206	193	1625	869	601	601	469	391	391	340	304
$\delta_4 = .40$		199	166	147	147	134	124	124	117	112	392	288	234	234	202	180	180	165	153
$\delta_4 = .45$		109	96	88	88	82	78	78	75	72	168	139	121	121	110	102	102	95	91

TABLE 4.7

Power of the Test of Non-inferiority Based on Relative Difference of Response Rates ($\alpha = 0.05$, $\beta = 0.20$, $\rho = 1$, $c_1 - (\mu_T + \mu_{\Delta_T}) = 0$)

			$c_1 - (\mu_R + \mu_{\Delta_R}) = -0.60$									$c_1 - (\mu_R + \mu_{\Delta_R}) = -0.80$								
			$\sigma_R^2 = \sigma_T^2$									$\sigma_R^2 = \sigma_T^2$								
			1.0			2.0			3.0			1.0			2.0			3.0		
δ_3	δ_4	$\sigma_{\Delta R}^2 = \sigma_{\Delta T}^2$	1.0	1.5	2.0	1.0	1.5	2.0	1.0	1.5	2.0	1.0	1.5	2.0	1.0	1.5	2.0	1.0	1.5	2.0
$\delta_3 = 25$	$\delta_4 = .35$		75.1	73.1	71.9	71.9	71.2	70.6	70.6	70.1	69.9	89.3	81.2	77.4	77.4	75.2	73.8	73.8	72.9	72.3
	$\delta_4 = .40$		97.0	94.6	92.8	92.8	91.4	90.2	90.2	89.2	88.5	100.0	99.7	98.6	98.6	97.1	95.7	95.7	94.5	93.4
	$\delta_4 = .45$		99.9	99.6	99.1	99.1	98.6	98.1	98.1	97.6	97.2	100.0	100.0	100.0	100.0	99.9	99.8	99.8	99.6	99.3
$\delta_3 = 30$	$\delta_4 = .35$		42.9	44.9	46.3	46.3	47.0	47.7	47.7	48.1	48.8	33.0	38.1	41.0	41.0	42.8	44.0	44.0	45.1	45.7
	$\delta_4 = .40$		71.9	70.5	69.9	69.9	69.1	68.6	68.6	68.3	68.3	79.1	75.5	73.5	73.5	72.1	71.1	71.1	70.6	70.0
	$\delta_4 = .45$		91.4	89.1	87.6	87.6	86.3	85.3	85.3	84.5	84.1	98.2	95.7	93.5	93.5	91.6	90.2	90.2	89.1	88.0
$\delta_3 = 35$	$\delta_4 = .35$		28.3	30.9	32.4	32.4	33.6	34.4	34.4	35.1	35.8	18.9	23.2	26.1	26.1	28.1	29.7	29.7	30.9	32.0
	$\delta_4 = .40$		49.3	50.2	50.5	50.5	50.9	51.0	51.0	51.2	51.5	46.4	47.7	48.6	48.6	49.2	49.7	49.7	50.1	50.6
	$\delta_4 = .45$		71.2	70.2	69.1	69.1	68.7	68.0	68.0	67.6	67.5	76.7	74.0	72.4	72.4	71.2	70.5	70.5	69.9	69.7
$\delta_3 = 40$	$\delta_4 = .35$		21.2	23.4	24.9	24.9	25.9	26.8	26.8	27.7	28.0	13.9	17.1	19.3	19.3	21.2	22.5	22.5	23.6	24.3
	$\delta_4 = .40$		35.9	37.4	38.3	38.3	38.9	39.4	39.4	40.3	40.1	30.6	33.2	34.6	34.6	36.0	36.9	36.9	37.6	37.8
	$\delta_4 = .45$		53.8	54.0	54.0	54.0	53.8	53.8	53.8	54.4	53.7	53.7	53.9	53.7	53.7	54.1	54.1	54.1	54.2	53.7
$\delta_3 = 45$	$\delta_4 = .35$		17.2	19.1	20.5	20.5	21.3	22.2	22.2	22.8	23.3	11.4	13.9	15.8	15.8	17.2	18.1	18.1	19.2	19.8
	$\delta_4 = .40$		27.9	29.6	30.8	30.8	31.4	32.2	32.2	32.6	32.9	22.7	25.1	26.9	26.9	28.1	28.7	28.7	29.8	30.0
	$\delta_4 = .45$		41.6	42.7	43.5	43.5	43.5	44.0	44.0	44.2	44.2	39.2	40.4	41.5	41.5	42.1	42.0	42.0	42.9	42.6

TABLE 4.8

Sample Sizes for Non-inferiority Testing Based on Absolute Difference and Relative Difference of Response Rates Defined by the Relative Difference (Y_{ij}) ($\alpha = 0.05$, $\beta = 0.20$, $\rho = 1$, $c_2 - \mu_{\Delta T} = 0$)

$\sigma^2_{\Delta R} = \sigma^2_{\Delta T}$	$c_2 - \mu_{\Delta R} = -0.30$				$c_2 - \mu_{\Delta R} = -0.40$				$c_2 - \mu_{\Delta R} = -0.50$				$c_2 - \mu_{\Delta R} = -0.60$			
	1.0	1.5	2.0	2.5	1.0	1.5	2.0	2.5	1.0	1.5	2.0	2.5	1.0	1.5	2.0	2.5
Absolute Difference																
$\delta_5 = .25$	173	130	111	101	329	201	157	135	836	351	238	189	4720	745	399	284
$\delta_5 = .30$	91	74	66	61	141	102	85	76	244	147	114	97	504	229	159	128
$\delta_5 = .35$	56	48	44	41	78	61	53	49	114	81	67	59	180	110	85	73
$\delta_5 = .40$	38	33	31	29	50	41	37	34	66	51	44	40	92	64	53	47
$\delta_5 = .45$	28	25	23	22	34	29	27	25	43	35	31	29	56	42	36	33
Relative Difference																
$\delta_6 = .35$	224	173	151	138	391	256	206	180	823	412	297	243	2586	754	458	344
$\delta_6 = .40$	124	104	94	88	180	137	117	106	279	186	151	132	478	266	199	166
$\delta_6 = .45$	78	68	63	60	102	83	75	69	136	104	90	81	189	132	109	96

TABLE 4.9

Power of the Test of Non-inferiority Based on Relative Difference of Response Rates ($\alpha = 0.05$, $\beta = 0.20$, $\rho = 1$, $c_2 - \mu_{\Delta_T} = 0$)

$\sigma^2_{\Delta R} + \sigma^2_{\Delta T}$		$c_2 - \mu_{\Delta R} = -0.30$				$c_2 - \mu_{\Delta R} = -0.40$				$c_2 - \mu_{\Delta R} = -0.50$				$c_2 - \mu_{\Delta R} = -0.60$			
		1.0	1.5	2.0	2.5	1.0	1.5	2.0	2.5	1.0	1.5	2.0	2.5	1.0	1.5	2.0	2.5
$\delta_5 = .25$	$\delta_6 = .35$	70.6	69.5	68.8	68.7	73.8	71.2	70.1	69.6	80.5	74.3	72.1	70.9	95.7	79.6	75.1	73.1
	$\delta_6 = .40$	90.2	87.4	85.7	84.9	95.7	91.6	89.2	87.7	99.6	96.2	93.1	91.0	100.0	99.4	97.0	94.6
	$\delta_6 = .45$	98.1	96.4	95.2	94.5	99.8	98.7	97.6	96.7	100.0	99.8	99.2	98.5	100.0	100.0	99.9	99.6
$\delta_5 = .30$	$\delta_6 = .35$	47.7	49.3	50.1	50.5	44.0	47.0	48.1	49.0	38.6	43.7	45.9	47.1	29.2	39.2	42.9	44.9
	$\delta_6 = .40$	68.6	67.7	67.3	66.8	71.1	69.4	68.3	67.7	75.2	71.4	69.9	68.9	81.9	74.6	71.9	70.5
	$\delta_6 = .45$	85.3	83.1	81.8	80.9	90.2	86.7	84.5	83.3	95.4	90.6	87.9	86.0	99.2	94.9	91.4	89.1
$\delta_5 = .35$	$\delta_6 = .35$	34.4	36.9	38.2	38.7	29.7	33.3	35.1	36.5	23.6	29.4	32.2	33.8	16.1	24.4	28.3	30.9
	$\delta_6 = .40$	51.0	52.0	52.5	52.4	49.7	50.8	51.2	51.8	47.8	49.9	50.6	50.9	45.3	48.2	49.3	50.2
	$\delta_6 = .45$	68.0	67.4	67.0	66.3	70.5	68.7	67.6	67.4	73.7	71.1	69.6	68.5	78.4	73.5	71.2	70.2
$\delta_5 = .40$	$\delta_6 = .35$	26.8	28.8	30.3	30.8	22.5	25.8	27.7	28.7	17.3	22.1	24.6	26.3	12.0	17.9	21.2	23.4
	$\delta_6 = .40$	39.4	40.5	41.6	41.7	36.9	39.0	40.3	40.7	33.2	36.6	38.2	39.3	29.0	33.6	35.9	37.4
	$\delta_6 = .45$	53.8	53.7	54.2	53.7	54.1	54.1	54.4	54.0	53.5	54.0	54.1	54.2	53.7	53.6	53.8	54.0
$\delta_5 = .45$	$\delta_6 = .35$	22.2	24.2	25.1	25.8	18.1	21.0	22.8	23.7	14.1	17.9	20.0	21.6	10.0	14.5	17.2	19.1
	$\delta_6 = .40$	32.2	33.7	34.1	34.6	28.7	31.0	32.6	33.1	25.2	28.6	30.3	31.7	21.4	25.6	27.9	29.6
	$\delta_6 = .45$	44.0	44.7	44.5	44.8	42.0	43.1	44.2	44.1	40.3	42.1	42.9	43.8	38.6	40.5	41.6	42.7

4.5 Development of Therapeutic Index Function

4.5.1 Introduction

In pharmaceutical/clinical development of a test treatment, clinical trials are often conducted to evaluate the effectiveness of the test treatment under investigation. In clinical trials, there may be multiple endpoints available for measurement of disease status and/or therapeutic effect of the test treatment under study (Williams et al., 2004; Filozof et al., 2017). In practice, it is usually not clear which study endpoint can best inform the disease status and can be used to measure the treatment effect. Thus, it is difficult to determine which endpoint should be used as the primary endpoint especially as these multiple potential primary endpoints may be correlated with some unknown correlation structures. Once the primary study endpoint has been selected, sample size requirement for achieving a desired power can then be performed. It, however, should be noted that different study endpoints might not translate one-to another although they might be highly correlated to one another. In other words, for a given clinical trial, some study endpoints may be achieved and some don't. In this case, it is of interest to know which study endpoint is telling the truth. It also should be noted that different study endpoints might result in different sample size requirements.

Typical examples for clinical trials with multiple endpoints would be cancer clinical trials. In cancer clinical trials, overall survival (OS), response rate (RR), and/or time to disease progression (TTP) are usually considered as primary clinical endpoints for evaluation of effectiveness of the test treatment under investigation in regulatory submissions. Williams et al. (2004) provided a list of oncology drug products approved by the FDA based on single endpoint, co-primary endpoints, and/or multiple endpoints between 1990 and 2002.

As can be seen from Table 4.10, a total of 57 oncology drug submissions were approved by the FDA between 1990 and 2002. Among the 57 applications, 18 were approved based on survival endpoint alone, while 18 were approved based on RR and/or TTP alone. About 9 submissions were approved based on RR plus tumor-related signs and symptoms (co-primary endpoints). More recently, Zhou et al. (2019) provided a list of oncology and hematology drug approval by the FDA between 2008 and 2016, and a total of 12 drugs were approved based on multiple endpoints. Table 4.10 and Figure 4.1 indicated that endpoint selection is key to the success of intended clinical trials. For example, for those 9 submissions in Table 4.10, if the selected endpoints were survival or TTP, the clinical trials may meet the study endpoints of RR plus tumor-related signs and symptoms but fail to meet the selected study endpoints.

As demonstrated in Table 4.10 and Figure 4.1, composite endpoints are commonly used where multiple outcomes are measured and combined to

TABLE 4.10

Endpoints Supporting Regular Approval of Oncology Drug Marketing Application, January 1, 1990, to November 1, 2002

Total	57
Survival	18
RR and/or TTP alone	18
(predominantly hormone treatment of breast cancer	13
or hematologic malignancies)	(9)
Tumor-related signs and symptoms	(4)
RR + tumor-related signs and symptoms	2
Tumor-related signs and symptoms alone	2
Disease-free survival (adjuvant setting)	2
Recurrence of malignant pleural effusion	1
Decreased incidence of new breast cancer occurrence	1
Decreased impairment creatinine clearance	
Decreased xerostomia	

Source: Williams, G. et al., *J. Biopharm. Stat.,* 14, 5–21, 2004.

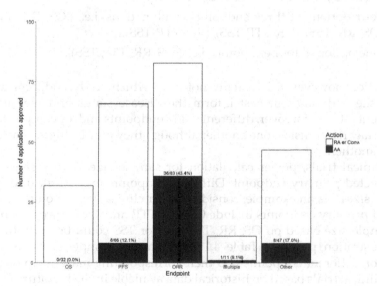

FIGURE 4.1

Number of applications approved by endpoint. The number of applications approved for a new indication in an oncology product by the US FDA during 2008 and 2016, grouped by primary endpoint of the trial that supported the application. Endpoints are abbreviated as follows: overall survival (OS), progression free survival (PFS), objective response rate (ORR), relapse-free survival (RFS), event-free survival (EFS), multiple endpoints other than a coprimary endpoint of overall survival and progression-free survival (Multiple), and other endpoints not included in the previous categories (Other). Types of approvals are abbreviated as follows: regular approval (RA), conversion to regular approval (Conv), and accelerated approval (AA). Data used were taken from the package inserts of the approved products and FDA records. (From Zhou, J. et al., *J. Natl. Cancer Inst.*, 111, 449–458, 2019.)

evaluate the treatment effect in clinical trials, especially for oncology drug developments. The adoption of composite endpoints can be due to that the primary endpoint is rare event or needs a long time to be observed, among other reasons. In oncology drug developments, commonly used composite endpoints may be comprised from the following four endpoints: (i) overall survival (OS); (ii) response rate (RR); (iii) time to disease progression (TTP) and; (iv) tumor-related signs and symptoms (TSS). Note that OS can be further divided into three categories: disease-free survival (DFS), progression-free survival (PFS), and relapse-free survival (RFS); yet for simplicity and without loss of generality, we won't differentiate them here. Assume all the four endpoints are used in constructing the composite endpoint(s). Depending on the number of composite endpoints, there can be up to a total of 15 possible combinations of them:

1. Four options of one endpoint, i.e., {OS, RR, TTP, TSS};
2. Six options of two endpoints combinations, i.e., {(OS, RR), (OS, TTP), (OS, TSS), (RR, TTP), (RR, TSS), (TTP, TSS)};
3. Four options of three endpoints combinations, i.e., {(OS, RR, TTP), (OS, RR, TSS), (OS, TTP, TSS), (RR, TTP, TSS)};
4. One option of four endpoints, i.e., {(OS, RR, TTP, TSS)}.

In practice, however, it is usually not clear which study endpoint and/or composite endpoint can best inform the disease status and measure the treatment effect. Moreover, different study endpoints and/or composite endpoints may not translate one another although they may be highly correlated to one another.

In clinical trials, power calculation for sample size is very sensitive to the selected primary endpoint. Different endpoints may result in different sample sizes. As an example, consider cancer clinical trials, commonly considered primary endpoints include OS, RR, TTP, and TSS. Power calculation for sample size based on OS, RR, TTP, and/or TSS could be very different. For illustration purpose, Table 4.11 summarizes sample sizes calculated based on different endpoints and their corresponding margins in oncology drug clinical trials based on historical data available in the literature (Motzer et al., 2019) and the margins are selected conventionally. From Table 4.11, it can be seen that different endpoints will result in different sample size requirements.

In this section, we intend to develop a therapeutic index based on a utility function to combine all study endpoints. The developed therapeutic index will fully utilize all the information collected via the available study endpoints for an overall assessment of the effectiveness of the test treatment under investigation. Statistical properties and performances of the proposed therapeutic index are evaluated both theoretically and via clinical trial simulations.

TABLE 4.11

Sample Size Calculation Based on Different Endpoints for One-sided Test of Non-inferiority with Significance Level $\alpha = 0.05$ and Expected Power $1 - \beta = 0.9$

Endpoint	OS	RR	TTP
H_a	$\theta > \delta$	$p_2 - p_1 > \delta$	$\theta > \delta$
Formula	$\dfrac{1}{\pi_1 \pi_2 p_0} \dfrac{(z_\alpha + z_\beta)^2}{(\ln\theta^* - \ln\delta)^2}$	$\left(\dfrac{p_1(1-p_1)}{\kappa} + p_2(1-p_2)\right)\left(\dfrac{z_\alpha + z_\beta}{p_2^* - p_1^* - \delta}\right)^2$	$\dfrac{1}{\pi_1 \pi_2 p_i} \dfrac{(z_\alpha + z_\beta)^2}{(\ln\theta^* - \ln\delta)^2}$
Margin (δ)	0.82	0.29	0.61
Other parameters	$p_0 = 0.14; \theta^* = 1$	$p_1 = 0.26; p_2 = 0.55;$ $p_2^* - p_1^* = 0$	$p_i = 0.4; \theta^* = 1$
Sample Size	6213	45	351

Note: 1. We assume a balanced design, i.e., $\kappa = 1$ and $\pi_2 = \pi_1 = \frac{1}{2}$. The sample size calculation formulas are obtained in Chow et al. (2018). The non-inferiority margin and other parameters are based on the descriptive statistics for the 560 patients with PD-L1-positive tumors in Motzer et al. (2019), where the margin is selected as the improvement of clinically meaningful difference.

2. H_a denotes the alternative hypothesis, δ is the non-inferiority margin, p_i is the response rate for sample i, π_1 and π_2 are the proportions of the sample size allocated to the two groups, $\theta = h_1/h_2$ is the hazard ratio, p_i is the overall probability of death occurring within the study period, p_0 is the overall probability of disease progression occurring within the study period, $\ln\theta$ is the natural logarithm of the hazard ratio, $\kappa = n_1/n_2$ is sample size ratio, $z_\alpha = \Phi^{-1}(1-\alpha)$ is $100(1-\alpha)\%$ quantile of the standard normal distribution.

4.5.2 Therapeutic Index Function

Let $e = (e_1, e_2, \cdots, e_J)'$ be the baseline clinical endpoints. The therapeutic index is defined as:

$$I_i = f_i(\omega_i, e), \quad i = 1, \cdots, K \tag{4.23}$$

where $\omega_i = (\omega_{i1}, \omega_{i2}, \cdots, \omega_{iJ})'$ is a vector of weights with ω_{ij} be the weight for e_j with respect to index I_i, $f_i(\cdot)$ is the therapeutic index function to construct therapeutic index I_i based on ω_i and e. Generally, e_j can be of different data types (e.g., continuous, binary, time to event) and ω_{ij} is pre-specified (or calculated by pre-specified criteria) and can be different for different therapeutic index I_i. Moreover, the therapeutic index function typically generates a vector of index $(I_1, I_2, \cdots, I_K)$; and if $K = 1$, it reduces to a single (composite) index. As an example, consider $I_i = \sum_{j=1}^{J} \omega_{ij} e_j$, then I_i is simply a linear combination of the endpoints; moreover, if $\omega_i = (\frac{1}{J}, \frac{1}{J}, \cdots \frac{1}{J})'$, then I_i is the average over all the endpoints.

Although e_j can be of different data types, we assume they are of the same type at this step. On one hand, we would like to investigate the predictability of I_i given that e_j can inform the disease (drug) status; one the other hand, we are also interested in the predictability of e_j given that I_i is informative. Particularly, we are interested in the following two conditional probabilities:

$$1. \quad p_{1ij} = \Pr(I_i \mid e_j), \quad i = 1, \cdots, K; j = 1, \cdots, J \tag{4.24}$$

and

$$2. \quad p_{2ij} = \Pr(e_j \mid I_i), \quad i = 1, \cdots, K; j = 1, \cdots, J \tag{4.25}$$

Intuitively, we would expect that p_{1ij} to be relatively large given that e_j is informative since I_i is a function of e_j; on the other hand, p_{2ij} could be small even if I_i is predictive since the information I_i contained may be attributed to another endpoint $e_{j'}$ rather other e_j. To derive Equations (4.24) and (4.25), we need to specify the weights ω_i, the distribution of e and the functions $f_i(\cdot)$ that are described more in details in the following subsections.

4.5.2.1 Selection of ω_i

One of the important concerns is how to select the weights ω_i. There might be various ways of specifying the weights and a reasonable one is based on the p-values. Specifically, denote $\theta_j, j = 1, \cdots, J$ as the treatment effect assessed by the endpoint e_j. Without loss of generality, θ_j is tested by the following hypotheses:

$$H_{0j} : \theta_j \le \delta_j \text{ versus } H_{aj}: \theta_j > \delta_j, \tag{4.26}$$

where $\delta_j, j = 1, \cdots, J$ are the pre-specified margins. Under some appropriate assumptions, we can calculate the p-value p_j for each H_{0j} based on the sample of e_j and the weights ω_i can be constructed based on $p = (p_1, p_2, \cdots, p_J)'$, i.e.,

$$\omega_{ij} = \omega_{ij}(p), \tag{4.27}$$

which is reasonable since each p-value indicates the significance of the treatment effect based on its responding endpoint; thus, it is possible to use all the information available to construct effective therapeutic indexes. Note that $\omega_{ij}(\cdot)$ should be constructed such that high value of ω_{ij} is with low value of p_j. For example, $\omega_{ij} = \frac{1}{p_j} / \sum_{j=1}^{J} \frac{1}{p_j}$.

4.5.2.2 Determination of $f_i(\cdot)$ and the Distribution of e

Another important issue is how to select the therapeutic index functions $f_i(\cdot)$. $f_i(\cdot)$ could be linear or nonlinear, or have an even more complicated form. We consider $f_i(\cdot)$ as linear here. Thus, (4.23) reduces to

$$I_i = \sum_{j=1}^{J} \omega_{ij} e_j = \sum_{j=1}^{J} \omega_{ij}(p) e_j, \ i = 1, \cdots, K. \tag{4.28}$$

Moreover, we need to specify the distribution of e. To simplify, assume e follows the multi-dimensional normal distribution $N(\theta, \Sigma)$, where

$$\theta = (\theta_1, \cdots, \theta_J)' \text{ and } \Sigma = (\sigma_{ij}^2)_{J \times J}$$

with

$$\sigma_{jj'}^2 = \sigma_j^2, j' = j \text{ and } \sigma_{jj'}^2 = \rho_{jj'} \sigma_j \sigma_{j'}, j' \ne j.$$

4.5.2.3 Derivation of $\Pr(I_i | e_j)$ and $\Pr(e_j | I_i)$

Suppose n subjects are independently and randomly selected from the population for the clinical trial. For each baseline endpoint e_j and hypothesis H_{0j}, a test statistic $\hat{e}_j$ is constructed based on the observations of the n subjects and the corresponding p-value p_j is calculated. e_j is informative is equivalent to $\hat{e}_j > c_j$ for some pre-specified critical value c_j pre-specified based on δ_j, significance level α and the variance of $\hat{e}_j$. The estimate of the therapeutic index I_i in (4.28) can be accordingly constructed as

$$\hat{I}_i = \omega_i{'}\hat{e} = \sum_{j=1}^{J} \omega_{ij}\hat{e}_j, \quad i = 1,\cdots,K, \tag{4.29}$$

where $\omega_i = (\omega_{i1},\omega_{i2},\cdots,\omega_{iJ}){'}$ and $\omega_{ij} = \omega_{ij}(p)$ is calculated based on the p-values on $p = (p_1,p_2,\cdots,p_J){'}$, and $\hat{e} = (\hat{e}_1,\hat{e}_2,\cdots,\hat{e}_J)$. $\hat{I}_i$ is informative if $\hat{I}_i > d_i$ for some pre-specified threshold d_i. Thus, (4.24) and (4.25) become

$$1.\ p_{1ij} = \Pr\left(\hat{I}_i > d_i \middle| \hat{e}_j > c_j\right), \quad i = 1,\cdots,K; j = 1,\cdots,J \tag{4.30}$$

and

$$2.\ p_{2ij} = \Pr\left(\hat{e}_j > c_j \middle| \hat{I}_i > d_i\right), \quad i = 1,\cdots,K; j = 1,\cdots,J \tag{4.31}$$

Without loss of generality, suppose $\hat{e}$ is the vector of sample means, then $\hat{e}$ follows the multi-dimensional normal distribution $N(\theta, \Sigma/n)$ based on the normality assumption of e. Moreover, $\hat{I}_i$ follows the normal distribution $N(\varphi_i, \eta_i^2/n)$, where

$$\varphi_i = \omega_i{'}\theta = \sum_{j=1}^{J} \omega_{ij}\theta_j \text{ and } \eta_i^2 = \omega_i{'}\Sigma\omega_i$$

Further, $(\hat{e}_j, \hat{I}_i){'}$ jointly follows a binormal distribution $N(\mu, \Gamma/n)$ where

$$\mu = (\theta_j, \varphi_i){'} \text{ and } \Gamma = \begin{pmatrix} \sigma_j^2 & 1_j{'}\Sigma\omega_i \\ 1_j{'}\Sigma\omega_i & \eta_i^2 \end{pmatrix} = \begin{pmatrix} \sigma_j^2 & \rho_{ji}^*\sigma_j\eta_i \\ \rho_{ji}^*\sigma_j\eta_i & \eta_i^2 \end{pmatrix},$$

where 1_j is an J dimensional vector of 0 except the jth item which equals 1 and thus

$$\rho_{ji}^* = 1_j{'}\Sigma\omega_i / (\sigma_j\eta_i) = \sum_{j'=1}^{J} \omega_{ij'}\sigma_{jj'}^2 / (\sigma_j\eta_i) = \sum_{j'=1}^{J} \omega_{ij'}\rho_{jj'}\sigma_j\sigma_{j'} / (\sigma_j\eta_i)$$

$$= \frac{1}{\eta_i}\sum_{j'=1}^{J} \omega_{ij'}\rho_{jj'}\sigma_{j'}.$$

Thus, the conditional probabilities (4.30) and (4.31) become

1.

$$p_{1ij} = \frac{\Pr\left(\hat{I}_i > d_i, \hat{e}_j > c_j\right)}{\Pr\left(\hat{e}_j > c_j\right)}$$

$$= \frac{1 - \Phi\left(\frac{\sqrt{n}(c_j - \theta_j)}{\sigma_j}\right) - \Phi\left(\frac{\sqrt{n}(d_i - \varphi_i)}{\eta_i}\right) + \Psi\left(\frac{\sqrt{n}(c_j - \theta_j)}{\sigma_j}, \frac{\sqrt{n}(d_i - \varphi_i)}{\eta_i}, \rho_{ji}^*\right)}{1 - \Phi\left(\frac{\sqrt{n}(c_j - \theta_j)}{\sigma_j}\right)} \quad (4.32)$$

2.

$$p_{2ij} = \frac{\Pr\left(\hat{I}_i > d_i, \hat{e}_j > c_j\right)}{\Pr\left(\hat{I}_i > d_i\right)}$$

$$= \frac{1 - \Phi\left(\frac{\sqrt{n}(c_j - \theta_j)}{\sigma_j}\right) - \Phi\left(\frac{\sqrt{n}(d_i - \varphi_i)}{\eta_i}\right) + \Psi\left(\frac{\sqrt{n}(c_j - \theta_j)}{\sigma_j}, \frac{\sqrt{n}(d_i - \varphi_i)}{\eta_i}, \rho_{ji}^*\right)}{1 - \Phi\left(\frac{\sqrt{n}(d_i - \varphi_i)}{\eta_i}\right)} \quad (4.33)$$

Moreover,

$$\frac{p_{2ij}}{p_{1ij}} = \frac{\Pr\left(\hat{e}_j > c_j\right)}{\Pr\left(\hat{I}_i > d_i\right)} = \frac{1 - \Phi\left(\frac{\sqrt{n}(c_j - \theta_j)}{\sigma_j}\right)}{1 - \Phi\left(\frac{\sqrt{n}(d_i - \varphi_i)}{\eta_i}\right)}, \quad (4.34)$$

where $\Phi(x)$ and $\Psi(x,y,\rho)$ denote the cumulative distribution functions for standard single variate normal and bivariate normal distributions respectively. Note that both conditional probabilities (4.32) and (4.33) depend on the parameters θ, Σ; the sample size n; the number of baseline endpoints J; the pre-specified weights ω_i; and the pre-specified thresholds c_j, d_i which further depend on the hypothesis testing margins δ_j and pre-specified type I error rate(s) among others. Intuitively, there are not simple formulas for (4.32) and (4.33) that can be derived directly. Although methods such as Taylor expansion may be employed to approximate (4.32) and (4.33) it is still nontrivial and could be quite complicated. However, note that $\Phi(x)$ is monotonic increasing, based on (4.34) we have

$$\frac{p_{2ij}}{p_{1ij}} < 1 \Leftrightarrow \frac{c_j - \theta_j}{\sigma_j} > \frac{d_i - \varphi_i}{\eta_i}. \quad (4.35)$$

Moreover, we assume $c_j = \delta_j + z_\alpha \dfrac{\sigma_j}{\sqrt{n}}$ conventionally and d_i is a linear combination of c_j, s, i.e.,

$$d_i = \sum_{j=1}^{J} \omega_{ij} c_j = \sum_{j=1}^{J} \omega_{ij} \delta_j + \frac{z_\alpha}{\sqrt{n}} \sum_{j=1}^{J} \omega_{ij} \sigma_j, \quad i = 1, \cdots, K, \tag{4.36}$$

Then, (4.35) can be further expressed as

$$\frac{p_{2ij}}{p_{1ij}} < 1 \Leftrightarrow \frac{c_j - \theta_j}{\sigma_j} > \frac{d_i - \varphi_i}{\eta_i}$$

$$\Leftrightarrow \left(1 - \frac{\sigma_j}{\eta_i} \omega_{ij}\right)\left(\Delta\theta - \frac{z_\alpha}{\sqrt{n}}\sigma_j\right) < \frac{\sigma_j}{\eta_i}\left(\Delta\theta_i^{(-j)} - \frac{z_\alpha}{\sqrt{n}}\sigma_i^{(-j)}\right), \tag{4.37}$$

where

$$\Delta\theta_j = \theta_j - \delta_j, \Delta\theta = \left(\Delta\theta_1, \Delta\theta_2, \cdots, \Delta\theta_J\right)',$$

$$\Delta\theta_i^{(-j)} = \omega_i^{(-j)'}\Delta\theta = \sum_{j' \neq j}^{J}\omega_{ij'}\Delta\theta_{j'},$$

$$\sigma_i^{(-j)} = \sum_{j' \neq j}^{J}\omega_{ij'}\sigma_{j'},$$

$\omega_i^{(-j)}$ equals ω_i except the ith item equals 0. To obtain more insights of (4.37), we assume $J = 2, K = 1$ and focus on $j = 1$ without loss of generality. Then the last inequation in (15) can be simplified as

$$\left(1 - \frac{\sigma_1}{\eta}\omega_1\right)\left(\Delta\theta_1 - \frac{z_\alpha}{\sqrt{n}}\sigma_1\right) < \frac{\sigma_1}{\eta}\omega_2\left(\Delta\theta_2 - \frac{z_\alpha}{\sqrt{n}}\sigma_2\right) \tag{4.38}$$

where ω_1 and ω_2 are the weights for the two endpoints respectively with $\omega_1 + \omega_2 = 1$, and $\eta = \sqrt{\omega_1^2\sigma_1^2 + 2\rho\omega_1\omega_2\sigma_1\sigma_2 + \omega_2^2\sigma_2^2}$ and ρ is the correlation coefficient of the two endpoints. Obviously, (4.15) depends on the variabilities of the endpoints and their correlation, the underlined effect sizes of both endpoints, the weights, and the sample size. We illustrate several special scenarios of (4.38) in Table 4.12.

From Table 4.12, we can see a remarkable situation that when $\rho = 1$, $\sigma_1 = \frac{1}{\tau}\sigma_2$, whether p_{1ij} is greater than p_{2ij} depends on whether the underlined effect size

TABLE 4.12

Illustrate Inequation (16) with Respective to Different Parameter Settings

Parameters	Inequation (16)
	$1-\dfrac{\sigma_1}{\sqrt{\omega_1^2\sigma_1^2+2\rho\omega_1\omega_2\sigma_1\sigma_2+\omega_2^2\sigma_2^2}}\,\omega_1\left(\Delta\theta_1-\dfrac{z_a}{\sqrt{n}}\sigma_1\right)<\dfrac{\sigma_1}{\sqrt{\omega_1^2\sigma_1^2+2\rho\omega_1\omega_2\sigma_1\sigma_2+\omega_2^2\sigma_2^2}}\,\omega_2\left(\Delta\theta_2-\dfrac{z_a}{\sqrt{n}}\sigma_2\right)$
$\omega_1=\omega_2=1/2$	$1-\dfrac{\sigma_1}{\sqrt{\sigma_1^2+2\rho\sigma_1\sigma_2+\sigma_2^2}}\left(\Delta\theta_1-\dfrac{z_a}{\sqrt{n}}\sigma_1\right)<\dfrac{\sigma_1}{\sqrt{\sigma_1^2+2\rho\sigma_1\sigma_2+\sigma_2^2}}\left(\Delta\theta_2-\dfrac{z_a}{\sqrt{n}}\sigma_2\right)$
$\rho=0$	$1-\dfrac{\omega_1\sigma_1}{\sqrt{\omega_1^2\sigma_1^2+\omega_2^2\sigma_2^2}}\left(\Delta\theta_1-\dfrac{z_a}{\sqrt{n}}\sigma_1\right)<\dfrac{\omega_2\sigma_1}{\sqrt{\omega_1^2\sigma_1^2+\omega_2^2\sigma_2^2}}\left(\Delta\theta_2-\dfrac{z_a}{\sqrt{n}}\sigma_2\right)$
$\rho=1$	$1-\dfrac{\omega_1\sigma_1}{\omega_1\sigma_1+\omega_2\sigma_2}\left(\Delta\theta_1-\dfrac{z_a}{\sqrt{n}}\sigma_1\right)<\dfrac{\omega_2\sigma_1}{\omega_1\sigma_1+\omega_2\sigma_2}\left(\Delta\theta_2-\dfrac{z_a}{\sqrt{n}}\sigma_2\right)$
$\sigma_1=\sigma_2$	$1-\dfrac{\omega_1}{\sqrt{\omega_1^2+2\rho\omega_1\omega_2+\omega_2^2}}\left(\Delta\theta_1-\dfrac{z_a}{\sqrt{n}}\sigma_1\right)<\dfrac{\omega_2}{\sqrt{\omega_1^2+2\rho\omega_1\omega_2+\omega_2^2}}\left(\Delta\theta_2-\dfrac{z_a}{\sqrt{n}}\sigma_2\right)$
$\omega_1=\omega_2=1/2$ $\rho=0$	$1-\dfrac{\sigma_1}{\sqrt{\sigma_1^2+\sigma_2^2}}\left(\Delta\theta_1-\dfrac{z_a}{\sqrt{n}}\sigma_1\right)<\dfrac{\sigma_1}{\sqrt{\sigma_1^2+\sigma_2^2}}\left(\Delta\theta_2-\dfrac{z_a}{\sqrt{n}}\sigma_2\right)$

(Continued)

TABLE 4.12 (Continued)

Illustrate Inequation (16) with Respective to Different Parameter Settings

Parameters	Inequation (16)
$\omega_1 = \omega_2 = 1/2$ $\rho = 1$	$\left(1-\dfrac{\sigma_1}{\sigma_1+\sigma_2}\right)\left(\Delta\theta_1 - \dfrac{z_a}{\sqrt{n}}\sigma_1\right) < \dfrac{\sigma_1}{\sigma_1+\sigma_2}\left(\Delta\theta_2 - \dfrac{z_a}{\sqrt{n}}\sigma_2\right)$
$\omega_1 = \omega_2 = 1/2$ $\sigma_1 = \sigma_2$	$\left(1-\dfrac{1}{\sqrt{2+2\rho}}\right)\Delta\theta_1 - \dfrac{1}{\sqrt{2+2\rho}}\Delta\theta_2 < \left(1-\dfrac{2}{\sqrt{2+2\rho}}\right)\dfrac{z_a}{\sqrt{n}}\sigma_1$
$\rho = 0,\ \sigma_1 = \sigma_2$	$\Delta\theta_1 < \dfrac{\sqrt{\omega_1^2+\omega_2^2}+\omega_1}{\omega_2}\Delta\theta_2 - \dfrac{\sqrt{\omega_1^2+\omega_2^2}+\omega_1-\omega_2}{\omega_2}\dfrac{z_a}{\sqrt{n}}\sigma_1$
$\omega_1 = \omega_2 = 1/2$ $\rho = 0,\ \sigma_1 = \sigma_2$	$\Delta\theta_1 < (\sqrt{2}+1)\Delta\theta_2 - \dfrac{\sqrt{2}z_a}{\sqrt{n}}\sigma_1$
$\omega_1 = 1/3,\ \omega_2 = 2/3$ $\rho = 0,\ \sigma_1 = \sigma_2$	$\Delta\theta_1 < \dfrac{\sqrt{5}+1}{2}\Delta\theta_2 - \dfrac{\sqrt{5}-1}{2}\dfrac{z_a}{\sqrt{n}}\sigma_1$
$\rho = 1,\ \sigma_1 = \dfrac{1}{\tau}\sigma_2$	$\Delta\theta_1 < \dfrac{1}{\tau}\Delta\theta_2$

$\Delta\theta_1$ is less than $\frac{1}{\tau}\Delta\theta_2$ only, regardless of the weights. For other situations, the relation between p_{1ij} and p_{2ij} varies for different combinations of weights, variabilities and correlations, underlined effect sizes, and sample sizes.

4.6 Concluding Remarks

In clinical trials, it is not uncommon that a study is powered based on expected absolute change from baseline of a primary study endpoint but the collected data are analyzed based on relative change from baseline (e.g., percent change from baseline) of the primary study endpoint, or the collected data are analyzed based on the percentage of patients who show some improvement (i.e., responder analysis). The definition of a responder could be based on either absolute change from baseline or relative change from baseline of the primary study endpoint. It is very controversial in terms of the interpretation of the analysis results, especially when a significant result is observed based on a study endpoint (e.g., absolute change from baseline, relative change from baseline, or the responder analysis) but not on the other study endpoint (e.g., absolute change from baseline, relative change from baseline, or responder analysis). Based on the numerical results of this study, it is evident that the power of the test can be decreased drastically when the study endpoint is changed. However, when switching from a study endpoint based on absolute difference to the one based on relative difference, one possible way to maintain the power level is to modify the corresponding non-inferiority margin, as suggested by the results given in Section 4.4.

In clinical trials, selection of appropriate study endpoints is critical for an accurate and reliable evaluation of effectiveness of a test treatment under investigation. In practice, however, there are usually multiple endpoints available for measurement of disease status and/or therapeutic effect of the test treatment under study. For example, in cancer clinical trials, overall survival, response rate, and/or time to disease progression are usually considered as primary clinical endpoints for evaluation of effectiveness of the test treatment under investigation. Once the study endpoints have been selected, sample size requirement for achieving a desired power can then be determined. However, different study endpoints may result in different sample size requirement. In practice, it is usually not clear which study endpoint can best inform the disease status and measure the treatment effect. Moreover, different study endpoints may not translate one another although they may be highly correlated one another. In this chapter, we intend to develop a therapeutic index based on a utility function to combine all study endpoints. The developed therapeutic index will fully utilize all the information collected via the available study endpoints for an overall assessment of the effectiveness of the test treatment under investigation. Statistical properties and performances of the proposed therapeutic index are evaluated both theoretically and via clinical trial simulations.

5

Non-inferiority/Equivalence Margin

5.1 Introduction

In clinical trials, it is unethical to not treat patients with critical/severe and/or life-threatening diseases such as cancer when approved and effective therapies such as standard of care or active control agents are available. In this case, an active control trial is often conducted for investigation of a new test treatment. The goal of an active control trial is to demonstrate that the test treatment is not inferior to or equivalent to the active control agent in the sense that the effect of the test treatment is not below some non-inferiority margin or equivalence limit when compared with the efficacy of the active control agent. In practice, there may be a need to develop a new treatment or therapy that is non-inferior (but not necessarily superior) to an established efficacious treatment due to the following reasons: (i) the test treatment is less toxic; (ii) the test treatment has a better safety profile; (iii) the test treatment is easy to administer; (iv) the test treatment is less expensive; (v) the test treatment provides better quality of life; (vi) the test treatment provides alternative treatment with some clinical benefits, e.g., generics or biosimilars. Clinical trials of this kind are referred to as non-inferiority trials. A comprehensive overview of design concepts and important issues that are commonly encountered in active control or non-inferiority trials can be found in D'Agostino et al. (2003).

For a non-inferiority trial, the concept is to reject the null hypothesis that a test treatment is inferior to a standard therapy or an active control agent and conclude that the difference between the test treatment and the active control agent is less than a clinically meaningful difference (non-inferiority margin) and hence the test treatment is at least as effective as (or not worsen than) the active control agent. The test treatment can then serve as an alternative to the active control agent. In practice, it, however, should be noted that unlike equivalence testing, non-inferiority testing is a one-sided equivalence testing which consists of the concepts of equivalence and superiority. In other words, superiority may be tested after the non-inferiority has been established. We conclude equivalence if fail to reject the null hypothesis of non-superiority. On the other hand, superiority may be concluded if the null hypothesis of non-superiority is rejected.

One of the major considerations in a non-inferiority trial is the selection of the non-inferiority margin. A different choice of non-inferiority margin will have an impact on sample size requirement for achieving a desired power for establishment of non-inferiority. It should be noted that non-inferiority margin could be selected based on either absolute change or relative change of the primary study endpoint, which will affect the method for data analysis of the collected clinical data, and consequently may alter the conclusion of the clinical study. In practice, despite the existence of some studies (e.g., Tsong et al., 1999; Hung et al., 2003; Laster and Johnson, 2003; Phillips, 2003), there is no established rule or gold standard for determination of non-inferiority margins in active control trials until early 2000. In 2000, the International Conference on Harmonization (ICH) published a guideline to assist the sponsors for selection of an appropriate non-inferiority margin (ICH E10, 2000). The ICH E10 guideline suggests that a non-inferiority margin may be selected based on past experience in placebo control trials with valid design under conditions similar to those planned for the new trial and the determination of a non-inferiority margin should not only reflect uncertainties in the evidence on which the choice is based, but also be suitably conservative. Along this line, in 2010, the FDA also published draft guidance on non-inferiority clinical trials and recommends a couple of approaches for selection of non-inferiority margin (FDA, 2010a).

In addition to exploring the FDA's recommended approaches, the purpose of this Chapter is to propose alternative methods for selection of non-inferiority margin in non-inferiority trials. In the next section, the relationship between non-inferiority testing and equivalence testing in active control trials is briefly described. Also included in this section is the impact on sample size requirement for achieving a desired power for establishment of non-inferiority. Section 5.3 discusses regulatory requirements and non-inferiority hypothesis for a pre-specified non-inferiority margin. Various methods for selection of non-inferiority margins are reviewed in Section 5.4. An example is given in Section 5.5 to illustrate various methods for determination of non-inferiority margin. Brief concluding remarks are given in Section 5.6.

5.2 Non-inferiority versus Equivalence

In clinical trials, some investigators confound the concepts of non-inferiority testing and equivalence testing. Thus, the question whether an equivalence testing can be replaced with a non-inferiority (especially for equivalence testing for biosimilar products) has been raised and discussed previously. In this section, we will explore the relationship among non-inferiority, equivalence, and superiority, their corresponding testing hypotheses, and power calculations for sample size.

5.2.1 Relationship among Non-inferiority, Equivalence, and Superiority

To study the relationship among non-inferiority, equivalence and superiority, we first assume that the non-inferiority margin, equivalence limit, and superiority margin are the same. Let M denote the non-inferiority margin (also equivalence limit and superiority margin). Also, let μ_T and μ_S be the mean responses of the test treatment and standard therapy (active control agent), respectively. If we assume that an observed mean response on the right-hand side of μ_S is an indication of improvement, then the relationship among non-inferiority, equivalence and superiority is illustrated in Figure 5.1.

As it can be seen from Figure 5.1, if μ_T falls within the equivalence limit of $(\mu_S - M, \mu_S + M)$, the test treatment is considered equivalent to the active control agent. If $\mu_T < \mu_S - M$, then the test treatment is considered inferior to the active control agent. In other words, $\mu_S - M \le \mu_T$ is an indication that the test treatment is not inferior to the active control agent. It should be noted that in this case, the test treatment could be either equivalent to the active control agent if $\mu_T < \mu_S + M$ or superior to the active control agent if $\mu_S + M < \mu_T$. Thus, we could test for superiority once the non-inferiority has been established without paying any statistical penalty because it is a closed testing procedure. Thus, non-inferiority consists of the concepts of equivalence and superiority and equivalence can be established through testing for non-inferiority and testing for non-superiority. Both non-inferiority testing and superiority testing are considered one-sided equivalence testing.

To provide a better understanding of the relationship among non-inferiority, equivalence, and superiority, their corresponding hypotheses are given in Figure 5.2. It, however, should be noted that if an observed response less than μ_S is considered improvement, the hypotheses for testing non-inferiority and superiority need to be modified.

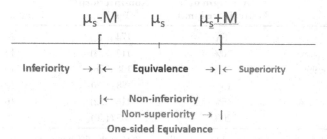

FIGURE 5.1
Relationship among non-inferiority, superiority, and equivalence.

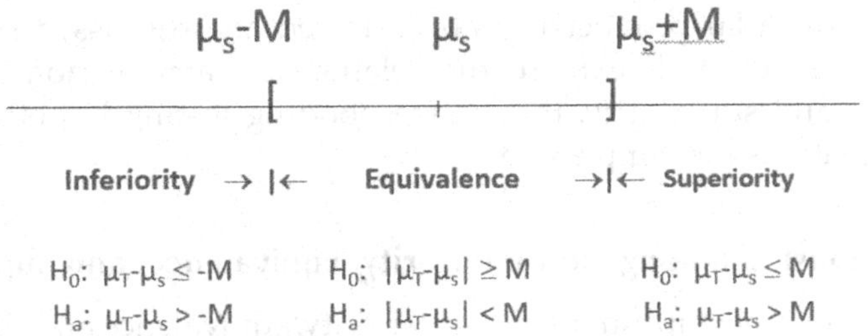

$$\mu_s\text{-M} \qquad \mu_s \qquad \mu_s\text{+M}$$

$$\underline{\qquad\qquad [\qquad\qquad\qquad\qquad]\qquad\qquad}$$

Inferiority → |←   Equivalence   →|← Superiority

$H_0: \mu_T\text{-}\mu_s \leq \text{-M}$ $H_0: |\mu_T\text{-}\mu_s| \geq M$ $H_0: \mu_T\text{-}\mu_s \leq M$

$H_a: \mu_T\text{-}\mu_s > \text{-M}$ $H_a: |\mu_T\text{-}\mu_s| < M$ $H_a: \mu_T\text{-}\mu_s > M$

FIGURE 5.2
Hypotheses for testing non-inferiority, equivalence, and superiority.

5.2.2 Impact on Sample Size Requirement

One of the major issues in non-inferiority or equivalence trials is the impact of the selected non-inferiority margin or equivalence limit on sample size requirement for achieving a desired power for establishing non-inferiority or equivalence. As an example, let μ_T and μ_s be the mean responses for the test treatment and the active control agent, respectively, and M be the non-inferiority margin or equivalence limit. For simplicity, consider power calculation for sample sizes for testing non-inferiority or equivalence of a test treatment as compared to an active control agent and assume that the non-inferiority margin is the same as the equivalence limit. For illustration purposes, we will focus on the study endpoint of binary responses. Based on formulas provided in Chow et al. (2008), sample size requirement for testing non-inferiority or equivalence for achieving an 80% power at the 5% level of significance for various combinations of μ_T and μ_s are summarized in Table 5.1.

As it can be seen from Table 5.1, testing non-inferiority (which is one-sided equivalence testing) requires fewer subjects. It, however, should be noted that the FDA requires non-inferiority testing be performed based on

TABLE 5.1

Sample Size[a] Requirement for Binary Responses

$\mu_T = \mu_s$[b]	NI Margin or Equivalence Limit	Non-inferiority Testing[c]	Equivalence Testing
$\mu_T = \mu_s \geq 90\%$	8%	174 (348)	241 (482)
	10%	112 (224)	155 (310)
$80\% \leq \mu_T = \mu_s < 90\%$	12%	138 (276)	191 (382)
	15%	88 (176)	122 (244)
$70\% \leq \mu_T = \mu_s < 80\%$	15%	116 (232)	160 (320)
	20%	65 (130)	90 (180)

[a] Power calculation was performed for achieving an 80% at the 5% level of significance.
[b] $\mu_T = \mu_s = 90\%$, 80%, and 70% were considered in this illustration.
[c] Non-inferiority testing was performed based on one-sided test at the 5% level of significance.

one side of a two-sided test at the 5% level of significance, which is equivalent to a one-sided test at the 5% level of significance. In other words, the FDA requires the significance level of 2.5% should be used when performing a one-sided non-inferiority testing. In this case, we need to increase sample size in order for achieving the same level of power for establishing non-inferiority. Also, Table 5.1 indicates that a narrower margin requires a much larger sample size for achieving the desired power for establishing non-inferiority. As a result, the selection of non-inferiority margin in very critical in non-inferiority trials.

5.3 Non-inferiority Hypothesis

5.3.1 Regulatory Requirements

International Conference on Harmonization Guideline—For selection of non-inferiority margin, the ICH guideline, *Guidance on Choice of Control Group and Related Design and Conduct Issues in Clinical Trials,* indicates that the selection of non-inferiority margin should be based on both statistical reasoning and clinical judgment. In addition, it should reflect uncertainties in the evidence of which the choice is based, and should be suitably conservative (ICH E10, 2000). From statistical point of view, the ICH E10 guideline suggests that the non-inferiority margin M should be chosen to satisfy at least the following two criteria:

Criterion 1. We want the ability to claim that the test therapy is non-inferior to the active control agent and is superior to the placebo (even though the placebo is not considered in the active control trial).

Criterion 2. The non-inferiority margin should be suitably conservative, i.e., variability should be taken into account.

It should be noted that a fixed M (i.e., it does not depend on any parameter) is rarely suitable under criterion 1. In addition, the selected margin should not be greater than the smallest effect size that the active drug would be reliably expected to have compared with placebo in the setting of a placebo-controlled trial.

United States Food and Drug Administration Guidance—Along this line, the FDA circulated guidance on non-inferiority clinical trials for comments (FDA, 2010a, 2018). Basically, this draft guidance consists of four parts: (i) a general discussion of regulatory, study design, scientific, and statistical issues associated with the use of non-inferiority studies when these are used to establish the effectiveness of a new drug; (ii) details of some of the issues such as the quantitative analytical and statistical approaches used to determine the non-inferiority margin for use in non-inferiority studies; (iii) Q&A of some commonly asked questions; and (iv) five examples of successful and unsuccessful efforts for determining non-inferiority margins and the conduct of non-inferiority studies.

In principle, the 2010 FDA draft guidance is very similar to the ICH E10 guideline. However, the 2010 FDA draft guidance provides more details regarding study design and statistical issues. The 2010 FDA draft guidance recommends two approaches be considered for determination of non-inferiority margin based on historical data of the approved active control agent.

5.3.2 Hypothesis Setting and Clinically Meaningful Margin

Let T, C and P denote the new or test treatment, the active protocol agent that has been demonstrated to be superior to a placebo, and the placebo, respectively. Thus, the relationship among T, C, P, and M (clinically meaningful margin) is illustrated in Figure 5.3 (a–c).

As it can be seen from Figure 5.3a, if T falls within $(C - M, C + M)$, we consider T and C are therapeutically equivalent assuming that the right side of C is improving and the left side of C is worsening. Thus, if T falls on the left-hand side of $C - M$, i.e., $T < C - M$ or $C - T > M$, we claim that T is inferior to C or C is superior to T. On the other hand, T is considered non-inferior to C if it falls

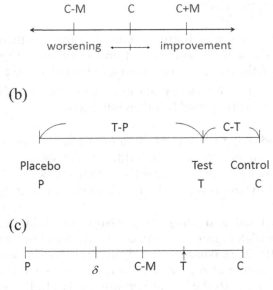

FIGURE 5.3
Relationship among T, C, P, and M. (a) Relationship between C and M. (b) Relationship among T, C, and P. (c) Relationship between M and δ.

on the right side of $C - M$, i.e., $C - M < T$ or $C - T < M$. In this case, hypotheses for testing non-inferiority between T and C can be described as follows.

$$H_0 : C - T > M \left(\text{or } C - M > T, T \text{ is inferior to } C \right);$$

$$H_a : C - T < M \left(\text{or } C - M < T, T \text{ is not inferior to } C \right).$$

Thus, we would reject the null hypothesis that T is inferior to C and conclude that the difference between T and C is less than a clinically meaningful non-inferiority margin (M) and hence T (test treatment) is at least as effective as (or not worsen than) C, the active control agent.

Figure 5.3b and 5.3c describes relationship among T, C, P, and M. If T is not inferior to C and is superior to P, then (i) $T > C - M$ or $T - C > -M$ and (ii) $T - P > \delta$, where $M \geq \delta$.

5.3.3 Retention of Treatment Effect in the Absence of Placebo

According to Figure 5.3b, Hung et al. (2003) proposed the concept of retention ratio, denoted by r, of the effect of the test treatment (i.e., $T - P$) and the effect of the active control agent (i.e., $C - P$) as compared to a placebo control regardless the presence of the placebo in the study. That is,

$$r = \frac{T - P}{C - P},$$

where r is a fixed constant between 0 and 1. Chow and Shao (2006) introduced the parameter of δ, which is the superiority margin as compared to the placebo. The relationship among P, T, C, δ, and M is illustrated in Figure 5.3c. At the worst possible scenario, we may select $M = \delta = T - P$. In this case, the retention rate becomes

$$r = \frac{T - P}{C - P} = \frac{\delta}{C - P} = \frac{M}{C - P},$$

This leads to

$$M = r(C - P).$$

Jones et al. (1996) suggests that $r = 0.5$ be chosen, while $r = 0.2$ is probably the most commonly employed for selection of non-inferiority margin without any clinical judgment or statistical reasoning. Thus, the selection of non-inferiority margin depends upon the estimation of the retention rate of the effect of the test treatment relative to the effect of the active control agent.

5.4 Methods for Selection of Non-inferiority Margin

5.4.1 Classical Method

In clinical trials, equivalent limits for therapeutic equivalence generally depend on the nature of the drug, targeted patient population, and clinical endpoints (efficacy and safety parameters) for the assessment of therapeutic effect. For example, for some drugs, such as topical antifungals or vaginal antifungals, which may not be absorbed in blood, the FDA proposed some equivalent limits for some clinical endpoints such as binary response (Huque and Dubey, 1990). As an example, for the study endpoint of cure rate, if the cure rate for the reference drug is greater than 95%, then a difference in cure rate within 5% is not considered a clinically important difference (see Table 5.2).

5.4.2 FDA's Recommendations

The 2010 FDA draft guidance recommends two non-inferiority margins, namely M_1 and M_2 should be considered. The 2010 FDA draft guidance indicated that M_1 is based on (i) the treatment effect estimated from the historical experience with the active control drug, (ii) assessment of the likelihood that the current effect of the active control is similar to the past effect (the constancy assumption), and (iii) assessment of the quality of the non-inferiority trial, particularly looking for defects that could reduce a difference between the active control and the new drug. Thus, M_1 is defined as the entire effect of the active control assumed to be present in the non-inferiority study

$$M_1 = C - P. \tag{5.1}$$

On the other hand, FDA indicates that M_2 is selected based on a clinical judgment which is never be greater than M_1 even if for active control drugs with small effects. It should be noted that a clinical judgment might argue that a larger difference is not clinically important. Ruling out that a difference

TABLE 5.2

Equivalence Limits for Binary Responses

Equivalence Limits (%)	Response Rate for the Reference Drug (%)
±20	50–80
±15	80–90
±10	90–95
±5	>95

between the active control and test treatment that is larger than M_1 is a critical finding that supports the conclusion of effectiveness. Thus, M_2 can be obtained as

$$M_2 = (1-\delta_0)M_1 = (1-\delta_0)(C-P),\qquad(5.2)$$

where

$$\delta_0 = 1-r = 1-\frac{T-P}{C-P} = \frac{C-T}{C-P}$$

is referred to as the ratio of the effect of the active control agent as compared to the test treatment and the effect of the active control agent as compared to the placebo. Thus, δ_0 becomes smaller if the difference between C and T decreases, i.e., T is close to C (the retention rate of T is close to 1). In this case, the FDA suggests a wider margin for the non-inferiority testing.

5.4.3 Chow and Shao's Method

By the 2010 FDA draft guidance, there are essentially two different approaches to analysis of the non-inferiority study: one is the fixed margin method (or the two confidence interval method) and the other one is the synthesis method. In the fixed margin method, the margin M_1 is based on estimates of the effect of the active comparator in previously conducted studies, making any needed adjustment for changes in trial circumstances. The non-inferiority margin is then pre-specified and it is usually chosen as a margin smaller than M_1 (i.e., M_2). The synthesis method combines (or synthesizes) the estimate of treatment effect relative to the control from the non-inferiority trial with the estimate of the control effect from a meta-analysis of historical trials. This method treats both sources of data as if they came from the same randomized trial to project what the placebo effect would have been had the placebo been present in the non-inferiority trial.

Following the idea of the ICH E10 (2000) that the selected margin should not be greater than the smallest effect size that the active control has, Chow and Shao (2006) introduced another parameter δ which is a superiority margin of the placebo $\delta > 0$) and assumed that the non-inferiority margin M is proportional to δ, i.e., $M = \lambda\delta$. Then, under the worst scenario, i.e., $T - C$ achieves its lower bound $-M$, then the largest possible M is given by $M = C - P - \delta$, which leads to

$$M = \frac{\lambda}{1+\lambda}(C-P),$$

where

$$\lambda = \frac{r}{1-r}.$$

It can be seen that if $0 < r \leq 1$, then $0 < \lambda \leq 1/2$.

To account for the variability of $C - P$, Chow and Shao suggested the non-inferiority margins, M_1 and M_2 be modified as follows, respectively,

$$M_3 = M_1 - \left(z_{1-\alpha} + z_\beta\right)SE_{C-T} = C - P - \left(z_{1-\alpha} + z_\beta\right)SE_{C-T}, \qquad (5.3)$$

where SE_{C-T} is the standard error of $\hat{C} - \hat{T}$ and $z_a = \Phi^{-1}(a)$ assuming that

$$SE_{C-P} \approx SE_{T-P} \approx SE_{C-T}.$$

Similarly, M_2 can be modified as follows

$$\begin{aligned}
M_4 = rM_3 &= r\left\{C - P - \left(z_{1-\alpha} + z_\beta\right)SE_{C-T}\right\} \\
&= \frac{\lambda}{1+\lambda}\left\{C - P - \left(z_{1-\alpha} + z_\beta\right)SE_{C-T}\right\}, \qquad (5.4) \\
&= \left(1 - \frac{1}{1+\lambda}\right)M_3,
\end{aligned}$$

where δ_0 is chosen to be $\dfrac{1}{1+\lambda}$ as suggested by Chow and Shao (2006).

5.4.4 Alternative Methods

Let C_L and C_U be the minimum and maximum effect of C when comparing with P. If the effect of the test treatment falls within the range of (C_L, C_U), we consider T is equivalent to C and superior to P. Consider the worst possible scenario that the effect of the active control falls on C_U, while the effect of the test treatment T falls on C_L. In this case, we may consider the difference between C_L and C_U and the non-inferiority margin. That is

$$M_5 = \hat{C}_U - \hat{C}_L. \qquad (5.5)$$

In addition, since the selection of M depends upon the choice of δ_0, in practice, δ_0 is often chosen as either $\delta_0 = 0.5$ $(r = 0.5)$ or $\delta_0 = 0.8$ $(r = 0.2)$. The non-inferiority margin becomes narrower when δ_0 closes to 1. Based on the above argument, at the worst possible scenario, δ_0 can be estimated by

$$\hat{\delta}_0 = \frac{\overline{C} - \overline{T}}{\overline{C} - \overline{P}} = 1 - \frac{\hat{T} - \hat{P}}{\hat{C} - \hat{P}} = 1 - \frac{\hat{C}_L}{\hat{C}_U}.$$

Thus,

$$M_6 = rM_1 = \left(1 - \frac{\hat{C}_L}{\hat{C}_U}\right)\left(\hat{C} - \hat{P}\right) \qquad (5.6)$$

5.4.5 An Example

A pharmaceutical company is interested in conducting a non-inferiority trial for evaluation of safety and efficacy (in terms of cure rate) of a test treatment intended for treating patients with certain diseases as compared to a standard of care treatment (or an active control agent). At the planning stage of the non-inferiority trial, the question regarding sample size requirement for achieving a desired power for establishing non-inferiority of the test treatment as compared to the active control agent is raised. Sample size calculation, however, depends upon the clinically meaningful difference (margin). A narrower margin will require a much larger sample size for achieving the desired power for establishing non-inferiority of the test treatment. For selection of non-inferiority margin, both the ICH guideline and the FDA guidance suggest that historical data of the active control agent as compared to placebo if available should be used for determination of the non-inferiority margin.

Historical data for comparing the active control agent with a placebo are summarized in Table 5.3. Since the response rate for the active control is $C = 61.4\%$, the classical method suggests a non-inferiority margin of 20% be considered. Also, from Table 5.4, the placebo effect is $P = 14.3\%$. Thus,

$$M_1 = C - P = 61.4\% - 14.3\% = 47.1\%.$$

The range of $C - P$ is given by (39.7%, 56.7%). If we assume that the retention rate is 70% (i.e., $\delta_0 = 1 - r = 0.3$), then $r = 1 - \delta_0 = 0.3$. This gives

$$M_2 = (1 - \delta_0) M_1 = 0.3 \times 47.1\% = 14.1\%.$$

TABLE 5.3

Summary Statistics of Historical Data

Active Control Agent	Year of Submission	N	Active Control (C)	Placebo Cure Rate (P)	Difference in Cure Rate
C_1	1984	279	63.1%	7.3%	55.8%
	1985	209	60.2%	4.0%	56.2%
C_2	1986	101	60.0%	14.0%	46.0%
C_3	1986	100	70.0%	13.3%	56.7%
	1986	108	55.1%	13.6%	41.5%
	1986	90	66.0%	18.6%	47.4%
	1988	137	58.7%	17.6%	41.1%
C_4	1982	203	60.2%	20.5%	39.7%
	1986	88	60.0%	16.7%	43.3%
	1988	97	60.9%	17.6%	43.3%
Mean			61.4%	14.3%	47.1%
SD			4.1%	5.2%	6.7%
Minimum			55.1%	4.0%	39.7%
Maximum			70.0%	20.5%	56.7%

Assuming that $SE_{C-P} \approx SE_{T-P} \approx SE_{C-T}$, we have $SE_{C-T} = 6.7\%$. This leads to

$$M_3 = M_1 - (z_{1-\alpha} + z_\beta)SE_{C-T}$$
$$= 47.1\% - (1.96 + 0.84) \times 6.7\%$$
$$= 27.7\%,$$

Consequently,

$$M_4 = \left(1 - \frac{1}{1+\lambda}\right)M_3 = 0.76 \times 27.7\% = 21\%.$$

For the proposed margin M_5, since the minimum effect and maximum effect of $C - P$ are given by $\hat{C}_L = 39.7\%$ and $\hat{C}_U = 56.7\%$, we have

$$M_5 = \hat{C}_U - \hat{C}_L = 56.7\% - 39.7\% = 17\%.$$

Also, since $\hat{\delta}_0 = \dfrac{39.7\%}{56.7\%} = 0.68$, $r = 1 - \hat{\delta}_0 = 1 - 0.68 = 0.32$. This leads to

$$M_6 = rM_1 = \left(1 - \frac{\hat{C}_L}{\hat{C}_U}\right)(\hat{C} - \hat{P}) = 0.32 \times 47.1\% = 15.1\%.$$

To provide a better understanding, these margins are summarized in Table 5.4. As it can be seen from Table 5.4, the margin ranges from 14.1% to 47.1% (the entire effect of the active control agent) with a median of 21%, which is close to the classical method. It should be noted that, prior to the publication of the 2010 draft guidance, the FDA recommended a non-inferiority margin of 15%, while the sponsor requested a non-inferiority margin of 20%.

TABLE 5.4

Non-inferiority Margins Suggested by Various Methods

Method	Suggested Non-inferiority Margin
Classical Methods	20.0%
Hung et al.'s suggestion with $r = 0.5$	23.1%
FDA's M_1 approach	47.1%
FDA's M_2 approach with $r = 0.3$[a]	14.1%
Chow and Shao's M_3 margin	27.7%
Chow and Shao's M_4 margin	21.0%
Proposed M_5 margin	17.0%
Proposed M_6 margin	15.1%

[a] Retention rate of 70%.

Note that considering $M = 0.5(C - P)$, a conservative estimate of C effect is obtained using the lower 95% confidence limit of 53.4%. Assuming a 14% therapeutic cure rate of placebo, a 34% therapeutic cure rate from T will maintain the retention ratio of $(T - P)/(C - P)$ for 50%.

5.4.6 Remarks

It should be noted that the above methods (except for the classical method) for determination of non-inferiority margin M is based on data observed from previous superiority studies comparing the active control agent and a placebo and data collected from superiority studies comparing the test treatment and the placebo if available. Thus, the selected margin is in fact an estimate rather than a fixed margin. In other words, the selected margin is a random variable whose statistical properties are unknown. In addition, since the selected non-inferiority margin has significant impact on power calculation for sample size, it is suggested that a sensitivity analysis be performed to carefully evaluate the potential impact of the selected margin on non-inferiority testing.

As indicated by the ICH guideline, the selection of a non-inferiority margin should take both clinical judgment and statistical reasoning into consideration. The 2010 FDA draft guidance, however, emphasizes on statistical reasoning based on historical data from previous superiority studies comparing the active control agent and the placebo. In practice, there is always discrepancy between the margin suggested by the investigator and the margin recommended by the FDA. In this case, it is suggested that medical/statistical reviewers be consulted who, one hopes, will be able to reach an agreement on the selection of the non-inferiority margin following the general principles as described in the FDA draft guidance.

5.5 Strategy for Margin Selection

For assessment of non-inferiority and/or equivalence/similarity, the selection of non-inferiority margin and/or equivalence limit (similarity margin) is critical. Too narrow a margin will require a much larger sample size for achieving the study objectives; while too wide a margin may increase the probability of wrongly accept bad products. Besides, the selected margin will have an impact on sample size required for achieving the study objectives. In practice, the sponsor tends to propose a wider margin, which often deviates from the margin recommended by the regulatory agency. The disagreement in margin selection could generate tremendous argument and discussion between the sponsor and the regulatory agency. To close up the gap between the sponsor's proposal and the regulatory agency's recommendation, Nei et al. (2019) proposed a strategy for selection of similarity margin in comparative clinical trials. Their proposed strategy is based on the evaluation of risk assessment of the sponsor's proposal assuming that the regulatory agency's recommended margin is the true margin, which can be summarized in the following steps:

Step 1: The sponsors are to identify historical studies available that are accepted by the FDA to determine similarity margin.

Step 2: Based on a meta-analysis that combines these identified historical studies, the similarity margin is determined; hence the corresponding sample size required for testing similarity. Power calculation for the required sample size is obtained based on the sponsor's proposed margin.

Step 3: At the same time, the FDA will propose a similarity margin by taking clinical judgment, statistical rationale, and regulatory feasibility into considerations.

Step 4: A risk assessment is then conducted for evaluation of the sponsor's proposed margin assuming that the FDA's proposal is true.

Step 5: The risk assessment is then reviewed by the FDA review team and communicated with the sponsor in order to reach agreement on the final similarity margin.

5.5.1 Criteria for Risk Assessment

In this section, we will focus on Step 4 from Nei et al. (2019)'s proposed strategy, which quantifies the risk of different margins based on several criteria. This will assist sponsors to adjust their margins according to the maximum risk that is allowed by FDA. Four criteria are considered regarding different aspects of the similarity test. Numerical derivations are given in the next section based on continuous endpoints (e.g. normality assumption). Let ϵ be the true difference between the proposed biosimilar product and its reference product, i.e., $\epsilon = \mu_B - \mu_R$, where μ_B and μ_R are the treatment effect of the biosimilar product and the reference product, respectively. We also assume that a positive value of ϵ means that the biosimilar product is more efficacious than the reference product in the selected efficacy endpoint. Let $\delta_{Sponsor}$ and δ_{FDA} be the sponsor's proposed margin and the FDA's recommended margin, respectively. In here, we assume $0 < |\epsilon| < \delta_{FDA} < \delta_{Sponsor}$.

Criterion 1: Sample Size Ratio (SSR)—When fixing the power of the similarity test, sample size is a decreasing function of similarity margin, i.e., the smaller the similarity margin is, the bigger the sample size is required. In clinical trial studies, large sample size corresponds to more costs on sponsors. The goal is to move sponsor proposed margin toward the FDA recommended margin, with a moderate increase in sample size while maintaining the power at the desired level. Let n_{FDA} be the sample size required to maintain $1-\beta$ power under δ_{FDA}, similarly for $n_{Sponsor}$. Define $SSR = \frac{n_{FDA}}{n_{Sponsor}}$ by the sample size ratio. Then sample size difference (SSD)

$$SSD = n_{FDA} - n_{Sponsor} = (SSR-1)\cdot n_{Sponsor}$$

can be viewed as the amount of the information lost for the use of a wider margin (i.e., $\delta_{Sponsor}$) assuming that δ_{FDA} is the true margin. By plotting the curve of sample size ratio, we can choose a threshold SSR_M which serves as a guideline for margin determination, say $105\%, 110\%, 115\%, 120\%$, which is corresponding to $5\%, 10\%, 15\%, 20\%$ loss based on $n_{Sponsor}$.

Criterion 2: Relative Difference in Power (RED)—When fixing sample size for the similarity test, power is an increasing function of similarity margin, i.e., the larger the margin is, the bigger the power is. Let $Power_{Sponsor}$ and $Power_{FDA}$ be the power under $\delta_{Sponsor}$ and δ_{FDA}, respectively. Since $\delta_{FDA} < \delta_{Sponsor}$, then $Power_{FDA} < Power_{Sponsor}$. This is due to the wider region of the alternative hypothesis under wider margin $\delta_{Sponsor}$ and because wider margins have smaller type II errors. Although we gain some power (smaller type II error rate) by using a wider margin, we also weaken the result (or say accuracy) when rejecting null hypothesis. Define $RED = Power_{Sponsor} - Power_{FDA}$. The quantity RED is the gain in power by scarifying accuracy, i.e., by using wider margin. In order to close up the gap between δ_{FDA} and $\delta_{Sponsor}$, we need to minimize RED. So we can set a threshold distance RED_M, say $0.05, 0.10, 0.15, 0.20$, between $Power_{FDA}$ and $Power_{Sponsor}$ to be the maximum power gain by using a wider margin.

Criterion 3: Relative Ratio in Power/Relative Risk (RR)—The power described in the last section is the probability of concluding biosimilarity. When given the same sample size under both FDA's and sponsor's margins,

$$Power_{FDA} < Power_{Sponsor}.$$

That is, the probability of concluding biosimilarity is bigger under $\delta_{Sponsor}$ than it is under δ_{FDA}. It means that among all the potential biosimilar products which would be considered biosimilar under the sponsor proposed margin, only a portion of them will be biosimilar under the FDA recommended margin. The rest are wrongly claimed as biosimilar by sponsors according to FDA's margin. This is regarded as a risk factor for using wider margins. Define RR as the probability which a product is not concluded as biosimilar under δ_{FDA} given that it is concluded as biosimilar under $\delta_{Sponsor}$,

$$RR = 1 - \frac{Power_{FDA}}{Power_{Sponsor}} = \frac{Power_{Sponsor} - Power_{FDA}}{Power_{Sponsor}}.$$

Under the FDA recommended margin, RR is the risk of wrongly concluding biosimilarity of a biosimilar drug using sponsor's margin. Furthermore, among all biosimilar drugs concluded using sponsor's margin, $100 \cdot RR$ of them would have been failed under the FDA recommended margin. Thus, RR is the risk of using sponsor proposed margin. Wider margins lead to larger risks. Thus, we may choose an appropriate margin by assuring that the risk is smaller than a maximum risk RR_M that is considered acceptable by the FDA (say 0.15). Let δ_M be the margin that corresponds to RR_M. We will derive an asymptotically analytical form of δ_M in the next section based on continuous endpoint.

Criterion 4: Inflation of Type I Error—Type I error inflation is the probability, assuming the smaller margin is the true difference, of rejecting a null hypothesis based on the wider margin in a study powered to rule out the wider margin (i.e., the type I error rate of the test under the "FDA" null). This will be greater than 5% and the degree of its inflation is probably relevant information. The inflation is also an increasing function of similarity margins. Bigger margins lead to larger inflations, i.e., larger type I error rates. We can set up a threshold value of type I error inflation for choosing the largest margin that is allowed.

5.5.2 Risk Assessment with Continuous Endpoints

In this section, both analytic and asymptotic forms of the four criteria proposed in the last section are derived. Without loss of generality, we only consider biosimilar products that have continuous endpoints. All calculations below can be derived in a similar fashion for biosimilar products with categorical endpoints.

Let $\delta > 0$ be the similarity margin and the null hypothesis of the similarity test is $H_0 : |\epsilon| \geq \delta$. Rejection of the null hypothesis implies similarity between the biosimilar product the reference product. For simplicity, we assume samples from both the biosimilar group and the reference group follow normal distributions with mean μ_B and μ_R, respectively and same unknown variance σ^2, which means the within-subject variances of both biosimilar and reference product are the same. That is,

$$x_1^B, x_2^B, \ldots, x_{n_B}^B \sim N\left(\mu_B, \sigma_B^2\right), x_1^R, x_2^R, \ldots, x_{n_R}^R \sim N\left(\mu_R, \sigma_R^2\right),$$

where n_B and n_R are sample sizes for the biosimilar group and the reference group. Let $\hat{\mu}_{BR} = \hat{\mu}_B - \hat{\mu}_R$ be the estimated treatment effect of the biosimilar product relative to the reference product with standard error of

$$\hat{\sigma}_{BR} = \hat{\sigma}\sqrt{1/n_B + 1/n_R},$$

where

$$\hat{\mu}_B = \frac{1}{n_B}\sum_{i=1}^{n_B} x_i^B, \ \hat{\mu}_R = \frac{1}{n_R}\sum_{i=1}^{n_R} x_i^R,$$

and

$$\hat{\sigma}^2 = \frac{1}{n_B + n_R - 2}\left\{\sum_{i=1}^{n_B}\left(x_i^B - \hat{\mu}_B\right)^2 + \sum_{i=1}^{n_R}\left(x_i^R - \hat{\mu}_R\right)^2\right\}.$$

Note that $\hat{\sigma}_{BR}$ depends on sample sizes (and treatment effects in some scenarios).

The rejection region for testing H_0 with statistical significance level α is

$$R = \left\{\frac{\hat{\mu}_{BR} + \delta}{\hat{\sigma}_{BR}} > t_{\alpha, n_B + n_R - 2}\right\}\bigcap\left\{\frac{\hat{\mu}_{BR} - \delta}{\hat{\sigma}_{BR}} < -t_{\alpha, n_B + n_R - 2}\right\}.$$

Thus, the power of the study is

$$Power = P\left(\frac{\hat{\mu}_{BR} + \delta}{\hat{\sigma}_{BR}} > t_{\alpha, n_B + n_R - 2} \text{ and } \frac{\hat{\mu}_{BR} - \delta}{\hat{\sigma}_{BR}} < -t_{\alpha, n_B + n_R - 2}\right).$$

$$\approx 1 - T_{n_B + n_R - 2}\left(t_{\alpha, n_B + n_R - 2}\left|\frac{\delta - \epsilon}{\sigma\sqrt{1/n_B + 1/n_R}}\right.\right) - T_{n_B + n_R - 2}\left(t_{\alpha, n_B + n_R - 2}\left|\frac{\delta + \epsilon}{\sigma\sqrt{1/n_B + 1/n_R}}\right.\right),$$

where $T_k(\cdot \mid \theta)$ is the cumulative distribution function of a non-central t-distribution with k degrees of freedom and the non-centrality parameter θ and $-\delta < \epsilon < \delta$ under H_a.

Sample size ratio (SSR)—Assume that $n_B = \kappa n_R$, the sample size n_R needed to achieve power $1 - \beta$ can be obtained by setting the power to $1 - \beta$. Since the power is larger than

$$1 - 2T_{n_B + n_R - 2}\left(t_{\alpha, n_B + n_R - 2}\left|\frac{\delta - |\epsilon|}{\sigma\sqrt{1/n_B + 1/n_R}}\right.\right),$$

a conservative approximation to the sample size n_R can be obtained by solving

$$T_{(1+\kappa)n_R - 2}\left(t_{\alpha, (1+\kappa)n_R - 2}\left|\frac{\sqrt{n_R}\left(\delta - |\epsilon|\right)}{\sigma\sqrt{1 + 1/\kappa}}\right.\right) = \frac{\beta}{2}.$$

When sample size n_R is sufficiently large, $t_{\alpha, (1+\kappa)n_R - 2} \approx z_\alpha$, and $T_{(1+\kappa)n_R - 2}(t \mid \theta) \approx \Phi(t - \theta)$, then

$$\frac{\beta}{2} = T_{(1+\kappa)n_R - 2}\left(t_{\alpha, (1+\kappa)n_R - 2}\left|\frac{\sqrt{n_R}\left(\delta - |\epsilon|\right)}{\sigma\sqrt{1 + 1/\kappa}}\right.\right) \approx \Phi\left(z_\alpha - \frac{\sqrt{n_R}\left(\delta - |\epsilon|\right)}{\sigma\sqrt{1 + 1/\kappa}}\right).$$

As a result, the sample size needed to achieve power $1 - \beta$ can be obtained by solving the following equation:

$$z_\alpha - \frac{\sqrt{n_R}\left(\delta - |\epsilon|\right)}{\sigma\sqrt{1 + 1/\kappa}} = z_{1-\beta/2} = -z_{\beta/2}.$$

This leads to

$$n_R = \frac{\left(z_\alpha + z_{\beta/2}\right)^2 \sigma^2 \left(1 + 1/\kappa\right)}{\left(\delta - |\epsilon|\right)^2}.$$

Thus

$$n_R^{FDA} = \frac{\left(z_\alpha + z_{\beta/2}\right)^2 \sigma^2 \left(1 + 1/\kappa\right)}{\left(\delta_{FDA} - |\epsilon|\right)^2}, \ n_R^{Sponsor} = \frac{\left(z_\alpha + z_{\beta/2}\right)^2 \sigma^2 \left(1 + 1/\kappa\right)}{\left(\delta_{Sponsor} - |\epsilon|\right)^2},$$

and with $\delta_{Sponsor} = \lambda\delta_{FDA}$, we have

$$SSR = \frac{n_{FDA}}{n_{Sponsor}} = \frac{\left(1 + \kappa\right)n_R^{FDA}}{\left(1 + \kappa\right)n_R^{Sponsor}} = \left(\frac{\lambda\delta_{FDA} - |\epsilon|}{\delta_{FDA} - |\epsilon|}\right)^2$$

and

$$\lambda_M = \sqrt{SSR_M} + \frac{|\epsilon|}{\delta}\left(1 - \sqrt{SSR_M}\right).$$

Relative difference in power (RED)—Let

$$B\left(\epsilon, \delta, n_R, \sigma, \kappa\right) = T_{(1+\kappa)n_R - 2}\left(t_{\alpha,(1+\kappa)n_R - 2} \left|\frac{\sqrt{n_R}\left(\delta + \epsilon\right)}{\sigma\sqrt{1 + 1/\kappa}}\right.\right),$$

based on the calculation above, we immediately have

$$RED = Power_{Sponsor} - Power_{FDA}$$

$$\approx B\left(\epsilon, \delta_{FDA}, n_R, \sigma, \kappa\right) - B\left(\epsilon, \lambda\delta_{FDA}, n_R, \sigma, \kappa\right) + B\left(-\epsilon, \delta_{FDA}, n_R, \sigma, \kappa\right)$$

$$- B\left(-\epsilon, \lambda\delta_{FDA}, n_R, \sigma, \kappa\right).$$

When n_R is sufficiently large, let

$$\tilde{\Phi}(\epsilon, \delta, n_R, \sigma, \kappa) = \Phi\left(z_\alpha - \frac{\sqrt{n_R}(\delta_{FDA} + \epsilon)}{\sigma\sqrt{1 + 1/\kappa}}\right),$$

by using the same approximation in last section,

$$RED \approx \left[\Phi\left(z_\alpha - \frac{\sqrt{n_R}(\delta_{FDA} + \epsilon)}{\sigma\sqrt{1 + \frac{1}{\kappa}}}\right) - \Phi\left(z_\alpha - \frac{\sqrt{n_R}(\lambda\delta_{FDA} + \epsilon)}{\sigma\sqrt{1 + \frac{1}{\kappa}}}\right)\right]$$

$$+ \left[\Phi\left(z_\alpha - \frac{\sqrt{n_R}(\delta_{FDA} - \epsilon)}{\sigma\sqrt{1 + 1/\kappa}}\right) - \Phi\left(z_\alpha - \frac{\sqrt{n_R}(\lambda\delta_{FDA} - \epsilon)}{\sigma\sqrt{1 + 1/\kappa}}\right)\right]$$

$$\approx \tilde{\Phi}(\epsilon, \delta_{FDA}, n_R, \sigma, \kappa) - \tilde{\Phi}(\epsilon, \lambda\delta_{FDA}, n_R, \sigma, \kappa) + \tilde{\Phi}(-\epsilon, \delta_{FDA}, n_R, \sigma, \kappa) -$$

$$\tilde{\Phi}(-\epsilon, \lambda\delta_{FDA}, n_R, \sigma, \kappa).$$

When plugging in the sample size $n_R^{Sponsor}$ which retains $1 - \beta$ power under $\delta_{Sponsor}$,

$$RED^\beta \approx \Phi\left[z_\alpha - \frac{\delta_{FDA} + \epsilon}{\lambda\delta_{FDA} - |\epsilon|}(z_\alpha + z_{\beta/2})\right] + \Phi\left[z_\alpha - \frac{\delta_{FDA} - \epsilon}{\lambda\delta_{FDA} - |\epsilon|}(z_\alpha + z_{\beta/2})\right] - \beta.$$

i. When $\epsilon > 0$, then

$$RED^\beta < 2\Phi\left[z_\alpha - \frac{\delta_{FDA} - \epsilon}{\lambda\delta_{FDA} - |\epsilon|}(z_\alpha + z_{\beta/2})\right] - \beta;$$

ii. When $\epsilon < 0$, then

$$RED^\beta < 2\Phi\left[z_\alpha - \frac{\delta_{FDA} + \epsilon}{\lambda\delta_{FDA} - |\epsilon|}(z_\alpha + z_{\beta/2})\right] - \beta.$$

Thus, combining the above two cases, we have

$$RED^\beta < 2\Phi\left[z_\alpha - \frac{\delta_{FDA} - |\epsilon|}{\lambda\delta_{FDA} - |\epsilon|}\left(z_\alpha + z_{\frac{\beta}{2}}\right)\right] - \beta := RED^{\beta+}.$$

For simplicity reason, we will use $RED^{\beta+}$ in the following discussion and write it as RED^β.

Relative ratio in power/relative risk (RR)—Let

$$S_{FDA} = \left\{ reject\, H_0\, when\, \delta = \delta_{FDA} \right\}$$

and

$$S_{Sponsor} = \left\{ reject\, H_0\, when\, \delta = \delta_{Sponsor} \right\}.$$

Since $\delta_{FDA} < \delta_{Sponsor}$, then $|\epsilon| \le \delta_{FDA}$ leads to $|\epsilon| \le \delta_{Sponsor}$. Therefore, rejecting H_0 under δ_{FDA} leads to the rejection of H_0 under $\delta_{Sponsor}$, which means $S_{FDA} \subseteq S_{Sponsor}$ and $S_{FDA} \cap S_{Sponsor} = S_{FDA}$. Define p_s be the probability of concluding biosimilarity under δ_{FDA} given concluding biosimilarity under $\delta_{Sponsor}$. Then based on the relationship between S_{FDA} and $S_{Sponsor}$, we have

$$p_s = Pr\left(conclude\, similarity\, under\, \delta_{FDA} | conclude\, similarity\, under\, \delta_{Sponsor} \right)$$

$$= \frac{Pr\left(reject\, H_0\, when\, \delta = \delta_{FDA} \right)}{Pr\left(reject\, H_0\, when\, \delta = \delta_{Sponsor} \right)} = \frac{Power_{FDA}}{Power_{Sponsor}}$$

$$\approx \frac{1 - B\left(\epsilon, \delta_{FDA}, n_R, \sigma, \kappa \right) - B\left(-\epsilon, \delta_{FDA}, n_R, \sigma, \kappa \right)}{1 - B\left(\epsilon, \delta_{Sponsor}, n_R, \sigma, \kappa \right) - B\left(-\epsilon, \delta_{Sponsor}, n_R, \sigma, \kappa \right)}$$

Thus, based on the definition of RR in Criterion 3, we have

$$RR = 1 - p_s \approx \frac{RED}{1 - B\left(\epsilon, \lambda\delta_{FDA}, n_R, \sigma, \kappa \right) - B\left(-\epsilon, \lambda\delta_{FDA}, n_R, \sigma, \kappa \right)}.$$

For large n_R we have

$$RR \approx \frac{RED}{1 - \hat{\Phi}\left(\epsilon, \lambda\delta_{FDA}, n_R, \sigma, \kappa \right) - \hat{\Phi}\left(-\epsilon, \lambda\delta_{FDA}, n_R, \sigma, \kappa \right)}.$$

Type I error inflation (TERI)—When assuming the smaller margin is the true difference, i.e., $\epsilon = \pm\delta_{FDA}$ and $\delta_{FDA} < \delta_{Sponsor}$, then type I error inflation is calculated as follows.

Type I Error $| \epsilon = \pm\delta_{FDA}$

$$= P\left(\frac{\hat{\mu}_{BR} + \delta_{Sponsor}}{\breve{\sigma}_{BR}} > t_{\alpha, n_B+n_R-2} \text{ and } \frac{\hat{\mu}_{BR} - \delta_{Sponsor}}{\breve{\sigma}_{BR}} < -t_{\alpha, n_B+n_R-2} | \epsilon = \pm\delta_{FDA} \right)$$

$$= 1 - T_{n_B+n_R-2}\left(t_{\alpha, n_B+n_R-2} | \frac{\delta_{Sposor} + \delta_{FDA}}{\sigma\sqrt{1/n_B + 1/n_R}} \right) - T_{n_B+n_R-2}\left(t_{\alpha, n_B+n_R-2} | \frac{\delta_{Sposor} - \delta_{FDA}}{\sigma\sqrt{1/n_B + 1/n_R}} \right)$$

$$= 1 - B\left(\delta_{FDA}, \lambda\delta_{FDA}, n_R, \sigma, \kappa \right) - B\left(-\delta_{FDA}, \lambda\delta_{FDA}, n_R, \sigma, \kappa \right)$$

For large sample we have

$$Inflation \approx 1 - \alpha - \phi\left(z_\alpha - \frac{\sqrt{n_R}\,(\lambda+1)}{\sigma\sqrt{1+1/\kappa}} \cdot \delta_{FDA} \right) - \phi\left(z_\alpha - \frac{\sqrt{n_R}\,(\lambda-1)}{\sigma\sqrt{1+1/\kappa}} \cdot \delta_{FDA} \right).$$

5.5.3 Numerical Studies

In this section, Numerical studies for all four criteria are conducted and risk curves are plotted. Based on the results, suggestions on choosing a reasonable threshold are discussed for different scenarios. The validity of large sample approximation is investigated for small sample sizes. During this section, type I error rate and type II error rate are fixed to be 0.05 and 0.2.

Sample size ratio (SSR)—It can be verified that

$$\sqrt{SSR} = \frac{\delta_{FDA}}{\delta_{FDA} - |\epsilon|}\lambda - \frac{|\epsilon|}{\delta_{FDA} - |\epsilon|}$$

To further investigate the relationship between SSR and λ, we consider $\sqrt{SSR}$ instead of SSR since $\sqrt{SSR}$ is a linear function of λ with $\frac{\delta_{FDA}}{\delta_{FDA} - |\epsilon|}$ as slope and $-\frac{|\epsilon|}{\delta_{FDA} - |\epsilon|}$ as intercept. The quantity $\sqrt{SSR}$ increases by a proportion of $\frac{\delta_{FDA}}{\delta_{FDA} - |\epsilon|}$. For example, if $\delta_{Sponsor}$ is 10% wider than δ_{FDA}, then $\sqrt{SSR}$ is increased by $\frac{0.1\delta_{FDA}}{\delta_{FDA} - |\epsilon|}$. So smaller values of $\delta_{FDA} - |\epsilon|$ leads to steeper lines. In other words, if δ_{FDA} is set to be closer to $|\epsilon|$, sample size of sponsor increases more rapidly when $\delta_{Sponsor}$ moves toward δ_{FDA}. We can observe this from the plot (Figure 5.4).

Let SSR_{cur} be the sample according to the current $\delta_{Sponsor}$. For a safe choice, we propose to use δ_{new} that corresponds to $SSR_{cur} - \Delta$, where Δ can range from 0.2 0.3 0.2 to 0.3. This will make the gap between δ_{FDA} and $\delta_{Sponsor}$ smaller. But this is not a universal choice. Different thresholds should be chosen based on a case-by-case basis. The plot the sample size difference (SSD) curve is given in Figure 5.5.

As it can be seen from Figure 5.5, SSD curve follows the same pattern as that of SSR curve.

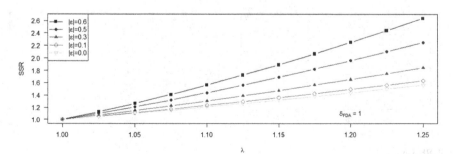

FIGURE 5.4
Plot of sample size ratio (SSR) curve.

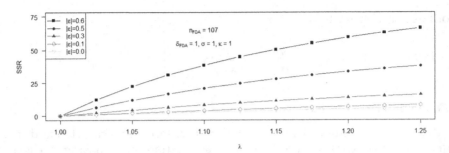

FIGURE 5.5
Plot of sample size difference (SSD) curve.

Relative difference in power (RED)—Since the large sample approxima-tion is used in deriving the asymptotic form of *RED*, we first investigate the validity of this approximation when sample size is small. As we can see from the four plots (Figure 5.6), when sample size of a single arm is 15, the approxi-mation is still close to the original. For sample size 30, two curves look identi-cal to each other. The condition of the normal approximation of t distribution is that the degree of freedom is greater than 30. In this case, $n_B + n_R - 2 > 30$, which is not difficult to satisfy in practice. For reasons of simplicity, we will use the asymptotic form instead of the original one in the following discus-sion. The rest of the parameters used in sample size comparison plot is set as follows, $\epsilon = -0.5$, $\delta_{FDA} = 1.0$, $\sigma = 1$, $\kappa = 1$. Since *RED* is symmetric about ϵ, we only plot when $\epsilon < 0$.

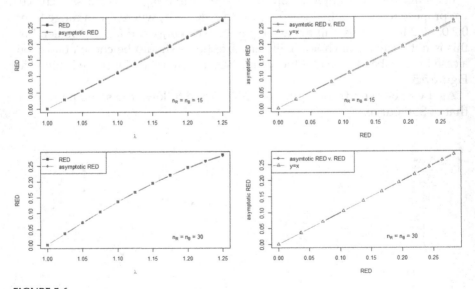

FIGURE 5.6
Plots of RED against λ and plots of asymptotic RED against RED with $n_R = n_B = 15$ and 30, respectively.

Next, to eliminate some of the parameters in *RED*, we rewrite it in terms of effect size (ES) $\Delta = -\epsilon/\sigma$ and $\Delta_{FDA} = \delta_{FDA}/\sigma$, and let $N = \sqrt{\frac{n_R}{1+1/\kappa}}$ be the sample size factor, then

$$RED \approx \Phi\left[z_\alpha - N\left(\Delta_{FDA} + \Delta\right)\right] - \Phi\left[z_\alpha - N\left(\lambda\Delta_{FDA} + \Delta\right)\right]$$

$$+ \Phi\left[z_\alpha - N\left(\Delta_{FDA} - \Delta\right)\right] - \Phi\left[z_\alpha - N\left(\lambda\Delta_{FDA} - \Delta\right)\right]$$

Figure 5.7 plots *RED* curves for 6 different ES_{FDA} values, which are corresponding to $0\%\,5\%\,10\%\,15\%\,20\%\,25\%$ increase from $ES = 0.5$. Large ES_{FDA} leads to steeper curve, i.e., the drastically increase in *RED* for smaller values of λ. So for large ES_{FDA}, narrowing the same portion of δ_{FDA} will yield more decrement in *RED* value. Therefore, under current parameter setting, for large ES_{FDA} we recommend choosing the margin (value of λ) such that *RED* is in the range of $(0.20,\ 0.40)$; for small ES_{FDA}, less than 0.20 is preferred. *RED* converges to $\Phi\left[z_\alpha - N\left(\Delta_{FDA} + \Delta\right)\right] + \Phi\left[z_\alpha - N\left(\Delta_{FDA} - \Delta\right)\right]$ when $\lambda \to \infty$; large sample size factor will result in quicker convergence.

The plot of *RED* curves for 6 different N values with $ES = 0.5$ is given in Figure 5.8.

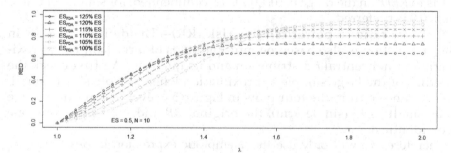

FIGURE 5.7
Plot of *RED* curves for 6 different ES_{FDA} values with $ES = 0.5$.

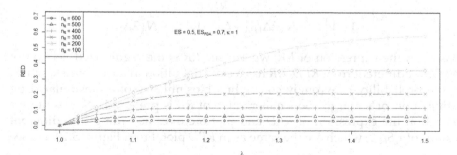

FIGURE 5.8
Plot of *RED* curves for 6 different N values with $ES = 0.5$.

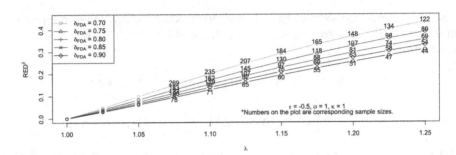

FIGURE 5.9
Plots of various δ_{FDA} values as λ increases.

Normally, the sample size used in sponsor's trial is required to maintain certain amount of power, such as 0.8. RED^{β} is used in this scenario. Figure 5.9 plots 5 different values of δ_{FDA} which is gradually increasing from ϵ. The sample sizes used for each value of λ here maintain $1-\beta$ power for sponsor's text. RED^{β} is different from RED, i.e., larger value of δ_{FDA} leads to slow growth of difference in power. For a large value of δ_{FDA}, λ which leads to RED^{β} in the range of $(0.1, 0.2)$ is recommended; for small value, less than 0.3 is preferred.

Relative ratio in power/relative risk (RR)—The definition of RR in the last section is also based on multiple steps of large sample approximation of noncentral t distribution and its quantile. We first check the validity of the large sample approximation when sample size is small. As we can see from the four plots in Figure 5.6, even when sample size is as small as 15 (single arm), the original RR and the asymptotic one look identical.

Therefore, we will only use the asymptotic expression in our following decision-making. The rest of the parameters used in sample size comparison plot is set as follows, $\epsilon = -0.5$, $\delta_{FDA} = 1.0$, $\sigma = 1$, $\kappa = 1$.

$$RR \approx \frac{RED}{1 - \Phi\left[z_{\alpha} - N\left(\lambda\Delta_{FDA} + \Delta\right)\right] - \Phi\left[z_{\alpha} - N\left(\lambda\Delta_{FDA} - \Delta\right)\right]}$$

Based on the expression of RR, we can see RR is the regularized version of RED. But different from RED, RR has a clear definition in terms of risk, which is the probability of wrongly concluding biosimilarity of a biosimilar drug using sponsor's margin. So smaller value of RR is preferred. We rewrite RR based on the expression of RED and plot 6 curves according to 6 different values of ES_{FDA}, which are the same as in RED plot. From Figure 5.10, we see

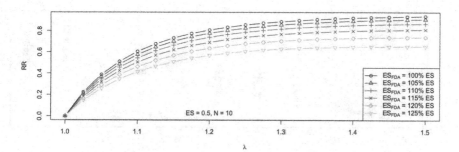

FIGURE 5.10
Plots *RED* curves with six different sample sizes n_R.

that a larger ES_{FDA} leads to smaller risk. *RR* converges to *RED* when $\lambda \to \infty$; large sample size factor will result in quicker convergence. Figure 5.10 plots *RED* curves with six different sample sizes n_R. As we can see larger sample size leads to lower risk.

When plugging in the sample size that retains power at $1 - \beta$, the shape of the following five are curves is almost identical to those in the RED^β plot (see Figure 5.11). This is because RR^β is approximately proportional to $\frac{RED^\beta}{1-\beta}$.

The threshold value of RR^β for selecting λ can be set as the threshold value of RED^β divided by $1 - \beta$ (see Figure 5.12).

Risk and sample size—Based on the expression of RR^β, relative risk is an increasing function of λ, hence, $\delta_{Sponsor}$. Furthermore, $n_R^{Sponsor}$ is a decreasing function of $\delta_{Sponsor}$. Minimizing risk is based on the increase of sample size, which leads to the increase of the cost of clinical trial for sponsor. So when

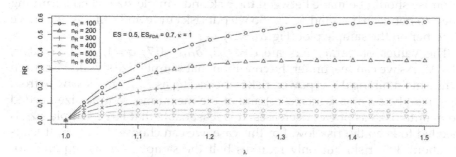

FIGURE 5.11
Plot of RR against λ.

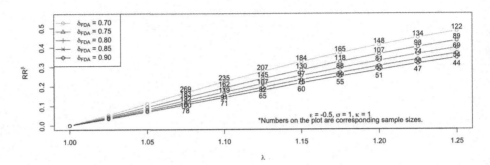

FIGURE 5.12
Plot of RR^β against λ.

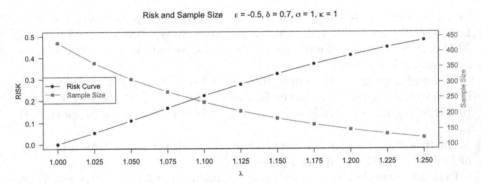

FIGURE 5.13
Plot of risk against sample size for the case where $\varepsilon = -0.5$, $\delta = 0.7$, $\sigma = 1$, $\kappa = 1$.

minimizing risk, sample size should be considered at the same time. A compromise should be made between the risk and sample size when shrinking the gap between δ_{FDA} and $\delta_{Sponsor}$. So we put risk curve and sample size curve together on the sample plot (Figure 5.13).

The values of parameters are $\varepsilon = -0.5$, $\delta_{FDA} = 0.7$, $\sigma = 1$, $\kappa = 1$, $\alpha = 0.05$, $\beta = 0.2$. As we can see, under the condition that sample size should be chosen while maintaining 0.8 as power, risk is increasing as $\delta_{Sponsor}$ moves away from δ_{FDA} (can be seen as λ increases) and at the meantime, sample size needed to maintain 0.8 as power decreases. It is reasonable that large sample size is needed to keep the risk low. For this case, we can choose $\lambda = 1.075$. It leads to about 30% risk, but only requires half the sample size as required for the FDA recommended margin. Figure 5.14 plots sample size against risk. The relationship is almost linear with a negative slope.

Type I error inflation—Figure 5.15 plots the type I error inflation when $\delta_{Sponsor}$ moves away from δ_{FDA}. The values of parameters are $\varepsilon = -0.5$, $\delta_{FDA} = 0.7$, $\sigma = 1$, $\kappa = 1$, $\alpha = 0.05$, $\beta = 0.2$. The sample size used in the plot is $n_R^{Sponsor}$, i.e.,

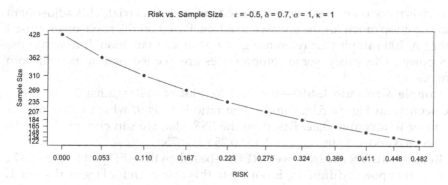

FIGURE 5.14
Plot of Type I error inflation against risk.

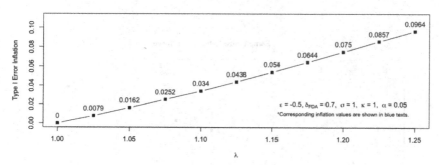

FIGURE 5.15
Plot of Type I error inflation against λ.

maintains 0.8 power. Only the asymptotic expression of type I error inflation is used here. Type I error rate can also be seen as a risk factor. Minimizing the inflation caused by wider margin is the goal here. In this case, about 50% of the inflation is acceptable, since type I error is 0.05 here. Thus, we can choose $\lambda = 1.15$ and $\delta_{Sponsor} = 0.805$.

5.5.4 An Example

In this section, we present a synthetic example to demonstrate the strategy proposed in this paper and four criteria in margin selection. Assume after the clinical trial, we observe the following settings $\hat{\mu}_B = 2.55$, $\hat{\mu}_R = 2.75$, $\hat{\sigma} = 1.35$, $n_B = 140$, $n_R = 200$, $\delta_{FDA} = 0.25$, $\delta_{Sponsor} = 0.35$. Based on the FDA recommended margin, the sample size required to maintain 0.8 power is around 274 for reference group (assuming true difference is zero in sample size calculation), which is more than the sample size used in the clinical trial. This adjustment may cost too much for the sponsor to adopt. Based on the sponsor-proposed margin, 200 samples for reference group is more than enough to retain the 0.8 power. Obviously, some compromises are needed here to benefit both parties.

Sample size ratio (SSR)—Figure 5.16. Plot of SSR against λ. As it can be seen from Figure 5.16, sample size ratio here is 9, which is too big for sponsor to accommodate. Based on the *SSR* plot, we can choose *RSS* to be between 3 and 4. Thus $\delta_{Sponsor} = 1.15 * 0.25 = 0.2875$.

Relative difference in power (RED)—Based on the *RED* plot (Figure 5.17), we see that power difference is not big in this case even for large values of λ. So $\delta_{Sponsor}$ does not need to move too much toward δ_{FDA}. Therefore, any value of λ between 1.15 and 1.20 is acceptable.

Relative ratio in power/relative risk (RR)—After regularization, the risk is more understandable than the previous two criteria. Consider that it has a clear meaning in terms of the probability of wrongly concluding

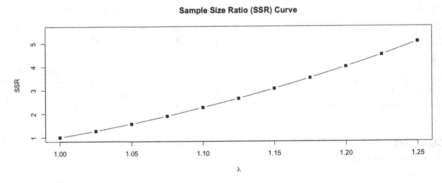

FIGURE 5.16
Plot of SSR against λ.

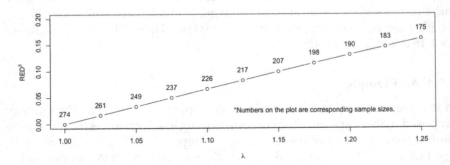

FIGURE 5.17
Plot of RED against λ.

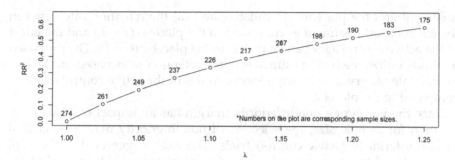

FIGURE 5.18
Plot of RR against λ.

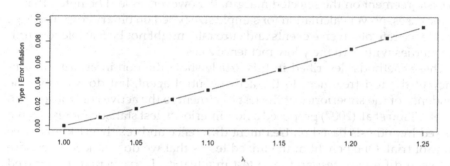

FIGURE 5.19
Plot of Type I error inflation against λ.

similarity. Figure 5.18 plots RR against λ. As it can be seen, anything larger than 40% may be too risky. So 40% can be our maximum risk, and $\lambda = 1.125$, $\delta_{Sponsor} = 0.281$.

Type I error inflation—Figure 5.19 plots Type I error inflation against λ. Since the significance level here is 0.05, after the inflation we want the significance level not to go beyond 0.10.1. Thus, the maximum inflation allowed is 0.05 and $\lambda = 1.15$, $\delta_{Sponsor} = 0.2875$.

5.6 Concluding Remarks

In this chapter, following similar ideas described in the 2010 FDA draft guidance on non-inferiority clinical trials, several alternative methods for selection of an appropriate non-inferiority margin are discussed. These methods were derived by taking into consideration: (i) the variability of the observed mean difference between the active control agent (C) and the placebo (P), the test treatment (T) and the active control agent, and the test treatment and the placebo (if available) and (ii) the retention rate between the effect of test treatment as compared to the placebo $(T - P)$ and the effect of the active control agent as compared to the placebo $(C - P)$. The proposed methods utilize median of estimates of the retention rates based on the historical data observed from superiority studies of the active control agent as compared to the placebo.

Since the width of a non-inferiority margin has an impact on power calculation for sample size, the selection of non-inferiority margin is critical in non-inferiority (active-control) trials. The FDA suggests that the 2010 FDA draft guidance on non-inferiority clinical trials should be consulted for selection of an appropriate non-inferiority margin. In addition, communications with medical and/or statistical reviewers are encouraged when there is disagreement on the selected margin. It, however, should be noted that in some cases, power calculation for sample size based on binary response (e.g., incidence rate of adverse events and cure rate) might not be feasible for clinical studies with extremely low incidence rates.

These methods described in this article show the non-inferiority in efficacy of the test treatment to the active control agent, but do not have the evidence of the superiority of the test treatment to the active control agent in safety. Tsou et al. (2007) proposed a non-inferiority test statistic for testing the mixed hypothesis based on treatment difference and relative risk for active control trial. One benefit of the mixed test is that we do not need to choose between difference test and ratio test in advance. In particular, this mixed null hypothesis consists of a margin based on treatment difference and a margin based on relative risk. Tsou et al.'s proposed mixed non-inferiority test not only preserves the type I error rate at desired level but also gives the similar power as that from the difference test or as that from the ratio test.

Based on risk assessment using the four proposed criteria, the proposed strategy can not only close up the gap between the sponsor proposed margin and the FDA recommended margin, but also select an appropriate margin by taking clinical judgment, statistical rationale, and regulatory feasibility into consideration. In this article, for simplicity, we focus on continuous endpoint. The proposed strategy with the four criteria can be applied to other data types such as discrete endpoints (e.g., binary response) and time-to-event data. In addition to evaluation of risk of sponsor's proposed margin, we can also assess the risk of FDA recommended margin assuming that the margin proposed by the sponsor is the true margin.

6

Missing Data

6.1 Introduction

Missing values or incomplete data are commonly encountered in clinical trials. One of the primary causes of missing data is the dropout. Reasons for dropout include, but are not limited to: refusal to continue in the study (e.g., withdrawal of informed consent); perceived lack of efficacy; relocation; adverse events; unpleasant study procedures; worsening of disease; unrelated disease; non-compliance with the study; need to use prohibited medication; and death (DeSouza et al., 2009). Following the idea of Little and Rubin (1987), DeSouza et al. (2009) provided an overview of three types of missingness mechanisms for dropouts. These three types of missingness mechanisms include (i) missing completely at random (MCAR), (ii) missing at random (MAR), and (iii) missing not at random (MNAR). Missing completely at random is referred to the dropout process that is independent of the observed data and the missing data. Missing at random indicates that the dropout process is dependent on the observed data but is independent of the missing data. For missing not at random, the dropout process is dependent on the missing data and possibly the observed data. Depending upon the missingness mechanisms, appropriate missing data analysis strategies can then be considered based on existing analysis methods in the literature. For example, commonly considered methods under MAR include (i) discard incomplete cases and analyze complete-case only, (ii) impute or fill-in missing values and then analyze the filled-in data, (iii) analyze the incomplete data by a method such as likelihood-based method (e.g., maximum likelihood, restricted maximum likelihood, and Bayesian approach), moment-based method (e.g., generalized estimating equations and their variants), and survival analysis method (e.g., Cox proportional hazards model) that does not require a complete data set. On the other hand, under MNAR, commonly considered methods are derived under pattern mixture models (Little, 1994), which can be divided into two types, parametric (see, e.g., Diggle and Kenward, 1994) and semi-parametric (e.g., Rotnitzky et al., 1998).

In practice, the possible causes of missing values in a study can generally be classified into two categories. The first category includes the reasons that are not directly related to the study. For example, a patient may be lost to follow-up because he/she moves out of the area. This category of missing values can be considered as missing completely at random. The second category includes the reasons that are related to the study. For example, a patient may withdraw from the study due to treatment-emergent adverse events. In clinical research, it is not uncommon to have multiple assessments from each subject. Subjects with all observations missing are called unit non-respondents. Because unit non-respondents do not provide any useful information, these subjects are usually excluded from the analysis. On the other hand, the subjects with some, but not all, observations missing are referred to as item non-respondents. In practice, excluding item non-respondents from the analysis is considered against the intent-to-treat (ITT) principle and, hence is not acceptable. In clinical research, the primary analysis is usually conducted based on ITT population, which includes all randomized subjects with at least post-treatment evaluation. As a result, most item non-respondents may be included in the ITT population. Excluding item non-respondents may seriously decrease power/efficiency of the study. Statistical methods for missing values imputation have been studied by many authors (see, e.g., Kalton and Kasprzyk, 1986; Little and Rubin, 1987; Schafer, 1997).

To account for item non-respondents, two methods are commonly considered. The first method is the so-called likelihood-based method. Under a parametric model, the marginal likelihood function for the observed responses is obtained by integrating out the missing responses. The parameter of interest can then be estimated by the maximum likelihood estimator (MLE). Consequently, a corresponding test (e.g., likelihood ratio test) can be constructed. The merit of this method is that the resulting statistical procedures are usually efficient. The drawback is that the calculation of the marginal likelihood could be difficult. As a result, some special statistical or numerical algorithms are commonly applied for obtaining the MLE. For example, the expectation–maximization (EM) algorithm is one of the most popular methods for obtaining the MLE when there are missing data. The other method for item non-respondents is imputation. Compared with the likelihood-based method, the method of imputation is relatively simple and easy to apply. The idea of imputation is to treat the imputed values as the observed values and then apply the standard statistical software for obtaining consistent estimators. However, it should be noted that the variability of the estimator obtained by imputation is usually different from the estimator obtained from the complete data. In this case, the formulas designed to estimate the variance of the complete data set cannot be used to estimate the variance of estimator produced by the imputed data. As an alternative, two methods are considered for estimation of its variability. One is based on Taylor's expansion. This method is referred to as the linearization method. The merit of the linearization method is that it requires less computation. However, the drawback is that its formula could be very complicated and/or not trackable. The other approach is based on re-sampling method (e.g., bootstrap and jackknife). The drawback of the re-sampling method is that it requires intensive computation. The merit is that it is very easy to apply. With the help of a fast-speed computer, the re-sampling method has become much more attractive in practice.

Note that imputation is very popular in clinical research. The simple imputation method of last observation carried forward (LOCF) at endpoint is probably the most commonly used imputation method in clinical research. Although the LOCF is simple and easy for implementation in clinical trials, its validity has been challenged by many researchers. As a result, the search for alternative valid statistical methods for missing values imputation has received much attention in the past decade. In practice, the imputation methods in clinical research are more diversified due to the complexity of the study design relative to sample survey. As a result, statistical properties of many commonly used imputation methods in clinical research are still unknown, while most imputation methods used in sample survey are well studied. Hence, the imputation methods in clinical research provide a unique challenge and also an opportunity for the statisticians in the area of clinical research.

In Section 6.2, statistical properties and the validity of the commonly used LOCF method is studied. Some commonly considered statistical methods for missing values imputation are described in the subsequent sections of this chapter. Some recent development and a brief concluding remark are given in Sections 6.4 and 6.5.

6.2 Missing Data Imputation

6.2.1 Last Observation Carried Forward

Last observation carried forward (LOCF) analysis at endpoint is probably the most commonly used imputation method in clinical research. For illustration purposes, one example is described below. Consider a randomized, parallel-group clinical trial comparing r treatments. Each patient is randomly assigned to one of the treatments. According to the protocol, each patient should undergo s consecutive visits. Let y_{ijk} be the observation from the kth subject in the ith treatment group at visit j. The following statistical model is usually considered.

$$y_{ijk} = \mu_{ij} + \varepsilon_{ijk}, \text{ where } \varepsilon_{ijk} \sim N(0, \sigma^2), \tag{6.1}$$

where μ_{ij} represents the fixed effect of the ith treatment at visit j. If there are no missing values, the primary comparison between treatments will be based on the observations from the last visit $(j = s)$ because this reflects the treatment difference at the end of the treatment period. However, it is not necessary that every subject complete the study. Suppose that the last evaluable visit is $j^* < m$ for the kth subject in the ith treatment group. Then the value of y_{ij^*k} can be used to impute t_{isk}. After imputation, the data at endpoint are analyzed by the usual ANOVA model. We will refer to the procedure described above as LOCF. Note that the method of LOCF is usually applied according to the ITT principle. The ITT population includes all randomized subjects. The LOCF is commonly employed in clinical research, although it lacks statistical justification. In what follows, its statistical properties and justification are studied.

6.2.1.1 Bias-variance Trade-off

The objective of a clinical study is usually to assess the safety and efficacy of a test treatment under investigation. Statistical inferences on the efficacy parameters are usually obtained. In practice, a sufficiently large sample size is required to obtain a reliable estimate and to achieve a desired power for establishment of the efficacy of the treatment. The reliability of an estimator can be evaluated by bias and by variability. A reliable estimator should have a small or zero bias with small variability. Hence, the estimator based on LOCF and the estimator based on completers are compared in terms of their bias and variability. For illustration purposes, we focus on only one treatment group with two visits. Assume that there are a total of $n = n_1 + n_2$ randomized subjects, where n_1 subjects complete the trial, while the remaining n_2 subjects only have observations at visit 1. Let y_{ik} be the response from the kth subject at the ith visit and $\mu_i = E(y_{ik})$. The parameter of interest is μ_2. The estimator based on completers is given by

$$\bar{y}_c = \frac{1}{n_1} \sum_{k=1}^{n_1} y_{i2k}.$$

On the other hand, the estimator based on LOCF can be obtained as

$$\bar{y}_{LOCF} = \frac{1}{n} \left(\sum_{i=1}^{n_1} y_{i2k} + \sum_{i=n_1+1}^{n} y_{i1k} \right).$$

It can be verified that the bias of $\bar{y}_c$ is 0 with variance $\sigma^2 = n_1$, while of the bias of $\bar{y}_{LOCF}$ is $n_2(\mu_1 - \mu_2)/n$ with variance $\sigma^2/(n_1 + n_2)$. As noted, although LOCF may introduce some bias, it decreases the variability. In a clinical trial with multiple visits, usually, $\mu_j \approx \mu_s$ if $j \approx s$. This implies that the LOCF is recommended if the patients withdraw from the study at the end of the study. However, if a patient drops out of the study at the very beginning, the bias of the LOCF could be substantial. As a result, it is recommended that the results from an analysis based on LOCF be interpreted with caution.

6.2.1.2 Hypothesis Testing

In practice, the LOCF is viewed as a pure imputation method for testing the null hypothesis of

$$H_0 : \mu_{1s} = \cdots = \mu_{rs},$$

where μ_{ij} are as defined in (6.1). Shao and Zhong (2003) provided another look at statistical properties of the LOCF under the above null hypothesis. More specifically, they partitioned the total patient population into s subpopulations according to the time when patients dropped out from the study. Note that in their definition, the patients who complete the study are considered a special case of "drop out" at the end of the study. Then μ_{ij} represents the population mean of the jth subpopulation under treatment i. Assume that the jth subpopulation under the ith treatment accounts for $p_i \times 100\%$ of the overall population under the ith treatment. They argued that the objective of the intend-to-treat analysis is to test the following hypothesis test

$$H_0 : \mu_1 = \cdots = \mu_r, \tag{6.2}$$

where

$$\mu_i = \sum_{j=1}^{s} p_{ij} \mu_{ij}.$$

Based on the above hypothesis, Shao and Zhong (2003) indicated that the LOCF bears the following properties:

1. In the special case of $r = 2$, the asymptotic $(n_i \to \infty)$ size of the LOCF under H_0 is $\leq \alpha$ if and only if

$$\lim \left(\frac{n_2 \tau_1^2}{n} + \frac{n_1 \tau_2^2}{n} \right) \leq \lim \left(\frac{n_1 \tau_1^2}{n} + \frac{n_2 \tau_2^2}{n} \right),$$

where

$$\tau_i^2 = \sum_{j=1}^{s} p_{ij} (\mu_{ij} - \mu_i)^2.$$

The LOCF is robust in the sense that its asymptotic size is α if $\lim(n_1 / n) = n_2 / n$ or $\tau_1^2 = \tau_2^2$. Note that, in reality, $\tau_1^2 = \tau_2^2$ is impractical unless $\mu_{ij} = \mu_i$ for all j. However, $n_1 = n_2$ (as a result $\lim(n_1/n) = n_2/n$) is very typical, in practice. The above observation indicates in such a situation $n_1 = n_2$ that LOCF is still valid.

2. When $r = 2$, $\tau_1^2 \neq \tau_2^2$, and $n_1 \neq n_2$, the LOCF has an asymptotic size smaller than α if

$$(n_2 - n_1)\tau_1^2 < (n_2 - n_1)\tau_2^2 \qquad (6.3)$$

or larger than α if the inequality sign in (6.3) is reversed.

3. When $r \geq 3$, the asymptotic size of the LOCF is generally not α except for some special case (e.g., $\tau_1^2 = \tau_2^2 = \cdots = \tau_r^2 = 0$).

Because the LOCF usually does not produce a test with asymptotic significance level α when $r \geq 3$, Shao and Zhong (2003) proposed the following testing procedure based on the idea of post-stratification. The null hypothesis H_0 should be rejected if $T > \chi_{1-\alpha,r-1}^2$, where $\chi_{1-\alpha,r-1}^2$ is a chi-square random variable with $r - 1$ degrees of freedom and

$$T = \sum_{i=1}^{r} \frac{1}{\hat{V}_i} \left(\bar{y}_{i\cdot\cdot} - \frac{\sum_{i=1}^{r} \bar{y}_{i\cdot\cdot} / \hat{V}_i}{\sum_{i=1}^{r} 1 / \hat{V}_i} \right)^2,$$

$$\hat{V}_i = \frac{1}{n_i(n_i - 1)} \sum_{j=1}^{s} \sum_{k=1}^{n_{ij}} (y_{ijk} - \bar{y}_{i\cdot\cdot})^2.$$

Under model (6.1) and the null hypothesis of (6.3), this procedure has the exact type I error α.

6.2.2 Mean/Median Imputation

Missing ordinal responses are also commonly encountered in clinical research. For those types of missing data, mean or median imputation is commonly considered. Let x_i be the ordinal response from the ith subject, where $i = 1, \ldots, n$. The parameter of interest is $\mu = E(x_i)$. Assume that x_i for $i = 1, \ldots, n_1 < n$ are observed and the rest are missing. Median imputation will impute the missing response by the median of the observed response (i.e., x_i, $i = 1, \ldots, n_1$). The merit of median imputation is that it can keep the imputed response within the sample space as the original response by appropriately defining the median. The sample mean of the imputed data set will be used as an estimator for the population mean. However, as the parameter of interest is population mean, median imputation may lead to biased estimates.

As an alternative, mean imputation will impute the missing value by the sample mean of the observed units, i.e., $(1/n_1)\sum_{i=1}^{n_1} x_i$. The disadvantage of the mean imputation is that the imputed value may be out of the original response sample space. However, it can be shown that the sample mean of the imputed data set is a consistent estimator of population mean. Its variability can be assessed by the jackknife method proposed by Rao and Shao (1987).

In practice, usually, each subject will provide more than one ordinal response. The summation of those ordinal responses (total score) is usually considered as the primary efficacy parameter. The parameter of interest is the population mean of the total score. In such a situation, mean/median imputation can be carried out for each ordinal response within each treatment group.

6.2.3 Regression Imputation

The method of regression imputation is usually considered when covariates are available. Regression imputation assumes a linear model between the response and the covariates. The method of regression imputation has been studied by various authors (see, e.g., Srivastava and Carter, 1986; Shao and Wang, 2002).

Let y_{ijk} be the response from the kth subject in the ith treatment group at the jth visit. The following regression model is considered:

$$y_{ijk} = \mu_i + \beta_i x_{ij} + \varepsilon_{ijk}, \tag{6.4}$$

where x_{ij} is the covariate of the kth subject in the ith treatment group. In practice, the covariates x_{ij} could be demographic variables (e.g., age, sex, and race) or patient's baseline characteristics (e.g., medical history or disease severity). Model (6.4) suggests a regression imputation method. Let $\hat{\mu}_i$ and $\hat{\beta}_i$ denote the estimators of μ_i and β_i based on complete data set, respectively. If y_{ijk} is missing, its predicted mean value $y_{ijk}^* = \hat{\mu}_i + \hat{\beta}_i x_{ij}$ is used to impute. The imputed values are treated as true responses and the usual ANOVA is used to perform the analysis.

6.3 Marginal/Conditional Imputation for Contingency

In an observational study, two-way contingency tables can be used to summarize two-dimensional categorical data. Each cell (category) in a two-way contingency table is defined by a two-dimensional categorical variable (A, B), where A and B take values in $\{1,...,a\}$ and $\{1,...,b\}$, respectively. Sample cell frequencies can be computed based on the observed responses of (A, B) from a sample of units (subjects). Statistical interest includes the estimation of cell probabilities and testing hypotheses of goodness of fit or the independence of the two components A and B. In an observational study, there can be more than one stratum. It is assumed that within a stratum, sampled units independently have the same probability π_A to have missing B and observed A, π_B to have missing A and observed B, π_C to have observed A and B. (The probabilities π_A, π_B, and π_C may be different in different imputation classes.) As units with both A and B missing are considered as unit non-respondent, they are excluded in the analysis. As a result, without loss of generality, it is assumed that $\pi_A + \pi_B + \pi_C = 1$. For a two-way contingency table, it is very important for an appropriate imputation method to keep imputed values in the appropriate sample space. Whether in calculating the cell probability or in testing hypotheses (e.g., testing independence or goodness of fit), the corresponding statistical procedures are all based on the frequency counts of a contingency table. If the imputed value is out of the sample space, additional categories will be produced which is of no practical meaning. As a result, two hot deck imputation methods are thoroughly studied by Shao and Wang (2002).

6.3.1 Simple Random Sampling

Consider a sampled unit with observed $A = i$ and missing B. Two imputation methods were studied by Shao and Wang (2002). The marginal (or unconditional) random hot deck imputation method imputes B by the value of B of a unit randomly selected from all units with observed B. The conditional hot deck imputation method imputes B by the value of B of a unit randomly selected from all units with observed B and $A = i$. All non-respondents are imputed independently.

After imputation, the cell probabilities p_{ij} can be estimated using the standard formulas in the analysis of data from a two-way contingency table by treating imputed values as observed data. Denote these estimators by $\hat{p}_{ij}^I$, where $i = 1,...,a$ and $j = 1,...,b$. Let

$$\hat{p}^I = \left(\hat{p}_{11}^I, ..., \hat{p}_{1b}^I, ..., \hat{p}_{a1}^I, ..., \hat{p}_{ab}^I \right)',$$

and

$$p = (p_{11}, ..., p_{1b}, ..., p_{a1}, ..., p_{ab})',$$

where $p_{ij} = P(A = i, B = j)$. Intuitively, marginal random hot deck imputation leads to consistent estimators of $p_{i \cdot} = P(A = i)$ and $p_{\cdot j} = P(B = j)$, but not p_{ij}. Shao and Wang (2002) showed that $\hat{p}^I$ under conditional hot deck imputation are consistent, asymptotically unbiased, and asymptotically normal.

Theorem 6.1: Assume that $\pi_C > 0$. Under conditional hot deck imputation,

$$\sqrt{n}(\hat{p}^I - p) \to_d N(0, MPM' + (1 - \pi_C)P),$$

where $P = diag\{p\} - pp'$ and

$$M = \frac{1}{\sqrt{\pi_C}}(I_{axb} - \pi_A diag\{p_{B|A}\}I_a \otimes U_b - \pi_B diag\{p_{A|B}\}U_a \otimes I_b,$$

$$p_{A|B} = (p_{11} / p_{\cdot 1}, ..., p_{1b} / p_{\cdot b}, ..., p_{a1} / p_{\cdot 1}, ..., p_{ab} / p_{\cdot b})',$$

$$p_{B|A} = (p_{11} / p_{1\cdot}, ..., p_{1b} / p_{1\cdot}, ..., p_{a1} / p_{a\cdot}, ..., p_{ab} / p_{a\cdot})',$$

where I_a denotes an a-dimensional identity matrix, U_b denotes a b-dimensional square matrix with all components being 1, and $\otimes$ is the Kronecker product.

6.3.2 Goodness-of-Fit Test

A direct application of Theorem 6.1 is to obtain a Wald-type test for goodness of fit. Consider the null hypothesis of the form $H_0 : p = p_0$, where p_0 is a known vector. Under H_0,

$$X_W^2 = n(\hat{p}^* - p_0^*)'\hat{\Sigma}^{*-1}(\hat{p}^* - p_0^*) \to_d \chi_{ab-1}^2,$$

where χ_v^2 denotes a random variable having the chi-square distribution with v degrees of freedom, $\hat{p}^*(p_0^*)$ is obtained by dropping the last component of $\hat{p}^I(p_0)$, and $\hat{\Sigma}^*$ is the estimated asymptotic covariance matrix of $\hat{p}^*$, which can be obtained by dropping the last row and column of $\hat{\Sigma}$, the estimated asymptotic covariance matrix of $\hat{p}^I$. Note that the computation of $\hat{\Sigma}^{*-1}$ is complicated, Shao and Wang (2002) proposed a simple correction of the standard Pearson chi-square statistic by matching the first-order moment, an approach developed by Rao and Scott (1987). Let

$$X_G^2 = n \sum_{j=1}^{b} \sum_{i=1}^{a} \frac{(\hat{p}_{ij}^I - p_{ij})^2}{p_{ij}}.$$

It is noted that under conditional imputation the asymptotic expectation of X_G^2 is given by

$$D = \frac{1}{\pi_C}(ab + \pi_A^2 a + \pi_B^2 b - 2\pi_A a - 2\pi_B b + 2\pi_A \pi_B + 2\pi_A \pi_B \delta) - \pi_C ab + (ab - 1).$$

Let $\lambda = D/(ab-1)$. Then the asymptotic expectation of X_G^2/λ is $ab-1$, which is the first-order moment of a standard chi-square variable with $ab-1$ degrees of freedom. Thus, X_G^2/λ can be used just like a normal chi-square statistic to test the goodness of fit. However, it should be noted that this is just an approximated test procedure that is not asymptotically correct. According to a Shao and Wang's simulation study, this test performs reasonably well with moderate sample sizes.

6.4 Test for Independence

Testing for the independence between A and B can be performed by the following chi-square statistic when there is no missing data

$$X^2 = n\sum_{j=1}^{b}\sum_{i=1}^{a}\frac{(\hat{p}_{ij}^I - \hat{p}_{i\cdot}\hat{p}_{\cdot j})^2}{\hat{p}_{i\cdot}\hat{p}_{\cdot j}} \to_d \chi_{(a-1)(b-1)}^2.$$

It is of interest to know what the asymptotic behavior of the above chi-square statistic is under both marginal and conditional imputation. It is found that, under the null hypothesis that A and B are independent, conditional hot deck imputation yields

$$X^2 \to_d (\pi_C^{-1} + 1 - \pi_C)\chi_{(a-1)(b-1)}^2$$

and marginal hot deck imputation yields

$$X_{MI}^2 \to_d \chi_{(a-1)(b-1)}^2.$$

6.4.1 Results Under Stratified Simple Random Sampling

When number of strata is small Stratified samplings are also commonly used in medical study. For example, a large epidemiology study is usually conducted by several large centers. Those centers are usually considered as strata. For those types of study, the number of strata is not very large; however, the sample size within each stratum is very large. As a result, imputation is usually carried out within each stratum. Within the hth stratum, we assume that a simple random sample of size n_h is obtained and samples across strata are obtained independently. The total sample size is $n = \sum_{h=1}^{H} n_h$, where H is the number of strata and n_h is the sample size in stratum h. The parameter of interest is the overall cell probability vector $p = \sum_{h=1}^{H} w_h p_h$, where w_h is the hth stratum weight. The estimator of p based on conditional imputation is given by $\hat{p}^I = \sum_{h=1}^{H} w_h \hat{p}^I_h$. Assume that $n_h = n \to p$ as $n \to \infty$, $h = 1, ..., H$. Then, a direct application of Theorem 6.1 leads to

$$\sqrt{n}(\hat{p}^I - p) \to_d N(0, \Sigma),$$

where

$$\Sigma = \sum_{h=1}^{H} \frac{w_h^2}{p_h} \Sigma_h$$

and Σ_h is the Σ in Theorem 6.1 but restricted to the hth stratum.

6.4.2 When Number of Strata Is Large

In a medical survey, it is also possible to have the number of strata (H) very large, while the sample size within each stratum is small. A typical example is that of a medical survey is conducted by family. Then each family can be considered as a stratum and all the members within the family become the samples from within this stratum. In such a situation, the method of imputation within stratum is impractical because it is possible that within a stratum, there are no completers. As an alternative, Shao and Wang (2002) proposed the method of imputation across strata under the assumption that $(\pi_{h,A}, \pi_{h,B}, \pi_{h,C})$, where $h = 1, ..., H$, is constant. More specifically, let $n_{h,ij}^C$ denote the number of completers in the hth stratum such that $A = i$ and $B = j$. For a sampled unit in the kth imputation class with observed $B = j$ and missing A, the missing value is imputed by i according to the conditional probability

$$p_{ij} \,|\, B, k = \frac{\sum_{h} w_h n_{h,ij}^C / n_h}{\sum_{h} w_h n_{h,\cdot j}^C / n_h}.$$

Similarly, the missing value of a sampled unit in the kth imputation class with observed $A = i$ and missing B can be imputed by j according to the conditional probability

$$p_{ij} \,|\, A, k = \frac{\sum_{h} w_h n_{h,ij}^C / n_h}{\sum_{h} w_h n_{h,i\cdot}^C / n_h}.$$

Note that $\hat{p}^I$ can be computed by ignoring imputation classes and treating imputed values as observed data. The following result establishes the asymptotic normality of $\hat{p}^I$ based on the method of conditional hot deck imputation across strata.

Theorem 6.2: Let $(\pi_{h,A}, \pi_{h,B}, \pi_{h,C}) = (\pi_A, \pi_B, \pi_C)$ for all h. Assume further that $H \to \infty$ and that there are constants c_j, for $j = 1, \ldots, 4$, such that $n_h \leq c_1$, $c_2 \leq Hw_h \leq c_3$, and $p_{h,ij} \geq c_4$ for all h. Then

$$\sqrt{n}(\hat{p}^I - p) \to_d N(0, \Sigma),$$

where Σ is the limit of

$$n\left(\sum_{h=1}^{H} \frac{w_h^2}{n_h}\Sigma_h + \Sigma_A + \Sigma_B\right).$$

6.5 Recent Development

6.5.1 Other Methods for Missing Data

As indicated earlier, depending upon the mechanisms of missing data, different approaches may be selected in order to address the medical questions asked. In addition to the methods described in the previous sections of this chapter, the methods that are commonly considered included the mixed effects model for repeated measures (MMRM), weighted and unweighted generalized estimating equations (GEE), multiple-imputation-base GEE (MI-GEE), and Complete-case (CC) analysis of covariance (ANCOVA). For recent development of missing data imputation, the Journal of Biopharmaceutical Statistics (JBS) has published a special issue on Missing Data – Prevention and Analysis (JBS, 19, No. 6, 2009, Ed. G. Soon). These recent development on missing data imputation are briefly summarized below.

For a time-saturated treatment effect model and an informative dropout scheme that depends on the unobserved outcomes only through the random coefficients, Kong et al. (2009) proposed a grouping method to correct the biases in the estimation of treatment effect. Their proposed method could improve the current methods (e.g., the LOCF and the MMRM) and give more stable results in the treatment efficacy inferences. Zhang and Paik (2009) proposed a class of unbiased estimating equations using a pair-wise conditional technique to deal with the generalized linear mixed model under benign non-ignorable missingness where specification of the missing model is not needed. The proposed estimator was shown to be consistent and asymptotically normal under certain conditions.

Moore and van der Laan (2009) applied targeted maximum likelihood methodology to provide a test that makes use of the covariate data that are commonly collected in randomized trials. The proposed methodology does not require assumptions beyond those of the log-rank test when censoring is uninformative. Two approaches based on this methodology are provided: (i) a substitution-based approach that targets treatment and time-specific survival from which the log-rank parameter is estimated, and (ii) directly targeting the log-rank parameter. Shardell and El-Kamary (2009), on the other hand, used the framework of coarsened data to motivate performing sensitivity analysis in the presence of incomplete data. The proposed method (under pattern-mixture models) allows departures from the assumption of coarsening at random, a generalization of missing at random, and independent censoring.

Alosh (2009) studied the missing data problem for count data by investigating the impact of missing data on a transition model, i.e., the generalized autoregressive model of order 1 for longitudinal count data. Rothmann et al. (2009) evaluated the loss to follow-up with respect to the intent-to-treat principle on the most important efficacy endpoints for clinical trials of anti-cancer biologic products submitted to the United States Food and Drug Administration from August 2005–October 2008 and provided recommendations in light of the results.

DeSouza et al. (2009) studied the relative performances of these methods for the analysis of clinical trial data with dropouts via an extensive Monte-Carlo study. The results indicate that the MMRM analysis method provides the best solution for minimizing the bias arising from missing longitudinal normal continuous data for small to moderate sample sizes under MAR dropout. For the non-normal data, the MI-GEE may be a good candidate as it outperforms the weighted GEE method.

Yan et al. (2009) discussed methods used to handle missing data in medical device clinical trials, focusing on tipping-point analysis as a general approach for the assessment of missing data impact. Wang et al. (2009) studied the performance of a biomarker predicting clinical outcome in large prospective study under the framework of outcome- and auxiliary dependent sub-sampling and proposed a semi-parametric empirical likelihood method to estimate the association between biomarker and clinical outcome. Nie et al. (2009) dealt with censored laboratory data due to assay limits by comparing a marginal approach and variance-component mixed effects model approach.

6.5.2 The Use of Estimand in Missing Data

An estimand is a parameter that is to be estimated in a statistical analysis. The term is used to more clearly distinguish the target of inference from the function to obtain this parameter, i.e., the estimator and the specific value obtained from a given data set, i.e., the estimate (Mosteller and Tukey, 1987).

To distinguish the terms of estimator and estimand, consider the following example. Let X be a normally distributed random variable with mean μ and variance σ^2. The variance is often estimated by sample variance s^2, which is an estimator of σ^2 and σ^2 is called the estimand. An estimand reflects what is to be estimated to address the scientific question of interest posed by a trial. In practice, the choice of an estimand involves population of interest, endpoint of interest, and measure of intervention effect. Measure of intervention effect may take into account the impact of post-randomization events such as dropouts, non-compliance, discontinuation of study, discontinuation of intervention, treatment switching, rescue medication, death and so on.

As indicated in NRC (2010), an estimand is closely linked to the purpose or objective of an analysis. It describes what is to be estimated based on the question of interest. In clinical trials, since an estimand is often free of the specific assumptions regarding missing data, it is reasonable to conduct sensitivity analyses using different estimators for the same estimand in order to test the robustness of inference to different assumptions of missing data mechanisms (ICH, 2017). The ICH-E9-R1 addendum on estimands and sensitivity analyses states that estimands are defined by the (i) target population; (ii) the outcome of interest; (iii) the specification of how post-randomization events (e.g., dropout, treatment withdrawal, noncompliance, and rescue medication) reflected in the research question; and (iv) the summary measure for the endpoint. All sensitivity analyses should address the same primary estimand for missing data. Also, all model assumptions that are varied in the sensitivity analysis should be in line with the estimand of interest. Note that sensitivity analyses can also be planned for secondary/exploratory estimands and aligned accordingly.

6.5.3 Statistical Methods Under Incomplete Data Structure

6.5.3.1 Introduction

In clinical trials, statistical inference is derived based on probability structure that relies on randomization. In case of missing data, statistical inference should be derived based on valid statistical methods developed under the structure of incomplete data rather than based on methods of missing data imputation. As an example, for illustration purpose, consider statistical methods for two-sequence, three-period crossover designs with incomplete data (Chow and Shao, 1997), which is described below.

A crossover design is a modified randomized block subject) design in which each subject receives more than one treatment at different periods. A crossover design allows a within-subject comparison between treatments (that is, each subject serves as his/her own control). The use of cross-over designs for clinical trials has had extensive discussion in the literature (for example, Brown, Jones, and Kenward). In particular, the standard two-sequence two-period cross-over design is viewed favorably by the FDA for assessment of bioequivalence between drug products. To assess the difference between two treatments, A and B, we describe the standard two-sequence, two-period crossover design as follows. Subjects are randomly assigned to each of two sequences of treatments. Subjects in sequence 1 receive treatment A at the first dosing period and then cross over to receive treatment B at the second dosing period, while subjects in sequence 2 receive treatments in the order of B and A at two dosing periods.

Let y_{kij} denote the response of the ith subject in the kth sequence at the jth period. Then, we can describe y_{kij} by the following model:

$$y_{kij} = \mu + p_j + q_k + t_{g(k,j)} + c_{h(k,j)} + r_{ki} + e_{kij},$$ (6.5)

where μ is the overall mean; p_j is the fixed effect of the jth period, $j = 1, 2$, and $p_1 + p_2 = 0$; q_k is the fixed effect of the kth sequence, $k = 1, 2$, and $q_1 + q_2 = 0$, $t_{g(k,j)}$ is the fixed treatment effect; $g(k,j) = A$ if $k = j$, $g(k,j) = B$ if $k \neq j$, and $t_A + t_B = 0$ $c_{h(k,j)}$ is the fixed carry-over effect of treatment A or B; $c_{h(1,1)} = c_{h(2,1)} = 0$; $h(1,2) = A$, $h(2,2) = B$, and $c_A + c_B = 0$; r_{ik} is the random effect of the ith subject in the kth sequence; $i = 1, \ldots, n_k$; and e_{kij} is a random measurement error. The carry-over effect c_h in (6.5) is the effect of a drug product that persists after the end of the dosing period. It differs from the treatment effect t_g, which is a direct treatment effect during the period in which the treatment is administered.

We can see from (6.5) that there are five independent fixed effect parameters in the model: μ, p_1, q_1, t_A, and c_A. In general, it is not possible to obtain unbiased estimators of these five parameters. If there is a sufficient washout between dosing periods, then we can ignore the carry-over effect, that is, $c_A = c_B = 0$, and we can estimate μ, p_1, q_1, and t_A using linear combinations of the four observed sample means

$$\bar{y}_{kj} = \frac{1}{n_k} \sum_{i=1}^{n_k} y_{kij},$$ (6.6)

where $j = 1, 2$, and $k = 1, 2$. Since the standard two-sequence, two-period crossover design does not provide unbiased estimates of treatment and carry-over effects when carry-over effects are present, it is recommended that one use a replicate crossover design. The simplest and the most commonly used replicate crossover design is the two-sequence, three-period crossover design (Chow and Liu, 1992a, 2013), which one can obtain by simply adding an additional period to the standard two-sequence two-period crossover design such that subjects in sequence 1 receive three treatments in the order of A, B and B, and subjects in sequence 2 receive three treatments in the order of B, A and A. The data from a two-sequence, three-period crossover design can still be described by model (6.5) except that $j = 1, 2, 3$ and $p_1 + p_2 + p_3 = 0$, $g(k,j) = A$ when $k = j$ or $(k,j) = (2,3)$, $g(k,j) = B$ otherwise, $c_{h(k,1)} = 0$, $h(1,2) = h(2,3) = A$ and $h(2,2) = h(1,3) = B$. There are six independent fixed effect parameters, which be estimated unbiasedly using linear combinations of the observed sample means $\bar{y}_{kj}$ defined in (6.6), $j = 1,2,3, k = 1, 2$ (Jones and Kenward, 1989; Chow and Liu, 1992a, 2008, 2013).

In clinical trials the data set is often incomplete for various reasons (protocol violations, failure of assay methods, lost to follow-up, etc.). Since there are three periods in a two-sequence three-period cross-over design, subjects are likely to drop out at the third period since they are required to return for tests more often. Also, due to cost or other administrative reasons, sometimes not all of the subjects receive treatments for the third period. One cannot directly apply standard statistical methods for a cross-over design to an incomplete or unbalanced data set. A simple and naive way to analyze an incomplete data set from a two-sequence, three-period crossover design is to exclude the data from subjects who do not receive all three treatments so that one can treat the data set as if it is from a two-sequence, three-period crossover design with smaller sample sizes. This, however, may result in a substantial loss in efficiency when the dropout rate is appreciable.

6.5.3.2 Statistical Methods for 2 × 3 Crossover Designs with Incomplete Data

The purpose of this section is to describe a statistical method proposed by Chow and Shao (1997) for analysis of incomplete or unbalanced data from a two-sequence, three-period crossover design. Chow and Shao's proposed method fully utilizes the data from subjects who completed at least two periods of study to obtain more efficient estimators as compared to the method of excluding data. Chow and Shao assumed that model (6.5) holds, $\{r_{ki}\}$ and $\{e_{kij}\}$ are mutually independent, and that the e_{kij}s are identically distributed as $N(0,\sigma_e^2)$. Also, it is assumed that the random effects r_{ki} are identically distributed as $N(0,\sigma_a^2)$. The normality assumption on the random effects is more restrictive than that on the random measurement errors.

Let us first consider the case where there are no missing data, under the normality assumptions on r_{ki} and e_{kij}, the maximum likelihood estimator of

$$\beta = \left(\mu, p_1, p_2, q_1, t_A, c_A \right)'$$

exists but its exact distribution may not have a known form and, therefore, an exact confidence interval for a component of β based on the maximum likelihood estimator may not exist. The ordinary least squares (LS) estimator of β as

$$\hat{\beta}_{LS} = A\bar{y} \tag{6.7}$$

where

$$
A = \frac{1}{2}
\begin{bmatrix}
\dfrac{1}{3} & \dfrac{1}{3} & \dfrac{1}{3} & \dfrac{1}{3} & \dfrac{1}{3} & \dfrac{1}{3} \\[2mm]
\dfrac{2}{3} & -\dfrac{1}{3} & \dfrac{1}{3} & \dfrac{2}{3} & -\dfrac{1}{3} & -\dfrac{1}{3} \\[2mm]
-\dfrac{1}{3} & \dfrac{2}{3} & \dfrac{1}{3} & \dfrac{1}{3} & \dfrac{2}{3} & -\dfrac{1}{3} \\[2mm]
\dfrac{1}{2} & \dfrac{1}{4} & \dfrac{1}{4} & -\dfrac{1}{2} & -\dfrac{1}{4} & \dfrac{1}{4} \\[2mm]
\dfrac{1}{2} & -\dfrac{1}{4} & -\dfrac{1}{4} & \dfrac{1}{2} & \dfrac{1}{4} & \dfrac{1}{4} \\[2mm]
0 & \dfrac{1}{2} & -\dfrac{1}{2} & 0 & -\dfrac{1}{2} & \dfrac{1}{2}
\end{bmatrix}
\tag{6.8}
$$

and

$$
\bar{y} = \begin{bmatrix} \bar{y}_1 \\ \bar{y}_2 \end{bmatrix}, \quad \bar{y}_k = \frac{1}{n_k}\sum_{i=1}^{n_k} y_{ki}, \quad y_{ki} = \begin{bmatrix} y_{ki1} & y_{ki2} & y_{ki3} \end{bmatrix}'.
$$

That is, components of the LS estimator are linear combinations of the sample means $\bar{y}_{kj}$. For example, the LS estimator of the treatment effect $\delta = t_A - t_B = 2t_A$ is

$$
\hat{\delta}_{LS} = \frac{1}{2}\bar{y}_{11} - \frac{1}{4}\bar{y}_{12} - \frac{1}{4}\bar{y}_{13} - \frac{1}{4}\bar{y}_{21} + \frac{1}{4}\bar{y}_{22} + \frac{1}{4}\bar{y}_{23}.
$$

Under the normality assumptions on r_{ki} and e_{kij}, we can obtain an exact confidence interval for any given component of β using (6.7) and (6.8) because any component of $\hat{\beta}_{LS}$ can be written in the form $c'\bar{y}_1 \pm c'\bar{y}_2$ for an appropriate three-dimensional vector c.

We now consider the case where there are missing data. Without loss of generality, we assume that in the kth sequence, the first m_{k1} subjects have data for all three periods, while the next $m_{k2} - m_{k1}$ subjects have data for periods 1 and 2 and the next $m_{k3} - m_{k2}$ subjects have data for periods 2 and 3, and the last $n_k - m_{k3}$ subjects have data for periods 1 and 3, where $0 \le m_{k1} \le m_{k2} \le m_{k3} \le n_k$ (subjects who have data for only one period are excluded). The sample sizes m_{k1} may be random and whether or not y_{kij} is missing may depend on the value of y_{kij}. Thus, it is difficult to make an inference on β since the joint distribution of y_{kij}'s (or the conditional joint distribution of y_{kij}'s, given m_{kl}'s) is unknown. It is very likely, however, that whether or not y_{kij} is missing is independent of the measurement error e_{kij} (that is, m_{kl}'s are related to the random subject effects r_{ki} only). If this is true, then we can make inferences on some components of b based on a transformed data set that is unrelated to the random subject effects (see, for example, Mathew and Sinha, 1992; Weerakkody and Johnson, 1992). More precisely, we may write model (6.5) as

$$y = X\beta + Zr + e$$

and consider the linear transformation Hy with $HZ = 0$, where X, Z and H are some suitably defined matrices, y, r and e are the vectors of y_{kij}'s, r_{ki}'s, and e_{kij}'s, respectively.

Since

$$Hy = HX\beta + He \tag{6.9}$$

the conditional distribution of Hy, given m_{kl}'s, is still normal if e is normal and independent of m_{kl}'s.

Under model (6.9), we usually cannot estimate all components of β. For many problems in clinical trials, bioavailability and bioequivalence studies, the primary parameters of interest are often the treatment effect $\delta = t_A - t_B$ and the carry-over effect $\gamma = c_A - c_B$. In the following we consider a special transformation H which is equivalent to taking within-subject differences (between two periods) and produces unbiased estimators of p_1, p_2, δ and γ. Consider the within-subject differences (obtained by taking differences between the two data points for subjects with one missing value, or taking differences between the data from the first two periods and the last two periods for subjects without missing data):

$$d_{1i1} = y_{1i1} - y_{1i2} = p_1 - p_2 + t_A - t_B - c_A + e_{1i1} - e_{1i2}, \ 1 \leq i \leq m_{12}$$

$$d_{1i2} = y_{1i2} - y_{1i3} = p_2 - p_3 + c_A - c_B + e_{1i2} - e_{1i3}, \ 1 \leq i \leq m_{11} \text{ or } m_{12} < i \leq m_{13}$$

$$d_{1i3} = y_{1i1} - y_{1i3} = p_1 = p_3 + t_A - t_B - c_B + e_{1i1} - e_{1i3}, \ m_{13} < i \leq n_1$$

$$d_{2i1} = y_{2i1} - y_{2i2} = p_1 - p_2 - t_A + t_B - c_B + e_{2i1} - e_{2i2}, \ 1 \leq i \leq m_{22}$$

$$d_{2i2} = y_{2i2} - y_{2i3} = p_2 - p_3 - c_A + c_B + e_{2i2} - e_{2i3}, \ 1 \leq i \leq m_{21} \text{ or } m_{22} < i \leq m_{23}$$

$$d_{2i3} = y_{2i1} - y_{2i3} = p_1 - p_3 - t_A + t_B - c_A + e_{2i1} - e_{2i3}, \ m_{23} < i \leq n_2.$$

Let d be the vector of these differences (arranged according to the order of the subjects). Then d is independent of r and Hy for some H satisfying $HZ = 0$ (we can obtain explicitly the matrix H, but it is unnecessary). Assuming that m_{kl}'s are independent of e, we obtain that

$$d \sim N\left(W\theta, \sigma_e^2 G\right) \qquad (6.10)$$

(conditional on m_{kl}'s), where

$$\theta = \left(p_1 - p_2, p_2 - p_3, \delta, \gamma\right)'$$

$$W = \begin{bmatrix} 1_{m_{11}} \otimes \begin{bmatrix} 1 & 0 & 1 & -\dfrac{1}{2} \\ 0 & 1 & 0 & 1 \end{bmatrix} \\[1.2em] 1_{m_{12}-m_{11}} \otimes \begin{bmatrix} 1 & 0 & 1 & -\dfrac{1}{2} \end{bmatrix} \\[1em] 1_{m_{13}-m_{12}} \otimes \begin{bmatrix} 0 & 1 & 0 & 1 \end{bmatrix} \\[1em] 1_{m_1-m_{13}} \otimes \begin{bmatrix} 1 & 1 & 1 & \dfrac{1}{2} \end{bmatrix} \\[1em] 1_{m_{21}} \otimes \begin{bmatrix} 1 & 0 & -1 & \dfrac{1}{2} \\ 0 & 1 & 0 & -1 \end{bmatrix} \\[1.2em] 1_{m_{22}-m_{21}} \otimes \begin{bmatrix} 1 & 0 & -1 & \dfrac{1}{2} \end{bmatrix} \\[1em] 1_{m_{23}-m_{22}} \otimes \begin{bmatrix} 0 & 1 & 0 & -1 \end{bmatrix} \\[1em] 1_{n_2-m_{23}} \otimes \begin{bmatrix} 1 & 1 & -1 & -\dfrac{1}{2} \end{bmatrix} \end{bmatrix}$$

and

$$G = \begin{bmatrix} I_{m_{11}} \otimes \begin{bmatrix} 2 & -1 \\ -1 & 2 \end{bmatrix} & 0 & 0 & 0 \\[1.5em] 0 & 2I_{m_1-m_{11}} & 0 & 0 \\[1em] 0 & 0 & I_{m_{21}} \otimes \begin{bmatrix} 2 & -1 \\ -1 & 2 \end{bmatrix} & 0 \\[1.5em] 0 & 0 & 0 & 2I_{n_2-m_{21}} \end{bmatrix}$$

where $\otimes$ is the Kronecker product, 1_v is the v-vector of ones, I_v is the identity matrix of order v, and 0 is the matrix of 0's of an appropriate order. Under model (6.10), the maximum likelihood estimator of θ is the weighted least squares estimator

$$\hat{\theta}_H = (W'G^{-1}W)^{-1}W'G^{-1}d.$$

the theory of least squares we immediately obtain the following estimator of the covariance matrix of $\hat{\theta}_H$

$$\hat{\sigma}_e^2(W'G^{-1}W)^{-1}$$

where

$$\hat{\sigma}_e^2 = \frac{d'\left[G^{-1}-G^{-1}W(W'G^{-1}W)^{-1}W'G^{-1}\right]d}{n_1 + m_{11} + n_2 + m_{21} - 4}$$

(the sum of squared residuals divided by the degrees of freedom). We can then construct an exact confidence interval for $l'\theta$ with a fixed vector l using the fact that

$$\frac{l'\hat{\theta}_H - l'\theta}{\hat{\sigma}_e \sqrt{\{l'(W'G^{-1}W)^{-1}l\}}}$$

has a t-distribution with $n_1 + m_{11} + n_2 + m_{21} - 4$ degrees of freedom.

6.5.3.3 A Special Case

As we discussed in the previous subsection, missing data are often a more serious problem in the third period of study. We now obtain simplified formulae for $\hat{\theta}_H$, $\hat{\sigma}_e^2$, and $(W'G^{-1}W)^{-1}$ in the important special case where there is no missing data in the first two periods: $1 \le m_{k1} \le m_{k2} = m_{k3} = n_k$. We assume $m_{k1} \ge 1$; otherwise the design becomes a two-sequence two-period crossover. Let $m_k = m_{k1}$

$$\bar{d}_{kj} = \frac{1}{n_k}\sum_{i=1}^{n_k}d_{kij} \quad \text{and} \quad \tilde{d}_{kj} = \frac{1}{m_k}\sum_{i=1}^{m_k}d_{kij}, \quad j=1,2, k=1,2$$

Then, it can be verified that

$$\hat{\theta}_H = \frac{1}{2}\begin{bmatrix} \bar{d}_{11} + \bar{d}_{21} \\[2mm] -\frac{1}{2}\bar{d}_{11} + \frac{1}{2}\tilde{d}_{11} + \tilde{d}_{12} - \frac{1}{2}\bar{d}_{21} + \frac{1}{2}\tilde{d}_{21} + \tilde{d}_{22} \\[2mm] \frac{3}{4}\bar{d}_{11} + \frac{1}{4}\tilde{d}_{11} + \frac{1}{2}\tilde{d}_{12} - \frac{3}{4}\bar{d}_{21} - \frac{1}{4}\tilde{d}_{21} - \frac{1}{2}\tilde{d}_{22} \\[2mm] -\frac{1}{2}\bar{d}_{11} + \frac{1}{2}\tilde{d}_{11} + \tilde{d}_{12} + \frac{1}{2}\bar{d}_{21} - \frac{1}{2}\tilde{d}_{21} - \tilde{d}_{22} \end{bmatrix} \tag{6.11}$$

$$\hat{\sigma}_e^2 = \frac{\displaystyle\sum_{k=1}^{2}\left[\frac{2}{3}\sum_{i=1}^{m_k}\left(d_{ki1}^2 + d_{ki2}^2 + d_{ki1}d_{ki2}\right) + \frac{1}{2}\sum_{i=m_k+1}^{n_k} d_{ki1}^2 - \frac{n_k}{2}\bar{d}_{k1}^2 - \frac{m_k}{6}\left(\tilde{d}_{k1} + 2\tilde{d}_{k2}\right)^2\right]}{n_1 + m_1 + n_2 + m_2 - 4},$$

$$\tag{6.12}$$

and the upper-triangle part of the symmetric matrix $(W'G^{-1}W)^{-1}$ is given by

$$\begin{bmatrix} \frac{1}{2}\left(\frac{1}{n_1} + \frac{1}{n_2}\right) & -\frac{1}{4}\left(\frac{1}{m_1} + \frac{1}{m_2}\right) & \frac{3}{8}\left(\frac{1}{m_1} - \frac{1}{m_2}\right) & -\frac{1}{4}\left(\frac{1}{m_1} - \frac{1}{m_2}\right) \\[2mm] & \frac{1}{8}\left(\frac{1}{n_1} + \frac{1}{n_2} + \frac{3}{m_1} + \frac{3}{m_2}\right) & -\frac{3}{16}\left(\frac{1}{n_1} - \frac{1}{n_2} - \frac{1}{m_1} + \frac{1}{m_2}\right) & \frac{1}{8}\left(\frac{1}{n_1} - \frac{1}{n_2} + \frac{3}{m_1} - \frac{3}{m_2}\right) \\[2mm] & & \frac{3}{32}\left(\frac{3}{n_1} + \frac{3}{n_2} + \frac{1}{m_1} + \frac{1}{m_2}\right) & -\frac{3}{16}\left(\frac{1}{n_1} + \frac{1}{n_2} - \frac{1}{m_1} - \frac{1}{m_2}\right) \\[2mm] & & & \frac{1}{8}\left(\frac{1}{n_1} + \frac{1}{n_2} + \frac{3}{m_1} + \frac{3}{m_2}\right) \end{bmatrix}$$

It is interesting to consider the case where $m_k = n_k$ (that is, no missing datum in all three periods). From (6.11) and the fact that $\bar{d}_{kj} = \hat{d}_{kj}$ when $m_k = n_k$

$$\hat{\theta}_H = \frac{1}{2}\begin{bmatrix} \bar{d}_{11} + \bar{d}_{21} \\[2mm] \bar{d}_{12} + \bar{d}_{22} \\[2mm] \bar{d}_{11} + \frac{1}{2}\bar{d}_{12} - \bar{d}_{21} - \frac{1}{2}\bar{d}_{22} \\[2mm] \bar{d}_{12} - \bar{d}_{22} \end{bmatrix}. \tag{6.13}$$

Comparing (6.7) and (6.8) with (6.13), we have

$$\hat{\theta}_H = \hat{\theta}_{LS} = B\hat{\beta}_{LS},$$

where

$$B = \begin{bmatrix} 0 & 1 & -1 & 0 & 0 & 0 \\ 0 & 1 & 2 & 0 & 0 & 0 \\ 0 & 0 & 0 & 0 & 2 & 0 \\ 0 & 0 & 0 & 0 & 0 & 2 \end{bmatrix}.$$

This raises two interesting points: (i) our method reduces to the least squares method when there is no missing datum; (ii) when no datum is missing, the least squares estimator $\hat{\theta}_{LS}$ does not depend on the random subject effects r_{ki}. Therefore, its properties do not rely on the normality assumption on r_{ki}.

6.5.3.4 An Example

A two-sequence, three-period crossover experiment was conducted to compare two treatments of a drug product in women who have a diagnosis of late luteal phase dysphoric disorder often referred to as marked premenstrual syndrome. After recording their daily symptoms for one to three months, patients received placebo for a full menstrual cycle, with dosing initiated around the times of menses. Before active treatments, each patient received a one-month washout period to remove possible responders to placebo from the study. Following the washout period, patients received double-blind treatment for two full menstrual cycles with either treatment A or treatment B. After two cycles, the patients were crossed over in a double-blind fashion to the alternative treatment for two full menstrual cycles. Then, the patients received a third double blind treatment using the second treatment medication, for a final two cycles.

The analysis of efficacy was based on the depression score, which is the sum of the responses to 13 symptoms in each patient's symptom checklist completed at each treatment period. The depression scores appear in Table 6.1. In this example, there is no missing data in the fiacy two periods; $n_1 = 32$, $n_2 = 36$, $m_1 = 24$, and $m_2 = 18$. The dropout rates are between 75% and 50%, respectively, for the two sequences.

We assume model (6.5) and focus on the treatment effect $\delta = t_A - t_B$ and the carry-over effect $\gamma = c_A - c_B$. Using (6.11), we have

$$\hat{\delta}_H = -3.65 \text{ and } \hat{\gamma}_H = -1.94.$$

An estimate of σ_e^2 calculated according to (6.12) is $\hat{\sigma}_e^2 = 60.62$. Using the formulae given in the previous subsection, the following 95% confidence intervals can be obtained:

$$\delta : (-6.13, -1.18) \text{ and } \gamma : (-5.17, 1.29).$$

The p-value for the two-sided t-test of $\delta = 0$ is 0.004, while the p-value for the two-sided t-test of $\gamma = 0$ is 0.234. Thus, we have strong reason to believe that in

TABLE 6.1

Depression Scores y_{kij}

Patient	Sequence	Period 1	Period 2	Period 3	Patient	Sequence	Period 1	Period 2	Period 3
1	1	20	22	26	35	2	26	18	15
2	1	18	38	22	36	2	21	23	23
3	1	49	49	53	37	2	35	26	38
4	1	26	41	35	38	2	13	18	15
5	1	30	23	22	39	2	13	13	26
6	1	14	18	15	40	2	24	16	13
7	1	38	20	50	41	2	23	30	18
8	1	30	33	31	42	2	25	36	29
9	1	20	13	16	43	2	18	29	17
10	1	13	15	16	44	2	33	34	24
11	1	21	25	32	45	2	45	21	35
12	1	27	34	28	46	2	36	16	15
13	1	13	24	17	47	2	36	36	26
14	1	20	20	16	48	2	21	39	34
15	1	34	37	36	49	2	33	25	29
16	1	25	32	27	50	2	28	13	21
17	1	42	37	40	51	2	47	24	*
18	1	18	22	18	52	2	17	16	*
19	1	15	45	31	53	2	42	50	*
20	1	22	40	47	54	2	19	31	*
21	1	37	22	28	55	2	25	26	*
22	1	22	32	52	56	2	24	21	*
23	1	10	23	25	57	2	19	34	*
24	1	32	35	46	58	2	47	35	*
25	1	16	21	*	59	2	40	26	*
26	1	36	54	*	60	2	43	33	*
27	1	39	43	*	61	2	28	47	*
28	1	40	46	*	62	2	34	14	*
29	1	29	41	*	63	2	22	16	*
30	1	17	16	*	64	2	21	23	*
31	1	46	28	*	65	2	29	17	*
32	1	52	27	*	66	2	42	28	*
33	2	26	29	21	67	2	44	59	*
34	2	38	21	27	68	2	15	34	*

* Missing value.

this example, the treatment effect is significant whereas the carry-over effect is not. Since the carry-over effect is not significant, one might wonder what would have occurred if one had used a standard two-period crossover design. For comparison and illustration, we drop the third period data from the patients who received all three treatments (that is, we treat the data as if they were from a standard two-period crossover design), and re-perform the analysis using standard methods for a two-sequence two-period crossover design. The resulting estimate of δ is -2.38; the 95% confidence interval for δ is $(-5.30, 0.53)$, and the p-value for the two-sided t-test of $\delta = 0$ is 0.108. Note that the length of the confidence interval based on two-period data is about 15% longer than that based on three-period (incomplete) data. More importantly, based on the two-period data, we can neither reject nor accept $\delta = 0$ with strong evidence and a statistician might conclude that there is a need for more experiments to detect whether or not there is a treatment effect. On the other hand, with the additional third period (incomplete) data, we have concluded that the treatment effect is significant. Also, we emphasize that one cannot assess the carry-over effect using a two-sequence two-period crossover design, and that $\gamma = 0$ is a necessary assumption for the use of such a design.

6.6 Concluding Remarks

One of the most controversial issues in missing data imputation is that reduction of power. In practice, it is often considered that the most worrisome impact of missing values on the inference for clinical trials is biased estimation of the treatment effect. As a result, little attention was given to the possible loss of power. In clinical trials, it is recognized that missing data imputation may inflate variability and consequently decrease the power. If there is a significant decrease in power, the intended clinical trial will not be able to achieve the study objectives as planned. This would be a major concern during the regulatory review and approval process.

In addition to the issue of the reduction of power, the following is a summary of controversial issues that present challenge to clinical scientists when applying missing data imputation in clinical trials:

1. When the data are missing, the data are missing. How can we make-up data for the missing data?
2. The validity of the method of LOCF for missing data imputation in clinical trials.
3. When there is a high percentage of missing values, missing data imputation could be biased and misleading.

For the first question; from a clinical scientist's point of view, if the data is missing, it is missing. Missing data imputation has been criticized by *using legal procedure (i.e., statistical model or procedure) to illegally make-up (i.e., impute) data* because (i) *missing is missing* and (ii) one cannot draw statistical inference based on imputed (i.e., predicted but *not observed*) data. Thus, one should not impute (or make-up) data in any way whenever possible—it is always difficult, if not impossible, to verify the assumptions behind the method/model for missing data imputation. However, from a statistician's point of view, we may be able to estimate the missing data based on information surrounding the missing data under certain statistical assumptions/models. Dropping subjects with incomplete data may not be a GSP.

For the second question, the method of LOCF for missing values has been widely used in clinical trials for years in practice although its validity has been challenged by many researchers and regulatory agencies such as the FDA. It is suggested that the method of LOCF for missing values should not be considered as the primary analysis for missing data imputation.

As for the third question, in practice, if the percentage of missing values exceeds a pre-specified number, it is suggested that missing data imputation should not be applied. This raises a controversial issue for selection of the criterion of the cut-off value for the percentage of missing value, which will preserve good statistical properties of the statistical inference derived based on the incomplete data set and imputed data.

In summary, missing values or incomplete data are commonly encountered in clinical research. How to handle the incomplete data is always a challenge to the statisticians in practice. Imputation as one of very popular methodology to compensate for the missing data is widely used in biopharmaceutical research. As compared to its popularity, however, its theoretical properties are far away from well understood. Thus, as indicated by Soon (2009), addressing missing data in clinical trials should focus missing data prevention and missing data analysis. Missing data prevention is usually done through the enforcement of GCP during protocol development and clinical operations personnel training for data collection. This will lead to reduced biases, increased efficiency, less reliance on modeling assumption and less need for sensitivity analysis. However, in practice, missing data cannot be totally avoided. Missing data often occur due to factors beyond the control of patients, investigators, and clinical project team.

7

Multiplicity

7.1 General Concepts

In clinical trials, one of the ultimate goals is to demonstrate that the observed difference of a given study endpoint (e.g., the primary efficacy endpoint) is not only of clinical importance (or a clinically meaningful difference) with statistical meaning (or of statistically significance). Statistical meaning is referred to as that the observed difference is not by chance alone and it is reproducible if we are to conduct a similar study under similar experimental conditions. In practice, the observed clinically meaningful difference that has achieved statistical significance is also known as a *statistical difference*. Thus, a statistical difference means that the difference is not by chance alone and it is reproducible. In drug research and evaluation, it is of interest to control the chance of false negative (or making type I error) and minimize the chance of false positive (or making type II error) at a pre-specified level of significance. As a result, based on a given study endpoint, controlling the overall type I error rate at a pre-specified level of significance for achieving a designed power (i.e., the probability of correctly detecting a clinically meaningful difference if such a difference truly exists) has been a common practice for sample size determination.

In practice, the investigator may consider more than one endpoint (say two study endpoints) as the primary study endpoints. In this case, our goal is to demonstrate that the observed differences of the two study endpoints are clinically meaningful differences with statistical meaning. In other words, the observed differences are not by chance alone and they are reproducible. In this case, the level of significance is necessarily adjusted for controlling the *overall* type I error rate at a pre-specified level of significance for multiple endpoints. This has raised the critical issue of multiplicity in clinical research and development. In clinical trials, *multiplicity* is usually referred to as multiple inferences that are made in simultaneous context (Westfall and Bretz, 2010). As a result, alpha adjustment for multiple comparisons is to make sure that the simultaneously observed differences are not by chance alone. In clinical trials, commonly seen multiplicity include comparison of (i) multiple treatments (dose groups), (ii) multiple endpoints, (iii) multiple time points, (iv) interim analyses, (v) multiple tests of the sample hypothesis, (vi) variable/model selection, and (vii) subgroup analyses.

In general, if there are k treatments, there are $k(k-1)/2$ possible pair-wise comparisons. In practice, two types of error rates are commonly considered (Lakshminarayanan, 2010). The first type of error rate is a comparison-wise error rate (CWE), which is a type I error rate for each comparison. That is, it is the probability of erroneously rejecting the null hypothesis between treatments involved in the comparison. The other type of error rate is an experiment-wise error rate (EWE) or family-wise error rate (FWER) which is the error rate associated with one or more type I errors for all comparisons included in the experiment. Thus, for k comparisons, CWE = α and FWER $= 1 - (1 - \alpha)^k$. As a result, the FWER could be much larger than the significance level associated with each test if multiple statistical tests are performed using the same data set. In practice, thus, it is of interest to control the FWER. In the past several decades, several procedures for controlling FWER have been suggested in the literature. These procedures can be classified into either single-step procedures or step-wise (e.g., step-up and step-down) procedures. Note that an alternative approach to multiplicity control is to consider the false discovery rate (FDR) (see, e.g., Benjamini and Hochberg, 1995).

In the next section, regulatory perspectives regarding multiplicity adjustment are discussed. Also included in this section are some commonly seen controversial issues of multiplicity in clinical trials. Section 7.3 provides a summary of commonly considered statistical methods for multiplicity adjustment for controlling the overall type I error rate. An example concerning a dose finding study is given in Section 7.4. A brief concluding remark is given in the last section of this chapter.

7.2 Regulatory Perspective and Controversial Issues

7.2.1 Regulatory Perspectives

Regulatory position regarding adjustment for multiplicity is not clear. In 1998, the ICH E9 published guidelines in "Statistical Principles in Clinical Trials." These guidelines have several comments reflecting concern over the multiplicity problem. The ICH E9 guidelines recommend that the analysis of clinical trial data may necessitate an adjustment to the Type I error. In addition, the ICH E9 suggests details of any adjustment procedure or an explanation of why adjustment is not thought to be necessary should be set out in the analysis plan. The European Agency for the Evaluation of Medicinal Products (EMEA), on the other hand, in its Committee for Proprietary Medicinal Products (CPMP) draft guidance Points to Consider on Multiplicity Issues in Clinical Trials indicates that multiplicity can have a substantial influence on the rate of false positive conclusions whenever there is an opportunity to choose the most favorable results from two or more analyses. The EMEA guidance also echoes the ICH recommendation for stating details of the multiple comparisons procedure in the analysis plan.

In 2017, the FDA published a draft guidance on *Multiple Endpoints in Clinical Trials Guidance for Industry,* which provides sponsors and review staff with the Agency's thinking about the problems posed by multiple endpoints in the analysis and interpretation of study results and how these problems can be managed in clinical trials for human drugs and biological products. The purpose of this guidance is to describe various strategies for grouping and ordering endpoints for analysis and applying some well-recognized statistical methods for managing multiplicity within a study in order to control the chance of making erroneous conclusions about a drug's effects. Basing a conclusion on an analysis where the risk of false conclusions has not been appropriately controlled can lead to false or misleading representations regarding a drug's effects.

As indicated by Snapinn (2017), despite recent advance in methods for handling multiple endpoints in clinical trials, some challenges remain. In this chapter I will discuss some of these challenges, including the following: (i) potential confusion surrounding the terminology used to describe the multiple endpoints; (ii) appropriate methods for assessing a treatment's effect on the components of a composite endpoint; (iii) advantages and disadvantages of fixed-sequence vs. alpha-splitting methods; (iv) the need to report adjusted p-values; and (v) situations where a single trial may be entitled to multiple sets of alpha.

7.2.2 Controversial Issues

When conducting clinical trials involving multiple comparisons, the following questions are always raised:

1. Why do we need to adjust for multiplicity?
2. When do we need to adjust for multiplicity?
3. How do we adjust for multiplicity?
4. Is the FWER well controlled?

To address the first question, it is suggested that the null/alternative hypotheses be clarified since the type I error rate and the corresponding power are evaluate under the null hypothesis and the alternative hypothesis, respectively.

Regarding the second question, it should be noted that adjustment for multiplicity is to ensure that the simultaneously observed differences are not by chance alone. For example, for evaluation of a test treatment under investigation, if regulatory approval is based on single endpoint, then no alpha adjustment is necessary. However, if regulatory approval is based on multiple endpoints, then α adjustment is a must in order to make sure that the simultaneously observed differences are not by chance alone and that they are reproducible. Conceptually, it is not correct that alpha needs to be adjusted if more than one statistical test (e.g., primary hypothesis and secondary hypothesis) is to be performed. Whether the α should be adjusted depends upon the null hypothesis (e.g., a single hypothesis with one primary endpoint or a composite hypothesis with multiple endpoints) to be tested. The interpretations of the test results for single null hypothesis and composite null hypothesis are different.

For questions (3) and (4), several useful methods for multiplicity adjustment are available in the literature (see, e.g., Hsu, 1996; Chow and Liu, 1998b; Westfall et al., 1999). These methods are single-step methods (e.g., Bonferroni's method), step-down methods (e.g., Holm's method), or step-up methods (e.g., Hochberg's method). In the next section, some commonly employed methods for multiplicity adjustment are briefly described.

As pointed out by Westfall and Bretz (2010), the commonly encountered difficulties surrounding multiplicity in clinical trials include (i) penalizing for doing more or good job (i.e., performing additional test), (ii) adjusting α for all possible tests conducted in the trial, and (iii) problems with determining the family of hypotheses to be tested. Penalizing for doing good job refers to adjustment for multiplicity in dose finding trials that include more dose groups than needed. Adjusting α for all possible tests conducted in the trial, although the α is controlled at the pre-specified level, is overkill because it is not the investigator's best interest to show that all of the observed differences simultaneously are not by chance alone. In practice, it can be very tricky to select the appropriate family of hypotheses (e.g., primary endpoints and secondary endpoints for efficacy or safety or both) for multiplicity adjustment for clinical evaluation of the test treatment under investigation.

It should be added that the most worrisome impact of multiplicity on the inference for clinical trials is not only the control of FWER—though that can be problematic—but also preserving the power for correctly detecting a clinically meaningful treatment effect. One of the most frustrating issues in multiplicity is having adequate control of FWER but may fail to achieve the desired power due to multiplicity.

7.3 Statistical Method for Adjustment of Multiplicity

As indicated earlier, commonly considered procedures or methods for controlling the FWER at some pre-specified level of significance can be classified into two categories: (i) single-step methods (e.g., Bonferroni correction) and

(ii) step-wise procedures, which include step-down methods (e.g., Holm's method) and step-up methods (e.g., Hochberg's method). In practice, commonly used procedures for controlling the FWER in clinical trials are classic multiple comparison procedures (MCP), which include Bonferroni, Tukey, and Dunnett procedures. These procedures and among others are briefly described below.

7.3.1 Bonferroni Method

Among the procedures mentioned above, the method of Bonferroni is probably the most commonly considered procedure for addressing multiplicity in clinical trials though it is somewhat conservative. Suppose there are k treatments and we are interested in testing the following hypothesis

$$H_0 : \mu_1 = \mu_2 = \cdots = \mu_k,$$

where μ_i, $i = 1, \ldots, k$ is the mean for the ith treatment. Let y_{ij}, $j = 1, \ldots, n_i$, $i = 1, \ldots k$ be the jth observation obtained in the ith treatment. Also, let $\bar{y}_i$ and

$$s^2 = \frac{\sum_{i=1}^{k} \sum_{j=1}^{n_i} (y_{ij} - \bar{y}_i)^2}{\sum_{i=1}^{k} (n_i - 1)}$$

be the least square mean for the ith treatment and an estimate of the variance obtained from an analysis of variance (ANOVA), respectively. Also, n_i is the sample size the ith treatment. We then reject the null hypothesis and in favor of the alternative hypothesis that the treatment means μ_i and μ_j are different for every $i \neq j$ if

$$\left| \bar{y}_i - \bar{y}_j \right| > t_{\alpha/2}(v) \left[s^2 (n_i^{-1} + n_j^{-1}) \right]^{1/2}, \tag{7.1}$$

where $t_{\alpha/2}(v)$ denotes a critical value for the t-distribution with $v = \Sigma(n_i - 1)$ degrees of freedom and an upper tail probability of $\alpha/2$. Bonferroni's method simply requires that if there are k inferences in a family, then all inferences should be performed at the α/k significance level rather than at the α level.

Note that the application of Bonferroni's correction to ensure that the probability of declaring one or more false positives is no more than α. However, this method is not recommended when there are many pairwise comparisons. In this case, the following multiple range test procedures are useful.

7.3.2 Tukey's Multiple Range Testing Procedure

Similar to (7.1), we can declare that the treatment means μ_i and μ_j are different for every $i \neq j$ if

$$\left| \bar{y}_i - \bar{y}_j \right| > q(\alpha, k, v) \left[s^2 \frac{(n_i^{-1} + n_j^{-1})}{2} \right]^{1/2}, \tag{7.2}$$

where $q(\alpha, k, v)$ is the studentized range statistic. This method is known as Tukey's multiple range test procedure. It should be noted that simultaneous confidence intervals on all pairs of mean differences $\mu_i - \mu_j$ can be obtained based on the following

$$P\left\{ \mu_i - \mu_j \in \bar{y}_i - \bar{y}_j \pm \left| q \right| \left[s^2 \frac{(n_i^{-1} + n_j^{-1})}{2} \right]^{1/2} \text{ for all } i \neq j \right\} = 1 - \alpha. \tag{7.3}$$

Note that Tables of critical values for the studentized range statistic are widely available. As an alternative to the Tukey's multiple range testing procedure, the following Duncan's multiple-range testing procedure is often considered. Duncan's multiple testing procedure leads us to conclude that the largest and smallest of the treatment means are significantly different if

$$\left| \bar{y}_i - \bar{y}_j \right| > q(\alpha_p, p, v) \left[\frac{MSE}{n} \right]^{1/2}, \tag{7.4}$$

where p is the number of averages, $q(\alpha_p, p, v)$ is the critical value from the studentized range statistic with an FWER of α_p.

7.3.3 Dunnett's Test

When comparing several treatments with a control, Dunnett's test is probably the most popular method. Suppose there are $k - 1$ treatment and one control. Denote by μ_i, $i = 1, \dots, k-1$ and μ_k be the mean of the ith treatment and the control, respectively. Further, supposes that the treatment groups can be described by the following balanced one-way analysis of variance model:

$$y_{ij} = \mu_i + \varepsilon_{ij}, i = 1, \dots, k; j = 1, \dots, n.$$

It is assumed that the ε_{ij} are normally distributed with mean 0 and unknown variance σ^2. Under this assumption, μ_i and σ^2 can be estimated. Consequently, one-sided and two-sided simultaneous confidence intervals for $\mu_i - \mu_k$ can be obtained.

For the one-sided simultaneous confidence interval of $\mu_i - \mu_k$, $i = 1, \dots, k-1$, the lower bound is given by

$$\hat{\mu}_i - \hat{\mu}_k - T\hat{\sigma}\sqrt{2/n}, \text{ for } i = 1, ..., k-1, \tag{7.5}$$

where $T = T_{k-1,v}\{\rho_{ij}\}(\alpha)$ satisfies

$$\int_0^\infty \int_{-\infty}^\infty \left[\Phi(z - \sqrt{2}Tu)\right]^{k-1} d\Phi(z)\gamma(u)du = 1 - \alpha,$$

where Φ is the distribution function of the standard normal. It should be noted that $T = T_{k-1,v}\{\rho_{ij}\}(\alpha)$ are the critical values of the distribution of $\max T_i$, where $T_1, T_2, ..., T_k$ multivariate t distributed with v degrees of freedom and correlation matrix $\{\rho_{ij}\}$.

For the two-sided simultaneous confidence interval $\mu_i - \mu_k, i = 1, ..., k-1$, the lower bound is given by

$$\hat{\mu}_i - \hat{\mu}_k \pm |h|\hat{\sigma}\sqrt{2/n}, \text{ for } i = 1, ..., k-1, \tag{7.6}$$

where $|h|$ satisfies

$$\int_0^\infty \int_{-\infty}^\infty \left[\Phi(z + \sqrt{2}|h|t) - \Phi(z - \sqrt{2}|h|t)\right]^{k-1} d\Phi(z)\gamma(t)dt = 1 - \alpha.$$

Similarly, $|h|$ are the critical values of the distribution of $\max T_i$, where $T_1, T_2, ..., T_k$ follow multivariate t distributed with v degrees of freedom and correlation matrix $\{\rho_{ij}\}$.

7.3.4 Closed Testing Procedure

In clinical trials involving multiple comparisons, as an alternative, the use of closed testing procedure has become very popular since introduced by Marcus et al. (1976). The closed testing procedure can be described as follows. First, form all intersections of elementary hypothesis H_i, then test all intersections using non-multiplicity adjusted tests. An elementary hypothesis H_i is then declared significant if all intersections that include the elementary hypothesis as a component of the intersection are significant. More specifically, suppose there is a family of hypotheses, denoted by $\{H_i, 1 \le i \le k\}$. Let $H_P = \cap_{j \in P} H_j$ where $P = \{1, 2, ..., k\}$. H_P is rejected if and only if every H_Q is rejected for all $Q \subset P$ assuming that an α-level test for each hypothesis H_P is available. Marcus et al. (1976) showed that this testing procedure controls the FWER.

In practice, the closed testing procedure is commonly employed in dose-finding study with several doses of a test treatment under investigation. As an example, consider the following family of hypotheses:

$$\{H_i : \mu_i - \mu_k \le 0, 1 \le i \le k-1\}$$

against one-sided alternatives, where the kth treatment group is the placebo group. Assuming that the sample sizes in the treatment groups are equal (say n) and the sample size for the placebo group is n_k. Let

$$\rho = \frac{n}{n + n_k}.$$

Then, the closed testing procedure can be carried out by the following steps:

> Step 1: Calculate T_i, the t statistics for $1 \le i \le k - 1$. Let the ordered t statistics be $T_{(1)} \le T_{(2)} \le \cdots \le T_{(k-1)}$ with their corresponding hypotheses denoted by $H_{(1)}, H_{(2)}, ..., H_{(k-1)}$.
>
> Step 2: Reject $H_{(j)}$ if $T_{(i)} > T_{i,v,\rho}(\alpha)$ for $i = k - 1, k - 2, ..., j$. If we fail to reject $H_{(j)}$, then conclude that $H_{(j-1)}, ..., H_{(1)}$ are also to be retained.

The closed testing procedures have been shown to be more powerful than the classic multiple comparisons procedures, such as the classic Bonferroni, Tukey, and Dunnett procedures. Note that the above step-down testing procedure is more powerful than that of the Dunnett's testing procedure given in (7.5). There is considerable flexibility in the choice of tests for the intersection hypotheses, leading to the wide variety of procedures that fall within the closed testing umbrella. In practice, a closed testing procedure generally starts with the global null hypothesis and proceeds sequentially towards intersection hypotheses involving fewer endpoints. However, it can begin with the individual hypotheses towards the globally null hypothesis.

7.3.5 Other Tests

In addition to the testing procedures described above, there are several tests (p-value based stepwise test procedures) that are also commonly considered in clinical trials involving multiple comparisons. These methods include, but are limited to, Simes method (see, e.g., Hochberg and Tamhane, 1987; Hsu, 1996; Sarkar and Chang, 1997), Holm's method (Holm, 1979), Hochberg's method (Hochberg, 1988), Hommel's method (Hommel, 1988), and Rom's method (Rom, 1990), and, which are briefly summarized below.

Simes' method is designed to reject global null hypothesis if $p_{(i)} \le i\alpha / m$ for at least one $i = 1, ..., m$. The adjusted p-value for the global hypothesis is given by

$$p = m \min\{p_{(1)} / 1, ..., p_{(m)} / m\}.$$

Note that Simes' method improves Bonferroni's method in controlling the global type I error rate under independence (Sarkar and Chang, 1997). One of the limitations of Simes' method is that it cannot be used to drawn inferences on individual hypothesis since it only tests the global hypothesis.

Holm's method is a sequentially rejective procedure, which sequentially contrasts ordered unadjusted p-values with a set of critical values and rejects a null hypothesis if the p-values and each of the smaller p-values are less than their corresponding critical values. Holm's method not only improves the sensitivity of Bonferroni's correction method to detect real differences, but also increases in power and provides a strong control of the FWER.

Hochberg's method applies exactly the same set of critical values as the Holm's method but performs the test procedure in a step-up fashion. The Hochberg's method enables to identify more significant endpoints and hence is more powerful than that of the Holm's method. In practice, the Hochberg's method is somewhat conservative when individual p-values are independent. In the case where the endpoints are negatively correlated, the FWER control is not guaranteed for all types of dependence among p-values (i.e., the size could potentially exceed α).

Following the principle of closed testing procedure and Simes' test, Hommel's method is a powerful sequentially rejective method that allows for inferences on individual endpoints. It is shown to be marginally more powerful than that of the Hochberg's method. However, the Hommel procedure also suffers from the disadvantage of not preserving the FWER. It does protect the FWER when the individual tests are independent or positively dependent (Sarkar and Chang, 1997).

Rom's method is a step-up procedure that is slightly more powerful than Hochberg's method. Rom's procedure controls the FWER at the α level under the independence of p-values. More details can be found in Rom (1990).

7.4 Gate-Keeping Procedures

7.4.1 Multiple Endpoints

Consider a dose-response study comparing m doses of a test drug to a placebo or an active control agent. Suppose that the efficacy of the test drug will be assessed using a primary endpoint and $s-1$ ordered secondary endpoints. Suppose that the sponsor is interested in testing null hypotheses of no treatment effect with respect to each endpoint against one-sided alternatives. Thus, there are a total of ms null hypotheses, which can be grouped into s families to reflect the ordering of the endpoints. Now, let y_{ijk} denote the measurement of the ith endpoint collected in the jth dose group from the k th patient, where $k = 1,...,n$, $i = 1,...,s$, and $j = 0$ (control), $1,...,m$. The mean of y_{ijk} is denoted by μ_{ij}. Also, let t_{ij} be the t-statistic for comparing the jth dose group to the control with respect to the ith endpoint. It is assumed that the t-statistics follow a multivariate t distribution. Furthermore, y_{ijk}'s are normally distributed. Denote by $\mathfrak{J}_i$ the family of null hypotheses for the ith endpoint, $i = 1,...,s$, i.e., $\mathfrak{J}_i = \{H_{i1} : \mu_{i0} = \mu_{i1},..., H_{im} : \mu_{i0} = \mu_{im}\}$. The s families of null hypotheses are tested in a sequential manner.

Family $\Im_1$ (the primary endpoint) is examined first and testing continues to Family $\Im_2$ (most important secondary endpoint) if at least one null hypothesis has been rejected in the first family. This approach is consistent with a regulatory view that findings with respect to secondary outcome variables are meaningful only when the primary analysis is significant. The same principle can be applied to the analysis of ordered secondary endpoints. Dmitrienko et al. (2006) suggest focusing on testing procedures that meet the following condition:

Condition A: Null hypotheses in $\Im_{i+1}$ can be tested only after at least one null hypothesis was rejected in $\Im_i, i = 1,...,s-1$. Secondly, it is important to ensure that the outcome of the multiple tests early in the sequence does not depend on the subsequent analyses;

Condition B: Rejection or acceptance of null hypotheses in $\Im_i$ does not depend on the test statistics associated with $\Im_{i+1},...,\Im_s, i = 1,...,s-1$. Finally, one ought to account for the hierarchical structure of this multiple testing problem and examine secondary dose–control contrasts only if the corresponding primary dose–control contrast was found significant;

Condition C: The null hypothesis $H_{ij}, i \geq 2$ can be rejected only if H_{1j} was rejected, $j = 1,...,m$. It is important to point out that the logical restrictions for secondary analyses in Condition C are caused only by the primary endpoint. This requirement helps clinical researchers streamline drug labeling and improves the power of secondary tests at the doses for which the primary endpoint was significant.

Within each of the s families, multiple comparisons can be carried out using the Dunnett's test as follows. Rejects H_{ij} if the corresponding t-statistic (t_{ij}) is greater than a critical value c for which the null probability of $\max(t_{i1},...,t_{im}) > c$ is α. Note that Dunnett's test protects the Type I error rate only within each family. Dmitrienko et al. (2006) extended Dunnett's test for controlling the family-wise error rate for all ms null hypotheses.

7.4.2 Gate-Keeping Testing Procedures

Dmitrienko et al. (2006) consider the following example to illustrate the process of constructing a gate-keeping testing procedure for dose-response studies. For simplicity, Dmitrienko et al. (2006) focus on the case where $m = 2$ and $s = 2$. In this example, it is assumed that the treatment groups are balanced with n patients per group. The four (i.e., $ms = 4$) null hypotheses are grouped into two ($s = 2$) families, i.e., $\Im_1 = \{H_{11}, H_{12}\}$ and $\Im_2 = \{H_{21}, H_{22}\}$. Note that $\Im_1$ consists of hypotheses for comparing low and high doses to placebo with respect to the primary endpoint, while) and $\Im_2$ contains hypotheses for comparing low and high doses to placebo with respect to the secondary endpoint.

Now let t_{11}, t_{12}, t_{21}, and t_{22} denote the t-statistics for testing H_{11}, H_{12}, H_{21}, and H_{22}. We can then apply the principle of the closed testing for constructing gate-keeping procedures. According to this principle, one first considers all possible non-empty intersections of the four null hypotheses (this family of 15 intersection hypotheses is known as the closed family) and then sets up tests for each intersection hypothesis. Each of these tests controls the Type I error rate at the individual hypothesis level and the tests are chosen to meet Conditions A, B and C described above. To define tests for each of the 15 intersection hypotheses in the closed family, let H denote an arbitrary intersection hypothesis and consider the following rules:

1. If H includes both primary hypotheses, the decision rule for H should not include t_{21} or t_{22}. This is done to ensure that a secondary hypothesis cannot be rejected unless at least one primary hypothesis was rejected (Condition A).

2. The same critical value should be used for testing the two primary hypotheses. This way, the rejection of primary hypotheses is not affected by the secondary test statistics (Condition B).

3. If H includes a primary hypothesis and a matching secondary hypothesis (e.g., $H = H_{11} \cap H_{21}$), the decision rule for H should not depend on the test statistic for the secondary hypothesis. This guarantees that H_{21} cannot be rejected unless H_{11} was rejected (Condition C).

Note that similar rules used in gate-keeping procedures based on the Bonferroni test can be found in Dmitrienko et al. (2003) and Chen et al. (2005). To implement these rules, it is convenient to utilize the decision matrix approach (Dmitrienko et al., 2003). For the sake of compact notation, we will adopt the following binary representation of the intersection hypotheses. If an intersection hypothesis equals H_{11}; it will be denoted by H^*_{1000}. Similarly, $H^*_{1100} = H_{11} \cap H_{12}$, $H^*_{1010} = H_{11} \cap H_{21}$, etc.

Table 7.1 (reproduced from Table I of Dmitrienko et al., 2006) displays the resulting decision matrix that specifies a rejection rule for each intersection hypothesis in the closed family. The three constants (c_1, c_2, and c_3) in Table 7.2 (reproduced from Table II of Dmitrienko et al., 2006) represent critical values for the intersection hypothesis tests. The values are chosen in such a way that, under the global null hypothesis of no treatment effect, the probability of rejecting each individual intersection hypothesis is α. Note that the constants are computed in a sequential manner (c_1 is computed first, followed by c_2, etc) and thus c_1 is the one-sided $100 \times (1 - \alpha)$th percentile of the Dunnett distribution with 2 and $3(n-1)$ degrees of freedom. Secondly, the other two critical values (c_2 and c_3) depend on the correlation between the primary and secondary endpoints, which is estimated from the data. Calculation of these critical values is illustrated later.

TABLE 7.1

Decision Matrix for a Clinical Trial With Two
Dose-Placebo Compassions and Two Endpoints
$(m = 2, s = 2)$

Intersection Hypothesis	Rejection Rule
H^*_{1111}	$t_{11} > c_1$ or $t_{12} > c_1$
H^*_{1110}	$t_{11} > c_1$ or $t_{12} > c_1$
H^*_{1101}	$t_{11} > c_1$ or $t_{12} > c_1$
H^*_{1100}	$t_{11} > c_1$ or $t_{12} > c_1$
H^*_{1011}	$t_{11} > c_1$ or $t_{22} > c_2$
H^*_{1010}	$t_{11} > c_1$
H^*_{1001}	$t_{11} > c_1$ or $t_{22} > c_2$
H^*_{1000}	$t_{11} > c_1$
H^*_{0111}	$t_{12} > c_1$ or $t_{21} > c_2$
H^*_{0110}	$t_{12} > c_1$ or $t_{21} > c_2$
H^*_{0101}	$t_{12} > c_1$
H^*_{0100}	$t_{12} > c_1$
H^*_{0011}	$t_{21} > c_1$ or $t_{22} > c_1$
H^*_{0010}	$t_{21} > c_3$
H^*_{0001}	$t_{22} > c_3$

Note: The test associated with this matrix rejects a null
hypothesis if all intersection hypotheses containing
it are rejected. For example, the test rejects H_{11} if
$H^*_{1111}, H^*_{1110}, H^*_{1101}, H^*_{1100}, H^*_{1011}, H^*_{1010}, H^*_{1001}$ and H^*_{1000} are
rejected.

TABLE 7.2

Critical Values for Individual Intersection Hypotheses in a
Clinical Trial with Two Dose-Placebo Comparisons and
Two Endpoints $(m = 2, s = 2)$

Correlation Between the Endpoints (ρ)	c_1	c_2	c_3
0.01	2.249	2.309	1.988
0.1	2.249	2.307	1.988
0.5	2.249	2.291	1.988
0.9	2.249	2.260	1.988
0.99	2.249	2.250	1.988

Source: Dmitrienko, A. et al., *Pharm. Stat.*, 5, 19–28, 2006.
Note: The correlation between the two endpoints (ρ) ranges between
0.01 and 0.99, overall one-sided Type I error probability is
0.025 and sample size per treatment group is 30 patients.

The decision matrix in Table 7.1 defines a multiple testing procedure that rejects a null hypothesis if all intersection hypotheses containing the selected null hypothesis were rejected. For example, H_{12} will be rejected if $H_{1111}^*, H_{1110}^*, H_{1101}^*, H_{1111}^*, H_{0111}^*, H_{0110}^*, H_{0101}^*$, and H_{0100}^* were all rejected. By the closed testing principle, the resulting procedure protects the family-wise error rate in the strong sense at the α level. It is easy to verify that the proposed procedure possesses the following properties and thus meets the criteria that define a gate-keeping strategy based on the Dunnett test:

1. The secondary hypotheses, H_{21} and H_{22} cannot be rejected when the primary test statistics, t_{11} and t_{12}; are non-significant (Condition A).

2. The outcome of the primary analyses (based on H_{11} and H_{12}) does not depend on the significance of the secondary dose–placebo comparisons (Condition B). In fact, the procedure rejects H_{11} if and only if $t_{11} > c_1$. Likewise, H_{12} is rejected if and only if $t_{12} > c_1$. Since c_1 is a critical value of the Dunnett test, the primary dose-placebo comparisons are carried out using the regular Dunnett test.

3. The null hypothesis H_{21} cannot be rejected unless H_{11} was rejected and thus the procedure compares the low dose to placebo for the secondary endpoint only if the corresponding primary comparison was significant. The same is true for the other secondary dose–placebo comparison (Condition C).

Under the global null hypothesis, the four statistics follow a central multivariate t-distribution. The three critical values in Table 7.1 can be found using the algorithm for computing multivariate t-probabilities proposed by Genz and Bretz (2002). Table 7.2 shows the values of c_1, c_2, and c_3 selected values of ρ (correlation between the two endpoints). It is assumed in Table 7.2 that the overall one-sided Type I error rate is 0.025 and the sample size per group is 30 patients.

The information presented in Tables 7.1 and 7.2 helps evaluate the effect of the described gate-keeping approach on the secondary tests. Suppose, for example, that the two dose-placebo comparisons for the primary endpoint are significant after Dunnett's adjustment for multiplicity ($t_{11} > 2.249$ and $t_{12} > 2.249$). A close examination of the decision matrix in Table 7.1 reveals that the null hypotheses in the second family will be rejected if their t-statistics are greater than 2.249. In other words, the resulting multiplicity adjustment ignores the multiple tests in the primary family.

However, if the low dose does not separate from placebo for the primary endpoint ($t_{11} \leq 2.249$ and $t_{12} > 2.249$), it will be more difficult to find significant outcomes in the secondary analyses. First of all, the low dose vs. placebo comparison is automatically declared non-significant. Secondly, the high dose will be significantly different from placebo for the secondary endpoint if $t_{22} > c_2$. Note that c_2, which lies between 2.250 and 2.309 when $0.01 \leq \rho \leq 0.99$ is greater than Dunnett's critical value $c_1 = 2.249$ (in general, $c_2 > c_1 > c_3$). The larger critical value is the price of sequential testing. Note, however, that the penalty becomes smaller with increasing correlation.

7.5 Concluding Remarks

When conducting clinical trial involving one or more doses (e.g., dose-finding study) or one or more study endpoints (e.g., efficacy versus safety endpoint), the first dilemma at the planning stage of the clinical trial is the establishment of a *family* of hypotheses *a priori* in the study protocol for achieving the study objective of the intended clinical trial. Based on the study design and various underlying hypotheses, clinical strategies are usually explored for testing various hypotheses for achieving the study objectives. One such set of hypotheses (e.g., drug versus placebo, positive control agent versus placebo, primary endpoint versus secondary primary endpoint) would help to conclude whether both the drug and positive control agent are superior to placebo or the drug is efficacious in terms of the primary endpoint, secondary primary endpoint or both. Under the family of hypotheses, valid multiple comparison procedures for controlling the over type I error rate should be proposed in the study protocol.

The other dilemma at the planning stage of the clinical trial is sample size calculation. A typical procedure is to obtain required sample size under either an analysis of variance (ANOVA) method or an analysis of covariance (ANCOVA) model based on an overall F-test. This approach may not be appropriate if the primary objective involves multiple comparisons. In practice, when multiple comparisons are involved, the method of Bonferroni is usually performed to adjust the type I error rate. Again, the Bonferroni's method is conservative and may require more patients than is actually needed. Alternatively, Hsu (1996) suggested a confidence interval approach as follows. Given a confidence interval approach with level of $1 - \alpha$, perform sample size calculations so that with a pre-specified power $1 - \beta$ ($< 1 - \alpha$), the confidence intervals will cover the true parameter value and be sufficiently narrow (Hsu, 1996).

As indicated, multiple comparisons are commonly encountered in clinical trials. Multiple comparisons may involve comparisons of multiple treatments (dose groups), multiple endpoints multiple time points, interim analyses, multiple tests of the sample hypothesis, variable/model selection, and subgroup analyses in a study. In this case, statistical methods for controlling error rates such as CWE, FWER, or FDR are necessary for multiple comparisons. The closed testing procedure is useful for addressing multiplicity issue in dose-finding studies. In the case when there are a large number of tests involved such as tests for safety data, it is suggested the method using FDR for controlling the overall type I error rate be considered.

From a statistical reviewer's point of view, Fritsch (2012) indicated that when dealing with the issue of multiplicity, one should carefully select the most appropriate hypotheses, i.e., choose *"need to have"* endpoints, but don't pile on *"nice to have"* endpoints, put the endpoints in the right families, carefully consider which hypotheses represent distinct claims, and ensure all *"claims"* are covered under the multiplicity control structure. In addition, one should ensure a good match between the study objectives and the multiplicity control methods by utilizing natural hierarchies (but avoid arbitrary ones) and taking the time to understand complex structures to ensure overall control of multiplicity.

8

Sample Size

8.1 Introduction

In clinical research and development, clinical trials are often conducted to scientifically evaluate safety and efficacy of a test treatment under investigation. For approval of a new test treatment or drug therapy, the United States Food and Drug Administration (FDA) requires that at least two adequate well-controlled clinical studies be conducted in humans to demonstrate substantial evidence of the effectiveness and safety of the drug product under investigation. Clinical data collected from adequate and well-controlled clinical trials are considered substantial evidence for evaluation of the safety and efficacy of the test treatment under investigation. Substantial evidence has the characteristics that (1) it is representative of target patient population, (2) it provides an accurate and reliable assessment of the test treatment, and (3) there is sufficient sample size for achieving a desired statistical power at a pre-specified level of significance. Sufficient sample size is often interpreted as the minimum sample size required for achieving the desired statistical power.

As indicated by Chow et al. (2017), sample size calculation can be performed based on either (i) precision analysis, which is to control type I error rate and maximum error of an estimate allowed; (ii) power analysis, which is to control type II error by achieving a desired probability of correctly detecting a clinically meaningful difference if such a difference truly exists; (iii) reproducibility analysis, which is to control both treatment effect (maintaining treatment effect) and variability; and (iv) probability statement, which is to ensure that the probability of observing certain events is less than some pre-specified values. Among these approaches, a pre-study power analysis is probably the most commonly employed method for sample size calculation in clinical research. In practice, sample size calculation in clinical trials are often classified into the categories of (i) sample size estimation or determination, which is ensure the estimated sample size imparts certain benefits (e.g., a desired power); (ii) sample size re-estimation, which is usually performed at interim analysis; (iii) sample size justification, which is to provide the level of assurance with selected sample size, and; (iv) sample size adjustment, which is to ensure that current sample size will achieve the study objectives with a desired power.

In practice, a typical process for sample size calculation is to estimate/determine a minimum sample size required for achieving study objectives (or correctly detecting a clinically meaningful difference or treatment effect of a given primary study endpoint if such a difference truly exists) with a desired power at a pre-specified level of significance under a valid study design. Sample size calculation is usually performed based on an appropriate statistical test that is derived under the null hypothesis (which reflects the study objectives) and the study design. Thus, information required for performing sample size calculation include (i) study objectives (e.g., test for equality, non-inferiority/equivalence, or superiority); (ii) study design (e.g., a parallel design, a crossover design, a group sequential design, or other designs such as adaptive designs); (iii) properties of the primary study endpoint(s) (e.g., continuous, discrete, or time-to-event data); (iv) the clinically meaningful difference being sought (e.g., non-inferiority margin or equivalence/similarity limit); (v) significance level (e.g., 1% or 5%); (vi) desired power (e.g., 80% or 90%); and (vii) other information such as stratification, 1:1 ratio or 2:1 ratio, or log-transformation. Thus, procedures used for sample size calculation could be very different from one another depending upon different study objectives/hypotheses and different data types under different study designs (see, e.g., Lachin and Foulkes, 1986; Lakatos, 1986, Wang and Chow, 2002a,b; Wang et al., 2002; Chow and Liu, 2008). For a good introduction and comprehensive summary, one can refer to Chow et al. (2008).

In the next section, classical pre-study power analysis for sample size calculation is briefly reviewed. Also included in this section is a summary table of formulas for sample size calculation for various data types of study endpoints under hypotheses testing for equality, non-inferiority/equivalence, and superiority. Section 8.3 reviews and studies the relationship among several clinical strategies for selection of derived study endpoints. The relationship between two one-sided tests procedure and confidence interval approach is examined in Section 8.4. Sample size calculation/allocation for multiple-stage adaptive designs is given in Section 8.5. Section 8.6 discusses sample size adjustment with protocol amendments. Sample size calculation for multi-regional or global clinical trials is given in Section 8.7. Some concluding remarks are given in Section 8.8.

8.2 Traditional Sample Size Calculation

In clinical trials, sample size is often determined for achieving a desired power at a pre-specified level of significance by evaluating test statistics (which is derived under the null hypothesis) under the alternative hypothesis (Chow et al., 2008). In clinical trials, the hypotheses that are of particular interest to the investigators include hypotheses for testing equality, non-inferiority/equivalence, and superiority, which are briefly described below.

Hypothesis testing for equality is a commonly employed approach for demonstration of the efficacy and safety of a test drug product. The purpose is first to show that there is a difference between the test drug and the control (e.g., placebo control). Second, it is to demonstrate that there is a desired power (say at least 80% power) for correctly detecting a clinically meaningful difference if such a difference truly exists. For hypothesis testing for non-inferiority, it is to show that the test drug is not inferior to or as effective as a standard therapy or an active agent. Hypothesis testing for non-inferiority is often applied to situations where (i) the test drug is less toxic, (ii) the test drug is easier to administer, and (iii) the test drug is less expensive. On the other hand, hypothesis testing for superiority is to show that the test drug is superior to a standard therapy or an active agent. It should be noted that the usual test for superiority is referred to test for *statistical* superiority rather than *clinical* superiority. In practice, testing for superiority is not preferred by the regulatory agencies unless there is some prior knowledge regarding the test drug. For hypotheses testing for equivalence, it is to show that the test drug can reach the same therapeutic effect as that of a standard therapy (or an active agent) or they are therapeutically equivalent. It should be noted that hypotheses testing for equivalence include hypotheses testing for bioequivalence and hypotheses testing for therapeutic equivalence. More details regarding difference between testing for bioequivalence and testing for therapeutic equivalence will be provided at later section. To provide a better understanding, Figure 8.1 displays the relationship among hypotheses for non-inferiority, superiority, and equivalence, where μ_T and μ_S are mean responses for the test product and standard therapy and δ is clinically meaningful difference (e.g., non-inferiority margin or equivalence limit).

In this section, for simplicity, we will focus on one-sample case where the primary response is continuous and the hypotheses of interest is to test whether there is a difference, i.e., $\varepsilon = \mu - \mu_0$, between the mean responses, where μ_0 is a pre-specified constant:

$$H_0 : \varepsilon = 0 \text{ versus } H_a : \varepsilon \neq 0. \tag{8.1}$$

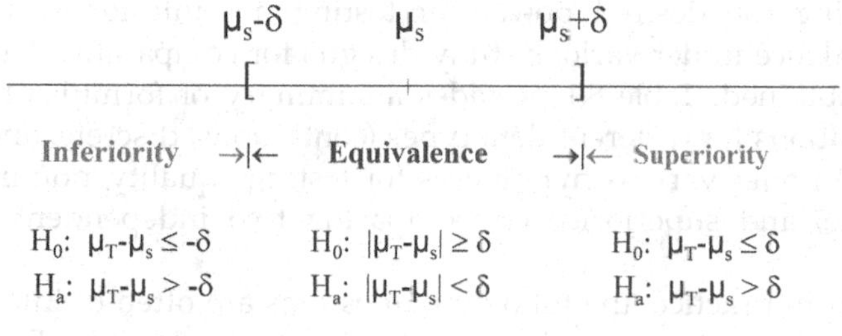

$\mu_s - \delta$	μ_s	$\mu_s + \delta$
Inferiority →|←	Equivalence	→|← Superiority
$H_0: \mu_T - \mu_s \leq -\delta$	$H_0: \|\mu_T - \mu_s\| \geq \delta$	$H_0: \mu_T - \mu_s \leq \delta$
$H_a: \mu_T - \mu_s > -\delta$	$H_a: \|\mu_T - \mu_s\| < \delta$	$H_a: \mu_T - \mu_s > \delta$

FIGURE 8.1
Relationship among non-inferiority, superiority, and equivalence.

When σ^2 is known, we reject the null hypothesis at the α level of significance if

$$\left| \frac{\bar{x} - \mu_0}{\sigma / \sqrt{n}} \right| > z_{\alpha/2},$$

where $\bar{x}$ is the sample mean and z_a is the upper ath quantile of a standard normal distribution. Under the alternative hypothesis that $\varepsilon \neq 0$, the power of the above test is given by

$$\Phi\left(\frac{\sqrt{n}\varepsilon}{\sigma} - z_{\alpha/2} \right) + \Phi\left(-\frac{\sqrt{n}\varepsilon}{\sigma} - z_{\alpha/2} \right), \tag{8.2}$$

where Φ is the cumulative standard normal distribution function. By ignoring a small value $\leq \alpha/2$, the power is approximately

$$\Phi\left(\frac{\sqrt{n}|\varepsilon|}{\sigma} - z_{\alpha/2} \right).$$

As a result, the sample size needed for achieving power $1 - \beta$ can be obtained by solving the following equation

$$\frac{\sqrt{n}|\varepsilon|}{\sigma} - z_{\alpha/2} = z_\beta.$$

This leads to

$$n = \frac{\left(z_{\alpha/2} + z_\beta \right)^2 \sigma^2}{\varepsilon^2} = \left[\frac{z_{\alpha/2} + z_\beta}{\theta} \right]^2, \tag{8.3}$$

where $\theta = \varepsilon/\sigma$ is the effect size adjusted for standard deviation. When σ^2 is unknown, it can be replaced by the sample variance s^2, which results in the usual one-sample t-test. Following the process, sample size required for achieving the desired power for testing non-inferiority, superiority, and equivalence under various study designs for comparative trials can be similarly obtained. Table 8.1 provides a summary of formulas for sample size calculations for different data types (continuous, discrete, and time-to-event data) under various hypotheses for testing equality, non-inferiority/equivalence, and superiority for comparing two independent treatment groups.

Note that in practice, the following questions are often of interest to the investigators or sponsors when conducting clinical trials. First, what's the impact on required sample size when switching from a two-sided test

TABLE 8.1

Formulas for Sample Size Calculation

Hypotheses	Continuous	Binary Responses	Time-to-Event Data
Equality $H_0:\varepsilon=0$ $H_a:\varepsilon\neq 0$	$n_1=kn_2$ $n_2=\dfrac{(z_{\alpha/2}+z_\beta)^2\,\sigma^2(1+1/k)}{(\mu_2-\mu_1)^2}$	$n_1=kn_2$ $n_2=\dfrac{(z_{\alpha/2}+z_\beta)^2}{(p_2-p_1)^2}\left[\dfrac{p_1(1-p_1)}{k}+p_2(1-p_2)\right]$	$n_1=kn_2$ $n_2=\dfrac{(Z_{\alpha/2}+Z_\beta)^2}{(\lambda_2-\lambda_1)^2}\left[\dfrac{\sigma^2(\lambda_1)}{k}+\sigma^2(\lambda_2)\right]$
Non-inferiority $H_0:\varepsilon\leq-\delta$ $H_a:\varepsilon>-\delta$	$n_1=kn_2$ $n_2=\dfrac{(z_\alpha+z_\beta)^2\,\sigma^2(1+1/k)}{\mu_2-\mu_1+\delta}$	$n_1=kn_2$ $n_2=\dfrac{(z_\alpha+z_\beta)^2}{(p_2-p_1+\delta)^2}\left[\dfrac{p_1(1-p_1)}{k}+p_2(1-p_2)\right]$	$n_1=kn_2$ $n_2=\dfrac{(Z_\alpha+Z_\beta)^2}{(\lambda_2-\lambda_1+\delta)^2}\left[\dfrac{\sigma^2(\lambda_1)}{k}+\sigma^2(\lambda_2)\right]$
Superiority $H_0:\varepsilon\leq\delta$ $H_a:\varepsilon>\delta$	$n_1=kn_2$ $n_2=\dfrac{(z_\alpha+z_\beta)^2\,\sigma^2(1+1/k)}{\mu_2-\mu_1-\delta}$	$n_1=kn_2$ $n_2=\dfrac{(z_\alpha+z_\beta)^2}{(p_2-p_1-\delta)^2}\left[\dfrac{p_1(1-p_1)}{k}+p_2(1-p_2)\right]$	$n_1=kn_2$ $n_2=\dfrac{(Z_\alpha+Z_\beta)^2}{(\lambda_2-\lambda_1-\delta)^2}\left[\dfrac{\sigma^2(\lambda_1)}{k}+\sigma^2(\lambda_2)\right]$
Equivalence $H_0:|\varepsilon|\geq\delta$ $H_a:|\varepsilon|<\delta$	$n_1=kn_2$ $n_2=\dfrac{(z_\alpha+z_{\beta/2})^2\,\sigma^2(1+1/k)}{\left(\delta-|\mu_2-\mu_1|\right)^2}$	$n_1=kn_2$ $n_2=\dfrac{(z_\alpha+z_{\beta/2})^2}{\left(\delta-|p_2-p_1|\right)^2}\left[\dfrac{p_1(1-p_1)}{k}+p_2(1-p_2)\right]$	$n_1=kn_2$ $n_2=\dfrac{(Z_\alpha+Z_{\beta/2})^2}{\left(\delta|\lambda_2-\lambda_1|\right)^2}\left[\dfrac{\sigma^2(\lambda_1)}{k}+\sigma^2(\lambda_2)\right]$

Note: ε is the difference between the true means, response rates, and hazard rates of a test drug and a control for the case of continuous, binary responses, and time-to-event data, respectively.

to a one-sided test? Second, will different ratio of treatment allocation (say 2:1 ratio) reduce required sample size and increase the probability of success? Third, can we determine/justify sample size based on effect size that one wishes to detect when there is little or no information regarding the test treatment? Partial answers (if not all) to the above questions can be obtained by carefully examining the formulas for sample size calculation given in Table 8.1.

8.3 Selection of Study Endpoints

When conducting clinical trials, appropriate clinical endpoints are often chosen in order to address the scientific and/or medical questions of interest (study objectives/hypotheses). For a given clinical study, the required sample size may be determined based on expected absolute change from baseline of a primary study endpoint but the collected data are analyzed based on relative change from baseline (e.g., percent change from baseline) of the primary study endpoint or based on the percentage of patients who show some improvement (i.e., responder analysis). The definition of a responder could be based on either absolute change from baseline or relative change from baseline of the primary study endpoint. In practice, it is not uncommon to observe a significant result on a study endpoint (e.g., absolute change from baseline, relative change from baseline, or the responder analysis) but not on other study endpoints (e.g., absolute change from baseline, relative change from baseline, or responder analysis). Thus, it is of interest to explore how an observed significant difference of a study endpoint (e.g., absolute change from baseline, relative change from baseline, or responder's analysis) can be translated to those of other study endpoints (e.g., absolute change from baseline, relative change from baseline, or responder's analysis).

The sample size required for achieving a desired power based on the absolute change could be very different from that obtained based on the percent change, or the percentage of patients who show an improvement based on the absolute change or relative change at α level of significance. Thus, the selection of an appropriate study endpoint has an immediate impact on the assessment of treatment effect. In practice, one of the most controversial issues regarding clinical endpoint selection is that which clinical endpoint is telling the truth. The other controversial issue is how to translate clinical results among the study endpoints. In what follows, we will make an attempt to answer these questions.

8.3.1 Translations among Clinical Endpoints

Suppose that there are two test treatments, namely, a test treatment (T) and a reference treatment (R). Denote the corresponding measurements of the ith subject in the jth treatment group before and after the treatment by W_{1ij} and W_{2ij}, where $j = T$ or R corresponds to the test and the reference treatment, respectively. Assume that the measurement W_{1ij} is lognormally distributed with parameters μ_j and σ_{1j}^2, i.e., $W_{1ij} \sim \text{lognormal}(\mu_j, \sigma_{1j}^2)$. Let $W_{2ij} = W_{1ij}(1 + \Delta_{ij})$, where Δ_{ij} denotes the percentage change after receiving the treatment. In addition, assume that Δ_{ij} is lognormal distributed with parameters $\mu_{\Delta j}$ and $\sigma_{\Delta j}^2$, i.e., $\Delta_{ij} \sim \text{lognormal}(\mu_{\Delta_j}, \sigma_{\Delta_j}^2)$. Thus, the difference and the relative difference between the measurements before and after the treatment are given by $W_{2ij} - W_{1ij}$ and $(W_{2ij} - W_{1ij}) / W_{1ij}$, respectively. In particular,

$$W_{2ij} - W_{1ij} = W_{1ij}\Delta_{ij} \sim \text{lognormal}\left(\mu_j, \mu_{\Delta_j}, \sigma_j^2 + \sigma_{\Delta_j}^2\right),$$

and

$$\frac{W_{2ij} - W_{1ij}}{W_{1ij}} \sim \text{lognormal}\left(\mu_{\Delta_j}, \sigma_{\Delta_j}^2\right).$$

To simplify the notations, define X_{ij} and Y_{ij} as $X_{ij} = \log(W_{2ij} - W_{1ij})$, $Y_{ij} = \log(\frac{W_{2ij} - W_{1ij}}{W_{1ij}})$. Then, both X_{ij} and Y_{ij} are normally distributed with means $\mu_j + \mu_{\Delta_j}$ and μ_{Δ_j}, $i = 1, 2, \ldots, n_j, j = T, R$, respectively.

Thus, possible derived study endpoints based on the responses observed before and after the treatment as described earlier include X_{ij}, the absolute difference between "before treatment" and "after treatment" responses of the subjects, Y_{ij}, the relative difference between "before treatment" and "after treatment" responses of the subjects, $r_{A_j} = \#\{x_{ij} > c_1, i = 1, \ldots, n_j\} / n_j$, the proportion of responders, which is defined as a subject whose absolute difference between "before treatment" and "after treatment" responses is larger than a pre-specified value c_1, $r_{R_j} = \#\{y_{ij} > c_1, i = 1, \ldots, n_j\} / n_j$, the proportion of responders, which is defined as a subject whose relative difference between "before treatment" and "after treatment" responses is larger than a pre-specified value c_2.

To define notation, for $j = T, R$, let $p_{A_j} = E(r_{A_j})$ and $p_{R_j} = E(r_{R_j})$. Given the above possible types of derived study endpoints, we may consider the following hypotheses for testing non-inferiority with non-inferiority margins determined based on either absolute difference or relative difference:

Case 1: Absolute difference

$$H_0 : \left(\mu_R - \mu_{\Delta_R}\right) - \left(\mu_T - \mu_{\Delta_T}\right) \geq \delta_1 \quad \text{vs.} \quad H_a : \left(\mu_R - \mu_{\Delta_R}\right) - \left(\mu_T - \mu_{\Delta_T}\right) < \delta_1 \qquad (8.4)$$

Case 2: Relative difference

$$H_0 : \left(\mu_{\Delta_R} - \mu_{\Delta_T}\right) \geq \delta_2 \quad \text{vs.} \quad H_a : \left(\mu_{\Delta_R} - \mu_{\Delta_T}\right) < \delta_2 \qquad (8.5)$$

Case 3: Absolute change in response rate (defined based on absolute difference)

$$H_0 : p_{A_R} - p_{A_T} \geq \delta_3 \text{ vs. } H_\alpha : p_{A_R} - p_{A_T} < \delta_3 \tag{8.6}$$

Case 4: Relative change in response rate (defined based on absolute difference)

$$H_0 : \frac{p_{A_R} - p_{A_T}}{p_{A_R}} \geq \delta_4 \text{ vs. } H_\alpha : \frac{p_{A_R} - p_{A_T}}{p_{A_R}} < \delta_4 \tag{8.7}$$

Case 5: Absolute change in response rate (defined based on relative difference)

$$H_0 : p_{R_R} - p_{R_T} \geq \delta_5 \text{ vs. } H_\alpha : p_{R_R} - p_{R_T} < \delta_5 \tag{8.8}$$

Case 6: Relative change in response rate (defined based on relative difference)

$$H_0 : \frac{p_{R_R} - p_{R_T}}{p_{R_R}} \geq \delta_6 \text{ vs. } H_\alpha : \frac{p_{R_R} - p_{R_T}}{p_{R_R}} < \delta_6 \tag{8.9}$$

8.3.2 Comparison of Different Clinical Strategies

Denote by X_{ij} the absolute difference between "before treatment" and "after treatment" responses of the ith subjects under the ith treatment, and by Y_{ij} the relative difference between "before treatment" and "after treatment" responses of the ith subjects under the jth treatment. Let $\bar{x}_{.j} = \frac{1}{n_j} = \sum_{i=1}^{n_j} x_{ij}$ and $\bar{y}_{.j} = \frac{1}{n_j} = \sum_{i=1}^{n_j} y_{ij}$ be the sample means of X_{ij} and Y_{ij} for the jth treatment group, $j = T, R$, respectively. Based on normal distribution, the null hypothesis in (8.4) is rejected at a level α of significance if

$$\frac{\bar{x}_{.R} - \bar{x}_{.T} + \delta_1}{\sqrt{\left(\dfrac{1}{n_T} + \dfrac{1}{n_R}\right)\left[\left(\sigma_T^2 + \sigma_{\Delta_T}^2\right) + \left(\sigma_R^2 + \sigma_{\Delta_R}^2\right)\right]}} > z_\alpha. \tag{8.10}$$

Thus, the power of the corresponding test is given as

$$\Phi\left(\frac{\left(\mu_T + \mu_{\Delta_T}\right) - \left(\mu_R + \mu_{\Delta_R}\right) + \delta_1}{\sqrt{\left(n_1^{-1} + n_2^{-1}\right)\left[\left(\sigma_T^2 + \sigma_{\Delta_R}^2\right) + \left(\sigma_R^2 + \sigma_{\Delta_R}^2\right)\right]}} - z_\alpha\right), \tag{8.11}$$

where $\Phi(.)$ is the cumulative distribution function of the standard normal distribution. Suppose that the sample sizes allocated to the reference and test treatments are in the ratio of r, where r is a known constant. Using these results, the required total sample size for the test the hypotheses (8.4) with a power level of $(1-\beta)$ is $N = n_T + n_R$, with

$$n_T = \frac{(z_\alpha + z_\beta)^2 (\sigma_1^2 + \sigma_2^2)(1+1/\rho)}{\left[(\mu_R + \mu_{\Delta R}) - (\mu_T + \mu_{\Delta T}) - \delta_1\right]^2},$$ (8.12)

$n_R = \rho n_T$ and z_u is $1-u$ quantile of the standard normal distribution.

Note that y_{ij} s are normally distributed. The testing statistic based on $\bar{y}_{.j}$ would be similar to the above case. In particular, the null hypothesis in (8.5) is rejected at a significance level α if

$$\frac{\bar{y}_{T.} - \bar{y}_{R.} + \delta_2}{\sqrt{\left(\dfrac{1}{n_T} + \dfrac{1}{n_R}\right)\left(\sigma_{\Delta T}^2 + \sigma_{\Delta R}^2\right)}} > z_\alpha.$$ (8.13)

The power of the corresponding test is given as

$$\Phi\left(\frac{\mu_{\Delta T} - \mu_{\Delta R} + \delta_2}{\sqrt{\left(n_T^{-1} + n_R^{-1}\right)\left(\sigma_{\Delta T}^2 + \sigma_{\Delta R}^2\right)}} - z_\alpha\right).$$ (8.14)

Suppose that $n_R = \rho n_T$, where r is a known constant. Then the required total sample size to test hypotheses (5) with a power level of $(1-\beta)$ is $(1+\rho)n_T$, where

$$n_T = \frac{(z_\alpha + z_\beta)^2 (\sigma_{\Delta T}^2 + \sigma_{\Delta R}^2)(1+1/\rho)}{\left[(\mu_R + \mu_{\Delta R}) - (\mu_T + \mu_{\Delta T}) - \delta_2\right]^2}.$$ (8.15)

For sufficiently large sample size n_j, r_{A_j} is asymptotically normal with mean p_{A_j} and variance $\frac{p_{A_j}(1-p_{A_j})}{n_j}$, $j = T, R$. Thus, based on Slutsky Theorem, the null hypothesis in (8.6) is rejected at an approximate α level of significance if

$$\frac{r_{AT} - r_{AR} + \delta_3}{\sqrt{\dfrac{1}{n_T} r_{AT}(1-r_{AT}) + \dfrac{1}{n_R} r_{AR}(1-r_{AR})}} > z_\alpha.$$ (8.16)

The power of the above test can be approximated by

$$\Phi\left(\frac{p_{A_T} - p_{A_R} + \delta_3}{\sqrt{n_T^{-1}p_{A_T}\left(1-p_{A_T}\right) + n_R^{-1}r_{A_R}\left(1-p_{A_R}\right)}} - z_\alpha\right). \tag{8.17}$$

If $n_R = \rho n_T$, where r is a known constant, the required sample size to test hypotheses (8.6) with a power level of $(1-\beta)$ is $(1+\rho)n_T$, where

$$n_T = \frac{\left(z_\alpha + z_\beta\right)^2\left[p_{A_T}\left(1-p_{A_T}\right) + p_{A_R}\left(1-p_{A_R}\right)/\rho\right]}{\left(p_{A_R} - p_{A_T} - \delta_3\right)^2}. \tag{8.18}$$

Note that, by definition, $p_{A_j} = 1 - \Phi\left(\frac{c_1 - (\mu_j + \mu_{\Delta j})}{\sqrt{\sigma_j^2 + \sigma_{\Delta j}^2}}\right)$, where $j = T, R$. Therefore, following similar arguments, the above results also apply to test hypotheses (8.8) with p_{A_j} replaced by $p_{R_j} = 1 - \Phi\left(\frac{c_2 - \mu_{\Delta j}}{\sigma_{\Delta j}}\right)$ and δ_3 replaced by δ_5. Thus, we have

$$H_0 : \left(1-\delta_4\right)p_{A_R} - p_{A_T} \geq 0 \quad \text{vs.} \quad H_1 : \left(1-\delta_4\right)p_{A_R} - p_{A_T} < 0. \tag{8.19}$$

Therefore, the null hypothesis in (8.7) is rejected at an approximate a level of significance if

$$\frac{r_{A_T} - \left(1-\delta_4\right)r_{A_R}}{\sqrt{\dfrac{1}{n_T}r_{A_T}\left(1-r_{A_T}\right) + \dfrac{\left(1-\delta_4\right)^2}{n_R}r_{A_R}\left(1-r_{A_R}\right)}} > z_\alpha. \tag{8.20}$$

Using normal approximation to the test statistic when both n_T and n_R are sufficiently large, the power of the above test can be approximated by

$$\Phi\left(\frac{p_{A_T} - \left(1-\delta_4\right)p_{A_R}}{\sqrt{n_T^{-1}p_{A_T}\left(1-p_{A_T}\right) + n_R^{-1}\left(1-\delta_4\right)^2 p_{A_R}\left(1-p_{A_R}\right)}} - Z_\alpha\right) \tag{8.21}$$

Suppose that $n_R = \rho n_T$, where r is a known constant. Then the required total sample size to test hypotheses (8.13), or equivalently (8.19), with a power level of $(1-\beta)$ is $(1+\rho)n_T$, where

$$n_T = \frac{\left(Z_\alpha + Z_\beta\right)^2\left[p_{A_T}\left(1-p_{A_T}\right) + \left(1-\delta_4\right)^2 p_{A_R}\left(1-p_{A_R}\right)/\rho\right]}{\left[p_{A_T} - \left(1-\delta_4\right)p_{A_R}\right]^2}. \tag{8.22}$$

Similarly, the results derived in (8.18) through (8.20) for the hypotheses (8.7) also apply to the hypotheses in (8.9) with p_{A_j} replaced by $p_{R_j} = 1 - \Phi\left(\frac{c_2 - \mu_{\Delta_j}}{\sigma_{\Delta_j}}\right)$ and δ_4 replaced by δ_6.

It should be noted that when conducting clinical trials, the sponsors always choose the clinical endpoints to their best interest. The regulatory agencies, however, require the primary clinical endpoint be specified in the study protocol. Positive results from other clinical endpoints will not be considered as the primary analysis results for regulatory approval. This, however, does not have any scientific or statistical justification for assessment of the treatment effect of the test drug under investigation.

8.4 Multiple-stage Adaptive Designs

Consider a clinical trial with K interim analyses. The final analysis is treated as the Kth interim analysis. Suppose that at each interim analysis, a hypothesis test is performed followed by some actions that are dependent of the analysis results. Such actions could be an early stopping due to futility/efficacy or safety, sample size re-estimation, modification of randomization, or other adaptations. In this setting, the objective of the trial can be formulated using a global hypothesis test, which is an intersection of the individual hypothesis tests from the interim analyses

$$H_0 : H_{01} \cap ... \cap H_{0K},$$

where $H_{0i}, i = 1, ..., K$ is the null hypothesis to be tested at the ith interim analysis. Note that there are some restrictions on H_{0i}, that is, rejection of any $H_{0i}, i = 1, ..., K$ will lead to the same clinical implication (e.g., drug is efficacious); hence all $H_{0i}, i = 1, ..., K$ are constructed for testing the *same* endpoint within a trial. Otherwise the global hypothesis cannot be interpreted.

In practice, H_{0i} is tested based on a sub-sample from each stage, and without loss of generality, assume H_{0i} is a test for the efficacy of a test treatment under investigation, which can be written as

$$H_{0i} : \eta_{i1} \geq \eta_{i2} \quad \text{versus} \quad H_{ai} : \eta_{i1} < \eta_{i2},$$

where η_{i1} and η_{i2} are the responses of the two treatment groups at the ith stage. It is often the case that when $\eta_{i1} = \eta_{i2}$, the p-value p_i for the sub-sample at the ith stage is uniformly distributed on $[0, 1]$ under H_0 (Bauer and Kohne, 1994). This desirable property can be used to construct a test statistic for multiple-stage seamless adaptive designs. As an example, Bauer and Kohne

(1994) used Fisher's combination of the *p*-values. Similarly, Chang (2007) considered a linear combination of the *p*-values as follows

$$T_k = \sum_{i=1}^{K} w_{ki} p_i, i = 1,\ldots,K,$$ (8.23)

where $w_{ki} > 0$ and K is the number of analyses planned in the trial. For simplicity, consider the case where $w_{ki} = 1$. This leads to

$$T_k = \sum_{i=1}^{K} p_i, i = 1,\ldots,K.$$ (8.24)

The test statistic T_k can be viewed as cumulative evidence against H_0. The smaller the T_k is, the stronger the evidence is. Equivalently, we can define the test statistic as $T_k = \sum_{i=1}^{K} p_i / K$, which can be viewed as an average of the evidence against H_0. The stopping rules are given by

$$\begin{cases} \text{Stop for efficacy} & \text{if } T_k \leq \alpha_k \\ \text{Stop for futility} & \text{if } T_k \geq \beta_k \\ \text{Continue} & \text{otherwise} \end{cases}$$ (8.25)

where T_k, α_k, and β_k are monotonic increasing functions of k, $\alpha_k < \beta_k$, $k = 1,\ldots,K-1$, and $\alpha_K = \beta_K$. Note that α_k, and β_k are referred to as the efficacy and futility boundaries, respectively. To reach the kth stage, a trial has to pass 1 to $(k-1)$th stages. Therefore, a so-called proceeding probability can be defined as the following unconditional probability:

$$\psi_k(t) = P\left(T_k < t, \alpha_1 < T_1 < \beta_1,\ldots,\alpha_{k-1} < T_{k-1} < \beta_{k-1}\right)$$

$$= \int_{\alpha_1}^{\beta_1} \cdots \int_{\alpha_{k-1}}^{\beta_{k-1}} \int_{-\infty}^{t} f_{T_1 \ldots T_k}(t_1,\ldots,t_k) dt_k dt_{k-1} \ldots dt_1,$$ (8.26)

where $t \geq 0$, t_i, $i = 1,\ldots,k$ is the test statistic at the ith stage, and $f_{T_1 \ldots T_k}$ is the joint probability density function. The error rate at the kth stage is given by

$$\pi_k = \psi_k(\alpha_k).$$ (8.27)

When efficacy is claimed at a certain stage, the trial is stopped. Therefore, the type I error rates at different stages are mutually exclusive. Hence, the experiment-wise type I error rate can be written as follows:

$$\alpha = \sum_{k=1}^{K} \pi_k. \tag{8.28}$$

Note that (8.26) through (8.28) are the keys to determine the stopping boundaries, which will be illustrated in the next sub-section with two-stage seamless adaptive designs. The adjusted p-value calculation is the same as the one in a classic group sequential design (see, e.g., Jennison and Turnbull, 2000). The key idea is that when the test statistic at the kth stage $T_k = t = \alpha_k$ (i.e., just on the efficacy stopping boundary), the p-value is equal to α spent $\sum_{i=1}^{k} \pi_i$. This is true regardless of which error spending function is used and consistent with the p-value definition of the classic design. The adjusted p-value corresponding to an observed test statistic $T_k = t$ at the kth stage can be defined as

$$p(t;k) = \sum_{i=1}^{k-1} \pi_i + \psi_k(t), k = 1, .., K. \tag{8.29}$$

This adjusted p-value indicates weak evidence against H_0, if the H_0 is rejected at a late stage because one has spent some α at previous stages. On the other hand, if the H_0 was rejected at an early stage, it indicates strong evidence against H_0 because there is a large portion of overall α that has not been spent yet. Note that p_i in (8.23) is the stage-wise naive (unadjusted) p-value from a sub-sample at the ith stage, while $p(t;k)$ are adjusted p-values calculated from the test statistic, which are based on the cumulative sample up to the kth stage where the trial stops, equations (8.28) and (8.29) are valid regardless how the p_i values are calculated.

An Example

Suppose that a clinical trial utilizing an adaptive group sequential design with one planned interim analysis is to be conducted to evaluate the safety and efficacy of a test treatment in treatment patient with certain disease. The sponsor would like to have the option for stopping the trial early due to either efficacy or futility and is able to control the overall type I error rate at the α level of significance. The study can be viewed as a two-stage adaptive design. Sample size calculation was performed based on the primary study endpoint of failure rate at 12 weeks post randomization. This study is powered to detect a clinically meaningful difference of 25% in failure rate between the test treatment and a placebo at the 5% level of significance assuming that the true placebo failure rate is 50%. Since there is an intention to stop the trial due to efficacy/futility at the end of Stage 1, sample size calculation was performed based on the method of individual p-values for a two-stage adaptive design proposed by Chang (2007).

At the end of the first stage, the following stopping rules based on individual p-values are considered:

Stop for efficacy if $T_1 \leq \alpha_1$;

Stop for futility if $T_1 > \beta_1$;

Continue with adaptation if $\alpha_1 < T_1 \leq \beta_1$,

where α_1 and β_1 $(\alpha_1 < \beta_1)$ are efficacy and futility boundaries, respectively, and T_1 is the test statistic (based on individual p-value) to be used at the first stage.

Note that after the review of the data by an independent data monitoring committee (DMC), additional adaptations such as dose modification and sample size re-estimation may be applied as suggested by the independent DMC if a decision to proceed is reached. As indicated by Chow and Chang (2006), for a two-stage adaptive design based on individual p-values, we have (see also, Chang, 2007)

$$\alpha = \alpha_1 + \alpha_2 (\beta_1 - \alpha_1).$$

Thus, for the proposed two-stage seamless adaptive design, we choose the efficacy and futility boundaries as follows

$$\alpha_1 = 0.005, \ \beta_1 = 0.40, \ \alpha_2 = 0.0506$$

for controlling the overall type I error rate at the 5% $(\alpha = 0.05)$ level of significance. Sample size calculation can then be performed accordingly.

8.5 Sample Size Adjustment with Protocol Amendments

In practice, for a given clinical trial, it is not uncommon to have 3–5 protocol amendments after the initiation of the clinical trial. One of the major impacts of many protocol amendments is that the target patient population may have been shifted during the process, which may have resulted in a totally different target patient population at the end of the trial. A typical example is the case when significant adaptation (modification) is applied to inclusion/exclusion criteria of the study. Denote by (μ, σ) the *target* patient population. After a given protocol amendment, the resultant (actual) patient population may have been shifted to (μ_1, σ_1), where $\mu_1 + \mu + \varepsilon$ is the population mean of the primary study endpoint and $\sigma_1 = C\sigma$ $(C > 0)$ is the population standard deviation of the primary study endpoint. The shift in target patient population can be characterized by

$$E_1 = \left|\frac{\mu_1}{\sigma_1}\right| = \left|\frac{\mu + \varepsilon}{C\sigma}\right| = |\Delta|\left|\frac{\mu}{\sigma}\right| = |\Delta|E,$$

where $\Delta = (1 + \varepsilon / \mu) / C$, and E and E_1 are the effect size before and after population shift, respectively. Chow et al. (2002) and Chow and Chang (2006) refer to Δ as a sensitivity index measuring the change in effect size between the actual patient population and the original target patient population.

As indicated in Chow and Chang (2006), the impact of protocol amendments on statistical inference due to shift in target patient population (moving target patient population) can be studied through a model that link the moving population means with some covariates (Chow and Shao, 2005). However, in many cases, such covariates may not exist or exist but not observable. In this case, it is suggested that inference on Δ be considered to measure the degree of shift in location and scale of patient population based on a mixture distribution by assuming the location or scale parameter is random (Chow et al., 2005).

In clinical trials, for a given target patient population, sample size calculation is usually performed based on a test statistic (which is derived under the null hypothesis) evaluated under an alternative hypothesis. After protocol amendments, the target patient population may have been shifted to an *actual* patient population. In this case, the original sample size may have to be adjusted in order for achieving the desired power for assessment of the treatment effect for the *original* patient population. As an example, consider sample size adjustment with protocol amendments based on covariate-adjusted model for testing non-inferiority hypothesis that

$$H_0 : p_{10} - p_{20} \leq -\delta \text{ versus } H_1 : p_{10} - p_{20} > -\delta,$$

where p_{10} and p_{20} are the response rate for a test treatment and an active control or placebo, respectively. Let $n_{Classic}$ and n_{Actual} be the sample size based on the original patient population and the actual patient population as the result of protocol amendments. Also, let $n_{Actual} = R n_{Classic}$, where R is the adjustment factor. Following the procedures described in Chow et al. (2008), sample sizes for both $n_{Classic}$ and n_{Actual} can be obtained. Let Y_{tij} and X_{tij} be the response and the corresponding relevant covariate for the jth subject after the ith amendment under the tth treatment $(t = 1, 2, i = 0, 1, ..., k, j = 1, 2, ..., n_{ti})$. For each amendment, patients selected by the same criteria are randomly allocated to either the test treatment $D_1 = 1$ or control treatment $D_2 = 0$ groups. In this particular case, the true mean values of the covariate for the two treatment groups are the same under each amendment. Therefore, the relationships between the binary response and the covariate for both treatment groups can be described by a single model,

$$p_{ti} = \frac{\exp\left(\beta_1 + \beta_2 D_t + \beta_3 v_i + \beta_4 D_t v_i\right)}{1 + \exp\left(\beta_1 + \beta_2 D_t + \beta_3 v_i + \beta_4 D_t v_i\right)}, t = 1, 2, i = 0, 1, ..., k.$$

Hence, the response rates for the test treatment and the control treatment are

$$P_{1i} = \frac{\exp\left(\beta_1, \beta_2 + (\beta_3 + \beta_4) v_i\right)}{1 + \exp\left(\beta_1 + \beta_2 + (\beta_3 + \beta_4) v_i\right)} \text{ and } p_{2i} = \frac{\exp\left(\beta_1 + \beta_3 v_i\right)}{1 + \exp\left(\beta_1 + \beta_3 v_i\right)}$$

respectively. Thus, the joint likelihood function of $\beta = (\beta_1, ..., \beta_4)^T$ is given by

$$\prod_{t=1}^{2} \prod_{i=0}^{k} \prod_{j=1}^{n_{ti}} \left[\left(\frac{\exp\left(\beta^T z^{(ti)}\right)}{1 + \exp\left(\beta^T z^{(ti)}\right)} \right)^{y_{tij}} \left(\frac{1}{1 + \exp\left(\beta^T z^{(ti)}\right)} \right)^{1 - y_{tij}} \times f_{\bar{X}_i}\left(\bar{x}_i\right) \right],$$

where $f_{\bar{X}_i}(\bar{x}_i)$ is the probability density function of $\bar{X}_i = \sum_{t=1}^{2} \sum_{j=1}^{n_{ti}} X_{tij}$ and $z^{(ti)} = (1, D_t, \bar{x}_i, D_t \bar{x}_i)^T$. The log likelihood function is then given by

$$l(\beta) = \sum_{t=1}^{2} \sum_{i=0}^{k} \sum_{j=1}^{n_{ti}} \left[y_{tij} \ln\left(\frac{\exp\left(\beta^T z^{(ti)}\right)}{1 + \exp\left(\beta^T z^{(ti)}\right)} \right) + (1 - y_{tij}) \ln\left(\frac{1}{1 + \exp\left(\beta^T z^{(ti)}\right)} \right) + \ln f_{\bar{X}_i}\left(\bar{x}_i\right) \right].$$

Given the resulting the maximum likelihood estimate $\hat{\beta} = (\hat{\beta}_1, ..., \hat{\beta}_4)^T$, we obtain the estimate of p_{10} and p_{20} as follows

$$\hat{p}_{10} = \frac{\exp\left(\hat{\beta}_1 + \hat{\beta}_2 + (\hat{\beta}_3 + \hat{\beta}_4) \bar{X}_0\right)}{1 + \exp\left(\hat{\beta}_1 + \hat{\beta}_2 + (\hat{\beta}_3 + \hat{\beta}_4) \bar{X}_0\right)}, \hat{p}_{20} = \frac{\exp\left(\hat{\beta}_1 + \hat{\beta}_3 \bar{X}_0\right)}{1 + \exp\left(\hat{\beta}_1 + \hat{\beta}_3 \bar{X}_0\right)}$$

Thus, we have

$$N_{Classic} = \frac{\left(Z_\alpha + Z_\gamma\right)^2}{\left(p_{10} - p_{20} + \delta\right)^2} \cdot \left[\frac{p_{10}(1 - p_{10})}{w} + \frac{p_{20}(1 - p_{20})}{1 - w} \right]$$

$$N_{Actual} = \frac{\left(Z_\alpha + Z_\gamma\right)^2 \tilde{V}_d}{\left(p_{10} - p_{20} + \delta\right)^2},$$

where w is the proportion of patients for the first treatment,

$$\tilde{V}_d = \left[g'(\beta) \right]^T \left(w \sum_{i=0}^{k} \rho_{1i}\, \mathbf{I}^{(1i)} + (1-w) \sum_{i=0}^{k} \rho_{2i}\, \mathbf{I}^{(2i)} \right)^{-1}$$

$$g'(\beta),\ w = n_1.\ /\ N,\ \rho_{ti} = n_{ti}\ /\ n_{t.},$$

and

$$g'(\beta) = \begin{pmatrix} p_{10}(1-p_{10}) - p_{20}(1-p_{20}) \\ p_{10}(1-p_{10}) \\ v_0\big(p_{10}(1-p_{10}) - p_{20}(1-p_{20})\big) \\ v_0\big(p_{10}(1-p_{10})\big) \end{pmatrix}.$$

Note that more details regarding formulas for sample size adjustment based on covariate-adjusted model for binary response endpoint and sample size adjustments based on random location shift and random scale shift can be found in Chow (2011).

8.6 Multi-regional Clinical Trials

As indicated by Uesaka (2009), the primary objective of a multi-regional bridging trial is to show the efficacy of a drug in all participating regions while also evaluating the possibility of applying the overall trial results to each region. To apply the overall results to a specific region, the results in that region should be consistent with either the overall results or the results from other regions. A typical approach is to show consistency among regions by demonstrating that there exists no treatment-by-region interaction. Recently, the Ministry of Health, Labor and Welfare (MHLW) of Japan published a guidance on Basic Principles on Global Clinical Trials that outlines the basic concepts for planning and implementation the multi-regional trials in a Q&A format. In this guidance, special consideration was placed on the determination of the number of Japanese subjects required in a multi-regional trial. As indicated, the selected sample size should be able to establish the consistency of treatment effects between the Japanese group and the entire group.

To establish the consistency of the treatment effects between the Japanese group and the entire group, it is suggested that the selected size should satisfy

$$P\left(\frac{D_J}{D_{All}} > \rho\right) \geq 1 - \gamma, \tag{8.30}$$

where D_J and D_{All} are the treatment effects for the Japanese group and the entire group, respectively. Along this line, Quan et al. (2010) derived closed form formulas for the sample size calculation/allocation for normal, binary and survival endpoints. As an example, the formula for continuous endpoint assuming that $D_J = D_{NJ} = D_{All} = D$, where D_{NJ} is the treatment effect for the non-Japanese subjects, is given below.

$$N_J \geq \frac{z_{1-\gamma}^2 N}{\left(z_{1-\alpha/2} + z_{1-\beta}\right)^2 \left(1 - \rho\right)^2 + z_{1-\gamma}^2 \left(2\rho - \rho^2\right)}, \tag{8.31}$$

where N and N_J are the sample size for the entire group and the Japanese group. Note that the MHLW recommends that ρ should be chosen to be either 0.5 or greater and γ should be chosen to be either 0.8 or greater in (8.30). As an example, if we choose $\rho = 0.5$, $\gamma = 0.8$, $\alpha = 0.05$, and $\beta = 0.9$, then $N_J / N = 0.224$. In other words, the sample size for the Japanese group has to be at least 22.4% of the overall sample size for the multi-regional trial.

In practice, $1 - \rho$ is often considered a non-inferiority margin. If ρ is chosen to be greater than 0.5, the Japanese sample size will increase substantially. It should be noted that the sample size formulas given in Quan et al. (2010) are derived under the assumption that there is no difference in treatment effects for the Japanese group and non-Japanese group. In practice, it is expected that there is a difference in treatment effect due to ethnic difference. Thus, the formulas for sample size calculation/allocation derived by Quan et al. (2010) are necessarily modified in order to take into consideration of the effect due to ethnic difference.

As an alternative, Kawai et al. (2008) proposed an approach to rationalize partitioning the total sample size among the regions so that a high probability of observing a consistent trend under the assumed treatment effect across regions can be derived, if the treatment effect is positive and uniform across regions in a multi-regional trial. Uesaka (2009) proposed new statistical criteria for testing consistency between regional and overall results, which do not require impractical sample sizes and discussed several methods of sample size allocation to regions. Basically, three rules of sample size allocation in multi-regional clinical trials are discussed. These rules include (1) allocating equal size to all regions, (2) minimizing total sample size, and (3) minimizing the sample size of a specific region. It should be noted that the sample size of a multi-regional trial may become very large when one wishes to ensure consistent results between region of interest and the other regions or between the regional results and the overall results regardless which rules of sample size allocation is used.

When planning a multi-regional trial, it is suggested that the study objectives should be clearly stated in the study protocol. Once the study objectives are confirmed, a valid study design can be chosen and the primary clinical endpoints can be determined accordingly. Based on the primary clinical endpoint, sample size required for achieving a desired power can then be calculated. Recent approaches for sample size determination in multi-regional trials developed by Kawai et al. (2008), Quan et al. (2010), and Ko et al. (2010) are all based on the assumption that the effect size is uniform across regions. For example, assume that we focus on the multi-regional trial for comparing a test product and a placebo control based on a continuous efficacy endpoint. Let X and Y be some efficacy responses for patients receiving the test product and the placebo control respectively. For convention, both X and Y are normally distributed with variance σ^2. We assume that σ^2 is known, although it can generally be estimated. Let μ_T and μ_P be the population means of the test and placebo, respectively, and let $\Delta = \mu_T - \mu_P$. Assume that effect size (Δ/σ) is uniform across regions. The hypothesis of testing for the overall treatment effect is given as

$$H_0 : \Delta \leq 0 \text{ versus } H_\alpha : \Delta > 0.$$

Let N denote the total sample size for each group planned for detecting an expected treatment difference $\Delta = \delta$ at the desired significance level α and with power $1 - \beta$. Thus,

$$N = 2\sigma^2 \left\{ \left(z_{1-\alpha} + z_{1-\beta} \right) / \delta \right\}^2,$$

where $z_{1-\alpha}$ is the $(1-\alpha)$th percentile of the standard normal distribution. Once N is determined, special consideration should be placed on the determination of the number of subjects from the Asian region in the multi-regional trial. The selected sample size should be able to establish the consistency of treatment effects between the Asian region and the regions overall. To establish the consistency of treatment effects between the Asian region and the entire group, it is suggested that the selected sample size should satisfy that the assurance probability of the consistency criterion, given that $\Delta = \delta$ and the overall result is significant at α level, is maintained at a desired level, say 80%. That is,

$$P_\delta \left(D_{Asia} \geq \rho D_{All} \mid Z > z_{1-\alpha} \right) > 1 - \gamma \tag{8.32}$$

for some pre-specified $0 < \gamma \leq 0.2$. Here Z represents the overall test statistic.

Ko et al. (2010) calculated the sample size required for the Asian region based on (8.32). For $\beta = 0.1$, $\alpha = 0.025$, and $\rho = 0.5$, the sample size for the Asian region has to be around 30% of the overall sample size to maintain the assurance probability of (8.32) at 80% level. On the other hand, by considering a two-sided test, Quan et al. (2010) derived closed form formulas for the sample size calculation for normal, binary and survival endpoints based on the consistency criterion. For examples, if we choose $\rho = 0.5$, $\gamma = 0.2$, $\alpha = 0.025$, and $\beta = 0.9$, then the Asian sample size has to be at least 22.4% of the overall sample size for the multi-regional trial.

It should be noted that the sample size determination given in Kawai et al. (2008), Quan et al. (2010), and Ko et al. (2010) are all derived under the assumption that the effect size is uniform across regions. In practice, it might be expected that there is a difference in treatment effect due to ethnic difference. Thus, the sample size calculation derived by Kawai et al. (2008), Quan et al. (2010), and Ko et al. (2010) may not be of practical use. More specifically, some other assumptions addressing the ethnic difference should be explored. For example, we may consider the following assumptions:

1. Δ is the same but σ^2 is different across regions;
2. Δ is different but σ^2 is the same across regions;
3. Δ and σ^2 are both different across regions.

Statistical methods for the sample size determination in multi-regional trials should be developed based on the above assumptions.

8.7 Current Issues

8.7.1 Is Power Calculation the Only Way?

In clinical trials comparing a test treatment (T) and a reference product (R) (e.g., a standard of care treatment or an active control agent), it can be verified that sample size formulae is a function of type I error rate (α), type II error rate (β) or power ($1 - \beta$), population mean difference (ε), clinically meaningful difference (δ), variability associated with the response (σ), and treatment allocation ratio (k) (see, e.g., Chow et al., 2017). Assuming that $n_T = n_R = n$ and $\sigma_T = \sigma_R = \sigma$, where n_T, σ_T, and n_R, σ_R are sample sizes and standard deviations for the test treatment and the reference product, respectively, sample size formulae can generally be expressed as follows:

$$n = f(\alpha, \beta, \varepsilon, \delta, \sigma, \text{and } k), \qquad (8.33)$$

in which $\varepsilon = \mu_T - \mu_R$, where μ_T and μ_R are the population means of the test treatment and the reference product, respectively. As indicated in Section 8.2, a typical approach (power calculation or power analysis for sample size calculation) for sample size determination is to fix α (control of type I error rate), ε (under the null hypothesis), δ (detecting the difference of clinical importance), σ (assuming that the variability associated with the response is known), and k (a fixed ratio of treatment allocation) and then select n for achieving a desired power of $1-\beta$ (Chow et al., 2017). In practice, it is not possible to select a sample size n and at the same time control all parameters given in (8.33) although many sponsors and most regulatory agencies intend to.

Under (8.33), sample size calculation can be divided into two categories: controlling one parameter and controlling two (multiple) parameters. For example, in the interest of maintaining clinically meaningful difference or margin, sample size can be selected by fixing all parameters except for δ. Sample size can also be determined by fixing all parameters except for σ and select an appropriate n for controlling the variability associated with the reference product. We will refer to these sample size calculations as sample size calculation controlling single parameter as described in (8.33). In some cases, the investigators may be interested in controlling two parameters such as δ and σ simultaneously for sample size calculation. Sample size calculation controlling δ and σ at the same time is also known as reproducibility analysis for sample size calculation. It should be noted that the required sample size will increase if one attempts to control more parameters at the same time for sample size calculation, which, however, may not be feasible in most clinical trials due to huge sample size required.

Thus, in clinical trials, power calculation is not the only way for sample size calculation. Based on (8.33), different approaches controlling single parameter or multiple parameters may be applied for sample size calculation. In addition, one may consider sample size calculation based on probability statement such as probability monitoring procedure for clinical studies with extremely low incidence rates or rare diseases drug development. Sample size determination based on probability monitoring procedure will be discussed in Chapter 18.

8.7.2 Instability of Sample Size

As discussed in Section 8.2, in clinical research, power analysis for sample size calculation is probably the most popular method for sample size determination. However, Chow (2011) indicated that sample size calculation based on estimate of σ^2/δ^2 might not be stable as expected. It can be verified that the asymptotic bias of $E(\hat{\theta} = s^2 / \hat{\delta}^2)$ is given by

$$E\left(\hat{\theta}\right) - \theta = N^{-1}\left(3\theta^2 - \theta\right) = 3N^{-1}\theta^2\left\{1 + o(1)\right\}.$$

Alternatively, it is suggested that the median of $s^2/\hat{\delta}^2$, i.e., $P(s^2/\hat{\delta}^2 \le \eta_{0.5}) = 0.5$ be considered. It can be shown that the asymptotic bias of the median of $s^2/\hat{\delta}^2$ is given by

$$\eta_{0.5} - \theta = -1.5N^{-1}\theta\{1 + o(1)\},$$

whose leading term is linear in θ. As it can be seen that bias of the median approach can be substantially smaller than the mean approach for a small sample size and/or small effect size. However, in practice, we usually do not know the exact value of the median of $s^2/\hat{\delta}^2$. In this case, a bootstrap approach in conjunction with a Bayesian approach may be useful.

8.7.3 Sample Size Adjustment for Protocol Amendment

In practice, it is not uncommon that there is a shift in target patient population due to protocol amendments. In this case, sample size is necessarily adjusted for achieving a desired power for correctly detecting a clinically meaningful difference with respect to the original target patient population. One of the most commonly employed approaches is to consider adjusting the sample size based on the change in effect size.

$$N_1 = \min\left\{N_{\max}, \max\left(N_{\min}, sign(E_0 E_1)\left|\frac{E_0}{E_1}\right|^{\alpha} N_0\right)\right\},$$

where N_0 and N_1 are the required original sample size before population shift and the adjusted sample size after population shift, respectively, $N_{\max}$ and $N_{\min}$ are the maximum and minimum sample sizes, α is a constant that is usually selected so that the sensitivity index $\Delta = \left|\frac{E_0}{E_1}\right|$ is within an acceptable range, and $sign(x) = 1$ for $x > 0$; otherwise $sign(x) = -1$.

The sensitivity index Δ regarding random target patient population due to protocol amendments and extrapolation is further studied in Chapter 10.

8.7.4 Sample Size Based on Confidence Interval Approach

As discussed in Chapter 3, the concepts of interval hypotheses testing and confidence interval approach for bioequivalence assessment for generics and similarity evaluation for biosimilars are different. The two one-sided tests (TOST) procedure for interval hypotheses testing is the official method recommended by the FDA. However, the 90% confidence interval approach, which is operationally equivalent to TOST under certain conditions, is often mix used. In this case, it is of interest to evaluate sample size requirement based on the 90% confidence interval. Base on the 90% confidence interval approach, sample size should be determined for achieving a desired probability that the constructed 90% confidence interval is totally within the equivalence limit or similarity margin. That is, an appropriate sample size should be selected for achieving a desired probability p based on the following probability statement:

$$p = \{90\%\text{CI} \sqsubset [\delta_L, \delta_U]\},$$

where $[\delta_L, \delta_U]$ is the bioequivalence limit or similarity margin. Noted that it can be verified that the above probability statement is not the same as that of the power function of the TOST for testing interval hypotheses based on $[\delta_L, \delta_U]$.

8.8 Concluding Remarks

In summary, sample size calculation plays an important role in clinical research. The purpose of sample size calculation is not only to ensure that there are sufficient or minimum required subjects enrolled in the study for providing substantial evidence of safety and efficacy of the test treatment under investigation, but also to identify any signals or trends—with some assurance—of other clinical benefits to the patient population under study. Since clinical trial is a lengthy and complicated process, standard procedures that fit all types of clinical trials conducted under different trial designs have been developed. In practice, under a valid study design, sample size is usually selected for achieving study objectives (such as detecting a clinically meaningful difference or treatment effect of a given primary study endpoint) at a pre-specified level of significance. Different types of clinical trials (e.g., non-inferiority/equivalence trials, superiority trials, dose-response trials, combinational trials, bridging studies, and vaccine clinical trials) are often conducted for different purposes of clinical investigations under different study designs (e.g., parallel-group design, crossover design, cluster randomized design, titration design, enrichment design, group sequential design, blinded reader design, designs for cancer research, and adaptive clinical trial designs). Thus, different procedures may be employed for achieving study objectives with certain desirable statistical inferences. Sample size calculation should be performed under an appropriate statistical test (derived under the null hypothesis) and evaluate under the alternative hypothesis for an accurate and reliable assessment of the clinically meaningful difference or treatment effect with a desired statistical power at a pre-specified level of significance GSP and GCP.

For clinical trials utilizing complex innovative designs such as multiple adaptive designs, statistical methods may not be well established. Thus, there may exist no formulae or procedures for sample size calculation/allocation at the planning stage of protocol development. In this case, it is suggested that a clinical trial simulation be conducted for obtaining the required sample size for achieving the study objectives with a desired power or statistical inference/assurance. Clinical trial simulation is a process that uses computers to mimic the conduct of a clinical trial by creating virtual patients and calculating (or predicting) clinical outcomes for each virtual patient based on pre-specified models. It should be noted that although clinical trial simulation does provide *"a"* solution (not *"the"* solution) to sample size calculation under complicated study designs, it is useful only when based on a well-established predictive model under certain assumptions, which are often difficult, if it is not impossible, to be verified. In addition, *"How to validate the assumed predictive model for clinical trial simulation?"* is often a major challenge to both investigators and biostatisticians.

9

Reproducible Research

9.1 Introduction

In clinical research and development, it is always a concern to the principal investigator that (i) the research finding does *not* reach statistical significance, i.e., it is purely by chance alone, and (ii) the significant research finding is *not* reproducible under the same experimental conditions with the same experimental units. Typical examples include (i) results from genomic studies for screening of relevant genes as predictors of clinical outcomes for building of a medical predictive model for critical and/or life-threatening diseases are often not reproducible and (ii) clinical results from two pivotal trials for demonstration of safety and efficacy of a test treatment under investigation are not consistent. In practice, it is then of particular interest to assess the validity/reliability and reproducibility of the research findings obtained from studies conducted in pharmaceutical and/or clinical research and development.

For genomic studies, thousands of genes are usually screened for selecting a handful of genes that are most relevant to clinical outcomes of a test treatment for treating some critical and/or life threatening diseases such as cancer. These identified genes, which are considered risk factors or predictors, will then be used for building a medical predictive model for the critical and/or life threatening diseases. A validated medical predictive model can definitely benefit patients with the diseases under study. In practice, it is not uncommon that different statistical methods may lead to different conclusions based on the same data set, i.e., different methods may select different group of genes that are predictive of clinical outcomes. The investigator often struggles with the situation that (i) which set of genes should be reported, and (ii) why the results are not reproducible. Some researchers attribute this to (i) the method is not validated and (ii) there is considerable variability (fluctuation) in data. Thus, it is suggested that necessary actions be taken to identify possible causes of variabilities and eliminate/control the identified variabilities whenever possible. In addition, it is suggested the method should be validated before it is applied to the clean and quality database.

For approval of a test drug product, the FDA requires two pivotal stud-ies be conducted (with the same patient population under the same study protocol) in order to provide substantial evidence of safety and efficacy of the test drug product under investigation. The purpose for two pivotal tri-als is to assure that the positive results (e.g., *p*-value is less than the nominal level of 5%) are *reproducible* with the same patient population under study. Statistically, there is higher probability of observing positive results of future study provided that positive results were observed in two independent tri-als as compare to that of observing positive results provided that positive results were observed in one single trial. In practice, however, it is a con-cern whether two positive pivotal trials can guarantee whether the positive results of future studies are reproducible if the study shall be repeatedly conducted with the same patient population.

In clinical research, it is often suggested that testing and/or statistical pro-cedure be validated for reducing possible deviation/fluctuation in research findings to increase the creditability of the research findings in terms of accuracy and reliability. This, however, does not address the question that whether the current observed research findings are reproducible if the study were conducted repeatedly under same or similar experimental conditions with the same patient population. In this chapter, we recommend the use of a Bayesian approach for assessment of the reproducibility of clinical research. In other words, the variability (or degree of fluctuation) in research findings is first evaluated followed by the assessment of reproducibility probability based on the observed variability (Shao and Chow, 2002). The suggested method provides certain assurance regarding the degree of reproducibility of the observed research findings if the study shall be conducted under the same experimental conditions and target patient population.

In the next section, the concept of reproducibility probability is briefly out-lined. Section 9.3 introduces the estimated power approach for assessment of reproducibility probability. Alternative methods for evaluation of reproduc-ibility probability are discussed in Section 9.4. Some applications are given in Section 9.5. Section 9.6 provides future perspectives regarding reproducible research in clinical development.

9.2 The Concept of Reproducibility Probability

In practice, reliability, repeatability, and reproducibility of research findings are related to various sources of variability such as intra-subject (experi-mental unit) variability, inter-subject variability, and variability due to subject-by-treatment interaction and so on during the pharmaceutical and/or clinical development process. To achieve the desired reliability, repeatability, and reproducibility of research findings, we will need to identify, eliminate, and control possible sources of variability. Chow and Liu (2013) classified possible sources of variability into four categories: (i) expected and controllable (e.g., a new equipment or technician); (ii) expected but uncontrollable (e.g., a new dose or treatment duration); (iii) unexpected but controllable (e.g., compliance); and (iv) unexpected and uncontrollable (e.g., pure random error). In pharmaceutical/clinical research and development, these sources of variability are often monitored through some variability (control) charts for statistical quality assurance and control (QA/QC) (see, e.g., Barrentine, 1991; JMP, 2012). The selection of acceptance limits, however, is critical to the success of these control charts. Following the idea of Shao and Chow (2002), Salah et al. (2017) proposed the concept of reproducibility based on an empirical power for evaluation of the degree of reliability, repeatability, and reproducibility which may be useful for determining the acceptance limits for monitoring reliability, repeatability, and reproducibility in variability control charts.

As mentioned in Section 9.1, for marketing approval of a new drug product, the FDA requires that at least two adequate and well-controlled clinical trials be conducted to provide substantial evidence regarding the safety and effectiveness of the drug product under investigation. The purpose of conducting the second trial is to study whether the observed clinical result from the first trial is reproducible on the same target patient population. Let H_0 be the null hypothesis that the mean response of the drug product is the same as the mean response of a control (for example, placebo) and H_a be the alternative hypothesis. An observed result from a clinical trial is said to be significant if it leads to the rejection of H_0. It is often of interest to determine whether clinical trials that produced significant clinical results provide substantial evidence to assure that the results will be reproducible in a future clinical trial with the same study protocol. Under certain circumstances, the FDA Modernization Act (FDAMA) of 1997 includes a provision (Section 115 of FDAMA) to allow data from one adequate and well-controlled clinical trial investigation and confirmatory evidence to establish effectiveness for risk/benefit assessment of drug and biological candidates for approval. Suppose that the null hypothesis H_0 is rejected if and only if $|T| > c$, where T is a test statistic and c is a positive critical value. In statistical theory, the probability of observing a significant clinical result when H_a is indeed true is referred to as the power of the test procedure. If the statistical model under H_a is a parametric model, then the power is

$$P\left(reject\, H_0 \mid H_a\right) = P(|T| > c \mid H_a) = P(|T| > c \mid \theta), \tag{9.1}$$

where θ is an unknown parameter or a vector of parameters under H_a. Suppose that one clinical trial has been conducted and the result is significant. What is the probability that the second trial will produce a significant result, that is, the significant result from the first trial is reproducible? Statistically, if the two trials are independent, the probability of observing a significant result from the second trial when H_a is true is still given by Equation (9.1), regardless of whether the result from the first trial is significant or not. However, information from the first clinical trial should be useful in the evaluation of the probability of observing a significant result in the second trial. This leads to the concept of reproducibility probability, which is different from the power defined by (9.1).

In general, the reproducibility probability is a person's subjective probability of observing a significant clinical result from a future trial, when he/she observes significant results from one or several previous trials. Goodman (1992) considered the reproducibility probability as the probability in (9.1) with θ replaced by its estimate based on the data from the previous trial. In other words, the reproducibility probability can be defined as an estimated power of the future trial using the data from the previous trial.

When the reproducibility probability is used to provide an evidence of the effectiveness of a drug product, the estimated power approach may produce a rather optimistic result. A more conservative approach is to define the reproducibility probability as a lower confidence bound of the power of the second trial. Perhaps a more sensible definition of reproducibility probability can be obtained by using the Bayesian approach. Under the Bayesian approach, the unknown parameter θ is a random vector with a prior distribution $\pi(\theta)$ assumed to be known. Thus, the reproducibility probability can be defined as the conditional probability of $|T| > c$ in the future trial, given the data set x observed from the previous trial(s), that is,

$$P(\lceil T \rceil > c|x) = \int P(\lceil T \rceil > c|\theta)\pi(\theta|x)d\theta, \tag{9.2}$$

where $T = T(y)$ is based on the data set y from the future trial and $\pi(\theta|x)$ is the posterior density of θ, given x.

9.3 The Estimated Power Approach

To study the reproducibility probability, we need to specify the test procedure, that is, the form of the test statistic T. In what follows, we consider several different study designs.

9.3.1 Two Samples with Equal Variances

Suppose that a total of $n = n_1 + n_2$ patients are randomly assigned to two groups, a treatment group and a control group. In the treatment group, n_1 patients receive the treatment (or a test drug) and produce responses $x_{11}, \ldots, x_{1n_1}$. In the control group, n_2 patients receive the placebo (or a reference drug) and produce responses $x_{21},\dots,x_{2n_2}$. This design is a typical two-group parallel design in clinical trials. Assume that x_{ij}'s are independent and normally distributed with means $\mu_i, i = 1, 2$, and a common variance σ^2. Suppose that the hypotheses of interest are

$$H_0 : \mu_1 - \mu_2 = 0 \text{ versus } H_a : \mu_1 - \mu_2 \neq 0. \tag{9.3}$$

Similar discussion applies for the case of a one-sided H_a. Consider the commonly used two-sample t-test that rejects H_0 if and only if $\lceil T \rceil > t_{0.975;n-2}$, where $t_{0.975;n-2}$ is the 97.5th percentile of the t-distribution with $n - 2$ degrees of freedom

$$T = \frac{\bar{x}_1 - \bar{x}_2}{\sqrt{\dfrac{(n_1 - 1)s_1^2 + (n_2 - 1)s_2^2}{n-2}}\sqrt{\dfrac{1}{n_1} + \dfrac{1}{n_2}}}, \tag{9.4}$$

and $\bar{x}_i$ and s_i^2 are the sample mean and variance, respectively, based on the data from the ith treatment group. The power of T for the second trial is

$$
\begin{aligned}
p(\theta) &= P\left(\lceil T(y) \rceil > t_{0.975;n-2}\right) \\
&= 1 - \mathcal{T}_{n-2}\left(t_{0.975;n-2} \mid \theta\right) + \mathcal{T}_{n-2}\left(-t_{0.975;n-2} \mid \theta\right),
\end{aligned}
\tag{9.5}
$$

where

$$\theta = \frac{\mu_1 - \mu_2}{\sigma\sqrt{\dfrac{1}{n_1} + \dfrac{1}{n_2}}}, \tag{9.6}$$

and $\mathcal{T}_{n-2}(\cdot \mid \theta)$ denotes the distribution function of the non-central t-distribution with $n-2$ degrees of freedom and the non-centrality parameter θ. Note that $p(\theta) = p(|\theta|)$. Values of $p(\theta)$ as a function of $|\theta|$ is provided in Table 9.1.

Replacing θ by its estimate $T(x)$, where T is defined by (9.4), reproducibility probability can be obtained as follows

$$\hat{P} = 1 - \mathcal{T}_{n-2}\left(t_{0.975;n-2} \mid T(x)\right) + \mathcal{T}_{n-2}\left(-t_{0.975;n-2} \mid T(x)\right), \tag{9.7}$$

which is a function of $|T(x)|$. When $|T(x)| > t_{0.975;n-2}$,

$$\hat{P} \approx \begin{cases} 1 - \mathcal{T}_{n-2}\left(t_{0.975;n-2} \mid T(x)\right) & \text{if } T(x) > 0 \\ \mathcal{T}_{n-2}\left(-t_{0.975;n-2} \mid T(x)\right) & \text{if } T(x) < 0 \end{cases} \tag{9.8}$$

TABLE 9.1

Values of the Power Function $p(\theta)$ in (9.5)

$|\theta|$	Total Sample Size							
	10	20	30	40	50	60	100	∞
1.96	0.407	0.458	0.473	0.480	0.484	0.487	0.492	0.500
2.02	0.429	0.481	0.496	0.504	0.508	0.511	0.516	0.524
2.08	0.448	0.503	0.519	0.527	0.531	0.534	0.540	0.548
2.14	0.469	0.526	0.542	0.550	0.555	0.557	0.563	0.571
2.20	0.490	0.549	0.565	0.573	0.578	0.581	0.586	0.594
2.26	0.511	0.571	0.588	0.596	0.601	0.604	0.609	0.618
2.32	0.532	0.593	0.610	0.618	0.623	0.626	0.632	0.640
2.38	0.552	0.615	0.632	0.640	0.645	0.648	0.654	0.662
2.44	0.573	0.636	0.654	0.662	0.667	0.670	0.676	0.684
2.50	0.593	0.657	0.675	0.683	0.688	0.691	0.697	0.705
2.56	0.613	0.678	0.695	0.704	0.708	0.711	0.717	0.725
2.62	0.632	0.698	0.715	0.724	0.728	0.731	0.737	0.745
2.68	0.652	0.717	0.735	0.743	0.747	0.750	0.756	0.764
2.74	0.671	0.736	0.753	0.761	0.766	0.769	0.774	0.782
2.80	0.690	0.754	0.771	0.779	0.783	0.786	0.792	0.799
2.86	0.708	0.772	0.788	0.796	0.800	0.803	0.808	0.815
2.92	0.725	0.789	0.805	0.812	0.816	0.819	0.824	0.830
2.98	0.742	0.805	0.820	0.827	0.831	0.834	0.839	0.845
3.04	0.759	0.820	0.835	0.842	0.846	0.848	0.853	0.860
3.10	0.775	0.834	0.849	0.856	0.859	0.862	0.866	0.872
3.16	0.790	0.848	0.862	0.868	0.872	0.874	0.879	0.884
3.22	0.805	0.861	0.874	0.881	0.884	0.886	0.890	0.895
3.28	0.819	0.873	0.886	0.892	0.895	0.897	0.901	0.906
3.34	0.832	0.884	0.897	0.902	0.905	0.907	0.911	0.916
3.40	0.844	0.895	0.907	0.912	0.915	0.917	0.920	0.925
3.46	0.856	0.905	0.916	0.921	0.924	0.925	0.929	0.932
3.52	0.868	0.914	0.925	0.929	0.932	0.933	0.936	0.940
3.58	0.879	0.923	0.933	0.937	0.939	0.941	0.943	0.947
3.64	0.889	0.931	0.940	0.944	0.946	0.947	0.950	0.953
3.70	0.898	0.938	0.946	0.950	0.952	0.953	0.956	0.959
3.76	0.907	0.944	0.952	0.956	0.958	0.959	0.961	0.965
3.82	0.915	0.950	0.958	0.961	0.963	0.964	0.966	0.969
3.88	0.923	0.956	0.963	0.966	0.967	0.968	0.970	0.973
3.94	0.930	0.961	0.967	0.970	0.971	0.972	0.974	0.977

Source: Shao, J. and Chow, S.C., *Stat. Med.*, 21, 1727–1742, 2002.

If T_{n-2} is replaced by the normal distribution and $t_{0.975;n-2}$ is replaced by the normal percentile, then (9.8) is the same as that in Goodman (1992) who studied the case where the variance σ^2 is known. Note that Table 9.1 can be used to find the reproducibility probability $\hat{P}$ in (9.7) with a fixed sample size n. As an example, if $T(x) = 2.9$ was observed in a clinical trial with $n = n_1 + n_2 = 40$, then the reproducibility probability is 0.807. If $T(x) = 2.9$ was observed in a clinical trial with $n = 36$, then an extrapolation of the results in Table 9.1 (for $n = 30$ and 40) leads to a reproducibility probability of 0.803.

9.3.2 Two Samples with Unequal Variances

Consider the problem of testing hypotheses (9.3) under the two-group parallel design without the assumption of equal variances. That is, x_{ij}'s are independently distributed as $N(\mu_i, \sigma_i^2), i = 1, 2$. When $\sigma_1^2 \neq \sigma_2^2$, there exists no exact testing procedure for the hypotheses in (9.3). When both n_1 and n_2 are large, an approximate 5% level test rejects H_0 when $|T| > z_{0.975}$, where

$$T = \frac{\bar{x}_1 - \bar{x}_2}{\sqrt{\dfrac{s_1^2}{n_1} + \dfrac{s_2^2}{n_2}}}. \tag{9.9}$$

Since T is approximately distributed as $N(\theta, 1)$ with

$$\theta = \frac{\mu_1 - \mu_2}{\sqrt{\dfrac{\sigma_1^2}{n_1} + \dfrac{\sigma_2^2}{n_2}}} \tag{9.10}$$

the reproducibility probability obtained by using the estimated power approach is given by

$$\hat{P} = \Phi\left(T(x) - z_{0.975}\right) + \Phi\left(-T(x) - z_{0.975}\right). \tag{9.11}$$

When the variances under different treatments are different and the sample sizes are not large, a different study design, such as a matched-pair parallel design or a 2×2 crossover design is recommended. A matched-pair parallel design involves m pairs of matched patients. One patient in each pair is assigned to the treatment group and the other is assigned to the control group. Let x_{ij} be the observation from the jth pair and the ith group. It is assumed that the differences $x_{1j} - x_{2j}$, $j = 1, \ldots, m$, are independent and identically distributed as $N(\mu_1 - \mu_2, \sigma_D^2)$. Then, the null hypothesis H_0 is rejected at the 5% level of significance if $|T| > t_{0.975;m-1}$, where

$$T = \frac{\sqrt{m}\left(\bar{x}_1 - \bar{x}_2\right)}{\hat{\sigma}_D} \tag{9.12}$$

and σ_D^2 is the sample variance based on the differences $x_{1j} - x_{2j}$, $j = 1,\ldots,m$. Note that T has the non-central t-distribution with $m-1$ degrees of freedom and the non-centrality parameter

$$\theta = \frac{\sqrt{m}\left(\mu_1 - \mu_2\right)}{\sigma_D}. \tag{9.13}$$

Consequently, the reproducibility probability obtained by using the estimated power approach is given by (9.7) with T defined in (9.12) and $n-2$ replaced by $m-1$.

Suppose that the study design is a 2×2 cross-over design in which n_1 patients receive the treatment at the first period and the placebo at the second period and n_2 patients receive the placebo at the first period and the treatment at the second period. Let x_{lij} be the normally distributed observation from the jth patient at the ith period and lth sequence. Then the treatment effect μ_D can be unbiasedly estimated by

$$\hat{\mu}_D = \frac{\bar{x}_{11} - \bar{x}_{12} - \bar{x}_{21} + \bar{x}_{22}}{2} \quad \sim N\left(\mu_D, \frac{\sigma_D^2}{4}\left(\frac{1}{n_1} + \frac{1}{n_2}\right)\right),$$

where $\bar{x}_{ij}$ is the sample mean based on $x_{lij}, j = 1,\ldots,n_l$ and $\sigma_D^2 = var\left(x_{l1j} - x_{l2j}\right)$. An unbiased estimator of σ_D^2 is

$$\hat{\sigma}_D^2 = \frac{1}{n_1 + n_2 - 2} \sum_{l=1}^{2}\sum_{j=1}^{n_l}(x_{l1j} - x_{l2j} - \bar{x}_{l1} + \bar{x}_{l2})^2,$$

which is independent of $\hat{\mu}_D$ and distributed as $\sigma_D^2 / \left(n_1 + n_2 - 2\right)$ times the chi-square distribution with $n_1 + n_2 - 2$ degrees of freedom. Thus, the null hypothesis $H_0 : \mu_D = 0$ is rejected at the 5% level of significance if $|T| > t_{0.975;n-2}$, where $n = n_1 + n_2$ and

$$T = \frac{\hat{\mu}_D}{\frac{\hat{\sigma}_D}{2}\sqrt{\frac{1}{n_1} + \frac{1}{n_2}}}. \tag{9.14}$$

Note that T has the non-central t-distribution with $n-2$ degrees of freedom and the non-centrality parameter

$$\theta = \frac{\mu_D}{\frac{\sigma_D}{2}\sqrt{\frac{1}{n_1} + \frac{1}{n_2}}}. \tag{9.15}$$

Consequently, the reproducibility probability obtained by using the estimated power approach is given by (9.7) with T defined by (9.14).

9.3.3 Parallel-Group Designs

Parallel-group designs are often adopted in clinical trials to compare more than one treatment with a placebo control or to compare one treatment, one placebo control and one active control. Let $a \geq 3$ be the number of groups and x_{ij} be the observation from the jth patient in the ith group, $j = 1, \ldots, n_i$, $i = 1, \ldots, a$. Assume that x_{ij}'s are independently distributed as $N(\mu_i, \sigma^2)$. The null hypothesis H_0 is then

$$H_0 : \mu_1 = \mu_2 = \ldots = \mu_a$$

which is rejected at the 5% level of significance if $T > F_{0.95;a-1,n-a}$, where $F_{0.95;a-1,n-a}$ is the 95th percentile of the F-distribution with $a - 1$ and $n - a$ degrees of freedom, $n = n_1 + \ldots + n_a$

$$T = \frac{SST/(a-1)}{SSE/(n-a)}, \tag{9.16}$$

$$SST = \sum_{i=1}^{a} n_i(\bar{x}_i - \bar{x})^2,$$

$$SSE = \sum_{i=1}^{a} \sum_{j=1}^{n_i} (x_{ij} - \bar{x}_i)^2,$$

$\bar{x}_i$ is the sample mean based on the data in the ith group, and $\bar{x}$ is the overall sample mean. Note that T has the non-central F-distribution with $a - 1$ and $n - a$ degrees of freedom and the non-centrality parameter

$$\theta = \sum_{i=1}^{a} \frac{n_i(\mu_i - \bar{\mu})^2}{\sigma^2}, \tag{9.17}$$

where $\bar{\mu} = \sum_{i=1}^{a} n_i \mu_i / n$. Let $\mathcal{F}_{a-1,n-a}(\cdot \mid \theta)$ be the distribution of T. Then, the power of the second clinical trial is

$$P(T(y) > F_{0.95;a-1,n-a}) = 1 - \mathcal{F}_{a-1,n-a}(F_{0.95;a-1,n-a} \mid \theta).$$

Thus, the reproducibility probability obtained by using the estimated power approach is

$$\hat{P} = 1 - \mathcal{F}_{a-1,n-a}(F_{0.95;a-1,n-a} \mid T(x)),$$

where $T(x)$ is the observed T based on the data x from the first clinical trial.

9.4 Alternative Methods for Evaluation of Reproducibility Probability

Since $\hat{P}$ in (9.7) or (9.11) is an estimated power, it provides a rather optimistic result. Alternatively, we may consider a more conservative approach, which considers a 95% lower confidence bound of the power as the reproducibility probability. In addition, we may also consider Bayesian approach for evaluation of reproducibility probability.

9.4.1 The Confidence Bound Approach

Consider first the case of the two-group parallel design with a common unknown variance σ^2. Note that $T(x)$ defined by (9.4) has the non-central t-distribution with $n-2$ degrees of freedom and the non-centrality parameter θ given by (9.6). Let $\mathcal{T}_{n-2}(\cdot \mid \theta)$ be the distribution function of $T(x)$ for any given θ. It can be shown that $\mathcal{T}_{n-2}(t \mid \theta)$ is a strictly decreasing function of θ for any fixed t. Consequently, a 95% confidence interval for θ is $\left(\hat{\theta}_-, \hat{\theta}_+\right)$, where $\hat{\theta}_-$ is the unique solution of $\mathcal{T}_{n-2}\left(T(x) \mid \theta\right) = 0.975$ and $\hat{\theta}_+$ is the unique solution of $\mathcal{T}_{n-2}\left(T(x) \mid \theta\right) = 0.025$. Then a 95% lower confidence bound for $|\theta|$ is

$$
|\hat{\theta}| = \begin{cases} \hat{\theta}_- & \text{if } \hat{\theta}_- > 0 \\ -\hat{\theta}_+ & \text{if } \hat{\theta}_+ < 0 \\ 0 & \text{if } \hat{\theta}_- \leq 0 \leq \hat{\theta}_+ \end{cases} \tag{9.18}
$$

and a 95% lower confidence bound for the power $p(\theta)$ in (9.5) is

$$
\hat{P}_- = 1 - \mathcal{T}_{n-2}\left(t_{0.975;n-2} \mid |\theta|_-\right) + \mathcal{T}_{n-2}\left(-t_{0.975;n-2} \mid |\theta|_-\right) \tag{9.19}
$$

If $|\theta|_- > 0$ and $\hat{P}_- = 0$ if $|\theta|_- = 0$. The lower confidence bound in (9.19) is useful when the clinical result from the first trial is highly significant. To provide a better understanding, values of the lower confidence bound $|\theta|_-$ corresponding to $|T(x)|$ values ranging from 4.5 to 6.5 are summarized in Table 9.2.

If $4.5 \leq |T(x)| \leq 6.5$ and the value of $|\theta|_-$ is found from Table 9.2, the reproducibility probability $\hat{P}_-$ (9.19) can be obtained from Table 9.1. For example, suppose that $|T(x)| = 5$ was observed from a clinical trial with $n = 30$. From Table 9.2, $|\theta|_- = 2.6$. Then, by Table 9.1, $\hat{P}_- = 0.709$. Consider the two-group parallel design with unequal variances σ_1^2 and σ_2^2. When both n_1 and n_2 are large, T given by (9.9) is approximately distributed as $N(\theta, 1)$ with θ given by (9.10). Hence, the reproducibility probability obtained by using the lower confidence bound approach is given by

TABLE 9.2

95% Lower Confidence Bound $|\hat{\theta}|_-$

$|T(x)|$	Total Sample Size							
	10	20	30	40	50	60	100	∞
4.5	1.51	2.01	2.18	2.26	2.32	2.35	2.42	2.54
4.6	1.57	2.09	2.26	2.35	2.41	2.44	2.52	2.64
4.7	1.64	2.17	2.35	2.44	2.50	2.54	2.61	2.74
4.8	1.70	2.25	2.43	2.53	2.59	2.63	2.71	2.84
4.9	1.76	2.33	2.52	2.62	2.68	2.72	2.80	2.94
5.0	1.83	2.41	2.60	2.71	2.77	2.81	2.90	3.04
5.1	1.89	2.48	2.69	2.80	2.86	2.91	2.99	3.14
5.2	1.95	2.56	2.77	2.88	2.95	3.00	3.09	3.24
5.3	2.02	2.64	2.86	2.97	3.04	3.09	3.18	3.34
5.4	2.08	2.72	2.95	3.06	3.13	3.18	3.28	3.44
5.5	2.14	2.80	3.03	3.15	3.22	3.27	3.37	3.54
5.6	2.20	2.88	3.11	3.21	3.31	3.36	3.47	3.64
5.7	2.26	2.95	3.20	3.32	3.40	3.45	3.56	3.74
5.8	2.32	3.03	3.28	3.41	3.49	3.55	3.66	3.84
5.9	2.39	3.11	3.37	3.50	3.58	3.64	3.75	3.94
6.0	2.45	3.19	3.45	3.59	3.67	3.73	3.85	4.04
6.1	2.51	3.26	3.53	3.67	3.76	3.82	3.94	4.14
6.2	2.57	3.34	3.62	3.76	3.85	3.91	4.03	4.24
6.3	2.63	3.42	3.70	3.85	3.94	4.00	4.13	4.34
6.4	2.69	3.49	3.78	3.93	4.03	4.09	4.22	4.44
6.5	2.75	3.57	3.86	4.02	4.12	4.18	4.32	4.54

Source: Shao, J. and Chow, S.C., *Stat. Med.*, 21, 1727–1742, 2002.

$$\hat{P}_- = \Phi\left(\left|T(x)\right| - 2z_{0.975}\right)$$

with T defined by (9.9).

For the matched-pair parallel design, T given by (9.12) has the non-central t-distribution with $m - 1$ degrees of freedom and the non-centrality parameter θ given by (9.13). Hence, the reproducibility probability obtained by using the lower confidence bound approach is given by (9.19) with T defined by (9.12) and $n - 2$ replaced by $m - 1$. Suppose now that the study design is the 2×2 cross-over design. Since T defined by (9.14) has the non-central t-distribution with $n - 2$ degrees of freedom and the non-centrality parameter θ given by (9.15), the reproducibility probability obtained by using the lower confidence bound approach is given by (9.19) with T defined by (9.14).

Finally, consider the parallel-group design, since T in (9.16) has the non-central F-distribution with $a - 1$ and $n - a$ degrees of freedom and the non-centrality parameter θ given by (9.17) and $\mathcal{F}_{a-1,n-a}(t\,|\,\theta)$ is a strictly decreasing function of θ, the reproducibility probability obtained by using the lower confidence bound approach is

$$\hat{P}_- = 1 - \mathcal{F}_{a-1,n-a}(F_{0.95;a-1,n-a} \mid \hat{\theta}_-),$$

where $\hat{\theta}_-$ is the solution of $\mathcal{F}_{a-1,n-a}\big(T(x) \mid \theta\big) = 0.95$.

9.4.2 The Bayesian Approach

Shao and Chow (2002) studied how to evaluate the reproducibility probability using Equation (9.1) under several study designs. When the reproducibility probability is used to provide an evidence of the effectiveness of a drug product, the estimated power approach may produce a rather optimistic result. A more conservative approach is to define the reproducibility probability as a lower confidence bound of the power of the second trial. Alternatively, a more sensible definition of reproducibility probability can be obtained by using the Bayesian approach. Under the Bayesian approach, the unknown parameter θ is a random vector with a prior distribution $\pi(\theta)$ assumed to be known. Thus, the reproducibility probability can be defined as the conditional probability of $|T| > c$ in the future trial, given the data set x observed from the previous trial, that is,

$$P(|T| > c \mid x) = \int P(|T| > c \mid \theta)\pi(\theta \mid x)d\theta,$$

where $T = T(y)$ is based on the data set y from the future trial and $\pi(\theta \mid x)$ is the posterior density of θ, given x. In practice, the reproducibility probability is useful when the clinical trials are conducted sequentially. It provides important information for regulatory agencies in deciding whether it is necessary to require the second clinical trial when the result from the first clinical trial is strongly significant.

Note that power calculation for required sample size for achieving a desired reproducibility probability at a pre-specified level of significance can be performed with appropriate selection of prior.

As discussed in Section 9.3, the reproducibility probability can be viewed as the posterior mean of the power function $p(\theta) = P(|T| > c \mid \theta)$ for the future trial. Thus, under the Bayesian approach, it is essential to construct the posterior density $\pi(\theta \mid x)$ in formula (9.2), given the data set x observed from the previous trial(s).

Consider first the two-group parallel design with equal variances, that is, x_{ij}'s are independent and normally distributed with means μ_1 and μ_2 and a common variance σ^2. If σ^2 is known, then the power for testing hypotheses in (9.3) is

$$\Phi(\theta - z_{0.975}) + \Phi(-\theta - z_{0.975})$$

with θ defined by (9.6). A commonly used prior for (μ_1, μ_2) is the non-informative prior $\pi(\mu_1, \mu_2) \equiv 1$. Consequently, the posterior density for θ is $N(T(x), 1)$, where

$$T = \frac{\overline{x}_1 - \overline{x}_2}{\sigma\sqrt{\dfrac{1}{n_1} + \dfrac{1}{n_2}}}$$

and the posterior mean given by (9.2) is

$$\int \left[\Phi(\theta - z_{0.975}) + \Phi(-\theta - z_{0.975}) \right] \pi(\theta \mid x)\, d\theta = \Phi\left(\frac{T(x) - z_{0.975}}{\sqrt{2}} \right)$$

$$+ \Phi\left(\frac{-T(x) - z_{0.975}}{\sqrt{2}} \right).$$

When $|T(x)| > z_{0.975}$, this probability is nearly the same as

$$\Phi\left(\frac{|T(x)| - z_{0.975}}{\sqrt{2}} \right),$$

which is exactly the same as that in formula (9.1) in Goodman (1992).

When σ^2 is unknown, a commonly used non-informative prior for σ^2 is the Lebesgue (improper) density $\pi(\sigma^2) = \sigma^{-2}$. Assume that the priors for μ_1, μ_2, and σ^2 are independent. The posterior density for (δ, u^2) is $\pi(\delta \mid u^2, x)\pi(u^2 \mid x)$, where

$$\delta = \frac{\mu_1 - \mu_2}{\sqrt{\dfrac{(n_1 - 1)s_1^2 + (n_2 - 1)s_2^2}{n - 2}}\sqrt{\dfrac{1}{n_1} + \dfrac{1}{n_2}}},$$

$$u^2 = \frac{(n - 2)\sigma^2}{(n_1 - 1)s_1^2 + (n_2 - 1)s_2^2},$$

$$\pi(\delta \mid u^2, x) = \frac{1}{u}\phi\left(\frac{\delta - T(x)}{u} \right),$$

in which ϕ is the density function of the standard normal distribution, T is given by (9.4), and $\pi(u^2 \mid x) = f(u)$ with

$$f(u) = \left[\Gamma\left(\frac{n-2}{2}\right) \right]^{-1} \left(\frac{n-2}{2}\right)^{(n-2)/2} u^{-n} e^{-(n-2)/2u^2}.$$

Since θ in (9.6) is equal to δ/u, the posterior mean of $p(\theta)$ in (9.5) is

$$\hat{P} = \int_0^\infty \left[\int_{-\infty}^\infty p\left(\frac{\delta}{u}\right) \phi\left(\frac{\delta - T(x)}{u}\right) d\delta \right] 2f(u) du, \qquad (9.20)$$

which is the reproducibility probability under the Bayesian approach. It is clear that $\hat{P}$ depends on the data x through the function $T(x)$.

The probability $\hat{P}$ in (9.20) can be evaluated numerically. A Monte Carlo method can be applied as follows. First, generate a random variate γ_j from the gamma distribution with the shape parameter $(n-2)/2$ and the scale parameter $2/(n-2)$, and generate a random variate δ_j from $N(T(x), u_j^2)$, where $u_j^2 = \gamma_j^{-1}$. Repeat this process independently N times to obtain $(\delta_j, u_j^2), j = 1, \ldots, N$. Then $\hat{P}$ in (9.20) can be approximated by

$$\check{P}_N = 1 - \frac{1}{N} \sum_{j=1}^N \left[T_{n-2}\left(t_{0.975;n-2} \left| \frac{\delta_j}{u_j} \right| \right) - T_{n-2}\left(-t_{0.975;n-2} \left| \frac{\delta_j}{u_j} \right| \right) \right]. \qquad (9.21)$$

Values of $\check{P}_N$ for $N = 10,000$ and some selected values of $T(x)$ and n are given in Table 9.3.

As it can be seen from Table 9.3, in assessing reproducibility, the Bayesian approach is more conservative than the estimated power approach, but less conservative than the confidence bound approach.

Consider the two-group parallel design with unequal variance and large $n_j's$. The approximate power for the second trial is

$$p(\theta) = \Phi(\theta - z_{0.975}) + \Phi(-\theta - z_{0.975}),$$

where

$$\theta = \frac{\mu_1 - \mu_2}{\sqrt{\dfrac{\sigma_1^2}{n_1} + \dfrac{\sigma_2^2}{n_2}}}.$$

Suppose that we use the non-inferiority prior density

$$\pi\left(\mu_1, \mu_2, \sigma_1^2, \sigma_2^2\right) = \sigma_1^{-2}\sigma_2^{-2}, \quad \sigma_1^2 > 0, \sigma_2^2 > 0.$$

Let $\tau_i^2 = \sigma_i^{-2}, i = 1,2$ and $\xi^2 = \left(n_1\tau_1^2\right)^{-1} + \left(n_2\tau_2^2\right)^{-1}$. Then, the posterior density $\pi(\mu_1 - \mu_2 \mid \tau_1^2, \tau_2^2, x)$ is the normal density with mean $\bar{x}_1 - \bar{x}_2$ and variance ξ^2 and the posterior density $\pi(\tau_1^2, \tau_2^2 \mid x) = \pi(\tau_1^2 \mid x)\pi(\tau_2^2 \mid x)$, where $\pi(\tau_i^2 \mid x)$ is the

TABLE 9.3

Reproducibility Probability Under the Bayesian Approach Approximated by Monte Carlo Simulation

$\|T(x)\|$	Total Sample Size							
	10	20	30	40	50	60	100	∞
2.02	0.435	0.482	0.495	0.501	0.504	0.508	0.517	0.519
2.08	0.447	0.496	0.512	0.515	0.519	0.523	0.532	0.536
2.14	0.466	0.509	0.528	0.530	0.535	0.543	0.549	0.553
2.20	0.478	0.529	0.540	0.547	0.553	0.556	0.565	0.569
2.26	0.487	0.547	0.560	0.564	0.567	0.571	0.577	0.585
2.32	0.505	0.558	0.577	0.580	0.581	0.587	0.590	0.602
2.38	0.519	0.576	0.590	0.597	0.603	0.604	0.610	0.618
2.44	0.530	0.585	0.610	0.611	0.613	0.617	0.627	0.634
2.50	0.546	0.609	0.624	0.631	0.634	0.636	0.640	0.650
2.56	0.556	0.618	0.638	0.647	0.648	0.650	0.658	0.665
2.62	0.575	0.632	0.654	0.655	0.657	0.664	0.675	0.680
2.68	0.591	0.647	0.665	0.674	0.675	0.677	0.687	0.695
2.74	0.600	0.660	0.679	0.685	0.686	0.694	0.703	0.710
2.80	0.608	0.675	0.690	0.702	0.705	0.712	0.714	0.724
2.86	0.629	0.691	0.706	0.716	0.722	0.723	0.729	0.738
2.92	0.636	0.702	0.718	0.730	0.733	0.738	0.742	0.752
2.98	0.649	0.716	0.735	0.742	0.744	0.748	0.756	0.765
3.04	0.663	0.726	0.745	0.753	0.756	0.759	0.765	0.778
3.10	0.679	0.738	0.754	0.766	0.771	0.776	0.779	0.790
3.16	0.690	0.754	0.767	0.776	0.781	0.786	0.792	0.802
3.22	0.701	0.762	0.777	0.790	0.792	0.794	0.804	0.814
3.28	0.708	0.773	0.793	0.804	0.806	0.809	0.820	0.825
3.34	0.715	0.784	0.803	0.809	0.812	0.818	0.828	0.836
3.40	0.729	0.793	0.815	0.819	0.829	0.830	0.838	0.846
3.46	0.736	0.806	0.826	0.832	0.837	0.839	0.847	0.856
3.52	0.745	0.816	0.834	0.843	0.845	0.846	0.855	0.865
3.58	0.755	0.828	0.841	0.849	0.857	0.859	0.867	0.874
3.64	0.771	0.833	0.854	0.859	0.863	0.865	0.872	0.883
3.70	0.778	0.839	0.861	0.867	0.870	0.874	0.884	0.891
3.76	0.785	0.847	0.867	0.874	0.882	0.883	0.890	0.898
3.82	0.795	0.857	0.878	0.883	0.889	0.891	0.898	0.906
3.88	0.800	0.869	0.881	0.891	0.896	0.899	0.904	0.913
3.94	0.806	0.873	0.890	0.897	0.904	0.907	0.910	0.919

Source: Shao, J. and Chow, S.C., *Stat. Med.*, 21, 1727–1742, 2002.

Note: Prior for $(\mu_1, \mu_2, \sigma^{-2}) = \sigma^{-2}$ with respect to the Lebesgue measure.

gamma density with the shape parameter $(n_i - 1)/2$ and the scale parameter $\frac{2}{[(n_i-1)s_j^2]}$, $i = 1,2$. Consequently, the reproducibility probability is the posterior mean of $p(\theta)$ given by

$$\hat{P} = \int \left[\Phi\left(\frac{\bar{x}_1 - \bar{x}_2}{\sqrt{2\xi}} - \frac{z_{0.975}}{\sqrt{2}} \right) + \Phi\left(-\frac{\bar{x}_1 - \bar{x}_2}{\sqrt{2\xi}} - \frac{z_{0.975}}{\sqrt{2}} \right) \right] \pi(\xi \mid x) d\xi,$$

where $\pi(\xi \mid x)$ is the posterior density of ξ constructed using $\pi(\tau_i^2 \mid x), i = 1,2$. The Monte Carlo method previously discussed can be applied to approximate $\hat{P}$.

Note that reproducibility probability under the Bayesian approach can be similarly obtained for the matched-pairs parallel design and the 2×2 crossover design described in the previous section.

Finally, consider the a-group parallel design, where the power is given by

$$p(\theta) = 1 - \mathcal{F}_{a-1,n-a}(F_{0.95;a-1,n-a} \mid \theta).$$

with θ given by (9.17). Under the non-informative prior

$$\pi\left(\mu_1, \ldots, \mu_a, \sigma^2\right) = \sigma^{-2}, \sigma^2 > 0$$

The posterior density $\pi\left(\theta \mid \tau^2, x\right)$, where $\tau^2 = \frac{SSE}{[(n-a)\sigma^2]}$, is the density of the non-central chi-square distribution with $a-1$ degrees of freedom and the non-centrality parameter $\tau^2(a-1)T(x)$. The posterior density $\pi(\sigma^2 \mid x)$ is the gamma distribution with the shape parameter $(n-a)/2$ and the scale parameter $\frac{2}{n-a}$. Consequently, the reproducibility probability under the Bayesian approach is

$$\hat{P} = \int_0^\infty \left[\int_0^\infty p(\theta)\pi\left(\theta \mid \tau^2, x\right) d\theta \right] \pi\left(\tau^2 \mid x\right) d\tau^2.$$

The reproducibility probability based on the Bayesian approach depends on the choice of the prior distributions. The non-informative prior we choose produces a more conservative reproducibility probability than that obtained using the estimated power approach, but is less conservative than that under the confidence bound approach. If a different prior such as an informative prior is used, a sensitivity analysis may be performed to evaluate the effects of different priors on the reproducibility probability.

9.5 Applications

9.5.1 Substantial Evidence with a Single Trial

An important application of the concept of reproducibility discussed in the previous sections is to address the following question:

> Is it necessary to conduct a second clinical trial when the first trial produces a relatively strong significant clinical result (for example, a relatively small p-value is observed), assuming that other factors (such as consistent results between centers, discrepancies related to gender, race and other factors, and safety issues) have been satisfactory addressed?

As mentioned in Section 9.1, the FDA Modernization Act of 1997 includes a provision (Section 115 of FDAMA) to allow data from one adequate and well-controlled clinical trial investigation and confirmatory evidence to establish effectiveness for risk/benefit assessment of drug and biological candidates for approval. This provision essentially codified an FDA policy that had existed for several years but whose application had been limited to some biological products approved by the Center for Biologic Evaluation and Research (CBER) of the FDA and a few pharmaceuticals, especially orphan drugs such as zidovudine and lamotrigine. A relatively strong significant result observed from a single clinical trial (say, p-value is less than 0.001) would have about 90% chance of reproducing the result in future clinical trials. Consequently, a single clinical trial is sufficient to provide substantial evidence for demonstration of efficacy and safety of the medication under study. In 1998, the FDA published a guidance which shed the light on this approach despite the fact that the FDA has recognized that advances in sciences and practice of drug development may permit an expanded role for the single controlled trial in contemporary clinical development (FDA, 1988).

Suppose it is agreed that the second trial is not needed if the probability for reproducing a significant clinical result in the second trial is equal to or higher than 90%. If a significant clinical result is observed in the first trial and the confidence bound $\hat{P}$ is equal to or higher than 90%, then we have 95% statistical assurance that, with a probability of at least 90%, the significant result will be reproduced in the second trial. As an example, under the two-group parallel design with a common unknown variance and $n = 40$, the 95% lower confidence bound $\hat{P}$ given in (9.19) is equal to or higher than 90% if and only if $|T(x)| \geq 5.7$, that is, the clinical result in the first trial is highly significant. Alternatively, if the Bayesian approach is applied to the same situation, the reproducibility probability in (9.20) is equal to or higher than 90% if and only if $|T(x)| \geq 3.96$.

9.5.2 Sample Size

When the reproducibility probability based on the result from the first trial is not higher than a desired level, the second trial must be conducted in order for obtaining substantial evidence of safety and effectiveness of the test drug under investigation. The results on the reproducibility probability discussed in the previous sections can be used to adjust the sample size for the second trial. If the sample size for the first trial was determined based on a power analysis with some initial guessing values of the unknown parameters, then it is reasonable to make a sample size adjustment for the second trial based on the results from the first trial. If the reproducibility probability is lower than a desired power level of the second trial, then the sample size should be increased. On the other hand, if the reproducibility probability is higher than the desired power level of the second trial, then the sample size may be decreased to reduce costs. In the following we illustrate the idea using the two-group parallel design with a common unknown variance.

Suppose that $\hat{P}$ in (9.7) is used as the reproducibility probability when $T(x)$ given in (9.4) is observed from the first trial. Let $\hat{\sigma}^2 = \frac{[(n_1-1)s_1^2+(n_2-1)s_2^2]}{n-2}$. For simplicity, consider the case where the same sample size $n^*/2$ is used for two treatment groups in the second trial, where n^* is the total sample size in the second trial. With fixed $\bar{x}_i$ and $\hat{\sigma}^2$ but a new sample size n^*, the T-statistic becomes

$$T^* = \frac{\sqrt{n^*}\left(\bar{x}_1 - \bar{x}_2\right)}{2\sigma}$$

and the reproducibility probability is $\hat{P}$ with T replaced by T^*. By letting T^* be the value to achieve a desired power, the new sample size n^* should be

$$n^* = \left(\frac{T^*}{T}\right)^2 / \left(\frac{1}{4n_1} + \frac{1}{4n_2}\right). \tag{9.22}$$

For example, if the desired reproducibility probability is 80%, then T^* needs to be 2.91 (Table 9.1). If $T = 2.58$ is observed in the first trial with $n = 30$ ($n_1 = n_2 = 15$), then $n^* \approx 1.27n \approx 38$ according to (9.22), that is, the sample size should be increased by about 27%. On the other hand, if $T = 3.30$ is observed in the first trial with $n = 30$ ($n_1 = n_2 = 15$), then $n^* \approx 0.78n \approx 24$, that is, the sample size can be reduced by about 22%.

9.5.3 Generalizability between Patient Populations

In clinical development, after the investigational drug product has been shown to be effective and safe with respect to a target patient population (for example, adults), it is often of interest to study a similar but different patient population (for example, elderly patients with the same disease under study or a patient population with different ethnic factors) to see how likely the clinical result is reproducible in the different population. This information is useful in regulatory submission for supplement new drug application (SNDA) (for example, when generalizing the clinical results from adults to elderly patients) and regulatory evaluation for bridging studies (for example, when generalizing clinical results from Gaussian to Asian patient population). For this purpose, we propose to consider the generalizability probability, which is the reproducibility probability with the population of a future trial slightly deviated from the population of the previous trial(s).

Consider a parallel-group design for two treatments with population means μ_1 and μ_2 and an equal variance σ^2. Other designs can be similarly treated. Suppose that in the future trial, the population mean difference is changed to $\mu_1 - \mu_2 + \varepsilon$ and the population variance is changed to $C^2\sigma^2$, where $C > 0$. The signal-to-noise ratio for the population difference in the previous trial is $|\mu_1 - \mu_2|/\sigma$, whereas the signal-to-noise ratio for the population difference in the future trial is

$$\frac{|\mu_1 - \mu_2 + \varepsilon|}{C\sigma} = \frac{|\Delta(\mu_1 - \mu_2)|}{\sigma},$$

where

$$\Delta = \frac{1 + \varepsilon/(\mu_1 - \mu_2)}{C} \tag{9.23}$$

is a measure of change in the signal-to-noise ratio for the population difference. For most practical problems, $|\varepsilon| < |\mu_1 - \mu_2|$ and, thus, $\Delta > 0$. Table 9.4 gives an example on the effects of changes of ε and C on Δ.

If the power for the previous trial is $p(\theta)$, then the power for the future trial is $p(\Delta\theta)$. Suppose that Δ is known. Under the frequentist approach, the generalizability probability is $\hat{P}_\Delta$, which is $\hat{P}$ given by (9.7) with $T(x)$ replaced by $\Delta T(x)$, or $\hat{P}_{\Delta-}$, which is $\hat{P}_-$ given in (9.19) with $|\hat{\theta}|_-$ replaced by $\Delta|\hat{\theta}|_-$. Under the Bayesian approach, the generalizability probability is $\hat{P}_\Delta$, which is $\hat{P}$ given by (9.20) with $p(\delta/u)$ replaced by $p(\Delta\delta/u)$. When the value of Δ is unknown, we may consider a set of Δ-values to carry out a sensitivity analysis. An example is given as follows.

A double-blind randomized trial was conducted in patients with schizophrenia for comparing the efficacy of a test drug with a standard therapy. A total of 104 chronic schizophrenic patients participated in this study. Patients were randomly assigned to receive the treatment of the test drug or the standard therapy for at least one year, where the test drug group has 56 patients and the standard therapy group has 48 patients. The primary clinical endpoint of this trial was the total score of Positive and Negative Symptom Scales (PANSS). No significant differences in demographics and baseline

TABLE 9.4

Effects of Changes in Mean and Standard Deviation

$\varepsilon / (\mu_1 - \mu_2)$	C	Range of Δ
<5%	0.8	1.188–1.313
	0.9	1.056–1.167
	1.0	0.950–1.050
	1.1	0.864–0.955
	1.2	0.792–0.875
	1.3	0.731–0.808
	1.4	0.679–0.750
	1.5	0.633–0.700
≥5% but <10%	0.8	1.125–1.375
	0.9	1.000–1.222
	1.0	0.900–1.100
	1.1	0.818–1.000
	1.2	0.750–0.917
	1.3	0.692–0.846
	1.4	0.643–0.786
	1.5	0.600–0.733
≥10% but <20%	0.8	1.000–1.500
	0.9	0.889–1.333
	1.0	0.800–1.200
	1.1	0.727–1.091
	1.2	0.667–1.000
	1.3	0.615–0.923
	1.4	0.571–0.857
	1.5	0.533–0.800

Source: Shao, J. and Chow, S.C., *Stat. Med.*, 21, 1727–1742, 2002.

characteristics were observed for baseline comparability. Mean changes from baseline in total PANSS for the test drug and the standard therapy are $\bar{x}_1 = -3.51$ and $\bar{x}_2 = 1.41$, respectively, with $s_1^2 = 76.1$ and $s_2^2 = 74.86$. The difference $\mu_1 - \mu_2$ is estimated by $\bar{x}_1 - \bar{x}_2 = -4.92$ and is considered to be statistically significant with $T = -2.88$, a p-value of 0.004, and a reproducibility probability of 0.814 under the estimated power approach or 0.742 under the Bayesian approach.

The sponsor of this trial would like to evaluate the probability for reproducing the clinical result for elderly patient population where Δ, the change in the signal-to-noise ratio, ranges from 0.75 to 1.2. The generalizability probabilities are given in Table 9.5. In this example, $|T|$ is not very large and, thus, a clinical trial is necessary. The generalizability probability can be used to determine the sample size n^* for such a clinical trial. As it can be seen from Table 9.5, if $\Delta = 0.9$ and the desired power (reproducibility probability)

TABLE 9.5

Generalizability Probability and Sample Size Requirement for Bridging Studies (under a Two-Group Parallel Design with $n_1 = 56$, $n_2 = 48$ and $T = -2.8$)

	Estimated Power Approach			Bayesian Approach		
		New Sample Size n^*			New Sample Size n^*	
Δ	$\hat{P}_\Delta$	70% Power	80% Power	$\hat{P}_\Delta$	70% Power	80% Power
1.20	0.929	52	66	0.821	64	90
1.10	0.879	62	80	0.792	74	102
1.00	0.814	74	96	0.742	86	118
0.95	0.774	84	106	0.711	98	128
0.90	0.728	92	118	0.680	104	140
0.85	0.680	104	132	0.645	114	154
0.80	0.625	116	150	0.610	128	170
0.75	0.571	132	170	0.562	144	190

Source: Shao, J. and Chow, S.C., *Stat. Med.*, 21, 1727–1742, 2002.

is 80%, then $n^* = 118$ under the estimated power approach and 140 under the Bayesian approach; if the desired power (reproducibility probability) is 70%, then $n^* = 92$ under the estimated power approach and 104 under the Bayesian approach. A sample size smaller than that of the original trial is allowed if $\Delta \geq 1$, that is, the new population is less variable.

The sample sizes n^* in Table 9.5 are obtained as follows. Under the estimated power approach

$$n^* = \frac{\left(\dfrac{T^*}{\Delta T}\right)^2}{\left(\dfrac{1}{4n_1} + \dfrac{1}{4n_2}\right)},$$

where T^* is the value obtained from Table 9.1 for which the reproducibility probability has the desired level (for example, 70% or 80%). Under the Bayesian approach, for each given Δ we first compute the value T_Δ^* at which the reproducibility probability has the desired level and then use

$$n^* = \frac{\left(\dfrac{T_\Delta^*}{T}\right)^2}{\left(\dfrac{1}{4n_1} + \dfrac{1}{4n_2}\right)}.$$

9.6 Future Perspectives

In practice, if a significant research finding is not reproducible, there is a reasonable doubt that the observed finding could be purely by chance alone and hence is not reliable. Statistically, a research finding is considered not creditable if it does not reach statistical significance (i.e., the observed finding is purely due to chance) and it is not reproducible under similar experimental conditions. To increase the creditability of the observed research findings, it is suggested that possible sources of bias and/or variability including (i) expected/controllable, (ii) expected but uncontrollable, (iii) unexpected but controllable, and (iv) unexpected/uncontrollable be identified, eliminated/minimized, and/or controlled whenever possible in order to increase reliability and the probability of reproducibility for an unbiased and reliable assessment of the test treatment under investigation in the pharmaceutical/clinical research and development process.

In summary, a non-reproducible research finding is not considered a valid science. The non-reproducible research finding may be biased and hence misleading in pharmaceutical/clinical research. It is then strongly suggested that given the observed research findings, the probability of reproducibility be assessed using the recommended method described in this chapter.

10

Extrapolation

10.1 Introduction

For marketing approval of a new drug product, the FDA requires that at least two adequate and well-controlled clinical trials be conducted to provide substantial evidence regarding safety and effectiveness of the drug product under investigation. The purpose of requiring at least two clinical studies is not only to assure the reproducibility of clinical findings, but also to provide valuable information regarding the generalizability of the clinical findings. Generalizability refers to one of the following situations. First, whether the clinical results of the original target patient population under study (e.g., adults) can be generalized to other similar but different patient populations (e.g., pediatrics or elderly). Second, whether a drug product newly developed or approved in one region (e.g., the US or the European Union) can be approved at another region (e.g., countries in the Asian-Pacific Region), given concerns that differences in ethnic factors could alter the safety and efficacy of the drug product in the new region. Third, for observational studies such as case-control studies, it is often of interest to determine whether the developed or established medical predictive model based on a database at one medical center can be applied to a different medical center that has similar database for patients with similar diseases under study. In many cases, generalizability of clinical results in the context of clinical research and development is also known as extrapolation of clinical results. In practice, since it is of interest to determine whether the observed clinical results from the original target patient population can be generalized to a similar but different patient population, we will focus on the first scenario. Statistical methods for assessment of generalizability of clinical results, however, can also be applied to the other scenarios.

Although the ICH E5 guideline establishes the framework for the acceptability of foreign clinical data, it does not clearly define the similarity in terms of dose response, safety, and efficacy between the original region and a new region. Shih (2001b) interpreted similarity as *consistency* among study centers by treating the new region as a new center of multicenter clinical trials. Under this definition, Shih proposed a method for assessment of consistency to determine whether the study is capable of bridging the foreign data to the new region. Alternatively, Shao and Chow (2002) proposed the concepts of reproducibility and *generalizability* probabilities for assessing bridging studies. In addition, Chow et al. (2002) proposed to assess similarity by analysis using a *sensitivity index*, which is a measure of population shift between the original region and the new region. For assessment of generalizability of clinical results from one population to another, Chow (2010) proposed to evaluate a so-called generalizability probability of a positive clinical result observed from the original patient population by studying the impact of shift in target patient through a model that links the population means with some covariates (see also, Chow and Shao, 2005; Chow and Chang, 2006). However, in many cases, such covariates may not exist, or exist but not be observable. In this situation, it is suggested that the degree of shift in location and scale of patient population be studied based on a mixture distribution by assuming the location or scale parameter is a random variable (Shao and Chow, 2002). The purpose of this chapter is to assess the generalizability of clinical results by evaluating the sensitivity index under three different model assumptions: that (i) shift in location parameter is a random variable; (ii) shift in scale parameter is a random variable, and (iii) shifts in both location and scale parameters are both random variables.

The remainder of this chapter is organized as follows. In Section 10.2, the concept of a sensitivity index for measuring the degree of population shift is briefly introduced. Section 10.3 discusses inference based on mixture distribution for the cases where (i) shift in location parameter is random and change in scale parameter is fixed, (ii) shift in location parameter is fixed and change in scale parameter is random, and (iii) both shift in location parameter and change in scale parameter are random. Statistical inferences of the effect size for the three possible scenarios are given in Section 10.4. In Section 10.5, an example concerning an asthma clinical trial under the three scenarios is presented. Brief concluding remarks are given in Section 10.6.

10.2 Shift in Target Patient Population

In clinical research, it is often of interest to generalize clinical results obtained from a given target patient population (or medical center) to a similar but different patient population (or another medical center). Denote the original target patient population by (μ_0, σ_0), where μ_0 and σ_0 are the population mean and population standard deviation, respectively. Similarly, denote the similar but different patient population by $(\mu_{new}, \sigma_{new})$. Since the two populations are similar but different, it is reasonable to assume that $\mu_{new} = \mu_0 + \varepsilon$ and $\sigma_{new} = C\sigma_0$ $(C > 0)$, where ε is referred to as the shift in location parameter (population mean) and C is the inflation factor of the scale parameter (population standard deviation). Thus, the (treatment) effect size adjusted for the standard deviation of population $(\mu_{new}, \sigma_{new})$ can be expressed as follows:

$$E_1 = \left| \frac{\mu_{new}}{\sigma_{new}} \right| = \left| \frac{\mu_0 + \varepsilon}{C\sigma_0} \right| = |\Delta| \left| \frac{\mu_0}{\sigma_0} \right| = |\Delta| E_0, \qquad (10.1)$$

where $\Delta = (1 + \varepsilon / \mu_0) / C$ and E_0 and E_1 are the respective effect sizes (of clinically meaningful importance) of the original target patient population and the similar but different patient population, respectively. The quantity Δ is usually referred to as a sensitivity index measuring the change in effect size between patient populations (see also, Shao and Chow, 2002; Chow and Chang, 2006).

As can be seen from (10.1), if $\varepsilon = 0$ and $C = 1$, $E_0 = E_1$. That is, the effect sizes of the two populations are identical. In this case, we claim that the results observed from the original target patient population (e.g., adults) can be generalized to the similar but different patient population (e.g., pediatrics or elderly). Applying the concept of bioequivalence assessment, we can claim that the effect sizes of the two patient populations are equivalent if the confidence interval of $|\Delta|$ is within (80%, 120%) of E_0, which is a criterion commonly considered in clinical research (Chow and Liu, 2008). However, it should be noted that there is a masking effect between the location shift ε and scale change C. In other words, shift in location parameter could be offset by the inflation or deflation of variability. As a result, the sensitivity index may remain unchanged while the target patient population has been shifted. Table 10.1 provides a summary of the impacts of various scenarios of location shift (i.e., change in ε) and scale change (i.e., change in Δ).

TABLE 10.1

Changes in Sensitivity Index

ε / μ (%)	Inflation of Variability Δ	Deflation of Variability Δ
−20	0.667	1.000
−10	0.750	1.125
−5	0.792	1.188
0	0.833	1.250
5	0.875	1.313
10	0.917	1.375
20	1.000	1.500

10.3 Assessment of Sensitivity Index

As indicated by Chow and Shao (2005), in many clinical trials, the effect size of the two populations could be linked by baseline demographics or patient characteristics if there is a relationship between the effect size and the baseline demographics and/or patient characteristics (e.g., a covariate vector). In practice, however, such covariates may not exist or exist but not be observable. In this case, the sensitivity index may be assessed by simply replacing ε and C with their corresponding estimates (Chow and Shao, 2005). Intuitively, ε and C can be estimated by

$$\hat{\varepsilon} = \hat{\mu}_{new} - \hat{\mu}_0 \text{ and } \hat{C} = \hat{\sigma}_{new} / \hat{\sigma}_0,$$

where $(\hat{\mu}_0, \hat{\sigma}_0)$ and $(\hat{\mu}_{new}, \hat{\sigma}_{new})$ are some estimates of (μ_0, σ_0) and $(\mu_{new}, \sigma_{new})$, respectively. Thus, the sensitivity index can be estimated by

$$\hat{\Delta} = \frac{1 + \hat{\varepsilon} / \hat{\mu}_0}{\hat{C}}.$$

In practice, the shift in location parameter ε and/or the change in scale parameter C could be random variables. If both ε and C are fixed, the sensitivity index can be assessed based on the sample means and sample variances obtained from the two populations. In real world problems, however, ε and C could be either fixed or random variables. In other words, there are three possible scenarios: (i) the case where ε is random and C is fixed, (ii) the case where ε is fixed and C is random, and (iii) the case where both ε and C are random. These possible scenarios are briefly described below.

10.3.1 The Case Where ε Is Random and C Is Fixed

To obtain estimates of the sensitivity index, we first consider the estimates of μ_{new} and σ_{new}. Chow et al. (2005) studied the case where $\mu_{new} = \mu_0 + \varepsilon$ is random variable and $\sigma_{new} = C\sigma_0$ is fixed constant.

Assume that x is conditional on μ, i.e., $x|_{\mu}$ follows a normal distribution $N(\mu, \sigma^2)$. That is, $x|_{\mu} \sim N(\mu, \sigma^2)$, where μ is distributed as $N(\mu_{\mu}, \sigma_{\mu}^2)$ and σ, μ_{μ}, and σ_{μ} are some unknown parameters. Let $\{x_{0i}, i = 1, \ldots, N_0\}$ be the responses observed from the original target patient population and $\{x_k, k = 1, \ldots, N_1\}$ be the responses observed from the similar but different population, that is, the sample size is n_1 in the similar but different population. If we split the data from the similar but different population into m segments of the same size n, then $\{x_k, k = 1, \ldots, N_1\}$ becomes $\{x_{ji}, j = 1, \ldots, m; i = 1, \ldots, n\}$. Combine $\{x_{0i}, i = 1, \ldots, n_0\}$ and $\{x_{ji}, j = 1, \ldots, m; i = 1, \ldots, n\}$, we have

$$\left\{ x_{ji}, i = 1, \ldots, n_0 \left(\text{if } j = 0 \right) n \left(\text{if } j > 0 \right); \ j = 0, 1, \ldots, m \right\}.$$

Note that $n_1 = \sum_{j=1}^{m} n = mn$. Under normality assumption, estimates of μ_0 and σ_0^2 can be obtained. Based on $x_{0i}, i = 1, \ldots, n_0$ the maximum likelihood estimates of μ_0 and σ_0^2 can be obtained as follows

$$\widehat{\mu}_0 = \frac{1}{n_0} \sum_{i=1}^{n_0} x_{0i}, \text{ and } \widehat{\sigma}_0^2 = \frac{1}{n_0 - 1} \sum_{i=1}^{n_0} (x_{0i} - \widehat{\mu}_0)^2. \tag{10.2}$$

Thus, the unconditional distribution of x is a mixed normal distribution (Chow et al., 2005) given below

$$\int N\left(x; \mu, \sigma^2\right) N\left(\mu; \mu_\mu, \sigma_\mu^2\right) d\mu = \frac{1}{\sqrt{2\pi\sigma^2}} \frac{1}{\sqrt{2\pi\sigma_\mu^2}} \int_{-\infty}^{\infty} e^{-\frac{(x-\mu)^2}{2\sigma^2} - \frac{(\mu-\mu_\mu)^2}{2\sigma_\mu^2}} d\mu,$$

where $x \in (-\infty, \infty)$, it can be verified that the above mixed normal distribution is a normal distribution with mean μ_μ and variance $\sigma_1^2 + \sigma_\mu^2$. In other words, x is distributed as $N\left(\mu_\mu, \sigma_1^2 + \sigma_\mu^2\right)$.

Theorem 10.1: Suppose that $X|_\mu \sim N(\mu, \sigma^2)$ and $\mu \sim N(\mu_\mu, \sigma_\mu^2)$, then we have

$$X \sim N\left(\mu_\mu, \ \sigma^2 + \sigma_\mu^2\right). \tag{10.3}$$

Proof: Consider the following characteristic function of a normal distribution $N(x, \mu, \sigma^2)$

$$\phi_0\left(w\right) = \frac{1}{\sqrt{2\pi\sigma^2}} \int_{-\infty}^{\infty} e^{iwt - \frac{1}{2\sigma^2}(t-\mu)^2} dt = e^{iw\mu - \frac{1}{2}\sigma^2 w^2}.$$

For distribution $X|_\mu \sim N(\mu, \sigma^2)$ and $\mu \sim N(\mu_\mu, \sigma_\mu^2)$, the characteristic function after exchanging the order of the two integrations is given by

$$\phi\left(w\right) = \int_{-\infty}^{+\infty} e^{iw\mu - \frac{1}{2}\sigma^2 w^2} N\left(\mu, \mu_\mu, \sigma_\mu\right) d\mu$$

$$= \int_{-\infty}^{+\infty} e^{iw\mu - \frac{\mu - \mu_\mu}{2\sigma_\mu^2} - \frac{1}{2}\sigma^2 w^2} d\mu$$

Note that

$$\int_{-\infty}^{+\infty} e^{iw\mu - \frac{\mu - \mu_\mu}{2\sigma_\mu^2}} d\mu = e^{iw\mu - \frac{1}{2}\sigma_\mu^2 w^2}$$

is the characteristic function of the normal distribution. It follows that

$$\phi(w) = e^{iw\mu_\mu - \frac{1}{2}(\sigma^2 + \sigma_\mu^2)w^2}.$$

According to the property of the character function, if $\phi(f_1) = \phi(f_2)$, there is $pdf(f_1) = pdf(f_2)$, where pdf is the probability density function, this leads to the characteristic function of $N(\mu_\mu, \sigma^2 + \sigma_\mu^2)$. This completes the proof.

Based on the above theorem, the maximum likelihood estimates of σ_1^2, μ_μ, and σ_μ^2 can be obtained from the following log-likelihood function

$$\ell\left(\mu_\mu, \sigma^2, \sigma_\mu^2\right) = -\frac{n}{2}\ln\left(2\pi\sigma^2\right) - \frac{m}{2}\ln\left(2\pi\sigma_\mu^2\right) - \frac{1}{2\sigma^2}\sum_{j=1}^{m}\sum_{i=1}^{n_j}\left(x_{ji} - \mu_j\right)^2$$

$$- \frac{1}{2\sigma_\mu^2}\sum_{j=0}^{m}\left[n_j\left(\mu_j - \mu_\mu\right)^2\right].$$

Thus, the maximum likelihood estimators are given by

$$\hat{\mu}_\mu = \frac{1}{m}\sum_{j=1}^{m}\hat{\mu}_j, \tag{10.4}$$

$$\hat{\sigma}_\mu^2 = \frac{1}{m}\sum_{j=1}^{m}(\hat{\mu}_j - \hat{\mu}_\mu)^2, \tag{10.5}$$

$$\hat{\sigma}^2 = \frac{1}{n_0}\sum_{j=1}^{m}\sum_{j=1}^{m}(x_{ji} - \hat{\mu}_j)^2, \tag{10.6}$$

where $\hat{\mu}_j = \frac{1}{n_j}\sum_{i=1}^{n_j}x_{ji}$.

Based on these maximum likelihood estimates, estimates of the shift parameter ε and the scale parameter C can be obtained as follows:

$$\hat{\varepsilon} = \hat{\mu}_\mu - \hat{\mu}_0 \text{ and } \hat{C} = \left(\hat{\sigma}_\mu^2 + \hat{\sigma}_1^2\right)/\hat{\sigma}_0^2,$$

respectively. Consequently, the sensitivity index can be estimated by simply replacing ε, μ, and C with their corresponding estimates, i.e., $\hat{\varepsilon}, \hat{\mu}_0$, and $\tilde{C}$.

10.3.2 The Case Where ε Is Fixed and C Is Random

We turn to the case where ε is fixed and C is random, that is the scale parameter is a random variable. Because the sample variance is usually distributed by a gamma distribution, we assume the scale parameter σ_{new}^2 distributed by an inverse gamma distribution. Similarly, the observed response variable $x\mid_{\sigma=\sigma_{new}}$ follows a normal distribution $N(\mu_1,\sigma^2)$ conditional on $\sigma \equiv \sigma_{new}$, that is

$$x\mid_\sigma \sim N(\mu,\sigma^2),$$

where $\sigma^2 \sim IG(\alpha,\beta)$ and μ, α and β are unknown parameters. Thus, we have the following results.

Theorem 10.2: Suppose that $x\mid_{\sigma=\sigma_{new}} \sim N(\mu,\sigma^2)$ and $\sigma^2 \sim IG(\alpha,\beta)$. Then, we have

$$x \sim f(x) = \frac{\Gamma\left(\alpha+\dfrac{1}{2}\right)}{\Gamma(\alpha)\sqrt{2\pi\beta}}\left[1+\frac{(x-\mu)^2}{2\beta}\right]^{-\left(\alpha+\frac{1}{2}\right)}, \tag{10.7}$$

where $\Gamma(\alpha)$ is the gamma function, which is defined as

$$\Gamma(\alpha) = \int_0^{+\infty} t^{\alpha-1}e^{-t}dt.$$

That is, x is a non-central t-distribution, where $\mu \in R$ is location parameter, $\sqrt{\beta/\alpha}$ is scale parameter and 2α is the degree of freedom in the t-distribution.

Proof:

$$f(x) = \int_0^{+\infty} f\left(x\mid\sigma^2\right)f\left(\sigma^2\right)d\sigma^2$$

$$= \int_0^{+\infty} \frac{\beta^\alpha}{\sqrt{2\pi\sigma^2}\Gamma(\alpha)\sigma^{2(\alpha+1)}}\exp\left\{-\frac{(x-\mu)^2+2\beta}{2\sigma^2}\right\}d\sigma^2$$

$$= \frac{\beta^\alpha}{\sqrt{2\pi}\Gamma(\alpha)}\int_0^{+\infty}\left(\frac{1}{\sigma^2}\right)^{\alpha+\frac{3}{2}}\exp\left\{-\frac{(x-\mu)^2+2\beta}{2\sigma^2}\right\}d\sigma^2$$

$$= \frac{\beta^\alpha}{\sqrt{2\pi}\Gamma(\alpha)}\int_0^{+\infty} t^{\alpha-\frac{1}{2}}\exp\left\{-\frac{(x-\mu)^2+2\beta}{2}t\right\}dt$$

$$= \frac{\Gamma\left(\alpha+\dfrac{1}{2}\right)}{\Gamma(\alpha)\sqrt{2\pi\beta}}\left[1+\frac{(x-\mu)^2}{2\beta}\right]^{-\left(\alpha+\frac{1}{2}\right)}$$

Thus, x follows a non-central t-distribution. Hence, we have

$$E(x) = \mu \text{ and } Var(x) = \beta / (\alpha - 1).$$

This completes the proof.

Based the above theorem, the maximum likelihood estimates of the parameters μ_1, α, and β can be obtained as follows. Suppose that the observations satisfy the following conditions.

1. $(x_{ji}|\mu,\sigma_j^2) \sim N(\mu,\sigma_j^2)$, $j = 1,...,m; i = 1,...,n_j$ and $\sigma_j^2, x_{j1},...,x_{jn_j}$ are independent identically distributed (*i.i.d*);
2. $\{x_{ji}, i = 1,...,n_j\}$, $j = 1, ...,m$ are independent;
3. $\sigma_j^2 \sim IG(\alpha,\beta), j = 1,...,m$.

For convenience, we combine the response variable $x = (x_{11},...,x_{1n_1},x_{21},...,x_{mn_m})$ and $\sigma_{new}^2 = (\sigma_1^2,...,\sigma_m^2)$ as the complete data z, with $z = (x, \sigma_{new}^2)$, and also combine the parameters to be estimated μ, α, β, as $\theta_1 = (\mu, \alpha, \beta)$. The estimator of the parameter vector θ_1 can be obtained by an expectation-maximization algorithm (EM algorithm) (see, e.g., Dempster, 1977; Lange, 1989; Liu, 1995). The detailed procedure is given as follows

$$\log f(z|\theta_1) = \log f\left(x_{ji}| \sigma_j^2\right) + \log f\left(\sigma_j^2| \alpha,\beta\right).$$

Thus, we have

$$\log f(z|\theta_1) = \sum_{j=1}^{m}\left(\frac{n_j}{2}+\alpha+1\right)\log\frac{1}{\sigma_j^2} + m\alpha\log\beta - m\log\Gamma(\alpha)$$

$$-\sum_{j=1}^{m}\frac{2\beta+\sum_{i=1}^{n_j}(x_{ji}-\mu)^2}{2\sigma_j^2}. \tag{10.8}$$

By (10.8) and by the EM algorithm, the E-step,

$$E\left(\frac{1}{\sigma_j^2}\bigg| x,\hat{\theta}_1^{(t)}\right) = \frac{\hat{\alpha}^{(t)}+n_j/2}{\hat{\beta}^{(t)}+\sum_{i=1}^{n_j}\left(x_{ji}-\hat{\mu}^{(t)}\right)^2/2},$$

$$E\left(\log\frac{1}{\sigma_j^2}\bigg| x,\hat{\theta}_1^{(t)}\right) = \phi\left(\hat{\alpha}^{(t)}+n_j/2\right) - \log\left(\hat{\beta}^{(t)}+\sum_{i=1}^{n_j}\left(x_{ji}-\hat{\mu}^{(t)}\right)^2/2\right),$$

where $\phi(*)$ is the gamma function, that is $\phi(*) = \frac{d}{dy}\ln(\Gamma(*))$. Consider

$$E\left(\frac{1}{\sigma_j^2}\Big|x,\hat{\theta}_1^{(t)}\right) \triangleq \hat{V}_{1j}^{(t)}, \quad E\left(\log\frac{1}{\sigma_j^2}\Big|x,\hat{\theta}_1^{(t)}\right) \triangleq \hat{V}_{2j}^{(t)},$$

and the expectation of the log-likelihood function

$$E\left(\log f\left(z\,|\,\hat{\theta}_1^{(t)}\right)\right) \triangleq Q\left(z\,|\,x,\hat{\theta}_1^{(t)}\right).$$

By the Newton-Raphson algorithm (Tjalling, 1995; Deuflhard, 2004), the $(t+1)^{\text{th}}$ iteration converges. Denote the estimates of the parameters after the $(t+1)^{\text{th}}$ iteration by

$$\hat{\theta} = \left(\hat{\mu}^{(t+1)}, \hat{\alpha}^{(t+1)}, \hat{\beta}^{(t+1)}\right),$$

where

$$\hat{\mu}^{(t+1)} = \sum_{j=1}^{m}\sum_{i=1}^{n_j} x_{ji}\hat{V}_{1j}^{(t)} \Big/ \sum_{j=1}^{m} n_j\hat{V}_{1j}^{(t)}, \tag{10.9}$$

$$\hat{\alpha}^{(t+1)} = \hat{\alpha}^{(t+1)} + \frac{\sum_{j=1}^{m}\hat{V}_{2j}^{(t)} - m\phi\left(\hat{\alpha}^{(t)}\right) + m\log\hat{\beta}^{(t)}}{m\phi'\left(\hat{\alpha}^{(t)}\right)}, \tag{10.10}$$

$$\hat{\beta}^{(t+1)} = m\hat{\alpha}^{(t+1)} \Big/ \sum_{j=1}^{m}\hat{V}_{1j}^{(t)}. \tag{10.11}$$

By the EM algorithm, the maximum likelihood estimates of μ, α and β can be obtained in (10.9–10.11). For simplicity, denote the estimates of the parameters as $\hat{\mu}, \hat{\alpha}$ and $\hat{\beta}$. Thus, the sensitivity index can be estimated as

$$\hat{\Delta} = \hat{\sigma}_0^2\left[1 + \left(\hat{\mu}_{new} - \hat{\mu}_0\right)/\hat{\mu}_0\right] / \hat{\sigma}_{new}^2, \tag{10.12}$$

where

$$\hat{\mu}_{new} = \hat{\mu} \text{ and } \hat{\sigma}_{new}^2 = \hat{\beta}/\left(\hat{\alpha} - 1\right).$$

10.3.3 The Case Where Both ε and C Are Random

Now, consider the case when both ε and C are random variables. We present a useful theorem. Assume that the clinical data based on the original population is distributed as (μ_0, σ_0^2) and the population mean μ and variance σ^2 in the new region are random variables, which follows a normal-scaled inverse gamma distribution, that is, $(\mu, \sigma^2) \sim N - \Gamma^{-1}(\mu_\mu, \upsilon, \alpha, \beta)$, in which both ε and C are random variables. We have the following results.

Theorem 10.3: Suppose that $x \mid_{\mu, \sigma^2} \sim N(\mu, \sigma^2)$ and $(\mu, \sigma^2) \sim N - \Gamma^{-1}(\mu_\mu, \nu, \alpha, \beta)$. Then, x is distributed as

$$f(x) = \frac{\Gamma(\alpha + 1/2)}{\Gamma(\alpha)\sqrt{2\pi\beta(\nu+1)/\nu}} \left[1 + \frac{\nu(x - \mu_\mu)^2}{2\beta(\nu+1)}\right]^{-\left(\alpha + \frac{1}{2}\right)},$$

where $\Gamma(\alpha)$ is the gamma function $\Gamma(\alpha) = \int_0^{+\infty} t^{\alpha-1} e^{-t} dt$, x is a non-central t-distribution with a location parameter μ_{new}, scale parameter $\beta(\nu+1)/\nu$ and the degree of freedom 2α, and the $Var(x) = \beta(\nu+1)/(\alpha\nu)$.

Proof:

$$f(x, \mu, \sigma^2) = \int_0^\infty \int_{-\infty}^{+\infty} f(x \mid \mu, \sigma^2) f(\mu, \sigma^2) d\mu d\sigma^2$$

$$= \int_0^\infty \int_{-\infty}^\infty \frac{1}{\sqrt{2\pi}\sigma} \exp\left\{-\frac{(x-\mu)^2}{2\sigma^2}\right\} \frac{\sqrt{\nu}}{\sigma\sqrt{2\pi}} \frac{\beta^\alpha}{\Gamma(\alpha)} \left(\frac{1}{\sigma^2}\right)^{\alpha+1} \exp\left\{-\frac{2\beta + \nu(\mu - \mu_\mu)^2}{2\sigma^2}\right\} d\mu d\sigma^2$$

$$= \frac{\sqrt{\nu}}{2\pi} \frac{\beta^\alpha}{\Gamma(\alpha)} \int_0^\infty \int_{-\infty}^\infty \exp\left\{-\left[\frac{(x-\mu)^2}{2\sigma^2} + \frac{2\beta + \nu(\mu - \mu_\mu)^2}{2\sigma^2}\right]\right\} d\mu d\sigma^2$$

$$= \frac{\sqrt{\nu}}{2\pi} \frac{\beta^\alpha}{\Gamma(\alpha)} \int_0^\infty \left(\frac{1}{\sigma^2}\right)^{\alpha+2} \exp\left\{-\frac{\upsilon[x - \mu_\mu]^2}{2(\upsilon+1)\sigma^2}\right\} \int_{-\infty}^\infty \exp\left\{-\frac{\left[\sqrt{1+\nu}\,\mu - \frac{x\nu\mu_\mu}{\sqrt{1+\nu}}\right]^2}{2\sigma^2}\right\} d\mu d\sigma^2$$

$$= \frac{\sqrt{\nu}}{\sqrt{2\pi(1+\nu)}} \frac{\beta^\alpha}{\Gamma(\alpha)} \int_0^\infty \left(\frac{1}{\sigma^2}\right)^{\alpha+3/2} \exp\left\{-\frac{2\beta + \upsilon[x - \mu_\mu]^2/(\upsilon+1)}{2\sigma^2}\right\} d\sigma^2$$

$$= \frac{\sqrt{\nu}}{\sqrt{2\pi\beta(1+\nu)}} \frac{\Gamma\left(\alpha + \frac{1}{2}\right)}{\Gamma(\alpha)} \left[1 + \frac{\upsilon[x - \mu_\mu]^2}{2\beta(\upsilon+1)}\right]^{-\left(\alpha + \frac{1}{2}\right)}.$$

Thus, the distribution of the response variable x follows a non-central t-distribution. This completes the proof.

The observed response variable $x = \{x_{ji}\}, j = 1,\ldots,m; i = 1,\ldots,n_j$ from N_2 patients in the similar but different population, where $\sum_{j=1}^{m} n_j = N_2$, the latent vector

$$\left(\mu,\sigma^2\right) = \left\{\left(\mu_j,\sigma_j^2\right), j = 1,\ldots,m\right\},$$

and the response variable x_{ji} satisfy the following conditions:

1. $(x_{ji}|\mu_j,\sigma_j^2) \sim N(\mu_j,\sigma_j^2)$, $j = 1,\ldots,m; i = 1,\ldots,n_j$, and given (μ_j,σ_j^2), $x_{j1},\ldots,x_{jnj}$ are i.i.d.;
2. $\{x_{ji}, i = 1,\ldots,n_j\}, j = 1,\ldots,m$ are independent;
3. $(\mu_j, \sigma^2_j) \sim N - \Gamma^{-1}(\mu_\mu,\nu,\alpha,\beta), j = 1,\ldots,m$.

We collect x with the random variables μ_1, σ_1^2 as the complete data $Z = (x,\mu_1,\sigma_1^2)$. The maximum likelihood estimates of $\theta_2 = \left(\mu_\mu,\nu,\alpha,\beta\right)$ can be derived by an EM algorithm, with the log-likelihood function as follows

$$\log f\left(z|\theta_2\right) = \log f\left(x|\mu,\sigma^2\right) + \log\left(\mu,\sigma^2|\theta_2\right)$$

$$= \sum_{j=1}^{m}\left(\frac{n_j+3}{2} + \alpha\right)\log\frac{1}{\sigma_j^2} - \sum_{j=1}^{m}\frac{2\beta + \nu\left(\mu_{1j} - \mu_\mu\right)^2}{2\sigma_j^2} - \sum_{j=1}^{m}\sum_{i=1}^{n_j}\frac{\left(x_{ji} - \mu_j\right)^2}{2\sigma_j^2}$$

$$- m\log\Gamma\left(\alpha\right) + \frac{m}{2}\log\nu + m\alpha\log\beta$$

conditional on the observed data x and the tth iterative estimates $\hat{\theta}_2^{(t)} = (\mu_\mu^{(t)},\nu^{(t)},\alpha^{(t)},\beta^{(t)})$, the expectation with regard to latent data μ and σ^2 is given by

$$E\left\{\frac{1}{\sigma_j^2}\middle|x,\hat{\theta}_2^{(t)}\right\} = \frac{\Gamma\left(\dfrac{n_j}{2} + \alpha^{(t)} + 1\right)}{\left(\hat{\beta}^{(t)} + \dfrac{1}{2}\sum_{i=1}^{n_j}\left(x_{ji} - \bar{x}_{j\cdot}\right)^2 + \dfrac{\hat{\nu}^{(t)}n_j}{2\left(n_j + \hat{\nu}^{(t)}\right)}\left(\bar{x}_{j\cdot} - \hat{\mu}_\mu^{(t)}\right)\right)\Gamma\left(\dfrac{n_j}{2} + \hat{\alpha}^{(t)}\right)}.$$

where $\bar{x}_{j\cdot} = \frac{1}{n_j}\sum_{i=1}^{n}x_{ji}$.

We define

$$E\left\{\frac{1}{\sigma_j^2}\middle|x,\hat{\theta}_{12}^{(t)}\right\} \triangleq \widehat{W}_{1j}^{(t)},$$

$$E\left\{\frac{\left(\mu_j - c\right)^2}{\sigma_j^2} \mid x, \hat{\theta}_2^{(t)}\right\} = \frac{1}{n_j + \hat{v}^{(t)}} + \left(\frac{\sum_{i=1}^{n} x_{ji} + v^{(t)}\hat{\mu}_\mu^{(t)}}{n_j + \hat{v}^{(t)}} - c\right)^2 \widehat{W}_{1j}^{(t)}$$

$$E\left\{\log\frac{1}{\sigma_j} \mid x, \hat{\theta}_2^{(t)}\right\} = \phi\left(\frac{n}{2} + \hat{\alpha}^{(t)}\right) - \ln\left(\hat{\beta}^{(t)} + \frac{1}{2}\sum_{i=1}^{n}\left(x_{ji} - \bar{x}_{j\cdot}\right)^2 + \frac{\hat{v}^{(t)}n_j}{2\left(n_j + \hat{v}^{(t)}\right)}\left(\bar{x}_{j\cdot} - \hat{\mu}_\mu^{(t)}\right)\right).$$

Also, denote

$$E\left\{\log\frac{1}{\sigma_j^2} \mid x, \hat{\theta}_2^{(t)}\right\} \triangleq \widehat{W}_{2j}^{(t)} \text{ and } E\left\{\frac{\left(\mu_j - c\right)^2}{\sigma_j^2} \mid x, \hat{\theta}_2^{(t)}\right\} \triangleq \widehat{W}_{3j}^{(t)}.$$

The maximum likelihood estimates of the vector θ_2 as follows,

$$\hat{\mu}_\mu^{(t+1)} = \sum_{j=1}^{m} \frac{\sum_{i=1}^{n_j} x_{ji} + \hat{v}^{(t)}\hat{\mu}_\mu^{(t)}}{n_j + \hat{v}^{(t)}} \Big/ \sum_{j=1}^{m} \widehat{W}_{1j}^{(t)}, \tag{10.13}$$

$$\hat{v}^{(t+1)} = m \Big/ \sum_{j=1}^{m} \widehat{W}_{3j}^{(t)} \tag{10.14}$$

$$\hat{\alpha}^{(t+1)} = \hat{\alpha}^{(t)} + \frac{\sum_{j=1}^{m} \widehat{W}_{2j}^{(t)} - m\phi\left(\hat{\alpha}^{(t)}\right) + m\log\hat{\beta}^{(t)}}{m\phi'\left(\hat{\alpha}^{(t)}\right)} \tag{10.15}$$

$$\hat{\beta}^{(t+1)} = m\hat{\alpha}^{(t+1)} \Big/ \sum_{j=1}^{m} \widehat{W}_{1j}^{(t)} \tag{10.16}$$

Without loss generality, assume that the $(t+1)$th iteration converges. In this case, the maximum likelihood estimates of the unknown constant vector θ_2 is given by

$$\left(\hat{\mu}_\mu^{(t+1)}, \hat{v}^{(t+1)}, \hat{\alpha}^{(t+1)}, \hat{\beta}^{(t+1)}\right).$$

As a result, when both ε and C are random, their estimates can be obtained

$$\hat{\varepsilon} = \hat{\mu}_{new} - \hat{\mu}_0 = \hat{\mu}_\mu^{(t+1)} - \hat{\mu}_0 \text{ and } \hat{C} = \hat{\sigma}_{new}^2 / \hat{\sigma}_0^2 = \hat{\beta}\left(\hat{v}+1\right) / \left(\hat{\alpha}\hat{v}\hat{\sigma}_0^2\right). \tag{10.17}$$

and the estimate of the sensitivity index can be obtained. If the estimate of the sensitivity index falls within a pre-specified interval, the similarity between the two regions can be confirmed.

By the estimate of the sensitivity index, the effect of the treatment in the new patient population may be assessed, and the generalizability of the treatment in the similar but different population can be obtained.

10.4 Statistical Inference

For the generalizability of the clinical trial data from the original target patient population to the similar patient population, based on the sensitivity index Δ, we have the following hypothesis,

$$H_0 : \Delta \leq 1-\delta \text{ or } \Delta \geq 1+\delta \text{ vs. } H_a : 1-\delta < \Delta < 1+\delta, \qquad (10.18)$$

where δ is the generalizability margin. Following the concept of bioequivalence, δ is usually chosen to be 20% (i.e., $\delta = 20\%$). We can claim that the effect sizes of the two patient populations are equivalent if the null hypothesis of (10.18) is rejected.

According to the expression of Δ given in Section 10.3, and given the maximum likelihood estimates $\hat{\mu}_0$ and, $\hat{\sigma}_0^2$ and $\hat{\mu}_{new}$ and, $\hat{\sigma}_{new}^2$, respectively, of the parameters in the original target patient population and of the similar but different population, we have the estimate of the sensitivity index,

$$\hat{\Delta} = \frac{\hat{\mu}_{new}}{\hat{\sigma}_{new}} \bigg/ \frac{\hat{\mu}_0}{\hat{\sigma}_0}$$

Using the nature log-transformation of $\hat{\Delta}$, hypotheses (10.18) are equivalent to the following hypotheses

$$H_{01}: \log \Delta \leq \log(1-\delta) \text{ or } \log \Delta \geq \log(1+\delta) \text{ vs. } H_{a1}: \log(1-\delta) < \log \Delta < \log(1+\delta)$$
$$(10.19)$$

the estimate of the log-transformation $\log\Delta$ is given by

$$\log\hat{\Delta} = \log\hat{\mu}_{new} - \log\hat{\sigma}_{new}^2 - \log\hat{\mu}_0 + \log\hat{\sigma}_0^2$$

and regarding to the property of MLE, as $N_1 \to \infty$, $\log\hat{\Delta}$ is the consistent estimate of $\log\Delta$.

In what follows, we will discuss the test of sensitivity index under the three scenarios that we mentioned earlier in the chapter.

10.4.1 The Case Where ε Is Random and C Is Fixed

In the case of that ε is random variable and C is fixed, the estimate of the log-transformation $\log\Delta$ is given by

$$\log\widehat{\Delta} = \log\widehat{\mu}_{new} - \frac{1}{2}\log\widehat{\sigma}^2_{new} - \log\widehat{\mu}_0 + \log\widehat{\sigma}^2_0$$

$$= \log\widehat{\mu}_\mu - \frac{1}{2}\log\left(\widehat{\sigma}^2_\mu + \widehat{\sigma}^2\right) - \log\widehat{\mu}_0 + \log\widehat{\sigma}^2_0$$

Assuming that

$$\frac{n_0}{\sum_{j=1}^m n_j} \to \gamma, \; N_0 \to \infty \text{ and } N_1 = \sum_{j=1}^m n_j \to \infty,$$

where $0 < \gamma < \infty$. A limiting result regarding the maximum likelihood estimate (Shao, 1999; Pfanzagl, 1994) can be obtained when N_0 and N_1 are sufficiently large. It can be verified that

$$\sqrt{N}\left[\begin{pmatrix}\widehat{\mu}_\mu \\ \widehat{\sigma}^2 \\ \widehat{\sigma}^2_\mu \\ \widehat{\mu}_0 \\ \widehat{\sigma}_0\end{pmatrix} - \begin{pmatrix}\mu_\mu \\ \sigma^2 \\ \sigma^2_\mu \\ \mu_0 \\ \sigma^2_0\end{pmatrix}\right] \xrightarrow{d} \left(\begin{pmatrix}0 \\ 0 \\ 0 \\ 0 \\ 0\end{pmatrix}, \Sigma_\varepsilon\right)$$

where d denotes convergence in distribution, $N = N_0 + N_1$, Σ_ε is the covariance matrix of the parameter vector, which is a block-diagonal matrix given by

$$\Sigma_\varepsilon = \begin{pmatrix} I_x^{-1}(\theta) & 0 \\ 0 & I_0(\mu_0,\sigma_0) \end{pmatrix},$$

where $I_x^{-1}(\theta)$ and $I_0(\mu_0,\sigma_0)$ are the covariance matrices of $\theta = (\mu,\sigma,\sigma_\mu)$ and (μ_0,σ_0), respectively.

Regarding to the property of the maximum likelihood estimate, $\log\widehat{\Delta}$ is the maximum likelihood estimate of $\log\Delta$, and by the multivariate delta method (Oehlert, 1992; Davison, 2003), it follows that asymptotically,

$$\sqrt{N}\left(\log\widehat{\Delta} - \log\Delta\right) \xrightarrow{d} N\left(0, \; \sigma^2_\varepsilon\right), \tag{10.20}$$

where d denote the convergence in distribution, and

$$\sigma_\varepsilon^2 = B_\varepsilon \Sigma_\varepsilon B_\varepsilon^T,$$

$$B_\varepsilon = \left(\frac{\partial \log \Delta}{\partial \mu_\mu}, \quad \frac{\partial \log \Delta}{\partial \sigma}, \quad \frac{\partial \log \Delta}{\partial \sigma_\mu}, \quad \frac{\partial \log \Delta}{\partial \mu_0}, \quad \frac{\partial \log \Delta}{\partial \sigma_0} \right),$$

$$= \left(\frac{1}{\mu_\mu}, \quad \frac{\sigma_\mu}{\sigma_\mu^2 + \sigma^2}, \quad \frac{\sigma}{\sigma_\mu^2 + \sigma^2}, \quad \frac{1}{\mu_0}, \quad \frac{1}{\sigma_0} \right)$$

where the estimates of parameters $\mu_\mu, \sigma, \sigma_\mu, \mu_0, \sigma_0$ are given in (10.4) through (10.6).

For testing the hypothesis, consider the test statistic

$$z_\varepsilon = \frac{\sqrt{N} \left(\log \hat{\Delta} - \log \Delta \right)}{\sigma_\varepsilon},$$

where ε is the significant level of the test. If $-z_{\varepsilon/2} < z_\varepsilon < z_{\varepsilon/2}$, the null hypothesis H_{01} is rejected and the generalizability in the similar but different population is concluded.

10.4.2 The Case Where ε Is Fixed and C Is Random

When ε is random variable and C is fixed, the estimate of the log-transformation $\log \Delta$ is,

$$\log \hat{\Delta} = \log \hat{\mu}_{new} - \frac{1}{2} \log \hat{\sigma}_{new}^2 - \log \hat{\mu}_0 + \frac{1}{2} \log \hat{\sigma}_0^2$$

$$= \log \hat{\mu} - \frac{1}{2} \log \left(\hat{\beta} \right) + \frac{1}{2} \log \left(\hat{\alpha} - 1 \right) - \log \hat{\mu}_0 + \frac{1}{2} \log \hat{\sigma}_0^2$$

Assuming that $\frac{N_0}{\sum_{j=1}^m n_j} \to \gamma$, $N_0 \to \infty$ and $N_1 \to \infty$, where $0 < \gamma < \infty$ and $N = N_0 + N_1$. Applying the property of the maximum likelihood estimate, we have

$$\sqrt{N} \left[\begin{pmatrix} \hat{\mu} \\ \hat{\beta} \\ \hat{\alpha} \\ \hat{\mu}_0 \\ \hat{\sigma}_0 \end{pmatrix} - \begin{pmatrix} \mu \\ \beta \\ \alpha \\ \mu_0 \\ \sigma_0 \end{pmatrix} \right] \xrightarrow{d} \begin{pmatrix} 0 \\ 0 \\ 0 \\ 0 \\ 0 \end{pmatrix}, \Sigma_C .$$

For the estimate of the covariance matrix Σ_C, the procedure is given in the Appendix of this chapter. As it can be seen from the Appendix, we have

$$\Sigma_C = \begin{pmatrix} I_x^{-1}(\theta_1) & 0 \\ 0 & I_0(\mu_0, \sigma_0) \end{pmatrix},$$

where $I_x^{-1}(\theta_1)$ is given in the appendix (Thomas, 1982).

In this case, the limiting results regarding the maximum likelihood estimators are obtained as the ratio of the sample sizes between the original population and the similar but different population. This ratio is finite whenever the numbers of observations from the two populations is sufficiently large. By the delta method, the asymptotical distribution of the log-transformation of the sensitivity index is as follows,

$$\sqrt{N}\left(\log\breve{\Delta} - \log\Delta\right) \xrightarrow{d} N\left(0, \ \sigma_{\breve{C}}^2\right), \tag{10.21}$$

where

$$\sigma_{\breve{C}}^2 = B_C \Sigma_C B_{\breve{C}}^T,$$

and

$$B_C = \left(\frac{\partial\log\Delta}{\partial\mu}, \ \frac{\partial\log\Delta}{\partial\beta}, \ \frac{\partial\log\Delta}{\partial\alpha}, \ \frac{\partial\log\Delta}{\partial\mu_0}, \ \frac{\partial\log\Delta}{\partial\sigma_0}\right)$$

$$= \left(\frac{1}{\mu_1}, \ \frac{1}{2\beta}, \ \frac{1}{2(\alpha-1)}, \ \frac{1}{\mu_0}, \ \frac{1}{\sigma_0}\right),$$

where the estimates of parameters μ_1, σ_μ, σ, μ_0, σ_0 are given in (10.8) through (10.10) and (10.2).

For testing the hypotheses, consider the test statistic

$$z_C = \frac{\sqrt{N}\left(\log\hat{\Delta} - \log\Delta\right)}{\sigma_\varepsilon}$$

at the significant level ε. if $-z_{\varepsilon/2} < z_C < z_{\varepsilon/2}$, the null hypothesis H_{01} is rejected. We then conclude the generalizability of clinical results.

10.4.3 The Case Where ε and C Are Random

When ε is random and C is fixed, the estimate of the log-transformation $\log\Delta$ is,

$$\log\hat{\Delta} = \log\hat{\mu}_{new} - \frac{1}{2}\log\hat{\sigma}_{new}^2 - \log\hat{\mu}_0 + \frac{1}{2}\log\hat{\sigma}_0^2$$

$$= \log\hat{\mu}_\mu - \frac{1}{2}\log\left(\hat{\beta}\right) - \frac{1}{2}\log\left(\hat{v}+1\right) + \frac{1}{2}\log\left(\hat{\alpha}\right) + \frac{1}{2}\log\left(\hat{v}\right) - \log\hat{\mu}_0 + \frac{1}{2}\log\hat{\sigma}_0^2$$

Assuming that $\frac{N_0}{\sum_{j=1}^{m} n_j} \to \gamma$, $N_0 \to \infty$ and $N_1 \to \infty$, where $0 < \gamma < \infty$, and $N = N_0 + N_1$ are sufficiently large. By the asymptotic normality of the maximum likelihood estimate, it can be verified that

$$\sqrt{N} \left[\begin{pmatrix} \hat{\mu}_1 \\ \hat{v} \\ \hat{\alpha} \\ \hat{\beta} \\ \hat{\mu}_0 \\ \hat{\sigma}_0 \end{pmatrix} - \begin{pmatrix} \mu_1 \\ v \\ \alpha \\ \beta \\ \mu_0 \\ \sigma_0 \end{pmatrix} \right] \overset{d}{\to} \begin{pmatrix} 0 \\ 0 \\ 0 \\ 0 \\ 0 \\ 0 \end{pmatrix}, \Sigma_{\varepsilon,C} .$$

The estimate of the covariance matrix $\Sigma_{\varepsilon,C}$, it is a bit complicated, so the procedure is given in the Appendix of this chapter. As it can be seen from the Appendix, we have

$$\Sigma_{\varepsilon,C} = \begin{pmatrix} I_x^{-1}(\theta_2) & 0 \\ 0 & I_0(\mu_0,\sigma_0) \end{pmatrix},$$

where $I_x^{-1}(\theta_2)$ and $I_0(\mu_0,\sigma_0)$ are given in the Appendix.

By the multivariate delta method, it follows, asymptotically, that

$$\sqrt{N} \left(\log\hat{\Delta} - \log\Delta \right) \overset{d}{\to} N\left(0, \ \sigma_{\varepsilon,C}^2 \right), \tag{10.22}$$

where

$$\sigma_{\varepsilon,C}^2 = B_{\varepsilon,C} \Sigma_{\varepsilon,C} B_{\varepsilon,C}^T,$$

and

$$B_{\varepsilon,C} = \left(\frac{\partial\log\Delta}{\partial\mu_\mu}, \ \frac{\partial\log\Delta}{\partial v}, \ \frac{\partial\log\Delta}{\partial\alpha}, \ \frac{\partial\log\Delta}{\partial\beta}, \ \frac{\partial\log\Delta}{\partial\mu_0}, \ \frac{\partial\log\Delta}{\partial\sigma_0} \right)$$

$$= \left(\frac{1}{\mu_\mu}, \ \frac{1}{2v} - \frac{1}{2(v+1)}, \ \frac{1}{2\alpha}, \ -\frac{1}{2\beta}, \ -\frac{1}{\mu_0}, \ \frac{1}{\sigma_0} \right),$$

where the estimates of parameters $\mu_\mu, v, \alpha, \beta$ and μ_0, σ_0 are given in (10.13) through (10.16) and (10.2).

For testing the hypothesis, consider the test statistic

$$Z_{\varepsilon,C} = \frac{\sqrt{N} \left(\log\hat{\Delta} - \log\Delta \right)}{\sigma_{\varepsilon,C}},$$

where ε is the significant level of the test. If $-z_{\varepsilon/2} < z_{\varepsilon,C} < z_{\varepsilon/2}$, the null hypothesis H_{01} is rejected and the generalizability of clinical results in the similar but different region is included.

10.5 An Example

To illustrate the proposed methods for analysis of the sensitivity index for addressing the generalizability of the observed clinical data from an original target patient population to a similar but different patient population, we consider the following example concerning an asthma clinical trial as described in Chow and Chang (2006).

A placebo-control clinical trial was conducted to evaluate the efficacy of an investigational drug product for treatment of patients with asthma. The primary study endpoint is the change in FEV1 (forced volume per second), which is defined to be the difference between the FEV1 after treatment and the baseline FEV1. Since the raw data are not available, without loss of generality, and for the sake of illustration, we simulated the asthma data according to summary statistics given in Table 10.2 (see also Chow and Chang, 2006). Next, taking parameter settings into account, 10,000 random samples were generated in each case for illustrating the proposed method.

10.5.1 Case 1: ε Is Random and C Is Fixed

According to the information from Chow and Chang (2006), the data of original target patient population were generated from the population $N(0.34, 0.15^2)$. There are 40 adult patients data points, and the data of similar but different target patient from the new patients $N(\mu_j, 0.143)$ and $\mu_j \sim N(0.32, 0.04)$, $j = 1, \ldots, 5$; we generate 8 data while $\mu_j = 0.3996, 0.3526, 0.3125, 0.3102, 0.3488$, respectively.

TABLE 10.2

Summary Statistics of an Asthma Trial

	Baseline FEV1 Range	Number of Patients	Baseline FEV1 Mean	FEV1 Change Mean	FEV1 Change SD
Test drug	1.5–2.0	9	1.86	0.31	0.14
	1.5–2.5	15	2.30	0.42	0.14
	1.5–3.0	16	2.79	0.54	0.16
Placebo	1.5–2.0	8	1.82	0.16	0.15
	1.5–2.5	16	2.29	0.19	0.13
	1.5–3.0	16	2.84	0.20	0.14

Source: Table 2.3 of Chow, S.C. and Chang, M., *Adaptive Design Methods in Clinical Trials*, Chapman and Hall/CRC Press, Taylor & Francis Group, New York, 2006.

By (10.4) through (10.6), the estimate value $\hat{\mu}_{new} = 0.3107$ and $\hat{\sigma}_{new} = 0.1370$ in the similar but different target patient, and $\hat{\mu}_0 = 0.3471$ and $\hat{\sigma}_0 = 0.1652$ in the adult patients population, so the estimate of the sensitivity index $\hat{\Delta} = 1.0791$. The result from clinical data in the original target population is likely to be generalizable to the new target population.

10.5.2 Case 2: ε Is Fixed and C Is Random

The data of original target patient population were generated from the population $N(0.34, 0.15^2)$. There are 40 adult patients data, and the data of similar but different target patient from the new population $N(0.32, \sigma_j^2)$ and $\sigma_j^2 \sim IG(8,1), j = 1, \ldots, 5$; we generated 8 data while $\sigma_j = 0.1490, 0.1005, 0.1655, 0.1234, 0.0890$ from $\sigma_j^2 \sim IG(8,1), j = 1, \ldots, 5$, respectively. By the (10.9) through (10.11), the estimate value $\hat{\mu}_{new} = 0.3136$ and $\hat{\sigma}_{new} = 0.1546$ in the similar but different target patient, and $\hat{\mu}_0 = 0.3280$ and $\hat{\sigma}_0 = 0.1546$ from the original adult patients, and the estimate of the sensitivity index $\hat{\Delta} = 1.0566$ between the two population. The result from clinical data in the original target population is likely to be generalizable to the new target population.

10.5.3 Case 3: ε and C Are Both Random

The data of original target patient population were generated from the population $N(0.34, 0.15^2)$. There are data sets for 40 adult patients, and the data of similar but different target patient from the new population $N(\mu_j, \sigma_j^2)$ with $\sigma_j^2 \sim IG(8,1), j = 1, \ldots, 5$; we generated 8 data while $\sigma_j = 0.1057, 0.1590, 0.0900, 0.1085, 0.1036$ from $\sigma_j^2 \sim IG(8,1), j = 1, \ldots, 5$ respectively, and the corresponding $\mu_j = 0.3417, -0.2132, 0.5486, 0.3410, 0.1179$. By the (10.13) through (10.16), the estimate value $\hat{\mu}_{new} = 0.3173$ and $\hat{\sigma}_{new} = 0.3024$ in the similar but different target patient, and $\hat{\mu}_0 = 0.3533$ and $\hat{\sigma}_0 = 0.1591$ from the original adult patients, and the estimate of the sensitivity index $\hat{\Delta} = 0.4726$ between the two population. The result from clinical data in the original target population is possibly not generalizable to the new target population in this procedure.

10.6 Concluding Remarks

In clinical research, it is often of interest to determine whether the observed clinical results can be generalized from the original target patient population (e.g., adults) to a similar but different patient population (e.g., pediatrics or elderly). Following the similar idea of population shift due to protocol amendments, analysis with covariate adjustment and the assessment of sensitivity index are the two commonly considered approaches for assessment of generalizability of clinical results. For the method of analysis with covariate adjustment, since such covariates may not exist or exist but not be observable, we focus on the assessment of generalizability through the evaluation of the sensitivity index. For the assessment of sensitivity index, in this chapter we have considered the cases where (i) ε is random variable and C is fixed, (ii) ε is fixed and C is random variable, and (iii) both ε and C are random variables. However, there are other cases such as (i) random split of the similar and different patient population (for assessment of shift in location parameter and change in scale parameter) and (ii) the number of segments in the split samples is also a random variable. The problems in analyzing these situations remain unsolved.

In addition, statistically it is also a challenge to clinical researchers when there are differences in demographics and/or patient characteristics. The imbalances between the original target patient population and the similar but different patient population could have a negative or positive impact on the generalizability of the clinical results. An illustration of the impact of ethnic factors on the responses to therapeutics is seen in the case of the epidermal growth factor receptor (EGFR) tyrosine kinase inhibitor gefitinib (Iressa). Recently, Iressa was approved in Japan and the United States for the treatment of non-small cell lung cancer (NSCLC). The EGFR is a promising target for anticancer therapy because it is more abundantly expressed in lung carcinoma tissue than in adjacent normal lung tissue. However, clinical trials have revealed significant variability in the response to gefitinib with higher responses observed in Japanese patients than in a predominantly European-derived population (27.5% vs. 10.4%, in a multi-institutional phase II trial; Fukuoka et al., 2003). Paez et al. (2004) also show that somatic mutations of the EGFR were found in 15 of 58 unselected tumors from Japan and 1 of 61 from the United States. Treatment with Iressa causes tumor regression in patients with NSCLC more frequently in Japan. Finally, the striking differences in the frequency of EGFR mutation and response to Iressa between Japanese and American patients raise general questions regarding variations in the molecular pathogenesis of cancer in different ethnic, cultural, and geographic groups. Recently, generalizability of a new treatment has attracted more and more attention from sponsors as well as regulatory authorities, the key issues such as when and how to address the geographic variations of efficacy and safety for the statistical inference are still some challenges.

Appendix

Consider the covariance-variance matrix of the parameter vector $\theta_1 = (\mu, \alpha, \beta)$. Thomas (1982) derived the observer information matrix that is employed when an EM algorithm is used to find maximum likelihood estimates in incomplete data problems. According to Louis's method, the information matrix of the vector $\theta_1 = (\mu, \alpha, \beta)$ is given by

$$I_x(\theta_1) = E_{\theta_1}\{B(z,\theta_1)\,|\,z \in R\} - E_{\theta_1}\{S(z,\theta_1)S^T(z,\theta_1)\,|\,z \in R\} + S^*(x,\theta_1)S^{*T}(x,\theta_1),$$

(A10.1)

where $S(z,\theta_1)$ are the gradient vectors of log-likelihood function $\log f(z\,|\,\theta_1)$, $B(z,\theta_1)$ are the negatives of the associated second derivative matrices. Of course, they need be evaluated only on the last iteration of the EM procedure, where $S^*(x,\theta_1) = E\{S(z,\ \theta_1)\,|\,z \in R\}$ is zero. Here

$$S^T(z,\theta_1) = \left(\sum_{j=1}^{m}\sum_{i=1}^{n_j}(x_{ji}-\mu)/\sigma_j^2, \sum_{j=1}^{m}\log\frac{1}{j} + m\log\beta - m\phi(\alpha), \frac{m\alpha}{\beta} - \sum_{j=1}^{m}\frac{1}{\sigma_j^2}\right)$$

(A10.2)

$$B(z,\theta_1) = \begin{pmatrix} \sum_{j=1}^{m}\dfrac{n}{\sigma_j^2} & 0 & 0 \\[2ex] 0 & m\phi'(\alpha) & -\dfrac{m}{\beta} \\[2ex] 0 & -\dfrac{m}{\beta} & \dfrac{m\alpha}{\beta^2} \end{pmatrix}$$

(A10.3)

and

$$S^{*T}(z,\theta_1) = \left(\sum_{j=1}^{m}\sum_{i=1}^{n_j}\widehat{V}_{1j}^{(t)}(x_{ji}-\mu), \sum_{j=1}^{m}\widehat{V}_{2j}^{(t)} + m\log\beta - m\phi(\alpha), \frac{m\alpha}{\beta} - \sum_{j=1}^{m}\widehat{V}_{1j}^{(t)}\right)$$

(A10.4)

substituting (A10.2)–(A10.4) to (A10.1), the information matrix $I_Y(\theta_1)$ of vector $\theta_1 = (\mu, \alpha, \beta)$ can be obtained. $I_x(\theta_1)$ can be inverted to find the covariance matrix of $\hat{\theta}$, that is,

$$Cov(\hat{\theta}_1) = I_x^{-1}(\hat{\theta}_1).$$

For the case 3 when the ε and C are random, the information matrix of the parameters vector θ_2 can be obtained by Thomas (1982) method, where

$$S^T(z,\theta_2) = \left(\sum_{j=1}^{m}\frac{v(\mu_j-\mu)}{\sigma_j^2}, -\sum_{j=1}^{m}\frac{(\mu_j-\mu)^2}{\sigma_j^2} + \frac{m}{2v}\sum_{j=1}^{m}\log\frac{1}{\sigma_j^2}\right.$$

$$\left. -m\phi(\alpha) + m\log\beta, -\sum_{j=1}^{m}\frac{1}{\sigma_j^2} + \frac{m\alpha}{\beta}\right),$$

(A10.5)

$$B(z,\theta_2)=\begin{pmatrix} \sum_{j=1}^{m}\dfrac{v}{\sigma_j^2} & -\sum_{j=1}^{m}\dfrac{(\mu_j-\mu)}{\sigma_j^2} & 0 & 0 \\[2ex] -\sum_{j=1}^{m}\dfrac{(\mu_j-\mu)}{\sigma_j^2} & \dfrac{m}{2v^2} & 0 & 0 \\[2ex] 0 & 0 & m\phi'(\alpha) & -\dfrac{m}{\beta} \\[2ex] 0 & 0 & -\dfrac{m}{\beta} & \dfrac{m\alpha}{\beta^2} \end{pmatrix} \tag{A10.6}$$

and

$$S^{*T}(x,\theta_2)=\left(\sum_{j=1}^{m}\left(\dfrac{\sum_{i=1}^{n}x_{ji}+v^{(t)}\mu^{(t)}}{n_j+v^{(t)}}-\mu_{new}\right)v\widehat{W}_{1i}^{(t)},-\sum_{j=1}^{m}\dfrac{\widehat{W}_{3j}^{(t)}}{2}+\dfrac{m}{2v}\right.,$$

$$\left.\sum_{j=1}^{m}\widehat{W}_{2j}^{(t)}-m\phi(\alpha)+m\log\beta,\ -\sum_{j=1}^{m}\widehat{W}_{1j}^{(t)}+\dfrac{m\alpha}{\beta}\right) \tag{A10.7}$$

Substituting (A10.5)–(A10.7) to (A10.1), the covariance-variance matrix of vector

$$\theta_2=\left(\mu_{new},v,\alpha,\beta\right)$$

can be obtained by the inverse matrix $I_{11}^{-1}(\hat{\theta}_2)$.

11

Consistency Evaluation

11.1 Introduction

In recent years, multi-regional (or multi-national) multi-center clinical trials have become very popular in global pharmaceutical/clinical development. The main purpose of a multi-regional clinical trial is not only to assess the efficacy of the test treatment over all regions in the trial, but also to bridge the overall effect of the test treatment to each of the region in the trial. Most importantly, a multi-regional clinical trial serves to shorten the time for drug development and regulatory development, submission, and approval around the world. Although multi-regional clinical trials provide the opportunity to fully utilize clinical data from all regions to support regional (local) registration, some critical issues such as regional differences (e.g., culture and medical practice/perception) and possible treatment-by-region interaction that may have an impact on the validity of the multi-regional trials inevitably occur.

In multi-regional trials for global pharmaceutical/clinical development, one of the commonly encountered critical issues is that clinical results observed from some regions (e.g., Asian-Pacific region) are inconsistent with clinical results from other regions (e.g., European Community) or global results (i.e., all regions combined). The inconsistency in clinical results between different regions (e.g., Asian-Pacific region and European Community and/or United States) could be due to differences in factors linked to ethnicity. In this case, the dose or dose regimen may require adjustment or a bridging study may be required before the data can be pooled for an overall combined assessment of the treatment effect. As a result, evaluation of consistency between specific regions (sub-population) and all regions combined (global population) is necessarily conducted before regional registration (e.g., Japan and China). It should be noted that different regions may have different requirements regarding sample sizes for studies conducted at specific regions in order to obtain registration (see, e.g., MHLW, 2007).

In practice, consistency between clinical results observed from a sub-population (specific region such as Japan or China) and the entire population (all regions combined) is often interpreted as similarity and/or equivalence between the two populations in terms of treatment effect (i.e., safety and/or efficacy). Along this line, several statistical methods including test for consistency (Shih, 2001b), assessment of consistency index (Tse et al., 2006), evaluation of sensitivity index (Chow et al., 2002), achieving reproducibility and/or generalizability (Shao and Chow, 2002), Bayesian approach (Hsiao et al., 2007; Chow and Hsiao, 2010), and the Japanese approach for evaluation of assurance probability (MHLW, 2007) have been proposed in the literature (see also, Liu et al., 2013). The purpose of this chapter is not only to provide an overview of these methods, but also to compare the relative performances of these methods through extensive clinical trial simulation (see also Ying et al., 2017).

In Section 11.2, some critical issues that are commonly encountered in multi-regional, multi-center trials are discussed. Several statistical methods for testing consistency or similarity/equivalence in clinical results observed from a sub-population as compared to that of the entire population are described in Section 11.3. Section 11.4 compares the performances of these statistical methods through extension simulation studies. In Section 11.5, an example concerning a multi-regional study involving a regional regulatory submission is discussed to illustrate the proposed statistical methods. Some concluding remarks are given in Section 11.6.

11.2 Issues in Multi-regional Clinical Trials

A multi-regional clinical trial (MRCT) is a trial conducted at more than one distinct region where the data collected from these regions are intended to be analyzed as a whole. Within a given region, the trial may be conducted as either a single-site study or a multi-center trial. In a MRCT, some practical issues are commonly encountered due to potential differences among centers within region and differences between regions.

11.2.1 Multi-center Trials

A multi-center trial is a trial conducted at more than one distinct center where the data collected from these centers are intended to be analyzed as a whole. At each center an *identical* study protocol is used. A multicenter trial is a trial with a center or site as a natural blocking or stratified variable that provides replications of clinical results. A multicenter trial should permit an overall estimation of the treatment difference for the targeted patient population across various centers. For a multi-center trial, the FDA suggests that individual center results should also be presented. In addition, the FDA suggests that statistical tests for homogeneity across centers (i.e., for detecting possible treatment-by-center interaction) be performed (FDA, 1988). Any extreme or contradictory results among centers should be noted and discussed. If there is no treatment-by-center interaction, the data can be pooled for analysis across centers.

Another issue is related to the use of central laboratory for testing of samples collected from different centers, which may have a significant impact on the assessment of efficacy and safety of the test treatment under investigation. A central laboratory provides a consistent assessment for laboratory tests. If a central laboratory is not used, the assessment of laboratory tests may differ from center to center depending on the equipment, analyst, and laboratory normal ranges used at that center. In such a case, possible confounding and interaction effects makes it difficult to combine the laboratory values obtained from the different centers for an unbiased assessment of the safety and efficacy of the test treatment.

For statistical analysis of data collected from a multi-center trial, if there is no evidence of treatment-by-center interaction, the data can be pooled for analysis for an overall assessment of the treatment effect across centers. Along this line, Nevius (1988) proposes a set of four conditions under which evidence from a multi-center trial would provide sufficient statistical evidence of efficacy. These conditions are summarized as follows:

1. The combined analysis shows significant results.
2. There is consistency over centers in terms of direction of results.
3. There is consistency over centers in terms of producing nominally significant results in centers with sufficient power.
4. Multiple centers show evidence of efficacy after adjustment for multiple comparisons.

11.2.2 Multi-regional, Multi-center Trials

Unlike a single multi-center trial within a given region, a multi-regional trial is much more complicated particularly when it incorporates several multi-center trials from different regions. In addition to those practical issues that are commonly seen in a multi-center trial, there are more critical issues that may have an impact on the assessment of treatment effect in global pharmaceutical/clinical development. These issues include, but are not limited to, potential differences in (1) study protocol (due to different regulatory requirements); (2) ethnic factors; (3) culture; and (4) medical practice or perception in different regions. Thus, in practice, it is suggested that critical issues such as representativeness, bias/variation control, heterogeneity (similarities and dis-similarities), consistency, and poolability across regions, which may have an impact on the validity of the multi-regional trial must be carefully evaluated in order to provide an overall assessment of the test treatment under investigation.

In practice, it is often a concern that only positive clinical results are observed in global population (i.e., all regions combined) and yet some regions (i.e., sub-populations) may fail to show positive results or show positive results with insufficient power. This inconsistency, if it is not purely due to chance, is concern for regional (local) registration. As inconsistency may be due to the fact that studies conducted at different regions may be conducted with similar but different (1) study protocols; (2) drug products and doses; (3) patient populations with different ethnic factors; (4) sample sizes; or (5) evaluability criteria. Regional regulatory agencies often request that such inconsistency be carefully evaluated during the regulatory review and approval process.

For evaluation of consistency, a null hypothesis of inconsistency is often tested. Many investigators then conclude the alternative hypothesis of consistency if we fail to reject the null hypothesis based on subgroup analysis (often with insufficient power for establishing non-inferiority or similarity between the sub-population and the entire population). It should be noted that regional regulatory agencies are not likely to accept this assessment with limited number of subjects available. On the other hand, the regional regulatory agencies may accept the conclusion that there is little evidence to demonstrate that the sub-population is inconsistent with the entire population. As a result, an additional clinical study may be required conducted on the sub-population in the region in order to confirm the consistency or similarity between the sub-population and the entire population.

11.3 Statistical Methods

As mentioned in the introduction section, several methods that have been proposed in the literature, in different contexts, could be useful for evaluation of consistency between a sub-population and the entire population in multi-regional studies. These methods are briefly outlined in the subsequent sections.

11.3.1 Test for Consistency

For a multi-regional clinical trial involving K regions, denote by $W = \{W_1, \ldots, W_K\}$ the results of the multi-regional trial, where W_i is the results observed from the ith region. To determine whether the local results from a given region is consistent with the global results, Shih (2001b) proposed to construct a predictive probability function, $P(v \mid W)$, where v is the study conducted at a given region. The predictive probability $P(v \mid W)$ provides a measure of the plausibility of v given the results of W. Thus, Shih's test for consistency can be summarized as follows (Shih, 2001b).

For a given region i, we first construct the predictive probability function, $P(W_i \mid W)$ of the results observed from the region. We then compare $P(W_i \mid W)$

with the plausibility of each of the results from other regions, i.e., $P(W_j| W)$, where $j \neq i$ and $j = 1,\ldots,K$. We conclude that the results of W_I is consistent with $W = \{W_1,\ldots,W_K\}$ if and only if

$$P(W_i| W) \geq \min\{P(W_j| W), j \neq i, j = 1,2,\ldots,K\}. \tag{11.1}$$

To accommodate the superiority test with the case of the smaller the better, the method can be modified as

$$W_i - W \leq \max\{W_j - W, j \neq i, j = 1,2,\ldots,K\}.$$

If the above inequality (11.1) holds, we conclude that the results of W_i is consistent with

$$W = \{W_1,\ldots,W_K\}.$$

To accommodate the superiority test with the case of the larger the better, the method can be modified as

$$W_i - W \geq \min\{W_j - W, j \neq i, j = 1,2,\ldots,K\}.$$

If the above inequality holds, we conclude that the results of W_i is consistent with

$$W = \{W_1,\ldots,W_K\}.$$

11.3.2 Assessment of Consistency Index

Let U and W be the measures of the primary study endpoint of a given MRCT from a sub-population (e.g., a specific region) and the global population (i.e., all regions combined), respectively, where $X = \log U$ and $Y = \log W$ follows normal distributions with means μ_X, μ_Y and variances V_X, V_Y, respectively. Similar to the idea of using $P(X < Y)$ to assess reliability in statistical quality control (Church and Harris, 1970; Enis and Geisser, 1971), Tse et al. (2006) proposed the following probability as an index to assess the consistency of clinical results observed from the sub-population as compared to the global population

$$p = P\left(1-\delta < \frac{U}{W} < \frac{1}{1-\delta}\right), \tag{11.2}$$

where $0 < \delta < 1$ and is defined as a limit that allows for consistency. Tse et al. (2006) referred p as the consistency index. Thus, p tends to 1 as δ tends to 1. For a given δ, if p is close to 1, the measures U and W are considered to be identical. It should be noted that a small δ implies the requirement of high degree of consistency between material U and material W. In practice, it may be difficult to meet this narrow specification for consistency. Under the normality assumption of $X = \log U$ and $Y = \log W$, (11.2) can be rewritten as

$$p = P(\log(1-\delta) < \log U - \log W < -\log(1-\delta))$$

$$= \Phi\left(\frac{-\log(1-\delta)-(\mu_X-\mu_Y)}{\sqrt{V_X+V_Y}}\right) - \Phi\left(\frac{\log(1-\delta)-(\mu_X-\mu_Y)}{\sqrt{V_X+V_Y}}\right).$$

where $\Phi(z_0) = P(Z < z_0)$ with Z being a standard normal random variable. Therefore, the consistency index p is a function of the parameters $\theta = (\mu_X, \mu_Y, V_X, V_Y)$, i.e. $p = h(\theta)$. Suppose that observations $X_i = \log U_i, i = 1, ..., n_X$ and $Y_i = \log W_i, i = 1, ..., n_Y$ are collected in an assay study. Then, using the invariance principle, the maximum likelihood estimator of p can be obtained as

$$\hat{p} = \Phi\left(\frac{-\log(1-\delta)-(\bar{X}-\bar{Y})}{\sqrt{\hat{V}_X+\hat{V}_Y}}\right) - \Phi\left(\frac{\log(1-\delta)-(\bar{X}-\bar{Y})}{\sqrt{\hat{V}_X+\hat{V}_Y}}\right), \qquad (11.3)$$

where $\bar{X} = \frac{1}{n_X}\sum_{i=1}^{n_X} X_i$, $\bar{Y} = \frac{1}{n_Y}\sum_{i=1}^{n_Y} Y_i$, $\hat{V}_X = \frac{1}{n_X}\sum_{i=1}^{n_X}(X_i-\bar{Y})^2$, and $\hat{V}_Y = \frac{1}{n_Y}\sum_{i=1}^{n_Y}(Y_i-\bar{Y})^2$. In other words, $\hat{p} = h(\hat{\theta}) = h(\bar{X}, \bar{Y}, \hat{V}_X, \hat{V}_Y)$. Furthermore, it can be easily verified that the following asymptotic result holds. Tse et al. (2006) showed $\hat{p}$ is asymptotically normal with mean $E(\hat{p})$ and variance $Var(p)$, where $E(p) = p + B(p) + o\left(\frac{1}{n}\right)$ and $Var(\hat{p}) = C(p) + o\left(\frac{1}{n}\right)$. The detailed expressions of $B(p)$ and $C(p)$ are given in Tse et al. (2006). In other words,

$$\frac{\hat{p}-E(\hat{p})}{\sqrt{var(\hat{p})}} \to N(0,1).$$

Hypotheses testing of the consistency index p can also be conducted based on the asymptotic normality of $\hat{p}$. Consider the following hypotheses

$$H_0 : p \le p_0 \text{ vs. } H_a : p > p_0.$$

In practice, we can choose p_0 as a pre-specified constant, say 0.8.

We would reject the null hypothesis and in favor of the alternative hypothesis of consistency. Under H_0, we have

$$\frac{\hat{p} - p_0 - B(\hat{p})}{\sqrt{var(\hat{p})}} \sim N(0, 1)$$

(11.4)

Thus, we reject the null hypothesis H_0 at the α level of significance if

$$\frac{\hat{p} - p_0 - B(\hat{p})}{\sqrt{var(\hat{p})}} > Z_\alpha.$$

This is equivalent to reject the null hypothesis H_0 when

$$\hat{p} > p_0 + B(\hat{p}) + Z_\alpha \sqrt{var(\hat{p})}.$$

For superiority test, we are probably more concerned with $p = P\left(\frac{U}{W} > 1 - \delta\right)$. Similar procedures for testing the consistency can be derived to suit this case. For simplicity, we do not give the details here.

11.3.3 Evaluation of Sensitivity Index

In a MRCT, denote the entire population (all regions combined) by (μ_0, σ_0), where μ_0 and σ_0 are the population mean and population standard deviation, respectively. Similarly, denote the sub-population (a or some specific regions) by (μ_1, σ_1). Since the two populations are similar but slightly different, it is reasonable to assume that $\mu_1 = \mu_0 + \varepsilon$ and $\sigma_1 = C\sigma_0$ ($C > 0$), where ε is referred to as the shift in location parameter (population mean) and C is the inflation factor of the scale parameter (population standard deviation). Thus, the (treatment) effect size adjusted for standard deviation of the sub-population (μ_1, σ_1) can be expressed as follows:

$$\delta_1 = \left|\frac{\mu_1}{\sigma_1}\right| = \left|\frac{\mu_0 + \varepsilon}{C\sigma_0}\right| = |\Delta|\left|\frac{\mu_0}{\sigma_0}\right| = |\Delta|\delta_0,$$

where $\Delta = \frac{1 + \varepsilon/\mu_0}{C}$ and δ_0 is the effect size adjusted for standard deviation of the entire population (μ_0, σ_0). Δ is referred to as a sensitivity index measuring the change in effect size between the sub-population and the entire population (Chow et al., 2002).

As it can be seen, if $\varepsilon = 0$ and $C = 1$, then $\delta_1 = \delta_0$. That is, the effect sizes of the two populations are identical. Applying the concept of bioequivalence assessment, we can claim that the effect sizes of the two patient populations are consistent if the confidence interval of $|\Delta|$ is within (80%, 120%).

In practice, the shift in location parameter (ε) and/or the change in scale parameter (C) could be random. If both ε and C are fixed, the sensitivity index can be assessed based on the sample means and sample variances obtained from the entire population and the sub-population. As indicated by Chow et al. (2002), ε and C can be estimated by $\hat{\varepsilon} = \hat{\mu}_1 - \hat{\mu}_0$ and $\hat{C} = \hat{\sigma}_1 / \hat{\sigma}_0$, respectively, where $(\hat{\mu}_0, \hat{\sigma}_0)$ and $(\hat{\mu}_1, \hat{\sigma}_1)$ are some estimates of (μ_0, σ_0) and (μ_1, σ_1) based on the entire population and the sub-population, respectively. Consequently, the sensitivity index Δ can be estimated by $\hat{\Delta} = \frac{1+\hat{\varepsilon}/\hat{\mu}_0}{\hat{C}}$ and the corresponding confidence interval can be obtained based on normal approximation (Chow et al., 2002; Lu et al., 2017).

In real world problems, however, ε and C could be either fixed or random variables. In other words, there are three possible scenarios: (i) the case where ε is random and C is fixed, (ii) the case where ε is fixed and C is random, and (iii) the case where both ε and C are random. Statistical inference of Δ under each of these possible scenarios has been studied by Lu et al. (2017).

Besides, we also give a simplified version: if $\hat{\Delta}$ is within (80%, 120%), the consistency is claimed. For superiority test, the criterion can be $\hat{\Delta} > 80\%$.

11.3.4 Achieving Reproducibility and/or Generalizability

Alternatively, Shao and Chow (2002) proposed the concept of reproducibility and generalizability probabilities for assessing consistency between specific sub-populations and the global population. If the influence of the ethnic factors is negligible, then we may consider the reproducibility probability to determine whether the clinical results observed in a patient population is reproducible in a different patient population. If there is a notable ethnic difference, the concept of generalizability probability can be used to determine whether the clinical results observed in a patient population can be generalized to a similar but slightly different patient population with notable differences in ethnic factors.

We proceed from the original definitions of reproducibility and generalizability probabilities, and proposed three modified methods for normal-based test procedure and power calculation. Versions for t-distribution based test procedure and power calculation can similarly be derived.

11.3.4.1 Specificity Reproducibility Probability for Inequality Test

Suppose we are concerned with the null hypothesis $H_0 : \mu_1 = \mu_2$ and alternative hypothesis $H_a : \mu_1 \neq \mu_2$, where μ_1 and μ_2 are means of a region and the global area. Denote the variances of the region and the global area by σ_1^2 and σ_2^2, respectively. Consider the case without the assumption of equal variance. When both sample sizes n_1 and n_2 are large, an approximate α level test rejects H_0 when $|T| > z_{1-\alpha/2}$, where

$$T = \frac{\bar{x}_1 - \bar{x}_2}{\sqrt{\left(\dfrac{s_1^2}{n_1} + \dfrac{s_1^2}{n_2}\right)}}$$

and $z_{1-\alpha/2}$ is the $1-\alpha/2$ quantile of the standard normal distribution. Since T is approximately distributed as $N(\theta,1)$ with

$$\theta = \frac{\mu_1 - \mu_2}{\sqrt{\left(\dfrac{\sigma_1^2}{n_1} + \dfrac{\sigma_2^2}{n_1}\right)}},$$

with a pre-specified level of β (say, 0.2), we get the two-sided $1-\beta$ confidence interval for θ as $(\hat{\theta}_- - \hat{\theta}_+) = \left(T - z_{1-\frac{\beta}{2}}, T + z_{1-\frac{\beta}{2}}\right)$. Then we get two types of one-sided $1-\beta$ confidence upper bound for $|\theta|$, denoted as $|\hat{\theta}|_+^1$ and $|\hat{\theta}|_+^2$: the first is $\max\{|\hat{\theta}_-|, |\hat{\theta}_+|\}$; the second is the non-negative value v satisfying $P\{|\theta| \le v \,|\, T\} = 1-\beta$, which is equivalent to the equation $\Phi(v-T) - \Phi(-v-T) = 1-\beta$, with $\Phi(\cdot)$ being the distribution function of the standard normal distribution. We define specificity reproducibility probability as

$$P\left\{|T| \le z_{1-\frac{\alpha}{2}} |\hat{\theta}|_+^j \right\}, j = 1, 2,$$

which is equal to $\Phi\left(z_{1-\frac{\alpha}{2}} - |\hat{\theta}|_+^j\right) - \Phi\left(-z_{1-\frac{\alpha}{2}} - |\hat{\theta}|_+^j\right)$. Compare the specificity reproducibility probability with a pre-specified constant (say, 0.8). If the specificity reproducibility probability is larger, we conclude that the consistency holds.

11.3.4.2 Superiority Reproducibility Probability

With respect to the superiority test with the case of the larger the better, consider null hypothesis $H_0 : \mu_1 \le q * \mu_2$ and alternative hypothesis $H_a : \mu_1 > q * \mu_2$, where q is a pre-specified constant (say 0.5). The notations here have the same meanings as those in the last paragraph. We have

$$T = \frac{\bar{x}_1 - q\bar{x}_2}{\sqrt{\left(\dfrac{s_1^2}{n_1} + \dfrac{q^2 s_1^2}{n_2}\right)}}$$

and

$$\theta = \frac{\mu_1 - q\mu_2}{\sqrt{\left(\dfrac{\sigma_1^2}{n_1} + \dfrac{q^2 \sigma_2^2}{n_1}\right)}}.$$

The rejection region is $\{T > z_{1-\alpha}\}$. The one-sided $1-\beta$ confidence lower bound for $|\theta|$, denoted as $\hat{\theta}_-$, is $T - z_{1-\beta}$. The superiority reproducibility probability is defined as

$$P\{T > z_{1-\alpha}|\hat{\theta}_-\} = \Phi(\hat{\theta}_- - z_{1-\alpha}) = \Phi(T - z_{1-\beta} - z_{1-\alpha}).$$

Compare the superiority reproducibility probability with a pre-specified constant (say, 0.8). If the superiority reproducibility probability is larger, we conclude that the consistency holds.

11.3.4.3 Reproducibility Probability Ratio for Inequality Test

Suppose we have the null hypothesis $H_0 : \mu_1 = \mu_2$ and alternative hypothesis $H_a : \mu_1 \neq \mu_2$, where μ_1 and μ_2 are means of the treatment and the reference for a specific region (or the global area). Denote by σ_1^2 and σ_2^2 the variances of the treatment and the reference for a specific region (or the global area). Consider the case without the assumption of equal variance. When both sample sizes n_1 and n_2 are large, an approximate α level test rejects H_0 when $|T| > z_{1-\alpha/2}$, where

$$T = \frac{\bar{x}_1 - \bar{x}_2}{\sqrt{\left(\dfrac{s_1^2}{n_1} + \dfrac{s_1^2}{n_2}\right)}}$$

and $z_{1-\alpha/2}$ is the $1-\alpha/2$ quantile of the standard normal distribution. Since T is approximately distributed as $N(\theta, 1)$ with

$$\theta = \frac{\mu_1 - \mu_2}{\sqrt{\left(\dfrac{\sigma_1^2}{n_1} + \dfrac{\sigma_2^2}{n_1}\right)}},$$

the reproducibility probability by using the estimated power approach is given by

$$\hat{P} = \Phi(T - z_{1-\alpha/2}) = \Phi(-T - z_{1-\alpha/2}).$$

Denote $\hat{P}_S$ and $\hat{P}_G$ as the reproducibility probabilities for a region and the global area, respectively. If $\hat{P}_S / \hat{P}_G \geq q$ where q is a pre-specified constant, then the consistency of the region's results and the global results is concluded.

In addition, the reproducibility probability using the confidence bound approach can be employed, as suggested in the literature by Shao and Chow (2002), with the same criterion of consistency test as seen above being applicable.

11.3.4.4 Reproducibility Probability Ratio for Superiority Test

Using techniques similar to the reproducibility probability ratio for inequality test, we can also consider the ratio for superiority test (with the case of the larger the better): $H_0 : \mu_1 \leq \mu_2$ versus $H_a : \mu_1 > \mu_2$. After obtaining the ratios for the region and the global area, the same criterion for testing consistency applies.

11.3.5 Bayesian Approach

For testing consistency or similarity between clinical results observed from a sub-population and those obtained from the global population, Chow and Hsiao (2010) proposed the following Bayesian approach via comparing the following hypotheses

$$H_0 : \Delta \leq 0 \quad \text{vs.} \quad H_a : \Delta > 0,$$

where $\Delta = \mu_S - \mu_G$, in which μ_S and μ_G are the population means of the sub-population and the global population, respectively. Let X_i and Y_j, $i = 1, 2, \ldots, n$ and $j = 1, 2, \ldots, N$, be some efficacy responses observed in the sub-population and the global population, respectively. For simplicity, assume that both X_is and Y_js are normally distributed with known variance σ^2. When σ^2 is unknown, it can generally be estimated by the sample variance. Thus, Δ can be estimated by

$$\hat{\Delta} = \bar{x} - \bar{y},$$

where $\bar{x} = \sum_{i=1}^{n} x_i / n$ and $\bar{y} = \sum_{j=1}^{N} y_j / N$. Under the following mixed prior information for Δ

$$\pi = \gamma \pi_1 + (1 - \gamma) \pi_2,$$

which is a weighted average of two priors, where $\pi_1 \equiv c$ is a non-informative prior and π_2 is a normal prior with mean θ_0 and variance σ_0^2, and γ is defined as the weight with $0 \leq \gamma \leq 1$. For the consistency test between a specific region and the global area, the choice of (θ_0, σ_0^2) can be derived from the values of $\hat{\Delta}$ from all regions except the specific region. Thus, the marginal density of $\hat{\Delta}$ is given by

$$m\left(\hat{\Delta}\right) = \gamma + (1 - \gamma) \frac{1}{\sqrt{2\pi \left(\sigma_0^2 + \tilde{\sigma}^2\right)}} \exp\left\{ -\frac{(\hat{\Delta} - \theta_0)^2}{2(\sigma_0^2 + \tilde{\sigma}^2)} \right\},$$

where $\tilde{\sigma}^2 = \sigma^2\left(\frac{1}{n}+\frac{1}{N}\right)$. Given the clinical data and prior distribution, the posterior distribution of Δ is

$$m\left(\Delta|\hat{\Delta}\right)=\frac{1}{m\left(\hat{\Delta}\right)}\left\{\gamma\frac{1}{\sqrt{2\pi}\,\tilde{\sigma}}\exp\left[-\frac{(\Delta-\hat{\Delta})^2}{2\tilde{\sigma}^2}\right]\right.$$

$$\left.+(1-\gamma)\frac{1}{2\pi\sigma_0\tilde{\sigma}}\exp\left[-\frac{(\Delta-\theta_0)^2}{2\sigma_0^2}-\frac{(\Delta-\hat{\Delta})^2}{2\tilde{\sigma}^2}\right]\right\}.$$

Thus, given the data and prior information, similarity (consistency) on efficacy in terms of a positive treatment effect for the sub-population can be concluded if the posterior probability of consistency (similarity) is

$$p_c = P\left(\mu_S - \mu_G > 0 | \text{clinical data and prior}\right) = \int_0^\infty \pi\left(\Delta|\hat{\Delta}\right)d\Delta > 1-\alpha.$$

For some pre-specified $0 < \alpha < 0.5$. In practice, α is determined such that the sub-population is generally smaller than 0.2 to ensure that posterior probability of consistency is at least 80%.

Based on discussion given in Chow and Hsiao (2010), the marginal density of $\hat{\Delta}$ can be re-expressed as

$$m\left(\hat{\Delta}\right)=\gamma+(1-\gamma)\frac{1}{\sqrt{2\pi\left(\sigma_0^2+2\sigma^2/N\right)}}\exp\left\{-\frac{(\hat{\Delta}-\theta_0)^2}{2(\sigma_0^2+2\sigma^2/N)}\right\}.$$

As a result, the posterior distribution of Δ is therefore given by

$$m\left(\Delta|\hat{\Delta}\right)=\frac{1}{m\left(\hat{\Delta}\right)}\left\{\gamma\frac{1}{\sqrt{4\pi\sigma^2/N}}\exp\left[-\frac{(\Delta-\hat{\Delta})^2}{4\sigma^2/N}\right]\right.$$

$$\left.+(1-\gamma)\frac{1}{\sqrt{8\pi\sigma_0^2\sigma^2/N}}\exp\left[-\frac{(\Delta-\theta_0)^2}{2\sigma_0^2}-\frac{(\Delta-\hat{\Delta})^2}{4\sigma^2/N}\right]\right\}$$

Consequently, we have

$$p_c = \int_0^\infty \pi\left(\Delta|\hat{\Delta}\right)d\Delta > 1-\alpha.$$

11.3.6 Japanese Approach

In a multi-regional trial, to establish the consistency of treatment effects between the Japanese population and the entire population under study, the Japanese MHLW suggested evaluating the relative treatment effect (i.e., treatment effect observed in Japanese population as compared to that of the entire population) to determine whether the Japanese population has achieved a desired assurance probability for consistency.

Let $\delta = \mu_T - \mu_P$ be the treatment effect of the test treatment under investigation. Denote by δ_J and δ_{All} the treatment effect of the Japanese population and treatment effect of the entire population (all regions), respectively. Let $\hat{\delta}_J$ and $\hat{\delta}_{All}$ be the corresponding estimators. For a multi-regional clinical trial, given δ, sample size ratio (i.e., the size of Japanese population as compared to the entire population) and the overall result is significant at the α level of significance, the assurance probability, p_a, of consistency is defined as

$$p_a = P_\delta\left(\hat{\delta}_J \geq \rho\hat{\delta}_{All}\,|\,Z\rangle z_{1-\alpha}\right) > 1 - \gamma,$$

where Z represents the overall test statistic, ρ is minimum requirement for claiming consistency, and $0 < \gamma \leq 0.2$ is a pre-specified desired level of assurance. The Japanese MHLW suggests that ρ be 0.5 or greater. The determination of ρ, however, should be different from product to product.

In practice, above required sample size may either not be easy to be achieved or not pre-planned. Instead, consistency maybe evaluated using above idea if the effect of a subgroup achieves a specified proportion (usually ≥50%) of the observed overall effect. This can be seen a simplified version of the Japanese approach.

11.3.7 The Applicability of Those Approaches

Considering the data type and which term the consistency test is based on, the approaches we have described above exhibit different levels of applicability. With respect to the data type, the normal data and binary data are the two widely used types. As for the term "consistency," the following two meanings can be considered: comparing the two effects directly, e.g., comparing $\mu_{T1} - \mu_{R1}$ and $\mu_{T2} - \mu_{R2}$ directly, where (μ_{Ti}, μ_{Ri}) are the treatment mean and the reference mean for the ith region, $i = 1, 2$; testing the consistency in terms of the original aim, e.g., testing the consistency between $\{\mu_{T1} - \mu_{R1} > 0\}$ and $\{\mu_{T2} - \mu_{R2} > 0\}$ when the original hypothesis is H_0: $\mu_T - \mu_R \leq 0$ versus H_a: $\mu_T - \mu_R > 0$. Table 11.1 displays the applicability of the approaches.

An example of direct comparison and original aim: assume a global clinical trial was carried out to compare the effect of a new drug and the effect of a reference drug globally and regionally. Denote μ_{GT}, μ_{GR}, μ_{ST} and μ_{ST} are the means of the new drug globally, the reference drug globally, the new drug

TABLE 11.1

Displays the Applicability of the Approaches

	Data Type		Aim	
Approach	Binary	Normal	Compare Directly	Original Aim
11.3.1 Shih	✔	✔	✔	✘
11.3.2 Tse	✘	✔	✔	✘
11.3.3 Sensitivity Index	✘	✔	✔	✘
11.3.4 Specificity reproducibility probability	✘	✔	✔	✘
11.3.4 Superiority reproducibility probability	✘	✔	✔	✘
11.3.4 Reproducibility probability ratio	✘	✔	✘	✔
11.3.5 Bayesian approach	✘	✔	✔	✔
11.3.6 Japanese approach	✔	✔	✔	✘

for a specific region and the reference drug for the specific region. We were concerned with the hypothesis testing

$$H_0 : \mu_{GT} - \mu_{GR} \geq 0 \quad \text{vs.} \quad H_a : \mu_{GT} - \mu_{GR} < 0$$

as well as

$$H_0 : \mu_{ST} - \mu_{SR} \geq 0 \quad \text{vs.} \quad H_a : \mu_{ST} - \mu_{SR} < 0$$

To test the consistency between the global effect $\mu_{GT} - \mu_{GR}$ and the specific region effect $\mu_{ST} - \mu_{SR}$, we may consider a variety of hypothesis testing. One is just to test $H_0 : \mu_{GT} - \mu_{GR} = \mu_{ST} - \mu_{SR}$ versus $H_a : \mu_{GT} - \mu_{GR} \neq \mu_{ST} - \mu_{SR}$, which can be seen as direct comparison and is not directly related to whether $\mu_{ST} - \mu_{SR}$ is smaller than 0 or not. Another one is directly related to $H_0 : \mu_{ST} - \mu_{SR} \geq 0$ versus $H_a : \mu_{ST} - \mu_{SR} < 0$, such as the Bayesian method introduced previously that is used to estimate the posterior probability of $\{\mu_{ST} - \mu_{SR} < 0\}$ using both the specific regional data and global data.

11.4 Simulation Study

11.4.1 The Case of the Matched-Pair Parallel Design with Normal Data and Superiority Test

First, we perform a simulation study for the matched-pair parallel design with normal data and superiority test. Assume the clinical trial with the aim of comparing the treatment effect mean μ_T and reference effect mean μ_R

(the smaller the better), consists of 80 patients from 10 regions with equal number of patients recruited in each region. Consider the superiority hypothesis test:

$$H_0 : \mu_T - \mu_R \geq 0 \text{ vs. } H_a : \mu_T - \mu_R < 0.$$

Presume $\mu_T - \mu_R$ is equal to -1 with the standard deviation being 3 globally for the power analysis. Given the pre-specified type I error level of $\alpha = 0.05$, 80 samples can achieve a power of 90%.

For the simulation parameter specification, let the standard deviations of the difference between treatment effect and reference effect be equal to 3 within each region. The values of $\mu_T - \mu_R$ for six regions is equal to -1, while the other four regions have the values of $-2, -1.5, -0.5$ and 0, respectively. The true standard deviations of the difference between treatment effect and reference effect for the global population are 3.04.

For each simulation, we simulated data for each region, combined them as the global data, and tested the superiority hypothesis both regionally and globally with the standard t-statistic method. Denoting $\overline{X}_S$ and $\overline{X}_G$ as the sample means for the region and global area respectively, the results for the subgroup analyses could be divided into the following 4 categories (see Table 11.2).

Seventeen approaches based on the methods discussed in Section 11.3 were selected to test the consistency. As the applicability of those methods are different, for each method, the version for the original aim would be used if the method applies to the original aim, otherwise the version for direct comparison would be adopted. Those methods are listed below with their abbreviations (Table 11.3).

We did altogether 100,000 repetitions of the simulation, with the following results recorded: the rejection rate of the null hypothesis $H_0 : \mu_T - \mu_R \geq 0$ regionally and globally (see Table 11.4), the frequency of the categories regionally and globally (see Table 11.5), consistency rates of each approach under different categories and in different regions (see Table 11.6). As the six regions have identical population parameters, we combined the results for them.

TABLE 11.2

Categories of Subgroup Analysis

Category	Superiority Trial
1	$\overline{X}_S < \overline{X}_G$ (met superiority)
2	$\overline{X}_G \leq \overline{X}_S < 0$ & upper bound of $1 - \alpha$ CI for the region using global CI width < 0
3	$\overline{X}_G \leq \overline{X}_S < 0$ but upper bound of $1 - \alpha$ CI for the region using global CI width ≥ 0
4	$\overline{X}_G < 0 \leq \overline{X}_S$

TABLE 11.3

List of Abbreviations

Approach	Choice of Parameters	Abbreviation
11.3.1 Shih		Shih Superiority
11.3.2 Tse	$p_0 = 0.8, \delta = 0.2, \alpha = 0.1$	Tse
11.3.2 Tse	R–R choice for p_0, $\delta = 0.2$, $\alpha = 0.1$	Tse Reference
11.3.2 Tse	$p_0 = 0.8, \delta = 0.2, \alpha = 0.1$ the version for superiority	Tse Sup
11.3.2 Tse	R–R choice for p_0, $\delta = 0.2$, $\alpha = 0.1$, the version for superiority	Tse Sup Reference
11.3.3 Sensitivity index	$p_0 = 0.8$, simple version	SenInd Simple
11.3.4 Specificity reproducibility probability	R–R choice for p_0, $\alpha = 0.05$, $\beta = 0.2$	RPspeP Reference
11.3.4 Superiority reproducibility probability	R–R choice for $p_0, q = 0.5, \alpha = 0.05, \beta = 0.2$	RPsupP Reference
11.3.4 Reproducibility probability ratio	$\alpha = 0.05, \beta = 0.2, p_0 = 0.8$, method of power estimation for inequality test	GenP
11.3.4 Reproducibility probability ratio	$\alpha = 0.05, \beta = 0.2, p_0 = 0.8$, method of confidence bound for inequality test	GenCB
11.3.4 Reproducibility probability ratio	$\alpha = 0.05, \beta = 0.2, p_0 = 0.8$, method of power estimation for superiority test	GenP Sup
11.3.4 Reproducibility probability ratio	$\alpha = 0.05, \beta = 0.2, p_0 = 0.8$, method of confidence bound for superiority test	GenCB Sup
11.3.5 Bayesian approach	$p_0 = 0.8$, select min $\{Pc(\gamma), \gamma = 0.1,...,1\}$	Bayesian Min Constant
11.3.5 Bayesian approach	R–R choice for p_0, select min $\{Pc(\gamma), \gamma = 0.1,...,1\}$	Bayesian Min Reference
11.3.5 Bayesian approach	$p_0 = 0.8$, select mean $\{Pc(\gamma), \gamma = 0.1,...,1\}$	Bayesian Mean Constant
11.3.5 Bayesian approach	R–R choice for p_0, select mean $\{Pc(\gamma), \gamma = 0.1,...,1\}$	Bayesian Mean Reference
11.3.6 Japanese approach	$p_0 = 0.8$, simple version	Japan Simple

TABLE 11.4

Passing Rate by Global and Regions

Passing Rate of Superiority Test	Global	Region 1–6	Region 7	Region 8	Region 9	Region 10
	0.902	0.214	0.521	0.357	0.114	0.050

The simulation results show that, for this scenario, the methods of the simple version for sensitivity index, superiority reproducibility probability, reproducibility probability ratio for superiority test, Bayesian approach and simple version for Japanese approach performed better than others, with overall consistency rate close to 65% (the rate of category 1 and 2).

TABLE 11.5

Frequencies by Global and Regions

Category	Category Frequency					
	Region 1–6	Region 7	Region 8	Region 9	Region 10	All
All	600000	100000	100000	100000	100000	1000000
1	300121	84111	68970	31077	15930	500209
2	99332	7583	12746	17283	14400	151344
3	96893	5223	10409	20144	19744	152413
4	103247	3055	7828	31393	49818	195341

TABLE 11.6

Consistency Rates of Each Approach under Different Categories and in Different Regions

Approach	Category	Consistency Rate					
		Region 1–6	Region 7	Region 8	Region 9	Region 10	All
Shih superiority	All	0.922	0.988	0.969	0.834	0.676	0.9
	1	1	1	1	1	1	1
	2	0.993	0.994	0.994	0.99	0.984	0.992
	3	0.95	0.965	0.961	0.936	0.905	0.944
	4	0.602	0.686	0.661	0.519	0.393	0.539
Tse	All	0	0	0	0	0	0
	1	0	0	0	0	0	0
	2	0	0	0	0	0	0
	3	0	0	0	0	0	0
	4	0	0	0	0	0	0
Tse reference	All	0.515	0.427	0.493	0.494	0.429	0.493
	1	0.515	0.405	0.474	0.54	0.553	0.494
	2	0.584	0.569	0.577	0.579	0.573	0.581
	3	0.557	0.554	0.551	0.551	0.534	0.553
	4	0.408	0.454	0.449	0.365	0.307	0.378
Tse sup	All	0	0.004	0.001	0	0	0.001
	1	0	0.004	0.001	0	0	0.001
	2	0	0	0	0	0	0
	3	0	0	0	0	0	0
	4	0	0	0	0	0	0
Tse sup reference	All	0.429	0.782	0.617	0.257	0.126	0.435
	1	0.829	0.921	0.877	0.784	0.731	0.845
	2	0.07	0.078	0.078	0.062	0.054	0.068
	3	0.014	0.022	0.017	0.011	0.009	0.013
	4	0	0.002	0	0	0	0

(Continued)

TABLE 11.6 (*Continued*)

Consistency Rates of Each Approach under Different Categories and in Different Regions

Approach	Category	Consistency Rate					
		Region 1–6	Region 7	Region 8	Region 9	Region 10	All
SenInd	All	0.589	0.882	0.759	0.399	0.227	0.58
simple	1	0.988	0.994	0.992	0.986	0.979	0.989
	2	0.494	0.53	0.516	0.458	0.425	0.487
	3	0.075	0.109	0.088	0.062	0.045	0.071
	4	0	0	0	0	0	0
RPspeP	All	0.623	0.433	0.57	0.573	0.436	0.575
reference	1	0.624	0.365	0.502	0.724	0.803	0.575
	2	0.959	0.976	0.968	0.948	0.933	0.957
	3	0.74	0.826	0.782	0.695	0.642	0.727
	4	0.189	0.297	0.247	0.138	0.092	0.16
RPsupP	All	0.545	0.864	0.727	0.352	0.189	0.541
reference	1	0.994	0.997	0.996	0.992	0.989	0.995
	2	0.258	0.299	0.281	0.226	0.197	0.253
	3	0.033	0.052	0.044	0.024	0.016	0.031
	4	0	0.001	0	0	0	0
GenP	All	0.668	0.898	0.795	0.554	0.503	0.676
	1	0.99	0.996	0.993	0.987	0.981	0.991
	2	0.586	0.615	0.597	0.555	0.537	0.581
	3	0.061	0.084	0.074	0.05	0.037	0.058
	4	0.377	0.296	0.331	0.446	0.524	0.423
GenCB	All	NA	NA	NA	NA	NA	NA
	1	NA	NA	NA	NA	NA	NA
	2	0.363	0.393	0.382	0.336	0.313	0.358
	3	NA	NA	NA	NA	NA	NA
	4	NA	NA	NA	NA	NA	NA
GenP sup	All	0.622	0.898	0.784	0.435	0.259	0.611
	1	0.995	0.998	0.996	0.993	0.991	0.995
	2	0.681	0.708	0.689	0.656	0.634	0.676
	3	0.075	0.104	0.087	0.063	0.048	0.072
	4	0	0	0	0	0	0
GenCB sup	All	0.575	0.875	0.749	0.385	0.219	0.568
	1	0.981	0.992	0.988	0.974	0.965	0.983
	2	0.465	0.492	0.483	0.433	0.412	0.459
	3	0.049	0.069	0.059	0.039	0.029	0.047
	4	0	0	0	0	0	0

(Continued)

TABLE 11.6 (*Continued*)

Consistency Rates of Each Approach under Different Categories and in Different Regions

Approach	Category	Consistency Rate					
		Region 1–6	Region 7	Region 8	Region 9	Region 10	All
Bayesian	All	0.549	0.849	0.718	0.37	0.214	0.545
min	1	0.917	0.959	0.94	0.895	0.878	0.925
constant	2	0.532	0.546	0.539	0.514	0.5	0.528
	3	0.015	0.015	0.014	0.012	0.012	0.014
	4	0	0	0	0	0	0
Bayesian	All	0.719	0.94	0.857	0.539	0.346	0.7
min	1	1	1	1	1	1	1
reference	2	0.945	0.962	0.953	0.933	0.915	0.942
	3	0.382	0.485	0.431	0.33	0.277	0.368
	4	0.004	0.009	0.009	0.002	0.001	0.003
Bayesian	All	0.59	0.873	0.751	0.412	0.251	0.583
mean	1	0.937	0.971	0.955	0.922	0.909	0.944
constant	2	0.7	0.721	0.698	0.689	0.694	0.699
	3	0.031	0.033	0.032	0.031	0.03	0.031
	4	0	0	0	0	0	0
Bayesian	All	0.749	0.95	0.877	0.576	0.384	0.728
mean	1	1	1	1	1	1	1
reference	2	0.99	0.993	0.993	0.99	0.985	0.99
	3	0.518	0.629	0.57	0.463	0.415	0.505
	4	0.007	0.017	0.014	0.004	0.002	0.005
Japan	All	0.579	0.882	0.755	0.385	0.214	0.571
simple	1	0.999	0.999	0.999	0.999	0.998	0.999
	2	0.442	0.516	0.483	0.395	0.355	0.436
	3	0.032	0.051	0.042	0.023	0.017	0.03
	4	0	0	0	0	0	0

11.4.2 The Case of the Two-Group Parallel Design with Normal Data and Superiority Test

We also performed a simulation study for two-group parallel design with normal data and superiority test. Assume the clinical trial with the aim of comparing the treatment effect mean μ_T and reference effect mean μ_R (the smaller the better), consists of 140 patients from 10 regions with equal number of patients recruited in each region and equal number of patients in each arm. Consider the superiority hypothesis test:

$$H_0 : \mu_T - \mu_R \geq 0 \text{ vs. } H_a : \mu_T - \mu_R < 0.$$

Presume $\mu_T - \mu_R$ is equal to -1 with the standard deviation of each arm being 2 globally for the power analysis. Given the pre-specified type I error level of $\alpha = 0.05$, 140 samples can achieve a power of 90%.

For the simulation parameter specification, let both standard deviations of the treatment arm and the reference arm be equal to 2 within each region. Let μ_R of each region is equal to 0. The values of $\mu_T - \mu_R$ for six regions is equal to -1, while the other four regions have the values of -2, -1.5, -0.5 and 0, respectively. The true standard deviations of the treatment arm and the reference arm for the global population are 2.06 and 2.

For each simulation, we simulated data for each region, combined them as the global data, and tested the superiority hypothesis both regionally and globally with the standard t-statistic method. Denoting $\bar{X}_S$ and $\bar{X}_G$ as the sample means of the difference between the treatment arm and reference arm for the region and global area, respectively, the results for the subgroup analyses could be divided into the following 4 categories (see Table 11.7).

Like the simulation study for the matched-pair parallel design, seventeen approaches based on the methods discussed in Section 11.3 were selected to test the consistency.

We did 100,000 repetitions of the simulation, with the following results recorded: the rejection rate of the null hypothesis $H_0 : \mu_T - \mu_R \geq 0$ regionally and globally (see Table 11.8), the frequency of the categories regionally and globally (see Table 11.9), consistency rates of each approach under different categories and in different regions (see Table 11.10). As the six regions have identical population parameters, we combined the results for them.

The simulation results show that, for this scenario, the methods of the simple version for sensitivity index, superiority reproducibility probability, reproducibility probability ratio for superiority test, Bayesian approach and

TABLE 11.7

Test Results by Category (Global and Regions)

Category	Superiority Trial
1	$\bar{X}_S < \bar{X}_G$ (met superiority)
2	$\bar{X}_G \leq \bar{X}_S < 0$ & upper bound of $1 - \alpha$ CI for the region using global CI width < 0
3	$\bar{X}_G \leq \bar{X}_S < 0$ but upper bound of $1 - \alpha$ CI for the region using global CI width ≥ 0
4	$\bar{X}_G < 0 \leq \bar{X}_S$

TABLE 11.8

Passing Rates of Superiority Test by Global and Regions

Passing Rate of Superiority Test	Global	Region 1–6	Region 7	Region 8	Region 9	Region 10
	0.898	0.223	0.547	0.378	0.116	0.051

TABLE 11.9

Frequencies by Global and Regions

	Category Frequency					
Category	Region 1–6	Region 7	Region 8	Region 9	Region 10	All
All	600000	100000	100000	100000	100000	1000000
1	299514	83715	69000	31172	16083	499484
2	98662	7776	12594	16829	14037	149898
3	97365	5444	10538	20031	19790	153168
4	103981	3032	7823	31856	49952	196644

TABLE 11.10

Consistency Rates of Each Approach under Different Categories and in Different Regions

		Consistency Rate					
Approach	Category	Region 1–6	Region 7	Region 8	Region 9	Region 10	All
Shih	All	0.923	0.988	0.968	0.834	0.674	0.9
superiority	1	1	1	1	1	1	1
	2	0.994	0.995	0.994	0.992	0.985	0.993
	3	0.951	0.958	0.957	0.937	0.903	0.944
	4	0.605	0.703	0.662	0.524	0.391	0.542
Tse	All	0	0	0	0	0	0
	1	0	0	0	0	0	0
	2	0	0	0	0	0	0
	3	0	0	0	0	0	0
	4	0	0	0	0	0	0
Tse	All	0.54	0.432	0.512	0.513	0.429	0.512
reference	1	0.54	0.397	0.481	0.586	0.599	0.513
	2	0.651	0.66	0.648	0.643	0.643	0.65
	3	0.6	0.621	0.62	0.591	0.573	0.598
	4	0.376	0.45	0.424	0.324	0.257	0.34
Tse sup	All	0	0.005	0.001	0	0	0.001
	1	0.001	0.007	0.002	0	0	0.002
	2	0	0	0	0	0	0
	3	0	0	0	0	0	0
	4	0	0	0	0	0	0
Tse sup	All	0.421	0.779	0.613	0.248	0.121	0.429
reference	1	0.832	0.926	0.881	0.78	0.728	0.848
	2	0.029	0.036	0.031	0.026	0.023	0.029
	3	0.005	0.011	0.006	0.004	0.003	0.005
	4	0	0	0	0	0	0

(*Continued*)

TABLE 11.10 (*Continued*)

Consistency Rates of Each Approach under Different Categories and in Different Regions

Approach	Category	Consistency Rate					
		Region 1–6	Region 7	Region 8	Region 9	Region 10	All
SenInd	All	0.603	0.89	0.773	0.412	0.24	0.593
simple	1	0.997	0.998	0.997	0.996	0.996	0.997
	2	0.564	0.621	0.591	0.526	0.493	0.558
	3	0.074	0.12	0.092	0.058	0.046	0.071
	4	0	0	0	0	0	0
RPspeP	All	0.607	0.419	0.551	0.554	0.416	0.558
reference	1	0.607	0.347	0.478	0.711	0.788	0.558
	2	0.963	0.979	0.973	0.951	0.934	0.961
	3	0.728	0.815	0.768	0.679	0.617	0.713
	4	0.157	0.266	0.214	0.112	0.072	0.132
RPsupP	All	0.552	0.868	0.734	0.36	0.196	0.547
reference	1	0.999	1	0.999	0.998	0.997	0.999
	2	0.29	0.35	0.315	0.257	0.228	0.286
	3	0.036	0.065	0.048	0.026	0.018	0.034
	4	0	0.001	0	0	0	0
GenP	All	0.664	0.896	0.795	0.549	0.499	0.672
	1	0.995	0.998	0.997	0.993	0.992	0.996
	2	0.576	0.616	0.598	0.551	0.528	0.573
	3	0.043	0.072	0.054	0.033	0.026	0.041
	4	0.375	0.277	0.325	0.434	0.518	0.417
GenCB	All	NA	NA	NA	NA	NA	NA
	1	NA	NA	NA	NA	NA	NA
	2	0.328	0.369	0.343	0.302	0.279	0.324
	3	NA	NA	NA	NA	NA	NA
	4	NA	NA	NA	NA	NA	NA
GenP sup	All	0.619	0.897	0.784	0.43	0.256	0.608
	1	0.998	0.999	0.999	0.996	0.997	0.998
	2	0.681	0.72	0.697	0.659	0.635	0.678
	3	0.053	0.085	0.067	0.041	0.034	0.051
	4	0	0	0	0	0	0
GenCB sup	All	0.572	0.874	0.748	0.382	0.217	0.565
	1	0.989	0.995	0.992	0.986	0.982	0.99
	2	0.442	0.489	0.462	0.415	0.391	0.438
	3	0.034	0.058	0.044	0.025	0.019	0.033
	4	0	0	0	0	0	0

(Continued)

TABLE 11.10 (*Continued*)

Consistency Rates of Each Approach under Different Categories and in Different Regions

Approach	Category	Region 1-6	Region 7	Region 8	Region 9	Region 10	All
				Consistency Rate			
Bayesian	All	0.563	0.858	0.732	0.383	0.224	0.557
min	1	0.936	0.969	0.953	0.92	0.911	0.942
constant	2	0.574	0.601	0.587	0.562	0.545	0.573
	3	0.005	0.008	0.006	0.005	0.004	0.005
	4	0	0	0	0	0	0
Bayesian	All	0.725	0.942	0.862	0.545	0.352	0.705
min	1	1	1	1	1	1	1
reference	2	0.971	0.98	0.977	0.963	0.95	0.969
	3	0.402	0.51	0.456	0.35	0.29	0.388
	4	0.004	0.01	0.007	0.002	0.001	0.003
Bayesian	All	0.602	0.88	0.764	0.422	0.259	0.593
mean	1	0.951	0.977	0.964	0.938	0.931	0.956
constant	2	0.753	0.774	0.76	0.748	0.751	0.754
	3	0.02	0.027	0.023	0.02	0.02	0.02
	4	0	0	0	0	0	0
Bayesian	All	0.752	0.951	0.88	0.579	0.388	0.731
mean	1	1	1	1	1	1	1
reference	2	0.996	0.997	0.998	0.995	0.994	0.996
	3	0.543	0.661	0.6	0.492	0.436	0.53
	4	0.007	0.016	0.011	0.004	0.002	0.005
Japan	All	0.578	0.88	0.755	0.385	0.217	0.57
simple	1	0.998	0.998	0.998	0.998	0.999	0.998
	2	0.447	0.522	0.484	0.403	0.365	0.441
	3	0.033	0.057	0.044	0.023	0.017	0.032
	4	0	0	0	0	0	0

simple version for Japanese approach performed better than others, with overall consistency rate close to 65% (the rate of category 1 and 2). The results are similar as those from matched-pair parallel design.

11.4.3 Remarks

In summary, six statistical methods are discussed and evaluated through simulations. Although there are limitations to the simulations, we have learned the following: For Shih's method, in order to make a meaningful comparison between regions using this method, the sample sizes for the regions should be similar and reasonably large. It is, in general, not recommended to be used for evaluating consistency in the country level results since the sample size varies a lot between countries in most MRCTs. Tse's method is based on equivalence idea. In the country level consistency evaluation, local results which are better than the global results will be considered consistent. Therefore, this method is not appropriately to be used for this type of consistency evaluation. In addition, it is difficult to select appropriate parameters for the method. Thus, this method is not recommended. Ying et al. (2017) also evaluated sensitivity index based on the simplified version (assuming the parameters are constant). It performs well. Although, in practice, the parameters are likely to be random, we would suggest using the simplified version since more assumptions need to be made to estimate the parameters that may or may not improve the performance for consistency evaluations. Other methods (reproducibility probability ratio for superiority test, GenP Sup, GenCB Sup, Bayesian approach using constant choice for p_0 and simple version of Japanese approach) perform well for consistency evaluations in general. Some of these approaches maybe more conservative than others, particularly when the sample size of the sub-population is small. Ying et al. (2017) suggested selecting two methods (one more-aggressive and one more-conservation) for consistency evaluation. For example, using Bayesian Mean Constant approach (more aggressive) and Bayesian Minimum Constant approach (more conservation). If the local results show consistency for both methods, then the evidence of consistency is strong. If the local results show inconsistency for both methods, then the evidence of consistency is weak. If the local results show consistency for one method but not the other method, then the evidence of consistency is not strong.

11.5 An Example

A phase 3, active (denoted as drug B) controlled, randomized (1:1 ratio), double-blind, parallel arm study to evaluate the efficacy and safety of a test drug (denoted as drug A) in preventing stroke in subjects with non-valvular atrial fibrillation. The primary objective is to demonstrate if the test drug A is non-inferior (NI with a NI margin of 1.40 for the hazard ratio of Drug A vs. Drug B) to the control drug B for the endpoint of adjudicated stroke events. The secondary objectives include superior test for the primary endpoint of stroke, all causes of death, and major bleeding. Thirty-eight countries (denoted as C1–C38) from 4 regions participated in the study. Different countries or regions have similar but different regulatory requirements (see Table 11.11).

TABLE 11.11

Examples of Different Regional Requirements

Regional Requirement	EU EMA	US FDA
Non-inferiority margin (HIV-infected patients)	10%–15%	10%–12%
Time points (HIV-RNA level)	16 weeks	24 weeks
Endpoint (Atirial Fibriliation)	Prevention of any recurrence	Delay in symptomatic recurrence
Endpoint (Hepatitus B Virus)	Combined composite of virological, histological and biochemical responses	Complete virological response at 52 weeks

TABLE 11.12

Categories of Subgroup Analysis

Category	Superiority Trial	Non-inferiority Trail
1	$HR_s < HR_o$ (met superiority)	$HR_s < HR_o$ (met NI)
2	$HR_o < HR_s < 1$ & upper bound of 95% CI for HR_s based on the overall number of events <1	$HR_o < HR_s < \Delta$ & upper bound of 95% CI for HR_s based on the overall number of events $<\Delta$
3	$HR_o < HR_s < 1$ but upper bound of 95% CI for HR_s based on the overall number of events >1	$HR_o < HR_s < \Delta$ but upper bound of 95% CI for HR_s based on the overall number of events $<\Delta$
4	$HR_o < 1 < HR_s$	$HR_o < HR_s$ & $HR_s > \Delta$

The results for the primary endpoint of stroke by country based on the intended treatment population (ITT) are summarized in Table 11.12 (for selected countries) and plotted in Figure 11.1.

Let us use an event-driving trial with two treatment groups as an example. Assuming the hazard ratio between test drug and control drug is HR_o for the overall population and HR_s for a subgroup. The type-I error rate is 5%. The trial is successful based on the overall population, i.e., the upper bound of 95% CI for the $HR_o < 1$ (if it is a superiority trial) or the upper bound of 95% CI for the $HR_o <$ non-inferiority (NI) margin denoted as Δ (if it is a NI trial). Then the results for the subgroup analyses could be divided into the following 4 categories (see Table 11.13).

For results in Category 1, it is clear that the subgroup performed better than the overall population. Consistency evaluation is usually not necessary.

For results in Category 2, it is likely all the statistical methods will show that no inconsistency can be found.

For results in Category 3, it is likely consistency cannot be shown by some statistical methods.

For results in Category 4, inconsistency is observed.

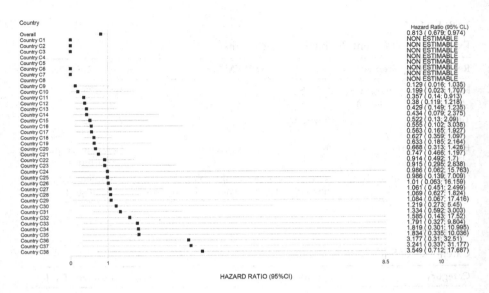

FIGURE 11.1
Forest plot of the results for the primary efficacy endpoint of the adjudicated stroke during the intended treatment period, overall and by country.

TABLE 11.13

The Results for the Primary Efficacy Endpoint of Adjudicated Stroke Events for Selected Sub-populations

Population	Total # of Subjects, N	Drug A: # Subjects with Events (Event Rate), n (%/year)	Drug B: # Subjects with Events (Event Rate), n (%/year)	HR of Drug A/Drug B (95% CI)
Overall	15421	183 (1.13)	230 (1.39)	0.81 (0.68, 0.97)
Country C18	1609	18 (0.96)	29 (1.52)	0.63 (0.36, 1.10)
Country C22	1255	16 (1.13)	16 (1.24)	0.91 (0.49, 1.70)
Country C28	752	25 (3.47)	23 (3.24)	1.07 (0.63, 1.82)
Country C31	830	12 (1.35)	9 (1.01)	1.33 (0.59, 3.00)
Country C34	289	3 (0.80)	2 (0.44)	1.82 (0.30, 10.99)

Please note that if there are only a few events in a subgroup, then no meaningful consistency evaluation can be done.

It is clear that no one will question the consistency for Country 18 since its HR is smaller than that for the overall population. Although the HR for Country 22 is less than 1 but the HR for Country 28 is greater than 1, both of them below to Category 2 since this is a non-inferiority trial (hazard ratio should be compared with the NI margin instead of 1). Result for Country 31 belongs to Category 3. Result for Country 34 belongs to Category 4. However, since the number of events is very small, the results are not meaningful.

Let us use the statistical methods described in the previous sections to evaluate the consistency for Country 28 and Country 31.

First of all, the HR for Country 28 is far less than the NI margin. If this result is observed for the overall population, then the NI claim will still be established.

The consistency evaluation for Countries 28 and 31 based on the Bayesian approach described in Section 3.5 are listed in Table 11.14. The results are as expected, i.e., the posterior probabilities of consistency for Country C28 are all >80% which indicates consistency. The posterior probabilities of consistency for Country C31 are not all >80% which indicates inconsistency may exist.

Some statistical methods such as the method described in the previous section for evaluating consistency may not work directly for the NI situations, e.g., the results like Countries C28 and C31. If we know the HR and its corresponding CI for the control drug vs. placebo, then an indirect comparison results between the test drug and placebo could be used to evaluate consistency. For example, the hazard ratio (HR_{BvsP}) and the corresponding 95% CI between the control Drug B and Placebo from historical trials are 0.38 (0.26, 0.56). Assuming HR_{BvsP} for Countries 28 and 31 is the same as that for the overall population, then the HRs between Drug A and placebo can be derived as 0.31, 0.41, and 0.51 for overall, Country C28, and Country 31, respectively. Then Country C28 reserves 85.7% of the overall effect (as measured by the relative risk reduction) and Country C31 reserves 71.5% of the overall effect. Based on the principal of Japanese guidelines, Both Country 28 and Country 31 results are consistent with the overall result.

TABLE 11.14

Values of Posterior Probability (P_c) for Various Values of Weight (γ) for Countries C28 and C31

γ	P_c for C28	P_c for C31
0.0	1.00	1.00
0.1	0.98	0.90
0.2	0.96	0.83
0.3	0.94	0.77
0.4	0.92	0.72
0.5	0.91	0.68
0.6	0.89	0.65
0.7	0.88	0.62
0.8	0.86	0.59
0.9	0.85	0.57
1.0	0.84	0.55

In addition, there are absence of treatment-by-region (Subgroup X vs. non-Subgroup X) interactions (the p-values for the interaction test are >0.15) for both countries C28 and C31.

Therefore, there is no indication that Country C28 is inconsistent with the overall population results from statistical point of view. However, such conclusion could not be made for Country 31.

11.6 Other Considerations/Discussions

In the above sections, we have mainly discussed the statistical methods for consistency evaluations. There are other areas for consideration of consistency evaluation. For example, if a numerical inconsistency was observed for the subgroup data, then we may wish to consider the following actions, taken from clinical point of view:

1. Consider the severity and subtype of the events if any, to see where the difference came from. For example, consider stroke types of hemorrhagic strokes and ischemic strokes for the example given in the previous section.

2. Conduct sensitivity analysis such as on-treatment analysis vs. ITT analysis which may probably help to locate the differences. For example, if the on-treatment result for a subgroup is consistent with the overall result but the ITT results are not, then check the post-treatment medications difference and check if there is a dose adjustment and stabilization period needed for the post-treatment medication which may increase some events of interest.

Since the subgroups are usually not powered in the MRCT, we are likely to observe numerical differences in the results for some subgroups. It is difficult to identify the causes for the difference between the subgroup and the overall population. Therefore, we suggest that, if a numerical difference is observed between a subgroup of interest and overall population and the sample size is not very small, then perform a complete consistency evaluation (possible difference in population, PK/PD, statistical tools, and clinical aspects) as discussed in this chapter. If no appropriate causes could be found to explain the numerical difference or to demonstrate inconsistency, then the conclusion can be drawn that is no inconsistency to be found. We may conclude that the numerical difference is likely due to a random chance, especially, when the sample size (or total events) for the subgroup is small. In such case, overall results should be used for the regional approval and/or more data could be obtained either pre- or after- regional approval.

11.7 Concluding Remarks

In this chapter, we have discussed several statistical methods for evaluating consistency between a specific region (i.e., sub-population) and all regions combined (i.e., global population or entire population) and compared different methods through simulations. We have also illustrated how to use those methods for non-inferiority situations through an example as it is more difficult to evaluate consistency for a non-inferiority situation. Statistical evaluation is only part of the consistency evaluation between a sub-population and the global population. Possible differences in population, PK/PD, and clinical aspects should also be evaluated. Since the sub-populations are usually not powered in the MRCT, it is likely that some numerical differences in the results for some sub-populations will be observed. It is difficult to identify the causes for the difference between the sub-population and the global population, especially when the sample size is small for the sub-population. Therefore, for evaluation of consistency in clinical results between a specific region and all regions combined, it is suggested that power calculation for sample size should take into consideration, to reveal the minimum sample size required at the specific region for achieving a desired level of consistency or assurance probability, as suggested by MHLW (2007) to ensure consistency in terms of treatment effect between a sub-population and the entire population.

12

Drug Products with Multiple Components—Development of TCM

12.1 Introduction

In recent years, as more and more innovative drug products are going off patent, the search for new medicines that treat critical and/or life-threatening diseases has become the center of attention of many pharmaceutical companies. As indicated by Chow and Liu (2000), pharmaceutical research and development is a lengthy and costly process. On average, it may take more than 12 years to bring a promising compound to the market. The probability of success, however, is usually very low. In the past several decades, tremendous effort was put on drug research and development, and yet only a handful of new drug products were approved by the regulatory agencies. As a result, an alternative approach for drug discovery is necessary. This leads to the study of the potential use of promising traditional Chinese medicines (TCM), especially for those intended for treating critical and/or life-threatening diseases. A TCM is defined as a Chinese herbal medicine developed for treating patients with certain diseases as diagnosed by the four major Chinese diagnostic techniques of inspection, auscultation and olfaction, interrogation, and pulse taking and palpation based on traditional Chinese medical theory of global dynamic balance among the functions/activities of all organs of the body. TCM is also known as drug products with multiple components (active ingredients).

Unlike evidence-based clinical research and development of a Western medicine (WM), clinical research and development of a TCM is usually experience-based with anticipated variability due to subjective evaluation of the disease under study. The use of TCM in humans for treating various diseases has a history of more than five thousand years, although no scientific documentation is available regarding clinical evidence of safety and efficacy of these TCMs.

In the past several decades, regulatory agencies of both China and Taiwan have debated which direction the TCM should take—Westernization or modernization. The Westernization of TCM is referred to the adoption of the typical (Western) process of pharmaceutical research and development for scientific evaluation of the safety and effectiveness of the TCM products under investigation, while the modernization of TCM is to evaluate the safety and effectiveness of TCM the Chinese way (i.e., different sets of regulatory requirements and evaluation criteria) scientifically. Although both China and Taiwan do attempt to build up an environment for the modernization of TCM, they seem to adopt the Westernization approach. As a result, in this chapter, we will place our emphasis on the Westernization of TCM.

In practice, it is a concern whether a TCM can be scientifically evaluated the Western way due to some fundamental differences between a WM and a TCM. These include differences in formulation, medical practice, drug administration, diagnostic procedure and criteria for evaluation, and flexibility. Under these differences, it is then of interest to the investigators regarding how to conduct a scientifically valid (i.e., an adequate and well-controlled) clinical trial for evaluation of the clinical safety and efficacy of the TCM under investigation. In addition, it is also of particular interest to the investigators as to how to translate an observed significant difference detected by the Chinese diagnostic procedure to a clinically meaningful difference based on some well-established clinical study endpoint. The purpose of this article is to provide some basic considerations regarding practical issues that are commonly encountered when conducting a TCM clinical trial the Western way.

In the next section, some fundamental differences between a WM and a TCM, which have an impact on the Westernization of TCM, are described. These fundamental differences include the concept of global dynamic balance/harmony among organs of the body (TCM) versus local site action (WM); subjective diagnostic techniques of inspection, auscultation and olfaction, interrogation, pulse taking and palpation (TCM) versus objectively clinical evaluation (WM); and personalized flexible dose with multiple components (TCM) versus fixed dose of single active ingredient (WM). Section 12.3 provides some basic considerations of TCM clinical trials. These basic considerations include study design, validation of a quantitative instrument developed for the four major TCM diagnostic techniques, the use/preparation of matching placebo, and sample size calculation. Some practical issues that are commonly encountered when conducting a TCM clinical trial are given in Section 12.4. Section 12.5 provides some recent development for assessment of TCM such as test for consistency in statistical quality control of raw material and/or final product, stability analysis, and calibration of Chinese diagnostic procedures against well-established study endpoints used for assessment of Western medicine. Some concluding remarks, including future strategy and recommendations in TCM research and development, are given in Section 12.6.

12.2 Fundamental Differences

As indicated earlier, the process for pharmaceutical research and development of Western medicines (WM) is well established, and yet it is a lengthy and costly process. This lengthy and costly process is necessary to ensure the efficacy, safety, quality, stability, and reproducibility of the drug product under investigation. For pharmaceutical research and development of a TCM, one may consider directly applying this well-established process to the TCM under investigation. However, this process may not be feasible due to some fundamental differences between a TCM and a WM. Some fundamental differences between a WM and a TCM are summarized in Table 12.1. These fundamental differences are briefly described below.

12.2.1 Medical Theory/Mechanism and Practice

TCM is a more than 5000-year-old holistic medical system encircling the entire scope of human experience. It combines the use of Chinese herbal medicines, acupuncture, massage, and therapeutic exercise such as Qigong (the practice of internal air) and Taigie for both treatment and prevention of disease. With its unique theories of etiology, diagnostic systems, and abundant historical literature, TCM itself consists of Chinese culture and philosophy, clinical practice experience, and the use of many medical herbs.

Chinese doctors believe how a TCM functions in the body is based on the eight principles, five-element theory, five Zang and six Fu, and information regarding channels and collaterals. Eight principles consist of Yin and Yang (i.e., negative and positive), cold and hot, external and internal, and Shi and Xu (i.e., weak and strong). The eight principles help Chinese doctors to differentiate syndrome patterns. For instance, people with Yin will develop disease in a negative, passive, and cool way (e.g., diarrhea and back pain), while people with Yang will develop disease in an aggressive, active, progressive, and warm way (e.g., dry eyes, tinnitus, and night sweats). The five elements

TABLE 12.1

Fundamental Differences Between a WM and a TCM

Description	Western Medicine	Traditional Chinese Medicine
Active ingredient	Single	Multiple
Dose	Fixed	Flexible
Diagnostic procedure	Objective; validated	Subjective; not validated
Therapeutic index	Well-established	Not well-established
Medical mechanism	Specific organs	Global dynamic balance/harmony among organs
Medical perception	Evidence-based	Experience-based
Statistics	Population	Individual

(earth, metal, water, wood, and fire) correspond to particular organs in the human body. Each element operates in harmony with the others.

The five Zang (or Yin organs) include heart (including the pericardium), lung, spleen, liver, and kidney, while the six Fu (or Yang organs) include gall bladder, stomach, large intestine, small intestine, urinary bladder, and three cavities (i.e., chest, epiastrium, and hypogastrium). Zang organs can manufacture and store fundamental substances. These substances are then transformed and transported by Fu organs. TCM treatments involve a thorough understanding of the clinical manifestations of Zang-Fu organ imbalance, and knowledge of appropriate acupuncture points and herbal therapy to rebalance or maintain the balance of the organs. The channels and collaterals are the representation of the organs of the body. They are responsible for conducting the flow of energy and blood through the entire body.

The elements of TCM can also help to describe the etiology of disease including six exogenous factors (i.e., wind, cold, summer, dampness, dryness, and fire), seven emotional factors (i.e., anger, joy, worry, grief, anxiety, fear, and fright), and other pathogenic factors. Once all of the information are collected and processed into a logical and workable diagnosis, the traditional Chinese medical doctor can determine the treatment approach.

Under the medical theory and mechanism described above, Chinese doctors believe that all of the organs within a healthy subject should reach the so-called global dynamic balance or harmony among organs. Once the global balance is broken at certain sites such as heart, liver, or kidney, some signs and symptoms then appear to reflect the imbalance at these sites. An experienced Chinese doctor usually assesses the causes of global imbalance before a TCM with flexible doses is prescribed to fix the problem. This approach is sometimes referred to as a personalized (or individualized) medicine approach.

12.2.1.1 Medical Practice

Different medical perceptions regarding signs and symptoms of certain diseases could lead to a different diagnosis and treatment for the diseases under study. For example, Chinese doctors could classify the signs and symptoms of Type 2 diabetic subjects as the disease of thirsty reduction. The disease of Type 2 diabetes is not recognized by Chinese medical literature, although they have the same signs and symptoms as the well-known disease of thirsty reduction. This difference in medical perception and practice has an impact on the diagnosis and treatment of the disease.

In addition, we tend to see therapeutic effect of WMs sooner than TCMs. Traditional Chinese medicines are often considered for patients who have chronic diseases or non-life-threatening diseases. For critical and/or life-threatening diseases such as cancer or stroke, TCMs are often used as the second-line or third-line treatment with no other alternative treatments. In many cases such as patients with later phase of cancer, TCMs are often used in conjunction with WMs without the knowledge of the primary care physicians.

12.2.2 Techniques of Diagnosis

The Chinese diagnostic procedure for patients with certain diseases consists of four major techniques, namely, inspection, auscultation and olfaction, interrogation, and pulse taking and palpation. All these diagnostic techniques aim mainly at providing an objective basis for differentiation of syndromes by collecting symptoms and signs from the patient. Inspection involves observing the patient's general appearance (strong or weak, fat or thin), mind, complexion (skin color), five sense organs (eye, ear, nose, lip, and tongue), secretions, and excretions. Auscultation involves listening to the voice, expression, respiration, vomit, and cough. Olfaction involves smelling the breath and body odor. Interrogation involves asking questions about specific symptoms and the general condition including history of the present disease, past history, personal life history, and family history. Pulse taking and palpation can help to judge the location and nature of a disease according to the changes of the pulse.

The Chinese diagnostic procedure of inspection, auscultation and olfaction, interrogation, and pulse taking and palpation is subjective, with large between-rater variability (i.e., variability from one Chinese doctor to another). This subjectivity and variability will have an impact not only on the patient's evaluability but also the prescribability of TCM, which will be further discussed below.

12.2.2.1 Objective versus Subjective Criteria for Evaluability

For evaluation of a WM, objective criteria based on some well-established clinical study endpoints are usually considered. For example, response rate (i.e., complete response plus partial response based on tumor size) is considered a valid clinical endpoint for evaluating clinical efficacy of oncology drug products. Unlike WMs, Chinese diagnostic procedure for evaluation of a TCM is very subjective. The use of a subjective Chinese diagnostic procedure has raised the following issues. First, it is a concern whether the subjective Chinese diagnostic procedure can accurately and reliably evaluate clinical efficacy and safety of the TCM under investigation. Thus, it is suggested that the subjective Chinese diagnostic procedure should be validated in terms of its accuracy, precision, and ruggedness before it can be used in TCM clinical trials. A validated Chinese diagnostic procedure should be able to detect a clinically significant difference if the difference truly exists. On the other hand, it is not desirable to wrongly detect a difference when there is no difference.

In clinical trials, evaluation is usually based on some validated tools (instruments) such as laboratory tests. Test results are then evaluated against some normal ranges for abnormality. Thus, it is suggested that the Chinese diagnostic procedure must be validated in terms of validity and reliability, and its false positive and false negative rates, before it can be used for evaluation of clinical efficacy and safety of the TCM under investigation.

12.2.3 Treatment

TCM prescriptions typically consist of a combination of several components. The combination is usually determined based on the medical theory of global dynamic balance (or harmony) among organs, and the observations from the Chinese diagnostic procedure. The use of Chinese diagnostic procedure is to find out what caused the imbalance among these organs. The treatment is to re-install the balance among these organs. Thus, the dose and treatment duration are flexible in order to achieve the balance. This concept leads to the concept of the so-called personalized (or individualized) medicine, which minimizes intra-subject variability.

12.2.3.1 Single Active Ingredient versus Multiple Components

Most Western medicines contain a single active ingredient. After drug discovery, an appropriate formulation (or dosage form) is necessarily developed so that the drug can be delivered to the site action in an efficient way. At the same time, an assay is necessarily developed to quantitate the potency of the drug. The drug is then tested on animals for toxicity, and humans (healthy volunteers) for pharmacological activities. Unlike the WMs, TCMs usually consist of multiple components with certain relative proportions among the components. As a result, the typical approach for evaluation of single active ingredient for WM is not applicable.

In practice, one may suggest evaluating the TCM component by component. However, this is not feasible due to the following difficulties. First, in practice, analytical methods for quantitation of individual components are often not tractable. Thus, the pharmacological activities of these components are not known. It should be noted that the component that comprises the major proportion of the TCM might not be the most active component. On the other hand, the component that has the least proportion of the TCM may be the most active component of the TCM. In practice, it is not known which relative proportions among these components can lead to the optimal therapeutic effect of the TCM. In addition, the relative component-to-component and/or component by food interactions are usually unknown, which may have an impact on the evaluation of clinical efficacy and safety of the TCM.

12.2.3.2 Fixed Dose versus Flexible Dose

Most WMs are usually administered in a fixed dose (say a 10 mg tablets or capsule). On the other hand, since a TCM consists of multiple components with possible varied relative proportions among the components, a Chinese doctor usually prescribes the TCM with different relative proportions of the multiple components based on the signs and symptoms of the patient according to his/her best judgment following a subjective evaluation based on the Chinese diagnostic procedure. Thus, unlike a WM that is prescribed as a fixed dose, a TCM is often prescribed as an individualized flexible dose.

The approach of WM with a fixed dose is a population approach to minimize the between subject (or inter-subject) variability, while the approach to TCM with an individualized flexible dose is to minimize the variability within each individual. In practice, it is a concern whether an individual flexible dose is compatible with a Western evaluation of the TCM. An individualized flexible dose depends heavily upon the Chinese doctor's subjective judgment, which may vary from one Chinese doctor to another. As a result, although an individualized flexible dose does minimize intra-subject variability, the variability from one Chinese doctor to another (i.e., the doctor-to-doctor or rater-to-rater variability) could be huge, and hence non-negligible.

12.2.4 Remarks

For the research and development of a TCM, before a TCM clinical trial is conducted, the following questions are necessarily asked.

1. Will the TCM clinical trial be conducted by Chinese doctors alone, Western clinicians alone, Western clinicians who have some background of Chinese herbal medicine alone, or both Chinese doctors and Western clinicians?
2. Will traditional Chinese diagnostic and/or trial procedures be used throughout the TCM clinical trial?
3. Upon approval, is the TCM intended for use by Chinese doctors or Western clinicians?

With respect to the first two questions, if the TCM clinical trial is to be conducted by Chinese doctors alone, the following questions arise. First, should the Chinese diagnostic procedure be validated in order to provide an accurate and reliable assessment of the TCM? In addition, it is of interest to determine how an observed difference obtained from the Chinese diagnostic procedure can be translated to the clinical endpoint commonly used in similar WM clinical trials with the same indication. These two questions can be addressed statistically by the calibration and validation of the Chinese diagnostic procedure with respect to some well-established clinical endpoints for evaluation of Western medicines. If the TCM clinical trial is to be conducted by Western clinicians or Western clinicians who have some background of Chinese herbal medicine, the standards and consistency of clinical results as compared to those WM clinical trials are ensured. However, the good characteristics of TCM may be lost during the process of the conduct of the TCM clinical trials. On the other hand, if the TCM clinical trial is to be conducted by both Chinese doctors and Western clinicians, difference in medical practice and/or possible disagreement regarding the diagnosis, treatment, and evaluation are major concerns.

For the third question, if the TCM is intended for use of Chinese doctors but it is employed by Western clinicians, difference in perception regarding how to prescribe the TCM is of great concern. The preparation of a package insert based on the clinical data could be a major issue not only to the sponsor but also to regulatory authorities. Similar comments apply to the situation where the TCM is intended for use of Western clinicians, but the trial is conducted by Chinese doctors.

As a result, it is suggested that the intention of use (i.e., labeling for the indication) be clearly evaluated when planning a TCM clinical trial. In other words, the sponsor needs to determine whether the TCM is intended for use of Western clinicians only, Chinese doctors only, or both Western clinicians and Chinese doctors at the planning stage of a TCM clinical trial, for an adequate package insert of the target diseases under study.

12.3 Basic Considerations

In this section, we describe some basic considerations that are necessary in order to ensure the success of a TCM clinical trial.

12.3.1 Study Design

To demonstrate clinical efficacy and safety of a TCM under investigation, like WMs, it is suggested that a randomized parallel-group, placebo-controlled clinical trial be conducted. However, it may not be ethical if the disease under study is critical and/or life threatening provided that a WM is available. Alternatively, a randomized placebo-control crossover clinical trial or a parallel-group design consisting of three arms (i.e., the TCM under study, a WM as an active control, and a placebo) is recommended. The three-arm, parallel-group design allows the establishment of non-inferiority/equivalence of the TCM as compared to the active control (WM) and the demonstration of the superiority of the TCM with respect to the placebo. One of the advantages of a crossover clinical trial is that a comparison within each individual can be made, although it will take a longer time to complete the study. Although a crossover design requires a smaller sample size as compared to a parallel-group design, there are some limitations for the use of crossover design. First, baselines prior to dosing may not be the same. Second, when a significant sequence effect is observed, we would not be able to isolate the effects of period effect, carry-over effects, and subject-by-treatment effect which are confounded to one another.

In many cases, factorial designs are used to evaluate the impact of specific components (with respect to the therapeutic effect) by fixing some of the components. For example, we may consider a parallel-group design comparing two treatment groups (one group is treated with the TCM with a specific component, and the other group is treated with the TCM without the specific component). The design of this kind may be useful to identify the most active component of the TCM with respect to the diseases under study. However, it does not address the possible drug-to-drug interactions among the components.

12.3.2 Validation of Quantitative Instrument

In TCM medical practice, a Chinese doctor usually collects information from the patient with a certain disease through the four subjective approaches as described in the previous section. The purpose of these subjective approaches is to collect information on various aspects of the disease under study such as signs, symptoms, patient's performance and functional activities, so a quantitative instrument with a large number of questions/items is necessary and helpful. For a simple analysis and an easy interpretation, these questions are usually grouped to form subscales, composite scores (domains) or overall score. The items (or subscales) in each subscale (or composite score) are correlated. As a result, the structure of responses to a quantitative instrument is multidimensional, complex, and correlated. As mentioned above, a standardized quantitative tool (instrument) is necessary to reduce variability from one Chinese doctor to another (prior to the conduct of a clinical trial).

Guilford (1954) discussed several methods such as Cronbach's α for measuring the reliability of internal consistency of a quantitative instrument. Guyatt et al. (1989) indicated that a quantitative instrument should be validated in terms of its validity, reproducibility, and responsiveness. Hollenberg et al. (1991) discussed several methods for validation of a quantitative instrument, such as consensual validation, construct validation, and criterion-related validation. There is, however, no gold standard as to how a quantitative instrument should be validated. In this chapter, we will focus on the validation of a quantitative instrument in terms of validity, reliability (or reproducibility), and responsiveness (see, e.g., Chow and Ki, 1994, 1996). As indicated in Chow and Shao (2002a), the validity of a quantitative instrument is the extent to which the instrument measures what is designed to measure. It is a measure of biasedness of the instrument. The biasedness of a quantitative instrument reflects the accuracy of the instrument. The reliability of a quantitative instrument measures the variability of the instrument, which directly relates to the precision of the instrument. On the other hand, the responsiveness of a quantitative instrument is usually referred to as the ability of the instrument to detect a difference of clinical significance within a treatment.

Hsiao et al. (2005) considered a specific design for calibration/validation of the Chinese diagnostic procedure. In the proposed study design, qualified subjects are randomly assigned to receive either a TCM or a WM. Each patient will be evaluated by a Chinese doctor and a Western clinician independently, regardless of which treatment group he/she is in. As a result, there are four groups of data, namely, (i) patients who receive TCM and evaluated by a Chinese doctor, (ii) patients who receive TCM but evaluated by a Western clinician, (iii) patients who receive WM but evaluated by a Chinese doctor, and (iv) patients who receive WM and evaluated by a Western clinician. Groups (iii) and (iv) are used to establish a standard curve for calibration between the TCM and the WM. Groups (i) and (ii) are then used to validate the Chinese diagnostic procedure based on the established standard curve.

12.3.3 Clinical Endpoint

Unlike WMs, the primary study endpoints for assessment of safety and effectiveness of a TCM are usually assessed subjectively by a quantitative instrument by experienced Chinese doctors. Although the quantitative instrument is developed by the community of Chinese doctors and is considered a gold standard for assessment of safety and effectiveness of the TCM under investigation, it may not be accepted by the Western clinicians due to fundamental differences in medical theory, perception, and practice. In practice, it is very difficult for a Western clinician to conceptually understand the clinical meaning of the difference detected by the subjective Chinese quantitative instrument. Consequently, whether the subjective quantitative instrument can accurately and reliably assess the safety and effectiveness of the TCM is always a concern to Western clinicians.

As an example, for assessment of safety and efficacy of a drug product for treatment of ischemic stroke, a commonly considered primary clinical endpoint is the functional status assessed by the so-called Barthel index. The Barthel index is a weighted functional assessment scoring technique composed of 10 items with a minimum score of 0 (functional incompetence) and a maximum score of 100 (functional competence). The Barthel index is a weighted scale measuring performance in self-care and mobility, which is widely accepted in ischemic stroke clinical trials. A patient may be considered a responder if his/her Barthel index is greater than or equal to 60. On the other hand, Chinese doctors usually consider a quantitative instrument developed by the Chinese medical community as the standard diagnostic procedure for assessment of ischemic stroke. The standard quantitative instrument is composed of six domains, which capture different information regarding patient's performance, functional activities, and signs and symptoms and status of the disease.

In practice, it is of interest to both Western clinicians and Chinese doctors how an observed clinically meaningful difference by the Chinese quantitative instrument can be translated to that of the primary study endpoint assessed by the Barthel index. To reduce the fundamental differences in medical theory/perception and practice, it is suggested that the subjective Chinese quantitative instrument be calibrated and validated with respect to that of the clinical endpoint assessed by the Barthel index before it can be used in TCM ischemic stroke clinical trials.

12.3.4 Matching Placebo

In clinical development, double-blind, placebo-control randomized clinical trials are often conducted for evaluation of the safety and effectiveness of a test treatment under investigation. To maintain blindness, a matching placebo should be identical to the active drug in all aspects of, size, color, coating, taste, texture, shape, and order except that it contains no active ingredient. In clinical trials, as advanced technique available for formulation, a matching placebo is not difficult to make because most Western medicines contain single active ingredient. Unlike Western medicines, TCMs usually consist of a number of components, which often have different taste. In TCM clinical trials, the TCM under investigation is often encapsulated. However, the test treatment will be easily unblinded if either the patient or Chinese doctor breaks the capsule. As a result, the preparation of matching placebo in TCM clinical trials plays an important role for the success of the TCM clinical trials.

12.3.5 Sample Size Calculation

In clinical trials, sample size is usually selected to achieve a desired power for detecting a clinically meaningful difference in one of the primary study endpoints for the intended indication of the treatment under investigation (see, e.g., Chow et al., 2002a; Chow et al., 2008). As a result, sample size calculation depends upon the primary study endpoint and the clinically meaningful difference that one would like to detect. Different primary study endpoints may result in very different sample sizes.

For illustration purpose, consider the example concerning a TCM for treatment of ischemic stroke, which was developed with more than 30 years clinical experience with humans. Suppose a sponsor would like to conduct a clinical trial to scientifically evaluate the safety and efficacy of the TCM the Western way as compared to an active control (e.g., aspirin). Thus, the intended clinical trial is a double-blind, parallel-group, placebo-control, randomized trial. The primary clinical endpoint is the response rate (a patient is considered a responder if his/her Barthel index is greater than or equal to 60) based on the functional status assessed by the Barthel index. Sample size calculation is performed based on the response rate after 4 weeks of treatment under the hypotheses of testing for superiority. As a result, a sample size of 150 patients per treatment group is required for achieving an 80% power for establishment of superiority of the TCM over the active control agent. Alternatively, we may consider the quantitative instrument developed by experienced Chinese doctors as the primary study endpoint for sample size calculation. Based on a pilot study, about 80% (79 out of 122) of ischemic stroke patients were diagnosed by one domain of the quantitative instrument. A patient is considered a responder if his/her domain score is greater than or equal to 7. Based on this primary study endpoint, a sample size of 90 per treatment group is required to achieve an 80% power for establishment of superiority.

The difference in sample size leads to the question of whether the use of the primary endpoint of response rate based on one domain of the Chinese quantitative instrument could provide substantial evidence of safety and effectiveness of the TCM under investigation.

12.4 TCM Drug Development

12.4.1 Statistical Quality Control Method for Assessing Consistency

Tse et al. (2006) proposed a statistical quality control (QC) method to assess a proposed consistency index of raw materials, which are from different resources and/or final product, and may be manufactured at different sites. The idea is to construct a 95% confidence interval for a proposed consistency index under a sampling plan. If the constructed 95% confidence lower limit is greater than a pre-specified QC lower limit, then we claim that the raw materials or final product has passed the QC and hence can be released for further process or use. Otherwise, the raw materials and/or final product should be rejected. For a given component (the most active component if possible), sampling plan is derived to ensure that there is a desired probability for establishing consistency between sites when truly there is no difference in raw materials or final products between sites. The statistical quality control method for assessment of consistency proposed by Tse et al. (2006) is described below.

Let U and W be the characteristics of the most active component among the multiple components of a TCM from two different sites, where $X = \log U$ and $Y = \log W$ follows normal distributions with means μ_X, μ_Y and variances V_X, V_Y, respectively. Similar to the idea of using $P(X<Y)$ to assess reliability in statistical quality control (Enis and Geisser, 1971; Church and Harris, 1970), we propose the following probability as an index to assess the consistency of raw materials and/or final product from two different sites

$$p = P\left(1 - \delta < \frac{U}{W} < \frac{1}{1-\delta}\right), \tag{12.1}$$

where $0 < \delta < 1$ and is defined as a limit that allows for consistency. We will refer p as the consistency index. Thus p tends to 1 as δ tends to 1. For a given δ, if p is close to 1, materials U and W are considered to be identical. It should be noted that a small δ implies the requirement of high degree of consistency between material U and material W. In practice, it may be difficult to meet this narrow specification for consistency. Under the normality assumption of $X = \log U$ and $Y = \log W$, (12.1) can be rewritten as

$$p = P\big(\log(1-\delta) < \log U - \log W < -\log(1-\delta)\big)$$

$$= \Phi\left(\frac{-\log(1-\delta)-(\mu_X - \mu_Y)}{\sqrt{V_X + V_Y}}\right) - \Phi\left(\frac{\log(1-\delta)-(\mu_X - \mu_Y)}{\sqrt{V_X + V_Y}}\right).$$

where $\Phi(z_0) = P(Z < z_0)$ with Z being a standard normal random variable. Therefore, the consistency index p is a function of the parameters $\theta = (\mu_X, \mu_Y, V_X, V_Y)$, i.e. $p = h(\theta)$. Suppose that observations $X_i = \log U_i$, $i = 1,...,n_X$ and $Y_i = \log W_i, i = 1,...,n_Y$ are collected in an assay study. Then, using the invariance principle, the maximum likelihood estimator (MLE) of p can be obtained as

$$\hat{p} = \Phi\left(\frac{-\log(1-\delta)-(\overline{X}-\overline{Y})}{\sqrt{\hat{V}_X + \hat{V}_Y}}\right) - \Phi\left(\frac{\log(1-\delta)-(\overline{X}-\overline{Y})}{\sqrt{\hat{V}_X + \hat{V}_Y}}\right), \qquad (12.2)$$

where $\quad \overline{X} = \dfrac{1}{n_X}\sum_{i=1}^{n_X} X_i, \quad \overline{Y} = \dfrac{1}{n_Y}\sum_{i=1}^{n_Y} Y_i, \quad \hat{V}_X = \dfrac{1}{n_X}\sum_{i=1}^{n_X}(X_i - \overline{X})^2, \quad$ and

$\hat{V}_Y = \dfrac{1}{n_Y}\sum_{i=1}^{n_Y}(Y_i - \overline{Y})^2$. In other words, $\hat{p} = h(\hat{\theta}) = h(\overline{X}, \overline{Y}, \hat{V}_X, \hat{V}_Y)$. Furthermore, it can be easily verified that the following asymptotic result holds.

Theorem 12.1: $\hat{p}$ as given in (12.2) is asymptotically normal with mean $E(\hat{p})$ and variance $Var(\hat{p})$. In other words,

$$\frac{\hat{p}-E(\hat{p})}{\sqrt{var(\hat{p})}} \to N(0,1), \qquad (12.3)$$

where $E(\hat{p}) = p + B(p) + o\left(\dfrac{1}{n}\right)$ and $Var(\hat{p}) = C(p) + o\left(\dfrac{1}{n}\right)$. The detailed expressions of $B(p)$ and $C(p)$ are given in the proof below.

Proof: Based on the definitions of $\overline{X}$ and $\hat{V}_X$, it is easy to show that

$$E(\overline{X}) = \mu_X, \quad E(\hat{V}_X) = \frac{n_X - 1}{n_X} V_X,$$

$$\text{var}\left(\overline{X}\right) = \frac{V_X}{n_X} \text{ and } \text{var}\left(\widehat{V}_X\right) = \frac{2(n_X - 1)}{n_X^2} V_X^2.$$

Similarly,

$$E\left(\overline{Y}\right) = \mu_Y, \ E\left(\widehat{Y}_Y\right) = \frac{n_Y - 1}{n_Y} V_Y,$$

$$\text{var}\left(\overline{Y}\right) = \frac{V_Y}{n_Y} \text{ and } \text{var}\left(\widehat{V}_Y\right) = \frac{2(n_Y - 1)}{n_Y^2} V_Y^2.$$

Applying expansion of $\hat{p}$ at p, we have

$$\hat{p} = p + \frac{\partial \hat{p}}{\partial \mu_X}\left(\overline{X} - \mu_X\right) + \frac{\partial \hat{p}}{\partial \mu_Y}\left(\overline{Y} - \mu_Y\right) + \frac{\partial \hat{p}}{\partial V_X}\left(\widehat{V}_X - V_X\right) + \frac{\partial \hat{p}}{\partial V_Y}\left(\widehat{V}_Y - V_Y\right)$$

$$+ \frac{1}{2}\left[\frac{\partial^2 \hat{p}}{\partial \mu_X^2}\left(\overline{X} - \mu_X\right)^2 + \frac{\partial^2 \hat{p}}{\partial \mu_Y^2}\left(\overline{Y} - \mu_Y\right)^2 + \frac{\partial^2 \hat{p}}{\partial V_X^2}\left(\widehat{V}_X - V_X\right)^2 + \frac{\partial^2 \hat{p}}{\partial V_Y^2}\left(\widehat{V}_Y - V_Y\right)^2\right] + \dots$$

The other second order partial derivatives are not considered because they will lead to expected values of order $O(n^{-2})$ or higher. Taking expectation,

$$E\left(\hat{p}\right) = p + \frac{1}{2}\left[\frac{\partial^2 \hat{p}}{\partial \mu_X^2}\frac{V_X}{n_X} + \frac{\partial^2 \hat{p}}{\partial \mu_Y^2}\frac{V_Y}{n_Y} + \frac{\partial^2 \hat{p}}{\partial V_X^2}\left(\frac{2V_X^2}{n_X}\right) + \frac{\partial^2 \hat{p}}{\partial V_Y^2}\left(\frac{2V_Y^2}{n_Y}\right)\right] + O\left(n^{-2}\right)$$

and

$$\text{var}\left(\hat{p}\right) = \left[\left(\frac{\partial \hat{p}}{\partial \mu_X}\right)^2\frac{V_X}{n_X} + \left(\frac{\partial \hat{p}}{\partial \mu_Y}\right)^2\frac{V_Y}{n_Y} + \left(\frac{\partial \hat{p}}{\partial V_X}\right)\left(\frac{2V_X^2}{n_X}\right) + \left(\frac{\partial \hat{p}}{\partial V_Y}\right)^2\left(\frac{2V_Y^2}{n_Y}\right)\right] + O\left(n^{-2}\right).$$

Therefore,

$$B(p) = \frac{1}{2}\left[\frac{\partial^2 \hat{p}}{\partial \mu_X^2}\frac{V_X}{n_X} + \frac{\partial^2 \hat{p}}{\partial \mu_Y^2}\frac{V_Y}{n_Y} + \frac{\partial^2 \hat{p}}{\partial V_X^2}\left(\frac{2V_X^2}{n_X}\right) + \frac{\partial^2 \hat{p}}{\partial V_Y^2}\left(\frac{2V_Y^2}{n_Y}\right)\right].$$

and

$$C(p) = \left[\left(\frac{\partial \hat{p}}{\partial \mu_X}\right)^2\frac{V_X}{n_X} + \left(\frac{\partial \hat{p}}{\partial \mu_Y}\right)^2\frac{V_Y}{n_Y} + \left(\frac{\partial \hat{p}}{\partial V_X}\right)^2\left(\frac{2V_X^2}{n_X}\right) + \left(\frac{\partial \hat{p}}{\partial V_Y}\right)^2\left(\frac{2V_Y^2}{n_Y}\right)\right].$$

For the sake of simplicity, denote

$$z_1 = \frac{\log(1-\delta)-(\mu_X-\mu_Y)}{\sqrt{V_X+V_Y}}, z_2 = \frac{-\log(1-\delta)-(\mu_X-\mu_Y)}{\sqrt{V_X+V_Y}}$$

and

$$\phi(z) = \frac{1}{\sqrt{2\pi}} \exp\left(-\frac{z^2}{2}\right).$$

Then after some algebra, the partial derivatives are given as

$$\frac{\partial \hat{p}}{\partial \mu_X} = -\frac{\partial \hat{p}}{\partial \mu_Y} = \left(\frac{-1}{\sqrt{V_X+V_Y}}\right)\left[\phi(z_2)-\phi(z_1)\right],$$

$$\frac{\partial \hat{p}}{\partial V_X} = \frac{\partial \hat{p}}{\partial V_Y} = \left(\frac{-1}{2\sqrt{V_X+V_Y}}\right)\left[z_2\phi(z_2)-z_1\phi(z_1)\right],$$

$$\frac{\partial^2 \hat{p}}{\partial \mu_X^2} = \frac{\partial^2 \hat{p}}{\partial \mu_Y^2} = \left(\frac{-1}{V_X+V_Y}\right)\left[z_2\phi(z_2)-z_1\phi(z_1)\right]$$

and

$$\frac{\partial^2 \hat{p}}{\partial V_X^2} = \frac{\partial^2 \hat{p}}{\partial V_Y^2} = \frac{1}{4(V_X+V_Y)^{3/2}}\left[\left(2z_2-z_2^3\right)\phi(z_2)-\left(2z_1-z_1^3\right)z_1\phi(z_1)\right].$$

This completes the proof.

Based on the result of Theorem 12.1, an approximate $(1-\alpha)100\%$ confidence interval for p, i.e., $\left(LL(\hat{p}), UL(\hat{p})\right)$, can be obtained. In particular,

$$LL(\hat{p}) = \hat{p}-B(\hat{p})-z_{\alpha/2}\sqrt{C(\hat{p})}, \text{ and } UL(\hat{p}) = \hat{p}-B(\hat{p})+z_{\alpha/2}\sqrt{C(\hat{p})}, \quad (12.4)$$

where z_α is the upper α percentile of a standard normal distribution.

For a valid statistical quality control process, a testing procedure is necessarily performed according to some pre-specified acceptance criteria under a sampling plan. In this section, we propose a statistical quality control (QC) method for assessing consistency of raw materials and/or final product of TCM. The idea is to construct a 95% confidence interval for a proposed consistency index described above under a sampling plan. If the constructed 95% confidence lower limit is greater than a pre-specified QC lower limit, then we claim that the raw material or final product has passed the QC and hence can be released for further processing or use. Otherwise, the raw materials and/or final product should be rejected. For a given component (the most active component if possible), sampling plan is derived to ensure that there is a desired probability for establishing consistency between sites when truly there is no difference in raw materials or final products between sites. In what follows, details regarding the choice of acceptance criteria, sampling plan and the corresponding testing procedure are briefly outlined.

12.4.1.1 Acceptance Criteria

In terms of consistency, we propose the following quality control (QC) criterion. If the probability that the lower limit $LL(\hat{p})$ of the constructed $(1-\alpha) \times 100\%$ confidence interval of p is greater than or equal to a pre-specified quality control lower limit, say, QC_L, exceeds a pre-specified number β (say $\beta = 80\%$), then we claim that U and W are consistent or similar. In other words, U and W are consistent or similar if $P\big(QC_L \leq LL(\hat{p})\big) \geq \beta$ where β is a pre-specified constant.

12.4.1.2 Sampling Plan

In practice, it is necessary to select a sample size to ensure that there is a high probability, say β, of consistency between U and W when in fact U and W are consistent. It is suggested that the sample size is chosen such that there is more than 80% chance that the lower confidence limit of p is greater than or equal to the QC lower limit, i.e., $\beta = 0.8$. In other words, the sample size is determined such that

$$P\left\{QC_L \leq LL\left(\hat{p}\right)\right\} \geq \beta. \tag{12.5}$$

Using (12.5), this leads to

$$P\left\{QC_L \leq \hat{p} - B\left(\hat{p}\right) - z_{\alpha/2}\sqrt{\mathrm{Var}\left(\hat{p}\right)}\right\} \geq \beta.$$

Thus,

$$P\left\{QC_L + z_{\alpha/2}\sqrt{\mathrm{Var}\left(\hat{p}\right)} - p \leq \hat{p} - p - B(p)\right\} \geq \beta.$$

This gives

$$P\left\{\frac{QC_L - p}{\sqrt{\mathrm{Var}\left(\hat{p}\right)}} + z_{\alpha/2} \leq \frac{\hat{p} - p - B(p)}{\sqrt{\mathrm{Var}\left(\hat{p}\right)}}\right\} \geq \beta.$$

Therefore, the sample size required for achieving a probability higher than β can be obtained by solving the following equation:

$$\frac{QC_L - p}{\sqrt{\mathrm{Var}\left(\hat{p}\right)}} + z_{\alpha/2} \leq -z_{1-\beta}. \tag{12.6}$$

Assuming that $n_X = n_Y = n$, then the common sample size is given by

$$n \geq \frac{(z_{1-\beta} + z_{\alpha/2})^2}{(p - QC_L)^2} \left\{ \left(\frac{\partial \hat{p}}{\partial \mu_X}\right)^2 V_X + \left(\frac{\partial \hat{p}}{\partial \mu_Y}\right)^2 V_Y + \left(\frac{\partial \hat{p}}{\partial V_X}\right)^2 \left(2V_X^2\right) + \left(\frac{\partial \hat{p}}{\partial V_Y}\right)^2 \left(2V_Y^2\right) \right\}. \tag{12.7}$$

The above result suggests that the required sample size will depend on the choices of $\alpha, \beta, V_X, V_Y, \mu_X - \mu_Y$, and $p - QC_L$. It is clear from the expression in (12.7) that larger sample size is required for smaller α and larger β, i.e., the interval is expected to have high confidence level $(1-\alpha)$ and high chance that the lower confidence limit is larger than QC_L. Furthermore, if we require the QC_L to be close to p, i.e., $p - QC_L$ is small, a relatively large sample size is required. The dependence of the sample size n on the other parameters V_X, V_Y, and $\mu_X - \mu_Y$ is relatively unclear because these parameters are linked to the corresponding partial derivatives. A numerical study is conducted to explore the pattern. Given the large number of parameters involved in Equation (12.7), it is impractical to list the value of n for all the parameters combinations. However, for illustration purpose, we only consider a certain combinations of parameter values in an attempt to explore the pattern of dependence of n on the parameters. For the sake of simplicity, define

$$S = \frac{1}{(p - QC_L)^2} \left\{ \left(\frac{\partial \hat{p}}{\partial \mu_X}\right)^2 V_X + \left(\frac{\partial \hat{p}}{\partial \mu_Y}\right)^2 V_Y + \left(\frac{\partial \hat{p}}{\partial V_X}\right)^2 \left(2V_X^2\right) + \left(\frac{\partial \hat{p}}{\partial V_Y}\right)^2 \left(2V_Y^2\right) \right\}.$$

Then, for given choices of α and β, the required sample size n is equal to $(z_{1-\beta} + z_{\alpha/2})^2 S$. In particular, in our study, $\delta = 0.10, 0.15$, and 0.20; $\mu_X - \mu_Y = 0.5, 1.0$, and 1.5; $p - QC_L = 0.02, 0.05$, and 0.08. V_X is chosen to be 1 and $V_Y = 0.2, 0.5, 1.0, 2.0$, and 5.0. For each combination of these parameters values, the corresponding value of S is listed in Table 12.2. Given the number of parameters involved and the complexity of the mathematical expression of S, it is not easy to detect a general pattern. However, in general, the results suggest that S increases as $\mu_X - \mu_Y$ decreases; and as the variances V_x and V_y differ more from each other. In other words, smaller sample size is required if the difference between the population means is large or the variability of the two sites are of similar magnitude.

As an illustration, if for a study with $\delta = 0.2, V_X = 1, V_Y = 0.5, \mu_X - \mu_Y = 1.0$, and an experiment expect $p - QC_L$ to be not larger than 0.05, then results in Table 12.2 suggests that $S = 3.024$. Suppose a probability higher than $\beta = 0.8$

TABLE 12.2

Values of $n/(z_{1-\beta}+z_{\alpha/2})^2$, Where n Is the Required Sample Size

		$d=0.10$			$d=0.15$			$d=0.20$		
		$\Delta=0.5$	$\Delta=1.0$	$\Delta=1.5$	$\Delta=0.5$	$\Delta=1.0$	$\Delta=1.5$	$\Delta=0.5$	$\Delta=1.0$	$\Delta=1.5$
$D=0.02$	$[V_Y]=0.2$	5.693	5.376	4.955	13.403	12.681	11.702	24.861	23.594	21.810
	0.5	4.518	4.289	4.196	10.655	10.134	9.921	19.820	18.901	18.520
	1.0	3.939	3.336	3.237	9.310	7.894	7.662	17.370	14.761	14.333
	2.0	4.231	2.962	2.226	10.020	7.021	5.280	18.756	13.163	9.906
	5.0	5.728	4.159	2.469	13.595	9.876	5.866	25.534	18.558	11.032
$D=0.05$	0.2	0.911	0.860	0.793	2.144	2.029	1.872	3.978	3.775	3.490
	0.5	0.723	0.686	0.671	1.705	1.622	1.587	3.171	3.024	2.963
	1.0	0.630	0.534	0.518	1.490	1.263	1.226	2.779	2.362	2.293
	2.0	0.677	0.474	0.356	1.603	1.123	0.845	3.001	2.106	1.585
	5.0	0.916	0.666	0.395	2.175	1.580	0.939	4.085	2.969	1.765
$D=0.08$	0.2	0.356	0.336	0.310	0.838	0.793	0.731	1.554	1.475	1.363
	0.5	0.282	0.268	0.262	0.666	0.633	0.620	1.239	1.181	1.158
	1.0	0.246	0.208	0.202	0.582	0.493	0.479	1.086	0.923	0.896
	2.0	0.264	0.185	0.139	0.626	0.439	0.330	1.172	0.823	0.619
	5.0	0.358	0.260	0.154	0.850	0.617	0.367	1.596	1.160	0.690

Notation: $\Delta=\mu_X-\mu_Y$, $D=p-QC_L$

at the $\alpha = 0.05$ level of significance is required, the corresponding required sample size is given by

$$n \geq (z_{1-0.8} + z_{0.05/2})^2 S = (0.842 + 1.96)^2 (3.024) = 23.74.$$

Thus, a sample of size at least 24 is required.

12.4.1.3 Testing Procedure

Hypotheses testing of the consistency index p can also be conducted based on the asymptotic normality of $\hat{p}$. Consider the following hypotheses

$$H_0 : p \leq p_0 \text{ versus } H_\alpha : p > p_0.$$

We would reject the null hypothesis and in favor of the alternative hypothesis of consistency. Under H_0, we have

$$\frac{\hat{p} - p_0 - B(\hat{p})}{\sqrt{\mathrm{var}(\hat{p})}} \sim N(0,1) \tag{12.8}$$

Thus, we reject the null hypothesis H_0 at the α level of significance if

$$\frac{\hat{p} - p_0 - B(\hat{p})}{\sqrt{\mathrm{var}(\hat{p})}} > Z_\alpha.$$

This is equivalent to reject the null hypothesis H_0 when

$$\hat{p} > p_0 + B(\hat{p}) + Z_\alpha \sqrt{\mathrm{var}(\hat{p})}.$$

Again, for illustration purpose, Table 12.3 provides critical values of the proposed test for consistency index for various combinations of the parameters. In particular, $\alpha = 0.1$, $p_0 = 0.75$, 0.85 and 0.9, $\delta = 0.10$ and 0.20; $\mu_X - \mu_Y = 0.5$, 1.0, and 1.5. V_X is chosen to be 1 and $V_Y = 0.2$, 0.5, 1.0, 2.0, and 5.0. Note that the critical value is closer to the corresponding p_0 either for larger sample size n, smaller δ, or smaller $\mu_X - \mu_Y$.

12.4.1.4 Strategy for Statistical Quality Control

In practice, raw materials, in-process materials, and/or final products at different sites are manufactured sequentially in batches or lots. As a result, it is important to perform statistical quality control on batches. A typical approach is to randomly select samples from several (consecutive) batches

TABLE 12.3

Critical Values of the Proposed Test for Consistency Index p_0

p_0	δ	V_Y	D = 0.5			D = 1.0			D = 1.5		
			$n = 15$	$n = 30$	$n = 50$	$n = 15$	$n = 30$	$n = 50$	$n = 15$	$n = 30$	$n = 50$
0.75	0.10	0.2	0.7695	0.7640	0.7609	0.7683	0.7632	0.7604	0.7680	0.7629	0.7601
		0.5	0.7673	0.7624	0.7597	0.7665	0.7619	0.7593	0.7665	0.7619	0.7593
		1.0	0.7662	0.7616	0.7590	0.7646	0.7605	0.7582	0.7645	0.7604	0.7581
		2.0	0.7668	0.7620	0.7594	0.7639	0.7600	0.7578	0.7620	0.7586	0.7567
		5.0	0.7697	0.7640	0.7609	0.7667	0.7619	0.7593	0.7628	0.7592	0.7572
	0.20	0.2	0.7907	0.7791	0.7727	0.7884	0.7777	0.7717	0.7878	0.7771	0.7712
		0.5	0.7863	0.7760	0.7703	0.7846	0.7749	0.7695	0.7847	0.7749	0.7695
		1.0	0.7839	0.7743	0.7689	0.7807	0.7721	0.7673	0.7805	0.7719	0.7671
		2.0	0.7853	0.7753	0.7697	0.7793	0.7710	0.7664	0.7754	0.7682	0.7642
		5.0	0.7915	0.7797	0.7731	0.7853	0.7752	0.7697	0.7771	0.7694	0.7651
0.85	0.10	0.2	0.8695	0.8640	0.8609	0.8683	0.8632	0.8604	0.8680	0.8629	0.8601
		0.5	0.8673	0.8624	0.8597	0.8665	0.8619	0.8593	0.8665	0.8619	0.8593
		1.0	0.8662	0.8616	0.8590	0.8646	0.8605	0.8582	0.8645	0.8604	0.8581
		2.0	0.8668	0.8620	0.8594	0.8639	0.8600	0.8578	0.8620	0.8586	0.8567
		5.0	0.8697	0.8640	0.8609	0.8667	0.8619	0.8593	0.8628	0.8592	0.8572
	0.20	0.2	0.8907	0.8791	0.8727	0.8884	0.8777	0.8717	0.8878	0.8771	0.8712
		0.5	0.8863	0.8760	0.8703	0.8846	0.8749	0.8695	0.8847	0.8749	0.8695
		1.0	0.8839	0.8743	0.8689	0.8807	0.8721	0.8673	0.8805	0.8719	0.8671
		2.0	0.8853	0.8753	0.8697	0.8793	0.8710	0.8664	0.8754	0.8682	0.8642
		5.0	0.8915	0.8797	0.8731	0.8853	0.8752	0.8697	0.8771	0.8694	0.8651

(Continued)

TABLE 12.3 (Continued)

Critical Values of the Proposed Test for Consistency Index p_0

p_0	δ	V_Y	D = 0.5			D = 1.0			D = 1.5		
			n = 15	n = 30	n = 50	n = 15	n = 30	n = 50	n = 15	n = 30	n = 50
0.90	0.10	0.2	0.9195	0.9140	0.9109	0.9183	0.9132	0.9104	0.9180	0.9129	0.9101
		0.5	0.9173	0.9124	0.9097	0.9165	0.9119	0.9093	0.9165	0.9119	0.9093
		1.0	0.9162	0.9116	0.9090	0.9146	0.9105	0.9082	0.9145	0.9104	0.9081
		2.0	0.9168	0.9120	0.9094	0.9139	0.9100	0.9078	0.9120	0.9086	0.9067
		5.0	0.9197	0.9140	0.9109	0.9167	0.9119	0.9093	0.9128	0.9092	0.9072
	0.20	0.2	0.9407	0.9291	0.9227	0.9384	0.9277	0.9217	0.9378	0.9271	0.9212
		0.5	0.9363	0.9260	0.9203	0.9346	0.9249	0.9195	0.9347	0.9249	0.9195
		1.0	0.9339	0.9243	0.9189	0.9307	0.9221	0.9173	0.9305	0.9219	0.9171
		2.0	0.9353	0.9253	0.9197	0.9293	0.9210	0.9164	0.9254	0.9182	0.9142
		5.0	0.9415	0.9297	0.9231	0.9353	0.9252	0.9197	0.9271	0.9194	0.9151

Notation: $\Delta = \mu_X - \mu_Y$

for testing. In this case, observations from the study would be subject to batch-to-batch variability. For the sake of administrative convenience, it is common to have equal number of observations from the batches. Consider the following model:

$$X_{ij} = \mu_X + A_i^X + \varepsilon_{ij}^X, i = 1,..., m_X; j = 1,..., n_X,$$

where A_i^X accounts for the batch-to-batch variability for the observations collected in site 1 and is normally distributed with mean 0 and variance σ_{b1}^2; m_X is the number of batches collected in the study at site 1 and ε_{ij}^X are normal random variables with mean 0 and variance σ_1^2. Similarly,

$$Y_{ij} = \mu_Y + A_i^Y + \varepsilon_{ij}^Y, i = 1,..., m_Y; j = 1,..., n_Y,$$

where A_i^Y accounts for the batch-to-batch variability for the observations collected in site 2 and is normally distributed with mean 0 and variance σ_{b2}^2; m_Y is the number of batches collected in the study at site 2 and ε_{ij}^Y are normal random variables with mean 0 and variance σ_2^2. Therefore, the total variability of the most active component at the two sites are given by var $X = V_X = \sigma_{b1}^2 + \sigma_1^2$ and var $Y = V_Y = \sigma_{b2}^2 + \sigma_2^2$, respectively. Furthermore, let

$$\bar{X}_{i\cdot} = \frac{1}{n_X} \sum_{j=1}^{n_X} X_{ij} \text{ and } \bar{X} = \frac{1}{m_X} \sum_{i=1}^{m_X} \bar{X}_{i\cdot}.$$

Then, the observed sums of squares are

$$SSA_1 = n_X \sum_{i=1}^{m_X} (\bar{X}_{i\cdot} - \bar{X})^2,$$

$$SSE_1 = \sum_{i=1}^{m_X} \sum_{j=1}^{n_X} (X_{ij} - \bar{X}_{i\cdot})^2$$

and

$$SST_1 = SSA_1 + SSE_1.$$

Following the results in Chow and Tse (1991), the MLE of σ_{b1}^2 and σ_1^2 are

$$
\hat{\sigma}_{b1}^2 = \begin{cases} \dfrac{1}{n_X}\left(\dfrac{1}{m_X}SSA_1 - \dfrac{1}{m_X(n_X-1)}SSE_1 \right) & \dfrac{1}{m_X}SSA_1 \geq \dfrac{1}{m_X(n_X-1)}SSE_1 \\ & \text{if} \\ 0 & \dfrac{1}{m_X}SSA_1 < \dfrac{1}{m_X(n_X-1)}SSE_1 \end{cases}
$$

(12.9)

and

$$
\hat{\sigma}_1^2 = \begin{cases} \dfrac{1}{m_X(n_X-1)}SSE_1 & \dfrac{1}{m_X}SSA_1 \geq \dfrac{1}{m_X(n_X-1)}SSE_1 \\ & \text{if} & . \\ \dfrac{1}{n_Xm_X}SST_1 & \dfrac{1}{m_X}SSA_1 < \dfrac{1}{m_X(n_X-1)}SSE_1 \end{cases}
$$

(12.10)

Furthermore, the MLE of the total variability V_X is given by $\widehat{V} = \frac{1}{n_Xm_X}SST_1$. The MLE of σ_{b2}^2, σ_2^2, and V_y, denoted by $\hat{\sigma}_{b2}^2$, $\hat{\sigma}_2^2$, and $\widehat{V}_Y$, respectively, can be obtained in a similar way using observations Y_{ij}. Comparison of the estimates $\hat{\sigma}_{b2}^2$ and $\hat{\sigma}_{b1}^2$ would give an idea of the magnitude of the batch-to-batch variability at the two sites.

12.4.1.5 Remarks

Note that the method proposed by Tse et al. (2006) only focuses on a single (i.e., the most active) component assuming that the most active component can be quantitatively identified among multiple active components. Following similar idea, Lu et al. (2007) extended their results to the case of two correlative components by considering p_1 and p_2, the consistency indices of the two most active components of a TCM from two different sites. Lu et al. (2007) proposed to define the consistency index of a TCM with two correlative components by $\min(p_1, p_2)$ and denote it by p, where

$$
p_i = P\left(1-\delta_i < \frac{U_i}{W_i} < \frac{1}{1-\delta_i} \right), 0 < \delta_i < 1, i = 1,2
$$

and δ_i is a limit that allows for consistency. Therefore, the consistency index p is a function of the parameters $\theta = (\mu_{X_1}, \mu_{X_2}, \mu_{Y_1}, \mu_{Y_2}, V_{X_1}, V_{X_2}, V_{Y_1}, V_{Y_2})$, i.e., $p = h(\theta)$. By invariance principle, the maximum likelihood estimator (MLE) of p_1 and p_2 is given by

$$\hat{p}_i = \Phi\left(\frac{-\log(1-\delta_i)-(\bar{X}-\bar{Y}_i)}{\sqrt{\hat{V}+\hat{V}}}\right) - \Phi\left(\frac{\log(1-\delta_i)-(\bar{X}-\bar{Y}_i)}{\sqrt{\hat{V}+\hat{V}}}\right), \qquad (12.11)$$

where $\Phi(z_0) = P(Z < z_0)$ with Z being a standard normal random variable,

$$\bar{X} = \frac{1}{n}\sum_{j=1}^{n}X_{ij}, \bar{Y}_i = \frac{1}{n}\sum_{j=1}^{n}Y_{ij},$$

and

$$\hat{V}_{X_i} = \frac{1}{n}\sum_{j=1}^{n}(X_{ij}-\bar{X})^2, \hat{V}_{Y_i} = \frac{1}{n}\sum_{j=1}^{n}(Y_{ij}-\bar{Y}_i)^2, \ i=1,2.$$

Thus, the MLE of the proposed consistency index p is given by $\hat{p} = \min\left(\hat{p}_1,\hat{p}_2\right)$. Furthermore, it can be verified that the following asymptotic result holds (see also Lu et al., 2007).

Theorem 12.2: $\log \hat{p}$ as given in (2.1) with mean $E(\log\hat{p})$ and variance $\mathrm{Var}(\log\hat{p})$, where $E(\log\hat{p}) = \log p + B(p) + o(n^{-1})$ and $\mathrm{Var}(\log\hat{p}) = C(p) + o(n^{-1})$. The detailed expressions of $B(p)$ and $C(p)$ are given in the appendix. Furthermore,

$$\frac{\log \hat{p} - \log p - B(\hat{p})}{\sqrt{C(\hat{p})}} \to N(0,1)$$

where $B(\hat{p})$ and $C(\hat{p})$ are estimates of $B(p)$ and $C(p)$ with the unknown population parameters $\theta = \left(\mu_{X_1},\mu_{X_2},\mu_{Y_1},\mu_{Y_2},V_{X_1},V_{X_2},V_{Y_1},V_{Y_2}\right)$ estimated by their corresponding MLE's $\hat{\theta} = \left(\bar{X}_1,\bar{X}_2,\bar{Y}_1,\bar{Y}_2,\hat{V}_{X_1},\hat{V}_{X_2},\hat{V}_{Y_1},\hat{V}_{Y_2}\right)$.

Proof: The details of the derivation of $B(p)$ and $C(p)$ are given in the Appendix. In particular,

$$B(p) = \frac{1}{np_k}\frac{\partial^2\hat{p}}{\partial V_{X_k}^2}\left(V_{X_k}^2 + V_{Y_k}^2\right) - \frac{1}{2np_k^2}\left[\left(\frac{\partial\hat{p}}{\partial\mu_{X_k}}\right)^2\left(V_{X_k}+V_{Y_k}\right) + 2\left(\frac{\partial\hat{p}}{\partial V_{X_k}}\right)^2\left(V_{X_k}^2 + V_{Y_k}^2\right)\right]$$

and

$$C(p) = \frac{1}{np_k^2}\left[\left(\frac{\partial \hat{p}}{\partial \mu_{X_k}}\right)^2 (V_{X_n} + V_{Y_n}) + 2\left(\frac{\partial \hat{p}}{\partial V_{X_k}}\right)^2 (V_{X_k}^2 + V_{Y_k}^2)\right],$$

where the subscript k is defined by $k = j$ if $\hat{p} = \hat{p}, j = 1$ or 2. Note that $B(p)$ converges to 0 as n tend to infinity. Thus, $\hat{p}$ is asymptotically unbiased. Since $\hat{\theta} = (\bar{X}_1, \bar{X}_2, \bar{Y}_1, \bar{Y}_2, \hat{V}_{X_1}, \hat{V}_{X_2}, \hat{V}_{Y_1}, \hat{V}_{Y_2})$ is asymptotically multivariate normally distributed and $\hat{p}$ is a function of $\hat{\theta}$, it follows from Serfling (1980) that

$$\frac{\log \hat{p} - E(\log \hat{p})}{\sqrt{var(\log \hat{p})}} \to N(0,1).$$

Using Slutsky's Theorem, it can be shown that

$$\frac{\log \hat{p} - \log p - B(\hat{p})}{\sqrt{C(\hat{p})}}$$

is asymptotically normal since $B(\hat{p})$ and $C(\hat{p})$ are consistent estimates of $B(p)$ and $C(p)$, respectively.

Based on the results given in Theorem 19.2, for a given level $0 < \alpha < 1$, an approximate $(1-\alpha) \times 100\%$ confidence interval for $\log p$, denote by $(LL(\log \hat{p}), UL(\log \hat{p}))$, can be obtained as follows,

$$LL(\log \hat{p}) = \log \hat{p} - B(\hat{p}) - z_{\alpha/2}\sqrt{C(\hat{p})}, \tag{12.12}$$

and

$$UL(\log \hat{p}) = \log \hat{p} - B(\hat{p}) + z_{\alpha/2}\sqrt{C(\hat{p})}, \tag{12.13}$$

where $Z_{\alpha/2}$ is the upper $\alpha/2$-percentile of the standard normal distribution. Consequently, an approximate $(1-\alpha) \times 100\%$ confidence interval for p, denote by $(LL(\hat{p}), UL(\hat{p}))$, is given as

$$\left(e^{LL(\log \hat{p})}, e^{UL(\log \hat{p})}\right). \tag{12.14}$$

12.4.2 Stability Analysis

In the pharmaceutical industry, stability analysis refers to a study conducted for determining the expiration dating period (shelf-life) of a drug product under appropriate storage conditions. The shelf-life of a drug is defined as the time interval at which the potency of the drug remains within the approved specification limit, e.g., the specification limit given in the United States Pharmacopedia (USP) and National Formulary (NF) (USP/NF, 2000). The US FDA requires that the shelf-life be indicated on the immediate container label for every drug product in the marketplace. While many drug products consist of a single active ingredient, there are drug products containing multiple active ingredients (see, e.g., Pong and Raghavarao, 2002). For example, as indicated in Chow and Shao (2007), Premarin (conjugated estrogens, USP) contains at least five active ingredients, estrone, equilin, 17α-dihydroequilin, 17α-estradiol, and 17β-dihydroequilin. Other examples include combinational drug products, such as the traditional Chinese medicines, which is known to contain multiple active components. For a drug product with multiple active ingredients, an ingredient-by-ingredient stability analysis may not be appropriate, since these active ingredients may have some unknown interactions. Chow and Shao (2007) proposed a statistical method for determining the shelf-life of a drug product with multiple active components or ingredients following similar idea as suggested by the FDA and assuming that these active components or ingredients are linear combinations of some factors. The method proposed by Chow and Shao (2007) is described below.

Let $y(t,k)$ be the potency of the kth component or ingredient at time t after the manufacture of a given drug product, $k = 1,...,p$. For ingredient k, its shelf-life is the time interval at which $E[y(t,k)]$ (the expectation of $y(t,k)$) remains within a specified limit, whereas the shelf-life for the drug product may be the time interval at which $E\left[f\left(y(t,1),...,y(t,p)\right)\right]$ remains within the specified limits, where f is a function (such as a linear combination of $y(t,1),...,y(t,p)$) that characterizes the impact of all active components or ingredients. In general, f is a vector-valued function with a dimension $q \leq p$.

If data are observed from $y(t,1),...,y(t,p)$ and the function f is a known function, then stability analysis can be made by using the transformed data $z(t) = f\left(y(t,1),...,y(t,p)\right)$. If the dimension of f is 1, then $z(t)$ can be treated as a single component or ingredient. If the dimension of f is $q > 1$, then one may define the shelf-life to be the minimum of the shelf-lives $\tau_1,...,\tau_q$ where τ_h is the shelf-life when the hth component or ingredient of $z(t)$ is treated as a single component or ingredient. One special case is where f is the identity function so that the shelf-life is the minimum of all shelf-lives corresponding to different components or ingredients $y(t,k), k = 1,...,p$.

In practice, however, f is typically unknown. Although the best way to estimate f is to fit a model between the y and z variables, it requires data observed from both y and z, which is not a common practice in pharmaceutical industry, because the variable z in many problems, such as the traditional Chinese medicines, is not clearly defined (see, e.g., Chow et al., 2006).

In this chapter, we assume that the components of z are linear combinations of the components of y and propose a method to establish the shelf-life. Note that the approach proposed by Chow and Shao (2006) is basically an application of the factor model in multivariate analysis (see, e.g., Johnson and Wichern, 1992).

12.4.2.1 Models and Assumptions

Let $y(t)$ denote the p-dimensional vector whose kth component is the potency of the kth component or ingredient at time t after the manufacture of a given drug product, $k = 1,...,p$. We assume that the drug potency is expected to decrease with time t. If $p = 1$, i.e., $y(t)$ is univariate, the current established procedure for determination of a shelf-life is to use the time at which a 95% lower confidence bound for the mean degradation curve $E[y(t)]$ intersects the acceptable lower product specification limit as specified in USP/NF (2000) (see also, FDA, 1987a; ICH, 1993). Let η be the vector whose kth component or ingredient is the lower product specification limit as specified in the USP/NF for the kth component or ingredient of $y(t)$. Assume that, for any t

$$y(t) - E[y(t)] = LF_t + \mathcal{E}_t, \tag{12.15}$$

where L is a $p \times q$ non-random unknown matrix of full rank, F_t and ε_t are unobserved independent random vectors of dimensions q and p, respectively, $E(F_t) = 0$, $\text{Var}(F_t) = I_q$ (the identity matrix of order q), $E(\varepsilon_t) = 0$, $\text{Var}(\varepsilon_t) = \Psi$, and ψ is an unknown diagonal matrix of order p. Note that model (12.15) with the assumptions on F_t and ε_t is the so-called orthogonal factor model (Johnson and Wichern, 1992). If ε_t is treated as a random error, then model (12.15) assumes that the p-dimensional component or ingredient vector $y(t)$ is governed by a q-dimensional unobserved vector F_t. Normally q is much smaller than p Let $z(t) = (L'L)^{-1}L'[y(t) - \eta]$. It follows from (12.15) that

$$z(t) - E[z(t)] = F_t + (L'L)^{-1}L'\varepsilon_t. \tag{12.16}$$

If L is known, then (23.16) suggests performing a stability analysis based on the transformed data observed from $z(t)$. In practice, since L is unknown, if we can estimate L based on model (12.15) and the observed data from $y(t)$, then we can carry out a stability analysis using the transformed $z(t)$ with L replaced by its estimate.

Let $x(t)$ be an s-dimensional covariate vector associated with $y(t)$ at time t. For example, $x(t) = (1,t)'\ (s = 2)$ or $x(t) = (1,t,t^2)'\ (s = 3)$. We assume the following model at any time t:

$$E[y(t) - \eta] = Bx(t), \text{Var}[y(t)] = \Sigma, i = 1,...,m, j = 1,...,n, \tag{12.17}$$

where B is a $p \times s$ matrix of unknown parameters and $\Sigma > 0$ is an unknown $p \times p$ positive definite covariance matrix. Since $z(t) = (L'L)^{-1}L'\left[y(t) - \eta\right]$, it follows from (12.17) that:

$$E\left[z(t)\right] = \gamma' x(t), i = 1,...,m, j = 1,...,n, \qquad (12.18)$$

where $\gamma = B'L(L'L)^{-1}$.

12.4.2.2 Shelf-Life Determination

Suppose that we independently observe data $y_{ij}, i = 1,...,m, j = 1,...,n$, where y_{ij} is the jth replicate of $y(t_i)$ and $t_1,...,t_m$ are designed time points for the stability analysis. Define

$$x_i = x(t_i), z_{ij} = (L'L)^{-1}L'\left(y_{ij} - \eta\right), i = 1,...,m, j = 1,...,n. \qquad (12.19)$$

Consider first the case of $q = 1$, i.e., Z_{ij} in (12.19) is univariate. If $z_{ij}'s$ are observed, then an approximate 95% lower confidence bound for $E[z(t)] = \gamma' x(t)$ is

$$l(r) = \hat{\gamma} x(t) - t_{0.95, mn-s} \hat{\sigma} \sqrt{D(t)}, \qquad (12.20)$$

where $\hat{\gamma}$ is the least squares estimator of γ in model (12.20) based on data $Z_{ij}'S$ and $x_i's, \hat{\sigma}^2$ is the usual sum of squared residuals divided by its degrees of freedom $mn - s$, $t_{0.95, mn-s}$ is the 95th percentile of the t-distribution with degrees of freedom $mn - s$, and

$$D(t) = \left[n \sum_{i=1}^{m} x(t)' x_i x_i' x(t)\right]^{-1}.$$

Hence, if $Z_{ij}'s$ are observed, a shelf-life according to the 1987 FDA guideline for stability (FDA, 1987a) is

$$\tau = \inf\left\{t : l(t) \leq 0\right\}. \qquad (12.21)$$

For TCM, $y_{ij}'s$, not $Z_{ij}'s$, are observed. Hence, the lower confidence bound $l(t)$ in (12.20) needs to be modified. Since $\gamma' = (L'L)^{-1}L'B$, we can obtain an estimator of γ in two steps. At the first step, we use model (12.17), observed data $y_{ij}'s$ and $x_i's$, and the multivariate linear regression to obtain a least squares estimator $\hat{B}$ of B. At the second step, we consider the orthogonal factor model (12.15) and apply the method of principal components to obtain an estimator $\hat{L}$ of L, using *data* $y_{ij} - \eta - \hat{B}x_i, i = 1,...,m, j = 1,...,n$. More precisely, $\hat{L}$ is the normalized eigenvector corresponding to the largest eigenvalue of the sample covariance matrix based on data $y_{ij} j - \hat{B}x_i, i = 1,...,m, j = 1,...,n$. Let $\hat{\gamma} = \hat{B}'\hat{L}(\hat{L}'\hat{L})^{-1}$.

The lower confidence bound in (12.20) is modified to

$$l(t) = \hat{\gamma}'x(t) - t_{0.95,mn-s}\sqrt{x(t)'Vx(t)}, \qquad (12.22)$$

where V is the jackknife variance estimator of $\hat{\gamma}$ (see, e.g., Shao and Tu, 1995), i.e.,

$$V = \frac{mn-1}{mn}\sum_{i=1}^{m}\sum_{j=1}^{n}(\hat{\gamma}_{i,j} - \hat{\gamma})(\hat{\gamma}_{i,j} - \hat{\gamma})',$$

where $\hat{\gamma}_{i,j}$ is the estimator of γ calculated using the same method as in the calculation of $\hat{\gamma}$ but with the (i,j)th data point deleted. The result for $q = 1$ is sufficient for applications with a small or moderate p. When p is large, Chow and Shao (2007) proposed the following procedure with $1 < q < p$. Let $\hat{B}$ be defined as before, λ_k be the kth largest eigenvalue of the sample covariance matrix based on $y_{ij}j - \eta - \hat{B}x_i, i = 1,\ldots,m, j = 1,\ldots,n$, and e_k be the normalized eigenvector corresponding to λ_k. Then, the estimator $\hat{L}$ of L is the $p \times q$ matrix whose kth column is $\lambda_k e_k, k = 1,\ldots,q$. The estimator of γ is still $\hat{\gamma} = \hat{B}'\hat{L}(\hat{L}'\hat{L})^{-1}$, which is an $s \times q$ matrix. Let $\hat{\gamma}_k$ be the kth column of $\hat{\gamma}, k = 1,\ldots,q$

$$l_k(t) = \hat{\gamma}'_k x(t) - t_{1-0.05/q,mn-s}\sqrt{x(t)'V_k x(t)} \qquad (12.23)$$

and

$$V_k = \frac{mn-1}{mn}\sum_{i=1}^{m}\sum_{j=1}^{n}(\hat{\gamma}_{k,i,j} - \hat{\gamma}_k)(\hat{\gamma}_{k,i,j} - \hat{\gamma}_k)',$$

where $\hat{\gamma}_{k,i,j}$ is the same as $\hat{\gamma}_k$ but calculated with the (i,j)th data point deleted. Then, $l_k(t), k = 1,\ldots,q$, are approximate 95% simultaneous lower confidence bounds for $\zeta_k(t), k = 1,\ldots,q$, where $\zeta_k(t)$ is the kth component of $E[z(t)] = \gamma'x(t)$. An approximate level 95% shelf-life for the drug product (when the sample size mn is large) is

$$\tau = \min_{k=1,\ldots,q}\tau_k'$$

where each τ_k is defined by the right-hand side of (12.21) with $l(t)$ replaced by $l_k(t)$ and is in fact a shelf-life for the kth component of z with confidence level $(1 - 0.05/q)\%$.

12.4.2.3 An Example

To illustrate the proposed method for determining the shelf-life of a drug product with multiple active ingredients, consider a stability study conducted for a traditional Chinese herbal medicine, which is newly developed for treatment of patients with rheumatoid arthritis. This medicine contains three active botanical components, namely, Herba Epimedii (HE), B extract, and C extract. Each of the three components has been used as herbal remedies since ancient times in China and is well documented in the Chinese Pharmacopeia. The proportions of each components are summarized below (Table 12.4).

To establish a shelf-life for this product, a stability study was conducted for a time period of 18 months under a testing condition of 25°C / 60% RH (relative humidity). The lower product specification limit for each component is 90%. Stability data (per cent of label claim) at each sampling time point for the three components are given in Table 12.5.

Since $p = 3$, we consider $q = 1$. Using the proposed procedure described in the previous sections, $l(t)$ in (12.22) for various t (in month) as follows (Table 12.6).

Hence, the estimated shelf-life for this product is 27 months.

TABLE 12.4

Components of a TCM

Component	Formulation
HE	60 mg
B	25 mg
C	25 mg
Excipient	90 mg
Total	200 mg

TABLE 12.5

Stability Data of a TCM

Component	0	Sampling	Time Point	(month)	12	18
		3	6	9		
HE	99.6	97.5	96.8	96.2	94.8	95.3
	99.7	98.3	97.0	96.0	95.1	94.8
	100.2	99.0	98.2	97.1	95.3	94.6
B	99.5	98.4	96.3	95.4	93.2	91.0
	100.5	98.5	97.4	94.9	94.5	92.1
	99.3	99.0	97.3	95.0	93.1	91.5
C	100.0	99.5	98.9	98.2	97.9	97.5
	99.8	99.4	99.0	98.5	98.0	97.9
	101.2	99.9	100.3	99.5	98.9	98.0

TABLE 12.6

$l(t)$ Values with Various t

t	19	20	21	22	23	24	25	26	27	28
$l(t)$	4.97	4.36	3.75	3.14	2.52	1.90	1.28	0.66	0.03	−0.60

12.4.2.4 Discussion

The statistical method for determining the shelf-life of a drug product with p active ingredients proposed by Chow and Shao (2007) assumed that these active ingredients are linear combinations of q factors. Since we propose to choose these factors using principal components, the first factor can be viewed as the primary active factor and the second factor can be viewed as the secondary active factor. We assume that active ingredients decrease with time. If one or more ingredients increase with time, then a transformation such as $g(y) = -y$ or $g(y) = 1/y$ may be applied. If p is small or moderate, then $q = 1$ is recommended. If P is large, then adding a few more factors may be considered. Since the principal components are orthogonal, adding more factors will not affect the previous selected factors (except that $t_{095, mn-s}$ changed to $t_{1-005/q, mn-s}$) so that one can compare the results in a sensitivity analysis. Finally, adding more factors always results in a more conservative procedure.

Note that in our proposed approach, we assume that there is no significant toxic degradant in the test drug product with multiple components. This is a reasonable assumption for most traditional Chinese medicine since multiple ingredients are to reduce toxicities when used in conjunction with primary therapy. However, in the case when toxic degradation products are detected, special attention should be paid to (i) identity (chemical structure), (ii) cross reference to information about biological effects and significance of concentration likely to be encountered, and (iii) indications of pharmacological action or inaction as indicated in the FDA guideline for stability analysis.

The approach proposed by Chow and Shao (2007) is useful when different ingredients degrade not independently of each other, which is the case for most traditional Chinese medicine. If multiple ingredients degrade independently, then an ingredient-by-ingredient analysis may be appropriate. If our approach is applied, then we will select $q = 1$ or $q = 2$ factors that are ingredients having the most variability.

12.4.3 Calibration of Study Endpoints in Clinical Development

When planning a clinical trial, it is suggested that the study objectives should be clearly stated in the study protocol. Once the study objectives are confirmed, a valid study design can be chosen and the primary clinical endpoints can be determined accordingly. For evaluation of treatment effect of a TCM, however, the commonly used clinical endpoint is usually not applicable due to the nature of the Chinese diagnostic procedures (CDP) as described earlier. The CDP is in fact an instrument (or questionnaire) that consists of a number of questions to capture the information regarding patient's activity, function, disease status and severity. As required by most regulatory agencies, such a subjective instrument must be validated before it can be used for assessment of treatment effect in clinical trials. However, without a reference marker, not only can the CDP not be validated, but also we do not know whether the TCM has achieved clinically significant effect at the end of the clinical trial. In this section, we will study the calibration and validation of the Chinese diagnostic procedure for evaluation of a TCM with respect to a well-established clinical endpoint for evaluation of a western medicine.

To address these issues described above, Hsiao et al. (2009) proposed a study design, which allows calibration and validation of a CDP with respect to a well-established clinical endpoint for WM (as a reference marker). Subjects will be screened based on criteria for western indication. Qualified subjects will be diagnosed by the CDP to establish baseline. Qualified subjects will then be randomized to receive either the test TCM or an active control (a well-established western medicine). Participated physicians including Chinese doctors and western clinicians will also be randomly assigned to either the TCM arm or the arm of western medicine (WM). As a result, this study design will result in three groups:

Group 1: Subjects who receive WM but evaluated by both a Chinese doctor and a Western clinician.

Group 2: Subjects who receive TCM and are evaluated by a Chinese doctor A.

Group 3: Subjects who receive TCM and are evaluated by a Chinese doctor B.

Group 1 can be used to calibrate the CDP against the well-established clinical endpoint, while Groups 2 and 3 can be used to validate the CDP based on the established standard curve for calibration.

12.4.3.1 Chinese Diagnostic Procedure

As indicated earlier, the diagnostic procedure for TCM consists of four major techniques, namely, inspection, auscultation and olfaction, interrogation, and pulse taking and palpation. All these diagnostic techniques aim mainly at providing objective basis for differentiation of syndromes by collecting symptoms and signs from the patient. Inspection involves observing the patient's general appearance (strong or weak, fat or thin), mind, complexion (skin color), five sense organs (eye, ear, nose, lip, and tongue), secretions, and excretions. Auscultation involves listening to the voice, expression, respiration, vomit, and cough. Olfaction involves smelling the breath and body odor. Interrogation involves asking questions about specific symptoms and the general condition including history of the present disease, past history, personal life history, and family history. Pulse taking and palpation can help to judge the location and nature of a disease according to the changes of the pulse. The smallest detail can have a strong impact on the treatment scheme as well as on the prognosis. While the pulse diagnosis and examination of the tongue receive much attention due to their frequent mention, the other aspects of diagnosis cannot be ignored.

After these four diagnostic techniques have been performed, the TCM doctor has to configure a syndrome diagnosis describing the fundamental substances of the body and how they function in the body based on the eight principles, five element theory, five Zang and six Fu, and information regarding channels and collaterals. Eight principles consist of Yin and Yang (i.e., negative and positive), cold and hot, external and internal, and Shi and Xu (i.e., weak and strong). Eight principles can help the TCM doctors to differentiate syndrome patterns. For instance, Yin people will develop disease in a negative, passive, and cool way (e.g., diarrhea and back pain), while Yang people will develop disease in an aggressive, active, progressive, and warm way (e.g., dry eyes, tinnitus, and night sweats). The five elements (earth, metal, water, wood, and fire) correspond to particular organs in the human body. Each element operates in harmony with the others.

Five Zang (or Yin organs) includes heart (including the pericardium), lung, spleen, liver, and kidney, while six Fu (or Yang organs) includes gall bladder, stomach, large intestine, small intestine, urinary bladder, and three cavities (i.e., chest, epiastrium, and hypogastrium). Zang organs can manufacture and store fundamental substances. These substances are then transformed and transported by Fu organs. TCM treatments involve a thorough understanding of the clinical manifestations of Zang-Fu organ imbalance and knowledge of appropriate acupuncture points and herbal therapy to rebalance the organs. The channels and collaterals are the representation of the organs of the body. They are responsible for conducting the flow of energy and blood through the entire body.

In addition to providing diagnostic information, these elements of TCM can also help to describe the etiology of disease including six exogenous factors (i.e., wind, cold, summer, dampness, dryness, and fire), seven emotional factors (i.e., anger, joy, worry, grief, anxiety, fear, and fright), and other pathogenic factors. Once all this information is collected and processed into a logical and workable diagnosis, the traditional Chinese medical doctor can determine the treatment approach.

12.4.3.2 Calibration

Let N be the number of patients collected in Group 1. For the data from Group 1, let x_j be the measurement of the well-established clinical endpoint of the jth patient. For simplicity, we assume that the measurement of well-established clinical endpoint is continuous. Suppose that the TCM diagnostic procedure consists of K items. Let z_{ij} denote the TCM diagnostic score of jth patient from the ith item, $i = 1, \ldots, K, j = 1, \ldots, N$. Let y_j represent the scale (or score) of the jth patient summarized from the K TCM diagnostic items. For simplicity, we assume that

$$y_J = \sum_{i=1}^{K} \sum_{J}^{N} z_{ij}.$$

Similar to calibration of an analytical method (cf. Chow and Liu (1995)), we will consider the following five candidate models:

Model 1: $y_j = \alpha + \beta x_j + e_j$,

Model 2: $y_j = \beta x_j + e_j$,

Model 3: $y_j = \alpha + \beta_1 x_J + \beta_2 x_j^2 + e_j$,

Model 4: $y_j = \alpha x_j^\beta e_j$,

Model 5: $y_j = \alpha e^{\beta x_j} e_j$,

where α, β, β_1, and β_2 are unknown parameters and *es* are independent random errors with $E(e_j) = 0$ and finite $Var(e_j)$ in models 1–3 and $E(\log(e_j)) = 0$ and finite $Var(\log(e_j))$ in models 4–5.

Model 1 is a simple linear regression model which is probably the most commonly used statistical model for establishment of standard curves for calibration. When the standard curve passes through the origin, model 1 reduces to Model 2. Model 3 indicates that the relationship between y and x is quadratic. When there is a nonlinear relationship between y and x, models 4 and 5 are useful. Note that both models 4 and 5 are equivalent to simple linear regression model after logarithm transformation. If all the above models can not fit the data, generalized linear models can be used.

By fitting an appropriate statistical model between these standards (well-established clinical endpoints) and their corresponding responses (TCM scores), an estimated calibration curve can be obtained. The estimated calibration curve is also known as the standard curve. For a given patient, his/her unknown measurement of well-established clinical endpoint can be determined based on the standard curve by replacing the dependent variable with its TCM score.

12.4.3.3 Validity

The validity itself is a measure of biasedness of the TCM instrument. Since a TCM instrument usually contains the four categories or domains, which in turn consist of a number of questions agreed by the community of the Chinese doctors, it is a great concern that the questions may not be the right questions to capture the information regarding patient's activity/function, disease status, and disease severity. We will use Group 2 to validate the CDP based on the previously established standard curve for calibration. Let X be the unobservable measurement of the well-established clinical endpoint which can be quantified by the TCM items, Z_i, $i = 1, ..., K$ based on the estimated standard curve in previous section. For convention, we assume that

$$X = (Y - \alpha) / \beta,$$

where $Y = \sum_{i=1}^{K} Z_i$. That is, model 1 was used for calibration. Suppose that X is distributed as a normal distribution with mean θ and variance τ^2. Let $Z = (Z_1, ..., Z_K)$. Again suppose Z follows a distribution with mean $\mu = (\mu_1, ..., \mu_K)'$ and variance Σ. To assess the validity, it is desired to see whether the mean of Z_i, $i = 1, ..., K$ is close to $(\alpha + \beta\theta)/K$. Let $\bar{\mu} = \frac{1}{K} \sum_{i=1}^{k} \mu_i$. Then $\theta = (\bar{\mu} - \alpha) / \beta$. Consequently, we can claim that the instrument is validated in terms of its validity if

$$|\mu_i - \bar{\mu}| < \delta, \forall i = 1, ..., K, \tag{12.24}$$

for some small pre-specified δ. To verify (12.24), we can consider construct a simultaneous confidence interval for $\mu_i - \bar{\mu}$. Assume that the TCM instrument is administered to N patients from Group 2. Let $\hat{\mu} = \frac{1}{N} \sum_{j=1}^{N} Z_j = \bar{Z}$. Then the $(1 - \alpha)100\%$ simultaneous confidence interval for $\mu_i - \bar{\mu}$ are given by

$$a_j'\hat{\mu} - \sqrt{\frac{1}{N} a_i' S a_i} T(\alpha, K, N - K) \le \mu_j - \bar{\mu} \le a_j'\hat{\mu} + \sqrt{\frac{1}{N} a_i' S a_i} T(\alpha, K, N - K),$$

$$i = 1, ..., K,$$

where

$$a_i' = \begin{pmatrix} -\dfrac{1}{K} \mathbf{1}_{i-1} \\[2mm] 1 - \dfrac{1}{K} \\[2mm] -\dfrac{1}{K} \mathbf{1}_{k-1} \end{pmatrix}$$

$$S = \frac{1}{N-1} \sum_{j=1}^{N} (\mathbf{Z}_j - \bar{\mathbf{Z}})(\mathbf{Z}_j - \bar{\mathbf{Z}})',$$

$$T^2(\alpha, K, N - K) = \frac{(N-1)K}{N-K} F(\alpha, K, N - K),$$

and

$$P\left(T^2\left(K,N-K\right) \le T^2\left(\alpha,K,N-K\right)\right) = 1-\alpha.$$

The Bonferroni adjustment of an overall α level might be conducted as follows:

$$\mathbf{a}_j'\hat{\mu} - \sqrt{\frac{1}{N}\mathbf{a}_i'S\mathbf{a}_i}\, T\left(\frac{\alpha}{2K},N-1\right) \le \mu_j - \bar{\mu} \le \mathbf{a}_j'\hat{\mu} + \sqrt{\frac{1}{N}\mathbf{a}_i'S\mathbf{a}_i}\, T\left(\frac{\alpha}{2K},N-1\right).$$

We can reject the null hypothesis that

$$H_0 : \left|\mu_i - \bar{\mu}\right| \ge \delta, \quad \forall i = 1,\ldots,K, \tag{12.25}$$

if any confidence interval falls completely within $(-\delta,\delta)$.

12.4.3.4 *Reliability*

The calibrated well-established clinical endpoints derived from the estimated standard curve are considered reliable if the variance of X is small. In this regard, we can test the hypothesis

$$H_0 : T^2 \le \Delta \text{ for some fixed } \Delta \tag{12.26}$$

to verify the reliability of estimating θ by X. We will use Group 2 to verify the reliability based on the previously established standard curve for calibration. Based on the estimated standard curve, we can derive that

$$\tau^2 = \frac{1}{\beta^2} \mathrm{Var}\left(\sum_{i=1}^{K} Z_i\right)$$

$$= \frac{1}{\beta^2} \mathbf{1}' \sum 1.$$

Note that the sample distribution of

$$\sum_{J=1}^{N} (X_J - \bar{X})^2 / \tau^2$$

has a chi-square distribution with $N-1$ degrees of freedom. According to Lehmann (1952; 1986), we can construct a $(1-\alpha)100\%$ one-sided confidence interval for τ^2 as follows

$$\tau^2 \geq \frac{\sum_{j=1}^{N}(X_i - \bar{X})^2}{\chi^2(\alpha, N-1)}$$

$$= \xi.$$

We can reject the null hypothesis of (12.26) and conclude that the items are not reliable in estimation of θ if $\xi > \Delta$.

12.4.3.5 Ruggedness

In addition to validity and reliability, an acceptable TCM diagnostic instrument should produce similar results on different raters. In other words, it is desirable to quantify the variation due to rater and the proportion of rater-to-rater variation to the total variation. We will use the one-way nested random model to evaluate instrument ruggedness (Chow and Liu (1995)). The one-way nested random model can be expressed as

$$x_{ij} = \mu + A_i + e_{j(i)}, \quad i = 1 \text{ (Group 2), 2 (Group 3)}; \ j = 1, \ldots, N,$$

where X_{ij} is the calibrated scale of the jth patient obtained from the ith rater, μ is the overall mean, A_i is the random effect due to the ith rater, and $e_{j(i)}$ is the random error of jth patient's scale nested within the ith rater. For the one-way nested random model, we need the following assumptions: A_i are i.i.d. normal with mean 0 and variance σ_A^2; $e_{j(i)}$ are i.i.d. normal with mean 0 and variance σ^2; A_i and $e_{j(i)}$ are mutually independent for all i and j (Searle et al. (1992)).

Let $\bar{X}_{i.} = \frac{1}{J}\sum_{j=1}^{N} X_{ij}$ and $\bar{X}.. = \frac{1}{2N}\sum_{i=1}^{2}\sum_{j=1}^{N} X_{ij} = \frac{1}{2}\sum_{i=1}^{2}\bar{X}_{i.}$. Also, let SSA and SSE denote the sum of squares of factor A and the sum of squares of errors respectively. In other words,

$$SSA = N\sum_{i=1}^{2}(\bar{X}_{i.} - \bar{X}..)^2$$

and

$$SSE = \sum_{i=1}^{2}\sum_{j=1}^{N}(X_{ij} - \bar{X}_{i.})^2.$$

Also let MSA and MSE denote mean squares for factor A and mean square error. Then MSA = SSA and MSE = SSE/[2(N − 1)]. As a result, the analysis of variance estimators of σ_A^2 and σ^2 can be obtained as follows:

$$\hat{\sigma}^2 = \text{MSE}$$

and

$$\hat{\sigma}_A^2 = \frac{MSA - MSE}{N}.$$

Note that $\hat{\sigma}_A^2$ is obtained from the difference between MSA and MSE, and thus it is possible to obtain a negative estimate for σ_A^2.

Three criteria can be used to evaluate instrument ruggedness. The first criterion is to compute the probability for obtaining a negative estimate of σ_A^2 given by

$$p(\hat{\sigma}_A^2 < 0) = P(F[1, 2(N-1)] < (F)^{-1}),$$

where $F[1, 2(N-1)]$ is a central F distribution with 1 and $2(N-1)$ degrees of freedom and

$$F = \frac{\sigma^2 + N\sigma_A^2}{\sigma^2}.$$

If $P(\hat{\sigma}_A^2 < 0)$ is large, it may suggest that $\sigma_A^2 = 0$. The second criterion is to test whether the variation due to factor A is significantly larger than zero:

$$H_0 : \sigma_A^2 = 0 \quad \text{versus.} \quad H_1 : \sigma_A^2 > 0. \tag{12.27}$$

The null hypothesis (12.27) is rejected at the α level of significance if

$$F_A > F_C = F(1, 2(N-1)),$$

where $F_A = MSA/MSE$. The third criterion is to evaluate the proportion of the variation due to factor A, which is defined as follows:

$$\rho_A = \frac{\sigma_A^2}{\sigma^2 + \sigma_A^2}.$$

By Searle et al. (1992), the estimator the $(1-\alpha)100\%$ confidence interval for σ_A^2 are given by

$$\hat{\rho}_A = \frac{MSA - MSE}{MSA + (N-1)MSE},$$

$$L_\rho = \frac{F_A / F_U - 1}{N + (F_A / F_U - 1)},$$

$$U_\rho = \frac{F_A / F_L - 1}{N + (F_A / F_L - 1)},$$

where $F_L = F(1 - 0.5\alpha, 1, 2(N - 1))$ and $F_U = F(0.5\alpha, 1, 2(N - 1))$.

It may be also desired to test whether or not the rater-to-rater variability is within an acceptable limit ω. In this case, Hsiao et al. (2007) have considered to test the following hypothesis

$$H_0 : \sigma_A^2 \geq \omega \quad \text{versus} \quad H_1 : \sigma_A^2 < \omega. \tag{12.28}$$

Since there exists no exact $(1-\alpha)100\%$ confidence interval for σ_A^2, we can derive the Williams-Tukey interval with a confidence level between $(1-2\alpha)100\%$ and $(1-\alpha)100\%$ which is given by (L_A, U_A), where

$$L_A = \frac{SSA(1 - F_U / F_A)}{N \chi_{UA}^2},$$

$$U_A = \frac{SSA(1 - F_L / F_A)}{N \chi_{LA}^2},$$

where $F_L = F(1-0.5\alpha, 1, 2(N-1))$ and $F_U = F(0.5\alpha, 1, 2(N-1))$ represent the $(1-0.5\alpha)$th and (0.5α)th upper quantiles of a central F distribution with 1 and $2(N-1)$ degrees of freedom, $\chi_{LA}^2 = \chi^2(1-0.5\alpha, 1)$ and $\chi_{UA}^2 = \chi^2(0.5\alpha, 1)$ are the $(1-0.5\alpha)$th and (0.5α)th upper quantiles of a central chi-square distribution with 1 degree of freedom, and $F_A = MSA/MSE$. The null hypothesis (19.28) is rejected at α level of significance if $U_A < \omega$.

12.5 Challenging Issues

Although TCM has a long history of being used in humans, no scientifically valid documentations are available. As indicated by the FDA, substantial evidence regarding safety and effectiveness of the test treatment under investigation can only be obtained by conducting adequate and well-controlled clinical trials. However, before the test treatment under investigation can be used in human, sufficient information regarding chemistry, manufacturing, and control (CMC), clinical pharmacology, and toxicology are necessary provided (see, e.g., Chow and Liu, 1995). Since most TCMs consist of multiple components with unknown pharmacological activities, valid information regarding CMC, clinical pharmacology, and toxicology are difficult to obtain. In what follows, these difficulties are briefly described.

12.5.1 Regulatory Requirements

Although the use of TCMs in humans has a long history, there have been no regulatory requirements regarding the assessment of safety and effectiveness of the TCMs until recently. For example, both regulatory authorities of China and Taiwan have published guidelines/guidances for clinical development of TCMs (see, e.g., MOPH, 2002; DOH, 2004a, 2004b). In addition, FDA also published a guidance for botanical drug products (FDA, 2004). These regulatory requirements for TCM research and development, especially for clinical development are very similar to well-established guidelines/guidances for pharmaceutical research and development for Western medicines. It is a concern whether these regulatory requirements and the corresponding statistical methods are feasible for research and development of TCM, based on the fact that there are so many fundamental differences in medical practice, drug administration, and diagnostic procedure. As a result, it is suggested that current regulatory requirements and the corresponding statistical methods should be modified in order to reflect these fundamental differences.

It is strongly recommended that regulatory requirements for development, review and approval process for Premarin (conjugated estrogens tablets, USP) be consulted because Premarin is a WM consisting of multiple components that are similar to a TCM (FDA, 1991; Liu and Chow, 1996). Premarin, which contains multiple components of estrone, equilin, 17α −dihydroequilin, 17α −estradiol, and 17β −dihydroequilin, is intended for treatment of moderate to severe vasomotor symptoms associated with the menopause. The experience with Premarin is helpful in developing appropriate guidelines/guidances for TCM drug products with multiple components.

12.5.2 Test for Consistency

As mentioned above, unlike most western medicines (WM), TCMs usually consist of a number of components. The pharmacological activities, interactions, and relative proportions of these components are usually unknown. In practice, TCM is usually prescribed subjectively by an experienced Chinese doctor. As a result, the actual dose received by each individual varies depending upon the signs and symptoms as perceived by the Chinese doctor. Although the purpose of this medical practice is to reduce the within-subject (or intra-subject) variability, it could also introduce non-negligible variability such as variations from component-to-component and from rater-to-rater (a Chinese doctor to another). Consequently, reproducibility or consistency of clinical results is questionable. Thus, how to ensure the reproducibility or consistency of the observed clinical results has become a great concern to regulatory agencies in the review and approval process. It is also a great concern to the sponsor of the manufacturing process. To address the question of reproducibility or consistency, a valid statistical quality control process on the raw materials and final product is suggested.

Tse et al. (2006) proposed a statistical quality control (QC) method to assess a proposed consistency index of raw materials, which are from different resources and/or final product, which may be manufactured at different sites. The consistency index is defined as the probability that the ratio of the characteristics (e.g., extract) of the most active component among the multiple components of a TCM from two different sites (locations) is within a limit of consistency. The consistency index closes to 1 indicates that the components from the two sites or locations are almost identical. The idea for testing consistency is to construct a 95% confidence interval for the proposed consistency index under a sampling plan. If the constructed 95% confidence lower limit is greater than a pre-specified QC lower limit, then we claim that the raw materials or final product has passed the QC and hence can be released for further process or use. Otherwise, the raw materials and/or final product should be rejected. More details regarding the statistical methods proposed by Tse et al. (2006) are given in the next section.

12.5.3 Animal Studies

The purpose of animal studies is not only to study possible toxicity in animals, but also to suggest an appropriate dose for use in humans, assuming that the established animal model is predictive of the human model. For a newly developed drug product, animal studies are necessary. However, for some well-known TCMs, which have been used in humans for years and have a very mild toxicity profile, it is questionable whether animal studies are necessary. It is suggested that all components of TCMs as described in Chinese Pharmacopoeia (CP) be classified into several categories depending upon their potential toxicities and/or safety profiles as a basis for regulatory requirements for animal studies. In other words, for some well-known TCM components such as ginseng, animal studies for testing toxicity may be waived depending upon past experience of human use, although health risks or side effects following the proper administration of designated therapeutic dosages were not recorded in human use. Note that the German regulatory Authority's herbal watchdog agency, commonly called Commission E, has conducted an intensive assessment of the peer-reviewed literature on some 300 common botanicals with respect to the quality of the clinical evidence and the uses for which the herb can be reasonably considered effective (PDR, 1998).

12.5.4 Shelf-Life Estimation

Most regulatory agencies require that the expiration dating period (or shelf-life) of a drug product must be indicated in the immediate container label before it can be released for use. To fulfill this requirement, stability studies are usually conducted in order to characterize the degradation of the drug product. For drug products with a single active ingredient,

statistical methods for determination of drug shelf-life are well established (e.g., FDA, 1987a; ICH, 1996c). However, regulatory requirements
for estimation of drug shelf-life for drug products with multiple components are not available.

Following the concept of estimating shelf-life for drug products with single active ingredient, two approaches are worthy considering. First, we may
(conservatively) consider the minimum of the shelf-lives obtained from each
component of the drug product. This approach is conservative, and yet may
not be feasible due to the fact that (i) not all of the components of a TCM can
be accurately and reliably quantitated, and (ii) the resultant shelf-life may be
too short to be useful (see, e.g., Pong and Raghavarao, 2002).

Alternatively, we may consider a two-stage approach for determination
of drug shelf-life. At the first stage, an attempt should be made to identify
the most active component(s) whenever possible. A shelf-life can then be
obtained based on the method suggested in the FDA and ICH guidelines.
At the second stage, the obtained shelf-life is adjusted based on the relationship and/or interactions of the most active ingredient(s) and other components. As an alternative, Chow and Shao (2005) proposed a statistical method
for determining the shelf-life of a TCM following a similar idea suggested
by the FDA, assuming that the components are linear combinations of some
factors.

12.5.5 Indication and Label

As indicated earlier, it is very important to clarify the intention for the
use of a TCM (by Chinese doctors alone, Western clinicians alone, or both
Chinese doctors and Western clinicians) once it is approved by the regulatory agencies. If a TCM is intended for use by Chinese doctors alone, the
clinical trials conducted for obtaining substantial evidence should reflect
medical theory of TCM and medical practice of Chinese doctors. The label
should provide sufficient information as to how to prescribe the TCM the
Chinese way. On the other hand, if the TCM under investigation is intended
for use by Western clinicians alone, patients under study should be evaluated based on clinical study endpoints for safety and efficacy the Western
way. Consequently, the label should provide sufficient information for pre-
scribing the TCM the western way. If the TCM is intended for both Western
clinicians and Chinese doctors, patients are necessarily evaluated by both
Western clinical study endpoints and Chinese diagnostic procedures (e.g.,
some standardized quantitative instrument) provided that the Chinese
diagnostic procedure has been calibrated and validated against the well-
established Western clinical endpoint. In this case, there is a clear understanding how an observed difference by Chinese diagnostic procedure can
be translated to a clinical effect which is familiar to Western clinicians, and
vice versa.

12.6 Recent Development

12.6.1 Introduction

In recent years, as more and more innovative drug products are going off patent, the search for new medicines that treat critical and/or life-threatening diseases such as cardiovascular diseases and cancer has become the center of attention of many pharmaceutical companies and research organizations such as National Institute of Health (NIH). This leads to the study of the potential use of promising traditional Chinese medicines (TCM), especially for critical and/or life-threatening diseases. Bensoussan et al. (1998) used randomized clinical trial (RCT) to assess the effect of Chinese herb medicine in treating the Irritable Bowel Syndrome. However, RCT is not in common use when studying TCM. There are fundamental differences between Western medicines and TCM in terms of diagnostic procedures, therapeutic indices, medical mechanism, medical theory and practice (Chow et al., 2006; Zhou et al., 2012). Besides, TCM often consists of multiple components with flexible dose.

Chinese doctors believe that all of the organs within a healthy subject should reach the so-called global dynamic balance and harmony among organs. Once the global balance is broken at certain sites such as heart, liver, or kidney, some signs and symptoms then appear to reflect the imbalance at these sites. An experienced Chinese doctor usually assesses the causes of global imbalance before a TCM with flexible doses is prescribed to fix the problem. This approach is sometimes referred to as a personalized (or individualized) medicine approach. In practice, TCM consider inspection, auscultation and olfaction, interrogation, and pulse taking and palpation as the primary diagnostic procedures. The scientific validity of these subjective and experience-based diagnostic procedures has been criticized due to lack of reference standards and anticipated large evaluator-to-evaluator (i.e., Chinese doctor-to Chinese doctor) variability. For a systematic discussion of the statistical issues of TCM, see Chow (2015).

In this chapter, we attempt to propose a unified approach to developing a composite illness index based on a number of indices collected from a given subject under the concept of global dynamic balance among organs. Dynamic balance among organs can be defined as follows. Following the concept of testing bioequivalence or biosimilarity, if the 95% confidence upper bound is less than some equivalent limit, we conclude that the treatment achieves dynamic balance among the organs of the subject hence is considered as efficacious. If we fail to reject the null hypothesis, we conclude that the treatment is not efficacious since there is still a signal of illness. In practice, these signals of illness can be grouped to diagnose specific diseases based on some pre-specified reference standards for diseases status of specific diseases which are developed based on indices related to specific organs (or diseases).

12.6.2 Health Index and Efficacy Measure

Let $X_T = (X_{T1}, \ldots, X_{Tk})$ and $X_H = (X_{H1}, \ldots, X_{Hk})$ be the k-dimensional health profiles of a subject who is under treatment of TCM and a subject who is a healthy control, respectively. The components X_{Ti} of a health profile could be either continuous or ordinal, and even if it is continuous its distribution may not be normal. The dimension of the health profile k is potentially high. Under this formulation, a TCM treatment is considered efficacious if the health profile of subjects who receive treatment is not significantly different from the one of a healthy subject, possibly age and gender matched if needed. To this end, we define illness index θ as a kind of pseudo distance of the profile of a treated subject and the profile of a healthy subject. Specifically, let μ_T and μ_H be the mean health profile of a treated subject and a healthy subject, respectively, and let Σ_T and Σ_H be the covariance matrix of a treated subject and a healthy subject, respectively. Let $\lambda_1(\Sigma)$ be the largest eigenvalue of a symmetric matrix Σ. Define the illness index as

$$\theta = \frac{(\mu_T - \mu_H)^T (\mu_T - \mu_H) + \lambda_1(\Sigma_T) - \lambda_1(\Sigma_H)}{\max\{\sigma_0^2, \lambda_1(\Sigma_H)\}}, \tag{12.29}$$

where σ_0^2 is a known constant.

The above definition of illness index has an interesting connection with principal component analysis. To see this, image that X_T and X_H are multivariate normal. Then, $(\mu_T - \mu_H)^T (\mu_T - \mu_H)$ is the squared mean of the first principal component of $X_T - X_H$, $\lambda_1(\Sigma_T)$ is the variance of the first principal component of X_T, and $\lambda_1(\Sigma_H)$ is the variance of the first principal component of X_H. A small θ implies two things. First, it implies a small difference between the mean health profiles μ_T and μ_H, relative to the maximal variance of a healthy profile X_H. Second, it implies that the maximal variance of the diseased profile X_T is at least not much greater than the maximal variance of the healthy profile X_H. Therefore, a small illness index can be viewed as evidence for efficacy of a treatment. The efficacy of a treatment could be formulated as a test for equivalence with the health profile of a healthy control. Let X_T denote the health profile of a diseased subject at the completion of the treatment in (12.29). The efficacy is claimed if a test rejects the following H_0

$$H_0 : \theta \geq \varepsilon \quad \text{versus} \quad H_1 : \theta < \varepsilon, \tag{12.30}$$

where ε is a known positive threshold.

12.6.3 Assessment of Efficacy

Let

$$\gamma = (\mu_T - \mu_H)^T (\mu_T - \mu_H) + \lambda_1(\Sigma_T) - \lambda_1(\Sigma_H) - \varepsilon \max\{\sigma_0^2, \lambda_1(\Sigma_H)\}. \tag{12.31}$$

Then, testing (12.30) is equivalent to testing

$$H_0 : \gamma \leq 0 \quad \text{versus} \quad H_1 : \gamma < 0. \tag{12.32}$$

We construct an asymptotic test via the duality between hypothesis test and confidence interval. Specifically, we construct a 95% approximate confidence upper bound for γ based on two independent random samples

$$X_{Ti} = (X_{Ti1},\ldots,X_{Tki})^T, i = 1,\ldots,n_1 \text{ and } X_{Hi} = (X_{Hi1},\ldots,X_{Hki})^T, i = 1,\ldots,n_2.$$

We then reject the null hypothesis H_0 if the 95% confidence upper bound is smaller than 0.

Let B be a $k \times k$ orthogonal matrix such that

$$B^T (q\Sigma_T + \Sigma_H)B = diag\{\eta_1,\ldots,\eta_k\},$$

where $q = n_2 / n_1$. Let

$$\upsilon = (\upsilon_1,\ldots,\upsilon_k)^T = B(\mu_T - \mu_H),$$

and

$$\hat{\upsilon} = (\hat{\upsilon}_1,\ldots,\hat{\upsilon}_k)^T = B(\hat{\mu}_T - \hat{\mu}_H).$$

Then, parameter γ can be written as

$$\gamma = \sum_{i=1}^k \upsilon_i^2 + \lambda_1(\Sigma_T) - \lambda_1(\Sigma_H) - \varepsilon \max\{\sigma_0^2, \lambda_1(\Sigma_H)\}.$$

It is easily seen that $\upsilon_i^2, i = 1,\ldots,k$, are independent and normally distributed as $N(\upsilon_i, \eta_i)$. Let $\hat{\eta}_i$ be the ith diagonal element of $B^T(q\hat{\Sigma}_T + \hat{\Sigma}_H)B$. Then, a 95% confidence upper bound for υ_i^2 is

$$\sum_{i=1}^k \upsilon_i^2 + 2z_{0.05}\sqrt{\frac{\sum_{i=1}^k \hat{\upsilon}_i^2 \hat{\eta}_i}{n_2}},$$

where $z_{0.05}$ is the 95%th percentile of the standard normal distribution.

Let $l_{1,T}$ be the largest eigenvalue of $\hat{\Sigma}_T$. Then, by Anderson and Hauck (1990), an asymptotic 95% confidence upper bound for $\lambda_1(\Sigma_T)$ is

$$\frac{l_{1,T}}{1 - z_{0.05}\sqrt{2/n_1}}.$$

Similarly, an asymptotic 95% confidence lower bound for $\lambda_1(\Sigma_H)$ is

$$\frac{l_{1,T}}{1 - z_{0.05}\sqrt{2/n_2}}.$$

If $\lambda_1(\Sigma_H) \ge \sigma_0^2$, then γ in (12.31) reduces to

$$\gamma = \sum_{i=1}^{k} \upsilon_i^2 + \lambda_1(\Sigma_T) - (1+\varepsilon)\lambda_1(\Sigma_H). \tag{12.33}$$

Since $\hat{\upsilon}_i's$, $l_{1,T}$, and $l_{1,H}$ are independent, then using the idea of Howe (1974) and Graybill and Wang (1980), we construct an approximate 95% confidence upper bound for γ as

$$\hat{\gamma}_{U,1} = \sum_{i=1}^{k} \hat{\upsilon}_i^2 + l_{1,T} - (1+\varepsilon)l_{1,H} + \sqrt{\Delta_1}, \tag{12.34}$$

where

$$\Delta_1 = \frac{4z_{0.05}^2 \sum_{i=1}^{k} \hat{\upsilon}_i^2 \hat{\eta}_i}{n_2} + \left(\frac{l_{1,T}}{1 - z_{0.05}\sqrt{\dfrac{2}{n_1}}} - l_{1,T}\right)^2 + (1+\varepsilon)^2\left(\frac{l_{1,H}}{1 - z_{0.05}\sqrt{2/n_2}} - l_{1,H}\right)^2$$

If $\lambda_1(\Sigma_H) < \sigma_0^2$, then γ in (12.31) reduces to

$$\gamma = \delta + \lambda_1(\Sigma_T) - \lambda_1(\Sigma_H) - \varepsilon\sigma_0^2. \tag{12.35}$$

An approximate 95% confidence upper bound for γ is given by

$$\hat{\gamma}_{U,2} = \sum_{i=1}^{k} \hat{\upsilon}_i^2 + l_{1,T} - l_{1,H} - \varepsilon\sigma_0^2 + \sqrt{\Delta_2}, \tag{12.36}$$

where

$$\Delta_2 = \frac{4z_{0.05}^2 \sum_{i=1}^{k} \hat{\upsilon}_i^2 \hat{\eta}_i}{n_2} + \left(\frac{l_{1,T}}{1 - z_{0.05}\sqrt{\dfrac{2}{n_1}}} - l_{1,T}\right)^2 + \left(\frac{l_{1,H}}{1 - z_{0.05}\sqrt{2/n_2}} - l_{1,H}\right)^2.$$

For testing (12.30), it is equivalent to testing (12.32). Cheng et al. (2019) proposed the following testing rule: reject H_0 if $l_{1,H} \ge \sigma_0^2$ and $\hat{\gamma}_{U,1} < 0$, or $l_{1,H} < \sigma_0^2$ and $\hat{\gamma}_{U,2} < 0$. Specifically, one may claim that testing rule

$$\phi = I\left(l_{1,H} \ge \sigma_0^2 \text{ and } \hat{\gamma}_{U,1} < 0\right) + I\left(l_{1,H} < \sigma_0^2 \text{ and } \hat{\gamma}_{U,2} < 0\right) \tag{12.37}$$

has an approximate level of 0.05 for testing H_0.

When the sample size is small, the above test ϕ tends to be aggressive in rejecting the null hypothesis. Zheng et al. (2019) recommended replacing $z_{0.05}$ by $z_{0.025}$ in practice unless the sample size is very large.

12.6.4 Remarks

Zheng et al. (2019) proposed a method to assess the efficacy of a Chinese medicine treatment based on the illness index and the idea of bioequivalence and biosimilarity concept. The approach we take is a population bioequivalence approach. It would be of interest to develop method using an individual bioequivalence approach.

In this section, we assume that the health profile has a multivariate normal distribution. When part or all the components of the health profile are not normally distributed, the proposed method will be valid asymptotically when $\min\{n_1, n_2\}$ tends to infinity. In this section, we consider the case that k is moderately large. When k is very large, that is, when $k \gg \max\{n_1, n_2\}$, it is suggested that methods for sparse principal component analysis, should be used and the corresponding 95% confidence upper bound could be constructed via the bootstrap method.

The illness index defined in this section measures the closeness between a profile of a treated subject and a file of a healthy control using both mean and variance information. However, we did not model the causal relationship among the k components $X_1, \ldots, X_k$. Per TCM theory, alternative model incorporating causal information could be considered. Future research is thus warranted.

12.7 Concluding Remarks

As indicated earlier, a TCM is defined as a Chinese herbal medicine developed for treating patients with certain diseases as diagnosed by the four major techniques of inspection, auscultation and olfaction, interrogation, and pulse taking and palpation based on traditional Chinese medical theory of global balance among the functions/activities of all organs of the body. When conducting a TCM clinical trial, it is suggested that the fundamental differences between a WM and a TCM, as described in Section 2, should be evaluated carefully for a valid and unbiased assessment of the safety and effectiveness of the TCM under investigation.

One of the key issues in TCM research and development is to clarify the difference between Westernization of TCM and modernization of TCM. For Westernization of TCM, we follow regulatory requirements at critical stages of the process for pharmaceutical development including drug discovery, formulation, laboratory development, animal studies, clinical development, manufacturing process validation and quality control, regulatory submission, review, and process despite the fundamental differences between WM and TCM. For modernization of TCM, it is suggested that regulatory requirements should be modified in order to account for the fundamental differences between WM and TCM. In other words, we still ought to be able to see if TCM is really working with modified regulatory requirements using Western clinical trials as a standard for comparison.

In practice, it is recognized that WMs tend to achieve the therapeutic effect sooner than that of TCMs for critical and/or life-threatening diseases. TCMs are found to be useful for patients with chronic diseases or non-life-threatening diseases. In many cases, TCMs have shown to be effective in reducing toxicities or improving safety profile for patients with critical and/or life-threatening diseases. As a strategy for TCM research and development, it is suggested that (i) TCM be used in conjunction with a well-established WM as a supplement to improve its safety profile and/or enhance therapeutic effect whenever possible, and (ii) TCM should be considered as the second line or third line treatment for patients who fail to respond to the available treatments. However, some sponsors are interested in focusing on the development of TCM as a dietary supplement due to: (i) the lack or ambiguity of regulatory requirements; (ii) the lack of understanding of the medical theory/mechanism of TCM; (iii) the confidentiality of non-disclosure of the multiple components; and (iv) the lack of understanding of pharmacological activities of the multiple components of TCM.

Since TCM consists of multiple components which may be manufactured from different sites or locations, the post-approval consistency in quality of the final product is both a challenge to the sponsor and a concern to the regulatory authority. As a result, some post-approval tests, such as tests for content uniformity, weight variation, and/or dissolution and (manufacturing) process validation, must be performed for quality assurance before the approved TCM can be released for use.

13

Adaptive Trial Design

13.1 Introduction

In the past several decades, it has been recognized that increasing spending of biomedical research does not reflect an increase of the success rate of pharmaceutical/clinical research and development. The low success rate of pharmaceutical/clinical development could be due to following factors: (i) there is a diminished margin for improvement that escalates the level of difficulty in proving drug benefits; (ii) genomics and other new sciences have not yet reached their full potential; (iii) mergers and other business arrangements have decreased candidates; (iv) easy targets are the focus as chronic diseases are harder to study; (v) failure rates have not improved; and (vi) rapidly escalating costs and complexity decreases the willingness/ability to bring many candidates forward into the clinic (Woodcock, 2005). In early 2000, the FDA established the Critical Path Initiative to assist the sponsors in (i) identifying possible causes of failure, (ii) providing resolutions, and (iii) increasing the efficiency and the probability of success in pharmaceutical research and development. In its 2004 *Critical Path Report*, the FDA presented its diagnosis of the scientific challenges underlying the medical product pipeline problems. Two years later, the FDA released a *Critical Path Opportunities List* that calls for advancing innovative trial designs by using prior experience or accumulated information in trial design. Many researchers interpret these actions as an encouragement for using innovative adaptive design methods in clinical trials, while some researchers believe them to amount to a recommendation for the use of Bayesian approach for assessment of treatment effect in pharmaceutical/clinical development. The purpose of adaptive design methods in clinical trials is to provide the flexibility to the investigator for identifying best (optimal) clinical benefit of the test treatment under study in a timely and efficient fashion without undermining the validity and integrity of the intended study.

The concept of adaptive design can be traced back to 1970s when adaptive randomization (play-the-winner) was introduced as a class of designs for sequential clinical trials (Wei, 1978). Most adaptive design methods in clinical research and development fall into the categories of adaptive randomization (see, e.g., Efron, 1971; Lachin, 1988; Atkinson and Doney, 1992; Rosenberger et al., 2001; Hardwick and Stout, 2002); group sequential designs with the flexibility for stopping a trial early due to safety, futility, and/or efficacy (see, e.g., Lan and DeMets, 1987; Wang and Tsiatis, 1987; Lehmacher and Wassmer, 1999; Posch and Bauer, 1999; Liu et al., 2002); and flexible sample size re-estimation at interim for achieving the desired statistical power by controlling the overall type I error rate at a pre-specified level of significance (see, e.g., Cui et al., 1999; Chung-Stein et al., 2006; Chow et al., 2008). The use of adaptive design methods for modifying the trial procedures and/ or statistical methods of ongoing clinical trials based on accrued data has been practiced for years in pharmaceutical/clinical research and development. Adaptive design methods in clinical research are very attractive to pharmaceutical/clinical scientists for the following reasons. First, it reflects medical practice in real world. Second, it is ethical with respect to both efficacy and safety (toxicity) of the test treatment under investigation. Third, it is not only flexible but also efficient in the early phase of clinical development. However, one concern is whether the p-value or confidence interval approach for evaluating the treatment effect obtained after the modification is correct or reliable. Another concern is that the use of adaptive design methods in a clinical trial may lead to a totally different trial that is unable to address scientific/medical questions that the trial is intended to answer.

In recent years, the potential use of adaptive design methods in clinical trials has attracted much attention. For example, the Pharmaceutical Research and Manufacturers of America (PhRMA) and Biotechnology Industry Organization (BIO) have established adaptive design working groups and proposed/published white papers regarding strategies, methodologies, and implementations for regulatory consideration (see, e.g., Gallo et al., 2006; Chang, 2007). However, there is no universal agreement in terms of definition, methodologies, applications, and implementations. In addition, many journals have also published special issues on adaptive design for evaluation the potential use of adaptive trial design methods in clinical research and development. These scientific journals include, but are not limited to, *Biometrics* (Vol. 62, No. 3); *Statistics in Medicine* (Vol. 25, No. 19); *The Journal of Biopharmaceutical Statistics* (Vol. 15, No. 4 and Vol. 17, No. 6); *The Biometrical Journal* (Vol. 48, No. 4); and *Pharmaceutical Statistics* (Vol. 5, No. 2). In addition, many professional conferences/meetings have devoted special sessions for discussion of the feasibility, applicability, efficiency, validity, and integrity of the potential use of the innovative adaptive design methods in clinical trials in the past several years. For example, the FDA/Industry Statistics Workshop has offered adaptive sessions and workshops from industrial, academic, and regulatory perspectives consecutively between 2006 and 2008. More details regarding the use of adaptive design methods in clinical trials can be found in the books by Chow and Chang (2006) and Chang (2007).

The purpose of this chapter is not only to provide a comprehensive summarization of the issues that are commonly encountered when applying/

implementing the adaptive design methods in clinical research but also to include recently development such as the role of the independent data safety monitoring board and sample size estimation/allocation, justification, and adjustment when implementing a much more complicated adaptive design in clinical trials. In Section 13.2, commonly employed adaptations and the resultant adaptive designs are briefly described. Also included in this section are regulatory and statistical perspectives regarding the use of adaptive design methods in clinical trials. The impact of protocol amendments, challenges of by design adaptations, and obstacles of retrospective adaptations when applying adaptive design methods in clinical trials are described in Section 13.4. Some trial examples and strategies for clinical development are discussed in Sections 13.5 and 13.6, respectively. Brief concluding remarks are given in Section 13.7.

13.2 What Is Adaptive Design?

In clinical trials, it is not uncommon to modify trial procedures and/or statistical methods during the conduct of clinical trials based on the review of accrued data at interim. The purpose is not only to efficiently identify clinical benefits of the test treatment under investigation but also to increase the probability of success of the intended clinical trial. Factors that influence trial procedures include eligibility criteria; study dose; treatment duration; study endpoints; laboratory testing procedures; diagnostic procedures; criteria for evaluability; and the assessment of clinical responses. Statistical method considerations include randomization scheme; study design selection; study objectives/hypotheses; sample size calculation; data monitoring and interim analysis; and statistical analysis plan and/or methods for data analysis. In this chapter, we will refer to the adaptations (changes or modifications) made to the trial and/or statistical procedures as the adaptive design methods. Thus, an adaptive design is defined as a design that allows adaptations to trial and/ or statistical procedures of the trial after its initiation without undermining the validity and integrity of the trial (Chow et al., 2005). In one of their publications, with the emphasis of the feature of by design adaptations only (rather than ad hoc adaptations), the PhRMA Working Group on Adaptive Design refers to an adaptive design as a clinical trial design that uses accumulating data to decide on how to modify aspects of the study as it continues, without undermining the validity and integrity of the trial (Gallo et al., 2006).

On the other hand, FDA defines an adaptive design clinical study as a study that includes a prospectively planned opportunity for modification of one or more specified aspects of the study design and hypotheses based on analysis of data (usually interim data) from subjects in the study (FDA, 2010b, 2018). The FDA's definition, however, has been criticized on the basis that (i) it is not flexible because only prospective adaptations are allowed, (ii) it does not reflect real practice (e.g., protocol amendments), and most importantly, (iii) it does not mention validity and integrity of clinical studies utilizing adaptive trial designs. Note that in many cases, an adaptive design is also known as a flexible design (EMEA, 2002, 2006).

13.2.1　Adaptations

An adaptation is referred to as a modification or a change made to trial procedures and/or statistical methods during the conduct of a clinical trial. Adaptations that are commonly employed in clinical trials can be classified into the categories of prospective adaptation, concurrent (or ad hoc) adaptation, and retrospective adaptation. Prospective adaptations include, but are not limited to, adaptive randomization; stopping a trial early due to safety, futility, or efficacy at interim analysis; dropping the losers (or inferior treatment groups); sample size re-estimation; and so on. Thus, prospective adaptations are usually referred to as by design adaptations as described in the PhRMA white paper (Gallo et al., 2006). Concurrent adaptations are usually referred to as any ad hoc modifications or changes made as the trial continues. Concurrent adaptations include, but are not limited to, modifications in inclusion/exclusion criteria; evaluability criteria; dose/regimen and treatment duration; changes in hypotheses and/or study endpoints; and so on. Retrospective adaptations are usually modifications and/or changes made to a statistical analysis plan prior to database lock or the unblinding of treatment codes. In practice, prospective, ad hoc, and retrospective adaptations are implemented by study protocol, protocol amendments, and statistical analysis plan with regulatory reviewer's consensus, respectively.

13.2.2　Types of Adaptive Designs

Based on the adaptations employed, commonly considered adaptive designs in clinical trials include, but are not limited to (i) adaptive randomization design, (ii) group sequential design, (iii) N-adjustable (or flexible sample size re-estimation) design, (iv) drop-the-losers design, (v) adaptive dose finding design, (vi) biomarker-adaptive design, (vii) adaptive treatment-switching design, (viii) adaptive-hypothesis design, (ix) adaptive seamless (e.g., phase I/II or phase II/III) trial design, and (x) multiple adaptive design. These adaptive designs are briefly described below.

13.2.2.1　Adaptive Randomization Design

An adaptive randomization design is a design that allows modification of randomization schedules based on varied and/or unequal probabilities of treatment assignment in order to increase the probability of success. As a result, an adaptive randomization design is sometimes referred to as a play-the-winner design since it will increase the probability of success. Commonly applied adaptive randomization procedures include treatment-adaptive randomization (see, e.g., Efron, 1971; Lachin, 1988), covariate-adaptive randomization, and response-adaptive randomization (Rosenberger et al., 2001; Hardwick and Stout, 2002).

Although an adaptive randomization design could increase the probability of success, it may not be feasible for a large trial or a trial with a relatively longer treatment duration because the randomization of a given subject depends on the response of the previous subject. A large trial or a trial with a relatively longer treatment duration utilizing adaptive randomization design will take a much longer time to complete. Besides, the randomization schedule may not be available prior to the conduct of the study. Moreover, statistical inference on treatment effect is often difficult to obtain due to the complexity of the randomization scheme. In practice, a statistical test is often difficult, if not impossible, to obtain due to complicated probability structure as the result of adaptive randomization, which has also limited the potential use of adaptive randomization design in practice.

13.2.2.2 Group Sequential Design

A group sequential design is a design that allows for prematurely stopping a trial due to safety, futility/efficacy, or both with options of additional adaptations based on results of interim analysis. Many researchers refer to a group sequential design as a typical adaptive design because some adaptations may be applied after the review of interim results of the study, such as stopping the trial early due to safety, efficacy, and/or futility. In practice, various stopping boundaries based on different boundary functions for controlling an overall type I error rate are available in the literature (see, e.g., Lan and DeMets, 1987; Wang and Tsiatis, 1987; Jennison and Turnbull, 2000, 2005; Rosenberger et al., 2001; Chow and Chang, 2006). In recent years, the concept of two-stage adaptive design has led to the development of the adaptive group sequential design (see, e.g., Cui et al., 1999; Lehmacher and Wassmer, 1999; Posch and Bauer, 1999; Liu et al., 2002).

It should be noted that when additional adaptations such as adaptive randomization, dropping the losers, and/or adding additional treatment arms (in addition to the commonly considered adaptations such as stop the trial early due to safety, efficacy and/or futility and sample size re-estimation, and so on) are applied to a typical group sequential design after the review of the interim results, the resultant group sequential design is usually referred to as an adaptive group sequential design. In this case, the standard methods for the typical group sequential design may not be appropriate. In addition, it may not be able to control the overall type I error rate at the desired level of 5% if (i) there are additional adaptations (e.g., changes in hypotheses and/or study endpoints) and/or (ii) there is a shift in target patient population due to additional adaptations or protocol amendments.

13.2.2.3 Flexible Sample Size Re-estimation (SSRE) Design

A flexible sample size re-estimation (or N-adjustable) design is referred to as an adaptive design that allows for sample size adjustment or re-estimation based on the observed data at interim. In clinical research, it can be verified that the selected sample size is a function of type I error rate, type II error rate (or power), treatment effect (or clinically meaningful difference), and variability associated with the response. A typical approach (power analysis) is to select a sample size that will achieve a desired power by fixing other parameters such as type I error rate, the clinically meaningful difference, and the variability associated with the response. In practice, it is impossible to select a sample size that controls all parameters. For a flexible SSRE design, sample size adjustment or re-estimation could be done in either a blinding or unblinding fashion based on the criteria of maintaining anticipated treatment effect-size, controlling variability within a tolerable limit, achieving a desired conditional power, and/or reaching certain degree of reproducibility probability (see, e.g., Cui et al., 1999; Woodcock, 2005; Chung-Stein et al., 2006; Chow et al., 2007). Thus, sample size in a flexible SSRE trial design is a random variable. Sample size re-estimation suffers from the same disadvantage as the original power analysis for sample size calculation prior to the conduct of the study because it is performed by treating estimates of the study parameters, which are obtained based on data observed at interim, as true values. Note that the criteria of maintain treatment effect size, controlling variability, and achieving conditional power are considered one-parameter problem, while the criterion of reaching reproducibility is a two-parameter problem.

According to an informal communication with the medical/statistical reviewers at the FDA, it is not a good clinical/statistical practice to start with a small sample number and then perform sample size re-estimation (adjustment) at interim while ignoring the clinically meaningful difference that one wishes to detect for the intended clinical trial. It should be noted that the observed difference at interim based on a small number of subjects may not be of statistically significance (i.e., it may be observed by chance alone). In addition, there is variation associated with the observed difference that is an estimate of the true difference. Thus, standard methods for sample size re-estimation based on the observed difference with a limited number of subjects may be biased and misleading. To overcome these problems, in practice, a sensitivity analysis (with respect to variation associated with the observed results at interim) for sample size re-estimation design is recommended.

13.2.2.4 Drop-the-Losers Design

A drop-the-losers design is a design that allows dropping the inferior treatment groups as well as adding additional (promising) arms. A drop-the-losers design is useful in early phase of clinical development especially when there are uncertainties regarding the dose levels (see, e.g., Bauer and Kieser, 1999; Brannath et al., 2003; Posch et al., 2005; Sampson and Sill, 2005). The selection criteria (including the selection of initial dose, the increment of the dose, and the dose range) and decision rules play important roles for drop-the-losers designs. Dose groups that are dropped may contain valuable information regarding dose response of the treatment under study. Typically, a drop-the-losers design is a two-stage design. At the end of the first stage, the inferior arms will be dropped based on some pre-specified criteria. The winners will then proceed to the next stage. In practice, the study is often powered for achieving a desired power at the end of the second stage (or at the end of the study). In other words, there may not be any statistical power for the analysis at the end of the first stage for dropping the losers (or).

In practice, it is not uncommon to drop the losers or pick up the winners based on so-called precision analysis (Chow et al., 2017). The precision approach is an approach based on the confidence level for achieving statistical significance. In other words, the decision will be made (i.e., to drop the losers) if the confidence level for observing a statistical significance (i.e., the observed difference is not by chance alone or it is reproducible with the pre-specified confidence level) exceeds a pre-specified confidence level. In addition, other criteria for dropping the losers such as predictive probability of success and probability of being the best dose (or treatment arm) are also commonly considered (Lee and Lin, 2016). Zheng and Chow (2019) compared relative performances of these criteria in terms of the probability of correctly identifying the most promising dose (treatment arm) under a two-stage adaptive trial design.

Note that in a drop-the-losers design, a general principle is to drop the inferior treatment groups or add promising treatment arms but at the same time it is suggested that a control group be retained for a fair and reliable comparison at the end of the study. It should be noted that dose groups that are dropped may contain valuable information regarding dose response of the treatment under study. In practice, it is also suggested that subjects who are assigned in the inferior dose groups should be switched to the better dose group for ethical consideration. Treatment switching in a drop-the-losers design could complicate statistical evaluation in the dose selection process. Note that some clinical scientists prefer the term pick-the-winners rather than drop-the-losers.

13.2.2.5 Adaptive Dose Finding Design

The purpose of an adaptive dose finding (e.g., escalation) design is multifold and includes (i) the identification whether there is a dose response, (ii) the determination of the minimum effective dose (MED) and/or the maximum tolerable dose (MTD), (iii) the characterization of dose response curve, and (iv) the study of dose ranging. The information obtained from an adaptive dose finding experiment is often used to determine the dose level for the next phase of clinical development (see, e.g., Bauer and Rohmel, 1995; Whitehead, 1997; Zhang et al., 2006). For adaptive dose finding design, the method of continual re-assessment method (CRM) in conjunction with Bayesian approach is usually considered (see, e.g., O'Quigley et al., 1990; O'Quigley and Shen, 1996; Chang and Chow, 2005). Mugno et al. (2004) introduced a non-parametric adaptive urn design approach for estimating a dose-response curve. For more details regarding PhRMA's proposed statistical methods, the reader should consult with a special issue recently published by *The Journal of Biopharmaceutical Statistics*, Vol. 17, No. 6. Note that a typical approach for adaptive dose finding design focuses on dose response (severe toxicity and/or tolerability) curve. In practice, it is suggested that both safety including mild-to-moderate toxicity and efficacy be considered.

Note that according to the ICH E4 guideline on Dose-response Information to Support Drug Registration, there are several types of dose-finding (response) designs: (i) randomized parallel dose-response designs; (ii) crossover dose-response design; (iii) forced titration design (dose escalation design); and (iv) optimal titration design (placebo-controlled titration to endpoint). Some commonly asked questions for an adaptive dose finding design include, but are not limited to: (i) how to select the initial dose; (ii) how to select the dose range under study; (iii) how to achieve statistical significance with a desired power with the least subjects; (iv) what the selection criteria and decision rules should be if one would like to make a decision based on safety, tolerability, efficacy and/or pharmacokinetic information; and (v) what the probability is of achieving the optimal dose. In practice, a clinical trial simulation and/or sensitivity analysis is often recommended to evaluate/address the above questions.

13.2.2.6 Biomarker-Adaptive Design

A biomarker-adaptive design is a design that allows for adaptations based on the response of biomarkers such as genomic markers. An adaptive biomarker design involves biomarker qualification and standard, optimal screening design, and model selection and validation. It should be noted that there is a gap between identifying biomarkers that associated with clinical outcomes and establishing a predictive model between relevant biomarkers and clinical outcomes in clinical development. For example, correlation between a biomarker and a true clinical endpoint makes a prognostic marker. However, correlation between a biomarker and a true clinical endpoint does not make a predictive biomarker. A prognostic biomarker informs the clinical outcomes, independent of treatment. They provide information about the natural course of the disease in individuals who have or have not received the treatment under study. Prognostic markers can be used to separate good- and poor-prognosis patients at the time of diagnosis. A predictive biomarker informs the treatment effect on the clinical endpoint (Chang, 2007).

A biomarker-adaptive design can be used to (i) select right patient popula tion (e.g., enrichment process for selection of a better target patient population), (ii) identify nature course of disease, and (iii) detect disease earlier. In clinical research and development, biomarker-adaptive design, which has led to the research of targeted clinical trials, is not only the key to the success of precision medicine, but also help in developing personalized medicine (Charkravarty et al., 2005; Chang, 2007; Wang et al., 2007).

13.2.2.7 Adaptive Treatment-Switching Design

An adaptive treatment-switching design is a design that allows the inves- tigator to switch a patient's treatment from an initial assignment to an alternative treatment if there is evidence of lack of efficacy or safety of the initial treatment (see, e.g., Branson and Whitehead, 2002; Shao et al., 2005). In cancer clinical trials, estimation of survival is a challenge when treatment- switching has occurred in some patients. A high percentage of subjects who switched due to disease progression could lead to change in hypotheses to be tested. In this case, sample size adjustments for achieving a desired power are necessary.

13.2.2.8 Adaptive-Hypotheses Design

An adaptive-hypotheses design refers to a design that allows modification or change in hypotheses based on interim analysis results (Hommel, 2001). Adaptive-hypotheses designs are often considered before database lock and/or prior to data unblinding, which are implemented by the develop- ment of statistical analysis plan (SAP). Typical examples include the switch from a superiority hypothesis to a non-inferiority hypothesis and the switch between the primary study endpoint and the secondary endpoints. The pur- pose for switching from a superiority hypothesis to a non-inferiority hypoth- esis is to increase the probability of the success of the clinical trial. A typical approach is to first establish non-inferiority and then test for superiority. In this way, we do not have to pay for statistical penalty due to the prin- ciple of closed testing procedure. The idea of switching the primary study endpoints and the secondary endpoints is also to increase the probability of success of clinical development. In practice, it is not uncommon to observe positive results in secondary endpoint while fail to demonstrate clinical ben- efit for the primary endpoints. In this case, there is a strong desire to switch the primary endpoints and the secondary endpoints whenever it is scientifi- cally, clinically, and regulatory justifiable.

It should be noted that, for the switch from a superiority hypothesis to a non-inferiority hypothesis, the selection of non-inferiority margin is critical and has an impact on sample size adjustment for achieving the desired power. According to the ICH guideline, the selected non-inferiority margin should be both clinical and statistical justifiable (ICH, 2000; Chow and Shao, 2006).

Regarding the switch between the primary endpoint and the secondary endpoints, there has been a tremendous debate about controlling the overall type I error rate at the 5% level of significance. Thus, as an alternative, it has been recommended by many researchers considering switching from the primary endpoint to either a co-primary endpoint or a composite endpoint. However, the optimal allocation of the alpha-spending function has raised another statistical/clinical/regulatory concern.

13.2.2.9 Seamless Adaptive Trial Design

An adaptive seamless trial design refers to a program that addresses several single trial objectives that are normally achieved through separate trials of clinical development. An adaptive seamless design is an adaptive seamless trial design that would use data from patients enrolled before and after the adaptation in the final analysis (see, e.g., Kelly et al., 2005b; Maca et al., 2006; Chow and Tu, 2009; Chow and Lin, 2015). Commonly considered adaptive seamless trials in clinical development include an adaptive seamless phase I/II design in early clinical development and an adaptive seamless phase II/III trial design in late phase clinical development.

An adaptive seamless phase II/III design is a two-stage design consisting of a learning or exploratory stage (phase IIb) and a confirmatory stage (phase III). A typical approach is to power the study for the phase III confirmatory phase and obtain valuable information with certain assurance using confidence interval approach at the phase II learning stage. Its validity and efficiency, however, has been challenged (see, e.g., Tsiatis and Mehta, 2003; Chow et al., 2007; Chow and Tu, 2009; Chow and Lin, 2015). Statistical methods for a combined analysis under the situation that the study objectives (and/or endpoints) are similar but different at different stages are studied in the literature (Chow and Lin, 2015; Filozof et al., 2017). One of the key assumptions is that there is a well-established relationship between study endpoints at different stages. In other words, study endpoints (e.g., biomarker, surrogate endpoint, or same clinical endpoint with shorter duration) at an earlier stage are predictive of study endpoints (e.g., clinical endpoint) at a later stage. More research regarding sample size estimation/allocation and statistical analysis for seamless adaptive designs with different study objectives and/or study endpoints for various data types (e.g., continuous, binary, and time-to-event) is needed.

13.2.2.10 Multiple Adaptive Design

Finally, a multiple adaptive design is any combinations of the above adaptive designs. Commonly considered multiple adaptive designs include (i) the combination of adaptive group sequential design, drop-the-losers design, and adaptive seamless trial design, and (ii) adaptive dose-escalation design with adaptive randomization (Chow and Chang, 2006; FDA, 2010b, 2018).

In practice, because statistical inference for a multiple-adaptation design is often difficult, it is suggested that a clinical trial simulation be conducted to evaluate the performance of the resultant multiple adaptive design at the planning stage.

When applying a multiple adaptive design, some frequently asked questions include (i) how to avoid/control potential operational biases which may be introduced due to various adaptations that apply to the trial; (ii) how to control the overall type I error rate at the 5%; (iii) how to determine the required sample size for achieving the study objectives with the desired power; and (iv) how to maintain the quality, validity, and integrity of the trial. The trade-off between the flexibility/efficiency and scientific validity/integrity needs to be carefully evaluated before a multiple adaptive design is implemented in clinical trials.

13.3 Regulatory/Statistical Perspectives

From a regulatory point of view, the use of adaptive design methods based on accrued data in clinical trials may introduce operational bias such as selection bias, method of evaluation, early withdrawal, and modification of treatment. Consequently, it may not be able to preserve the overall type I error rate at the pre-specified level of significance. In addition, p-values may not be correct and the corresponding confidence intervals for the treatment effect may not be reliable. Moreover, it may result in a totally different trial that is unable to address the medical questions that original study intended to answer. Li (2006) also indicated that commonly seen adaptations that have an impact on the type I error rate include, but are not limited to: (i) sample size adjustments at interim; (ii) sample size allocations to treatments; (iii) deletions of, additions to, or changes to treatment arms; (iv) shifts in target patient population such as changes in inclusion/exclusion criteria; (v) changes in statistical test strategy; (vi) changes in study endpoints; and (vii) changes in study objectives such as the switch from a superiority trial to a non-inferiority trial. As a result, it is difficult to interpret the clinically meaningful effect size for the treatments under study (Quinlan et al., 2006).

From a statistical point of view, major (or significant) adaptations to trial and/or statistical procedures could (i) introduce bias/variation to data collection, (ii) result in a shift in location and scale of the target patient population, and (iii) lead to inconsistency between hypotheses to be tested and the corresponding statistical tests. These concerns will not only have an impact on the accuracy and reliability of statistical inference drawn on the treatment effect but also present challenges to biostatisticians for development of appropriate statistical methodology for an unbiased and fair assessment of the treatment effect.

Note that although the flexibility of modifying study parameters is very attractive to clinical scientists, several regulatory questions/concerns arise. First, what level of modifications to the trial procedures and/or statistical procedures would be acceptable to the regulatory authorities? Second, what are the regulatory requirements and standards for review and approval process of clinical data obtained from adaptive clinical trials with different levels of modifications to trial procedures and/or statistical procedures of ongoing clinical trials? Third, has the clinical trial become a totally different clinical trial after the modifications to the trial procedures and/or statistical procedures for addressing the study objectives of the originally planned clinical trial? These concerns must be addressed by the regulatory authorities before the adaptive design methods can be widely accepted in clinical research and development.

13.4 Impact, Challenges, and Obstacles

13.4.1 Impact of Protocol Amendments

In practice, for a given clinical trial, it is not uncommon to have between three and five protocol amendments after the initiation of the clinical trial. One of the major impacts of many protocol amendments is that the target patient population may have been shifted during the process, which may have resulted in a totally different target patient population at the end of the trial. A typical example is the case when significant modifications are applied to inclusion/exclusion criteria of the study protocol. As a result, the resultant actual patient population following certain modifications to the trial procedures is a moving target patient population rather than a fixed target patient population. As indicated in Chow and Chang (2006), the impact of protocol amendments on statistical inference due to shift in target patient population (moving target patient population) can be studied through a model that link the moving population means with some covariates (Chow and Shao, 2005). Chow and Shao (2005) derived statistical inference for the original target patient population for simple cases.

13.4.2 Challenges in By Design Adaptations

In clinical trials, commonly employed prospective (by design) adaptations include stopping the trial early due to safety, futility, and/or efficacy, sample size re-estimation (adaptive group sequential design), dropping the losers (adaptive dose finding design), and combining two separate trials into a single trial (adaptive seamless design). These designs are typical multiple-stage designs with different adaptations. In this section, major challenges in analysis and design are described. Recommendations and future development for resolution are provided whenever possible.

The major difference between a classic multiple-stage design and an adaptive multiple-stage design is that an adaptive design allows adaptations after the review of interim analysis results. These by design adaptations may include sample size adjustment (re-assessment or re-estimation); stopping the trials due to safety or efficacy/futility; and dropping the losers (picking the winners). Note that commonly considered adaptive group sequential design, adaptive dose finding design, and adaptive seamless trial design are special cases of multiple-stage designs with different adaptations. In this section, we will discuss major challenges in design (e.g., sample size calculation) and analysis (controlling type I error rate under moving target patient population) of an adaptive multiple-stage design with $K - 1$ interim analyses.

A multiple-stage adaptive group sequential design is very attractive to sponsors in clinical development. However, major (or significant) adaptations such as modification of doses and/or change in study endpoints may introduce bias/variation to data collection as the trial continues. To account for these (expected and/or unexpected) biases/variation, statistical tests are necessary adjusted to maintain the overall type I error, and the related sample size calculation formulas have to be modified for achieving the desired power. In addition, the impact on statistical inference is not negligible if the target patient population has been shifted due to major or significant adaptations and/or protocol amendments. This has presented challenges to biostatisticians in clinical research when applying a multiple-stage adaptive design. In practice, thus, it is worthy pursuing the following specific directions: (i) derive valid statistical test procedures for adaptive group sequential designs assuming model, which relates the data from different interim analyses; (ii) derive valid statistical test procedures for adaptive group sequential designs assuming the random-deviation model; (iii) derive valid Bayesian methods for adaptive group sequential designs; and (iv) derive sample size calculation formulas for various situations. Tsiatis and Mehta (2003) showed that there exists an optimal (i.e., uniformly more powerful) design for any class of sequential design with a specified error spending function. It should be noted that adaptive designs do not require in general a fixed error spending function. One of major challenges for an adaptive group sequential design is that the overall type I error rate may be inflated when there is a shift in target patient population (Feng et al., 2007).

For adaptive dose-finding design, Chang and Chow's method can be improved by the following specific directions: (i) study the relative merits and disadvantage of their method under various adaptive methods, (ii) examine the performance of an alternative method by forming the utility first with different weights to the response levels and then modeling the utility, and (iii) derive sample size calculation formulas for various situations. Recall that an adaptive seamless phase II/III design is a two-stage design that consists of two phases namely a learning (or exploratory) phase and a confirmatory phase. One of the major challenges for designs of this kind is that different study endpoints are often considered at different stages for achieving different study objectives. In this case, the standard statistical methodology for assessment of treatment effect and for sample size calculation cannot be applied.

For a two-stage adaptive design, the aforementioned method by Chang (2007) can be applied. Chang's method, however like other stagewise combination methods, is valid under the assumption of constancy of the target patient populations, study objectives, and study endpoints at different stages.

13.4.3 Obstacles of Retrospective Adaptations

In practice, retrospective adaptations such as adaptive-hypotheses may encounter prior to database lock (or unblinding) and be implemented through the development of a statistical analysis plan. To illustrate the impact of retrospective adaptations, we first consider the situation where switching hypotheses between a superiority hypothesis and a non-inferiority hypothesis. For a promising test drug, the sponsor would prefer an aggressive approach for planning a superiority study. The study is usually powered to compare the promising test drug with an active control agent.

However, the collected data may not support superiority. Instead of declaring the failure of the superiority trial, the sponsor may switch from testing superiority to testing the following non-inferiority hypotheses. The margin is carefully chosen to ensure that the treatment of the test drug is larger than the placebo effect and, thus, declaring non-inferiority to the active control agent means that the test drug is superior to the placebo effect. The switch from a superiority hypothesis to a non-inferiority hypothesis will certainly increase the probability of success of the trial because the study objective has been modified to establishing non-inferiority rather than showing superiority. This type of switching hypotheses is recommended provided that the impact of the switch of statistical issues and inference (e.g., appropriate statistical methods) on the assessment of treatment effect is well justified.

13.5 Some Examples

In this section, we will present some examples for adaptive trial designs, which have been implemented in practice (Chow and Chang, 2008). These trial examples include (i) an adaptive dose escalation design for early phase cancer trials; (ii) a multiple-stage adaptive design for Non-Hodgkin's Lymphoma (NHL) trial; (iii) a phase IV drop-the-losers adaptive design for multiple myeloma trial; and (iv) a two-stage seamless phase I/II adaptive trial design for hepatitis C virus (HCV) trial, which are described below.

Example 13.1: Adaptive dose Escalation Design for Early Phase Cancer Trials

In a phase I dose escalation cancer trial, suppose the primary objective is to determine the maximum tolerable dose (MTD) of a radiation therapy in treating patients with recurrent or refractory of a certain cancer. The identified MTD will be considered as the optimal dose for subsequent clinical trials conducted for later phase clinical development. Based on the toxicity data from animal studies, the initial dose was chosen to be 0.5 mCi/kg. Dose range under study was selected from 0.5 to 4.5 mCi/kg.

A typical approach is to consider an algorithm-based design, i.e., the so-called 3 + 3 traditional escalation rule (TER). The traditional TER is to enter three patients at a new dose level and then enter another three patients when a DLT is observed. The assessment of the six patients is then performed to determine whether the trial should be stopped at the level or to escalate to the next dose level. The traditional 3 + 3 TER design is simple and easy to implement and yet it suffers the following drawbacks: (i) there is no room for dose de-escalation; (ii) there is no sample size justification; (iii) there is no need for further analysis of data; (iv) there is no objective estimation of MTD with statistical assurance; and (v) there is no sampling error and no confidence interval (i.e., statistical inference is unknown). In addition, the 3 + 3 TER with a pre-specified sequence Fibonacci sequence for dose escalation is found to be (i) inefficient and (ii) often under-estimate the MTD especially when the starting dose is too low. Alternatively, one may consider an adaptive type approach, which is to apply the continual re-assessment method (CRM) in conjunction with a Bayesian approach. For the method of CRM, the dose-response relationship is continually reassessed based on accumulative data collected from the trial. The next patient who enters the trial is then assigned to the potential MTD level.

For the dose escalation trial, the principal investigator provided a wish list for selection of the study design from either traditional 3 + 3 TER and CRM in conjunction with a Bayesian approach. The investigator's wish list includes (i) small size cohort for lower dose level; (ii) minimize the number of patients at lower dose groups; (iii) majority patients near the MTD; (iv) ideally, the last two dose cohorts under study; (v) flexibility for dose de-escalation; (vi) limited dose jump if CRM is used; (vii) higher probability of reaching the MTD; and (viii) smaller probability of overdosing. The FDA, however, proposed design selection based on the following criteria: (i) number of patients expected; (ii) number of DLT expected; (iii) probability of observing DLT prior to MTD; (iv) probability of correctly achieving the MTD; (v) probability of overdosing; and (vi) others considerations, such as flexibility of dose de-escalation.

TABLE 13.1

Summary of Simulation Results

Design	# Patients Expected (N)	# of DLT Expected	Mean MTD (SD)	Prob. of Selecting Correct MTD
"3 + 3" TER	15.23	2.8	1.94 (0.507)	0.392
"3 + 3" STER[a]	17.59	3.2	1.70 (0.499)	0.208
CRM[b]	13.82	3.2	2.33 (0.451)	0.696

Note: Based on 5,000 simulation runs.

[a] Allows dose de-escalation.

[b] Uniform prior was used.

In the interest of choosing an efficient design for the dose escalation trial, extensive simulation studies with 5,000 runs were conducted. A logistic toxicity model is chosen for the simulations. Initial dose was chosen to be 0.5 mCi/kg and the number of dose levels (or cohorts) considered is 6. The six dose levels considered are 0.5, 1, 1.6, 2.5, 3.5, and 4.7. For the option of de-escalation, a strict TER (STER) was also considered. DLT rate at MTD is assumed to be $1/3 = 33\%$. For the CRM design, a uniform prior is assumed and dose jump is not allowed.

In this example, as it can be seen from Table 13.1 that if the true MTD is 3.5 mCi/kg, the TER approach underestimates MTD (1.94 mCi/kg), whereas the Bayesian CRM adaptive method also slightly underestimates the MTD (2.33 mCi/kg). The average number of patients required is 15.2 and 13.8 for the TER approach and the Bayesian CRM adaptive method, respectively. From a safety perspective, the average number of DLTs is 2.8 and 3.2 per trial for the TER approach and the Bayesian CRM, respectively. The probability of correctly selecting the MTD is 39.2% and 69.6% for the TER and the Bayesian CRM, respectively. As a result, the Bayesian CRM adaptive method is preferable. More details regarding Bayesian CRM adaptive method and adaptive dose finding can be found in Chang and Chow (2005).

Example 13.2: Multiple-Stage Adaptive Design for NHL Trial

A phase III two parallel group Non-Hodgkin's Lymphoma trial was designed with three analyses. The primary endpoint is progression-free survival (PFS), the secondary endpoints are (i) overall response rate (ORR) including complete and partial response and (ii) complete response rate (CRR). The estimated median PFS is 7.8 months and 10 months for the control and test groups, respectively. Assume a uniform enrollment with an accrual period of 9 months and a total study duration of 23 months. The estimated ORR is 16% for the control group and 45% for the test group. The classic design with a fixed sample size of 375 subjects per group will allow for detecting a three-month difference in median PFS with 82% power at a one-sided significance level of $\alpha = 0.025$. The first interim analysis will be conducted on the first 125 patients/group based on ORR. The objective of the first IA is to modify the randomization. Specifically, if the difference in ORR (test-control), $\Delta_{ORR} > 0$ the enrollment will continue. If $\Delta_{ORR} \leq 0$, then the enrollment will stop. If the enrollment is terminated prematurely, there will be one final analysis for efficacy based on PFS and possible efficacy claimed on the secondary endpoints. If the enrollment continues, there will be a second interim analysis based on PFS. The second interim analysis will could lead to either claim efficacy or futility, or continue to the next stage with possible sample size re-estimation. For the final analysis of PFS, when the primary endpoint (PFS) is significant, the analyses for the secondary endpoints will be performed for the potential claim on the secondary endpoints. During the interim analyses, the patient enrollment will not stop.

Example 13.3: Drop-the-Losers Adaptive Design for Multiple Myeloma Trial

For phase IV study, the adaptive design can also work well. For example, a year after an oncology drug came on the market, physicians are using the drug with different combinations for treating patients with multiple myeloma (MM). However, there is a strong desire to know which combination is the best for the patient populations. Many physicians have their own experiences, but no one has convincing data. Therefore, the sponsor is planned a trial to investigate the optimal combination for the drug. In this scenario, we can use much smaller sample size than we would in phase III trials because the issue is focused on the type I control, as the drug is already approved. The issue is: given a minimum clinically meaningful difference (e.g., two weeks in survival), what is the probability of the trial will be able to identify the optimal combination? This can be done through simulations, the strategy is to start with about five combinations, drop inferior arms, and calculate the probability of selecting the best arm under different combinations of sample size at the interim and final stages. In this adaptive design, we will drop two arms based on the observed response rate, that is, the two arms with lowest observed response rates will be dropped at interim analysis and the rest three arms will be carried forward to the second stage.

Given the response rate 0.4, 0.45, 0.45, 0.5, and 0.6 for the five arms and 91% power, if a traditional design is used, there will be nine multiple comparisons (adjusted alpha = 0.0055 using Bonferroni method). The required sample size for the traditional design is 209 per group or a total of 1,045 subjects comparing a total of 500 subjects for the adaptive trial (250 at the first stage and 150 at the second stage). In cases where the null hypothesis is true (the response rates are the same for all the arms), it doesn't matter too much which arm is chosen as optimal.

Example 13.4: Two-Stage Seamless Phase II/III Adaptive Design for HCV Trial

A pharmaceutical company is interested in conducting a clinical trial utilizing a two-stage seamless adaptive design for evaluation of safety, tolerability, and efficacy of a test treatment as compared to a standard care treatment for treating subjects with hepatitis C virus (HCV) genotype 1 infection. The proposed adaptive trial design consists of two stages of dose selection and efficacy confirmation. The primary efficacy endpoint is the incidence of sustained virologic response (SVR), defined as an undetectable HCV RNA level (<10 IU/mL) at 24 weeks after treatment is complete (Study Week 72). This assessment will be made when the last subject in Stage 2 has completed the Study Week 72 evaluation. In both Stages 1 and 2, consider the incidence of the following primary efficacy variables of (i) rapid virologic response (RVR), that is, undetectable HCV RNA level at Study Week 4; (ii) early virologic response (EVR), that is, ≥2-log10 reduction in HCV RNA level at Study Week 12 compared with the baseline level; (iii) end-of-treatment response (EOT), that is, undetectable HCV RNA level at Study Week 48; and (iv) SVR, that is, undetectable HCV RNA level at Study Week 72 (24 weeks after treatment is complete).

Stage 1 is a four-arm randomized evaluation of three dose levels of continuous subcutaneous (SC) delivery of the test treatment compared with PegIntron (standard care) given as once weekly SC injections. All subjects will receive oral weight-based ribavirin. After all Stage 1 subjects have completed Study Week 12, an interim analysis will be performed. The interim analysis will provide information to enable selection of an active dose of the test treatment based on safety/tolerability, outcomes, and early indications of efficacy to proceed to testing for non-inferiority compared with standard of care in Stage 2. Depending upon individual response data for safety and efficacy, Stage 1 subjects will continue with their randomization assignments for the full planned 48 weeks of therapy, with a final follow-up evaluation at Study Week 72. Stage 2 will be a non-inferiority comparison of a selected dose and the same PegIntron active-control regimen used in Stage 1, both again given with oral ribavirin, for up to 48 weeks of therapy, with a final follow-up evaluation at Study Week 72. A second interim analysis of all available safety/tolerability, outcomes, and efficacy data from Stage 1 and Stage 2 will be performed when all Stage 2 subjects have completed Study Week 12. Depending upon individual response data for safety and efficacy, Stage 2 subjects will receive the full planned 48 weeks of treatment, with final follow-up at Study Week 72. A diagram of two-stage seamless adaptive trial design is illustrated in Figure 13.1 (see also, Chow and Lin, 2015).

For power analysis for sample size estimation, a total study enrollment of 388 subjects will be required (120 subjects or 30 subjects per arm for Stage 1 and 268 subjects or 134 subjects per arm for Stage 2) to account for a probable dropout rate of 15% as well as 2 planned interim analyses using the O'Brien-Fleming method. Stage 1 will enroll a total of 120 subjects divided equally among the four treatment arms to gather

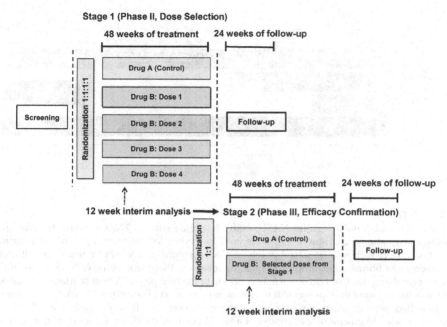

FIGURE 13.1
A diagram of 2-stage seamless adaptive trial design.

sufficient data to select an active dose of the test treatment to proceed to testing in Stage 2. Stage 2 will enroll an additional cohort of 268 subjects divided equally between its two treatment arms to provide a sufficient number of subjects to evaluate non-inferiority of continuous interferon delivery to standard-of-care interferon therapy, for an overall total of 306 subjects enrolled in Stage 1 and Stage 2 combined. Therefore, accounting for 15% attrition in each Stage, we expect that 164 subjects from the selected dose arms (30 from Stage 1 plus 134 additional subjects enrolled in Stage 2) and 164 subjects from the PegIntron active-control arms (30 from Stage 1 plus an additional 134 subjects enrolled in Stage 2) will need to be enrolled to meet the study objective of achieving an 80% power for establishing non-inferiority (with a non-inferiority margin of 15%) at the overall type I error rate of 5%.

Note that above two-stage seamless trial design is a combination of group sequential design, drop-the-losers design and seamless adaptive phase II/III design utilizing precision analysis (i.e., confidence interval approach) for decision-making on dose selection at the first stage. If we apply adaptive randomization for the second stage, the study design would be even more complicated. From regulatory point of view, it is important to ensure that (i) no operational biases are introduced during the conduct of the trial and (ii) the overall type I error rate is well controlled at the 5% level.

Recently, on August 9, 2016, the FDA allows a sponsor to conduct a phase II/III/IV seamless adaptive clinical trial for NASH (Non-Alcoholic

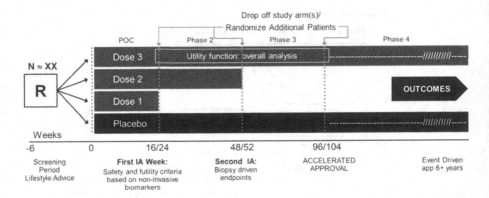

FIGURE 13.2
Phase 2/3/4 adaptive design for NASH study. A single seamless 2/3/4 adaptive trial design allows adaptations, continuous exposure, and long-term follow-up. Endpoints Ad interim analysis are reduction of at least 2 points in NAS, resolution of NASH by histology without worsening of fibrosis, and/or improvement in fibrosis without worsening of NASH. One (the most promising dose) or two doses may continue to the next phase. A post-marketing phase 4 with demonstration of improvement in clinical outcomes will lead to final marketing authorization. Because only one trial would lead to approval, a very small overall alpha (i.e., <0.001) is recommended to ensure proper control of a type I error. Abbreviations: IA, interim analysis; N, number of subjects per study arm; R, randomized patients.

SteatoHepatitis) studies. This is very encouraging to the sponsors. However, statistical methods for multiple-stage seamless adaptive trial design of this kind is not fully developed (see Figure 13.2). More research on this topic is necessary (Filozof et al., 2017).

Example 13.5: Targeted Clinical Trials/Biomarker-Adaptive Trial Design

Biomarker-adaptive trial design is often considered for identifying those patients most likely to respond to the test treatment under study through some validated diagnostic tests of biomarkers such as genomic markers. The process is referred to as the enrichment process in clinical trials, which has led to the concept of targeted clinical trials (Liu and Chow, 2008). As indicated by many researchers (see, e.g., Simon and Maitournam, 2004; Maitournam and Simon, 2005; Casciano and Woodcock, 2006; Dalton and Friend, 2006; Varmus, 2006), the disease targets at the molecular level can be identified by virtue of the data made available at the completion of the Human Genome Project (HGP). As a result, the importance of diagnostic tests for identification of molecular targets will increase as more targeted clinical trials will be conducted for the individualized treatment of patients (personalized medicine). For example, based on the risk of distant recurrence determined by a 21-gene Oncotype DXf0d2 breast cancer assay, patients with a recurrence score of 11–25 in the TAILORx (Trial Assigning Individualized Options for Treatment) trial sponsored by the United States National Cancer

Institute (NCI) are randomly assigned to receive either adjuvant chemo-therapy and hormonal therapy or adjuvant hormonal therapy alone.

Despite of different technical platforms employed in the diagnostic devices for molecular targets used in the trial, the assay falls in the category of the in vitro diagnostic multivariate index assays (IVDMIA) based on the selected differentially expressed genes for detection of the patients with the molecular targets (FDA, 2006a). In addition, to reduce the variation, the IVDMIAs do not usually use all genes during the development stage. Therefore, identification of the differentially expressed genes between different groups of patients is the key to the accuracy and reliability of diagnostic devices for molecular targets. Once the differentially expressed genes are identified, the next task is to search an optimal representation or algorithm that provides the best discrimination ability between the patients with molecular targets and those without the targets. The current validation procedure for diagnostic device is for the assay based on one analyte. However, the IVDMIAs are in fact the parallel assays based on the intensities of multiple analytes. As a result, the current approach to assay validation for one analyte may not be appropriate and is inadequate for validation of IVDMIAs.

With respect to the enrichment design for the targeted clinical trials, patients with positive diagnosis for the molecular targets are randomized to receive the test drug or the control. However, because no IVDMIA can provide the perfectly correct diagnosis, some patients with positive diagnosis may not actually have the molecular targets. Consequently, the treatment effect of the test drug for the patients with targets is underestimated. On the other hand, estimation of the treatment effect based on the data from the targeted clinical trials needs to take into consideration the variability associated with the estimates of accuracy of the IVDMIA such as positive predictive value and false positive rate obtained from the clinical effectiveness trials of the IVDMIA.

In general, there are three classes of targeted clinical trials. The first type is to evaluate the efficacy and safety of targeted treatment for the patients with molecular targets; Herceptin® clinical trials belong to this class. The second type of targeted clinical trials is to select the best treatment regimen for the patients based on the results of some tests for prognosis of clinical outcomes. The last type of the targeted clinical trials is to investigate the correlation of the treatment effect with variations of the molecular targets. Because the objectives of different targeted clinical trials vary, therefore the FDA's *Drug-Diagnostic Co-Development Concept Paper* proposed three different designs to meet different objectives of targeted clinical trials (FDA, 2005). These three designs are given in Figures 13.3 through 13.5.

Design A is the enrichment design, in which the only patients tested positively for identification of molecular targets are randomized either to receive the test drug or the concurrent control (Chow and Liu, 2003). The enrichment design is usually employed when there is a high degree of certainty that the drug response occurs only in the patients tested positively for the molecular targets and the mechanism of pathological pathways is clearly understood. Most of the Herceptin® phase III clinical trials used the enrichment design. However, as pointed out in

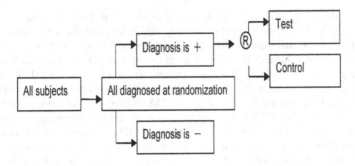

FIGURE 13.3
Design A for target clinical trials.

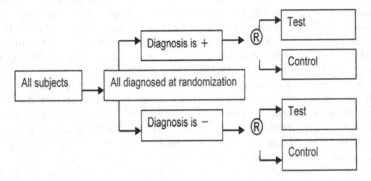

FIGURE 13.4
Design B for target clinical trials.

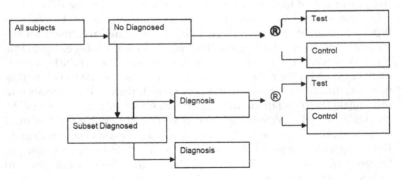

FIGURE 13.5
Design C for target clinical trials.

the FDA *Concept Paper*, the description of test sensitivity and specificity
will not be possible using this type of design without drug and placebo
data in the patients tested negative for the molecular targets. Design B
is a stratified randomized design and stratification factor is the results
of the test for the molecular targets. In other words, the patients are

stratified into two groups depending upon whether the diagnostic test is either positive or negative. Then a separate randomization is independently performed within each group to receive the test drug or concurrent control. The information of the test results for the molecular targets in Design C is primarily used as covariates and is not involved with randomization. Sometimes, only a part of the patients is tested for the molecular targets Design C is useful when the association of the treatment effect of the drug with the results of the diagnostic test needs to be further explored.

13.6 Strategies for Clinical Development

Clinical development of a new drug product is a lengthy and costly process, which includes phases I to III clinical development (prior to regulatory review and approval) and phase IV clinical development (post-approval). For life-threatening diseases or diseases with unmet medical need (rare diseases), this lengthy clinical development process is not acceptable. Such cases call for the use of adaptive design methods in clinical trials in order to shorten the development process (speed) without compromising the safety and efficacy of the drug product under investigation (validity) by maximizing the power for identify best clinical benefit of the drug product under investigation with limited number of subjects (efficiency). As a result, many adaptive design methods are developed for archiving the ultimate goals of validity, efficiency, and speed in clinical development. In practice, however, many other important factors beyond statistical components may have an impact on the development process. These factors include, but are not limited to, patient enrollment, treatment durations, and time required for regulatory review/approval.

Commonly considered strategies for the use of adaptive design methods in clinical development include, but are not limited to, adaptive dose finding, adaptive seamless phase I/II in early clinical development, and adaptive seamless phase II/III in late phase clinical development. These strategies help not only to shorten the development time in a more efficient way but also to increase the probability of success in clinical development. As an example, let us consider the development of a new drug product for the rare disease with unmet medical need, multiple myeloma (MM). A traditional approach is to conduct a typical phase I study for identifying the maximum tolerated dose (MTD), which is often considered as the optimal dose for the late phase of clinical development. In practice, there are several options that we can run this study with several doses or dose schedules. For example, this trial can be run with parallel dose groups or sequentially with the test drug as a single or add-on agent. The traditional approach will not be able to provide the entire spectrum of the dose response of the test drug. On the other hand, if we consider an adaptive approach, it allows us to use more options at the beginning and drop some of the arms (options) if they are either too toxic or ineffective (activity too low or requires very high dose that makes the treatment very costly—biologic product may be very costly). In addition, because the treatment arms are often correlated, the information such as dose limiting toxicity (DLT) from different arms can be synchronized using methods such as Bayesian hierarchical model (Chang and Chow, 2005).

Similar approaches can be used for phase II trials, phase I/II, or phase II/III seamless design. In this approach, the interim endpoint may be a marker such as response rate or time to disease progression. Some of the inferior arms can be dropped at interim. If some of the arms are promising, we can apply different randomization scheme to assign more patients to the superior arms (play-the-winner). In this case, not only will the total cost not increase, but additionally we will increase the chance of success because the increase of sample size for the superior treatment arms is most promising. In addition, the schedules for the trial are similar because the total number of patients remains unchanged. This adaptive design strategy can be applied to phase III trials, but with much less arms. For phase IV studies, the adaptive design can also be applied. However, the situation is different because the drug has been approved during phase IV clinical development. The focus of phase IV clinical development will be the safety rather than efficacy.

13.7 Concluding Remarks

As indicated earlier, although the use of adaptive design methods in clinical trials is motivated by their flexibility and efficiency, many researchers are not convinced and still challenge their validity and integrity (Tsiatis and Mehta, 2003). As a result, many discussions concern the flexibility, efficiency, validity, and integrity of these methods. Li (2006) suggested a couple of principles to be used when one implements an adaptive design in a clinical trial: (i) adaptation should not alter trial conduct and (ii) type I errors should be preserved. Beyond these principles, some basic considerations such as dose/dose regimen, study endpoints, treatment duration, and logistics should be carefully evaluated for feasibility (Quinlan et al., 2006). To maintain the validity and integrity of an adaptive design with complicated adaptations, it is strongly suggested that an independent data monitoring committee (IDMC) be established. In practice, IDMCs have been widely used in group sequential design with adaptations of stopping a trial early and sample size re-estimation. The role and responsibility of an IDMC for a clinical trial using adaptive design should be clearly defined. An IDMC usually conveys very limited information to investigators or sponsors about treatment effects, procedural conventions, and statistical methods with recommendations in order to maintain the validity and integrity of the study.

When applying adaptive design methods in clinical trials, the feasibility of certain adaptations such as changes in study endpoints/hypotheses should be carefully evaluated to prevent from any possible misuse and abuse of the adaptive design methods. For a complicated multiple adaptive design, an independent data monitoring committee should be established to ensure the integrity of the study. Clinical trial simulation provides *a* solution, not *the* solution, for a complicated multiple adaptive design. In practice, deciding how to validate the assumed predictive model for clinical trial simulation is a major challenge to both investigators and biostatisticians.

In February 2010, the FDA published a draft guidance, *Adaptive Design Clinical Trials for Drug and Biologics*. The FDA has expressed its intention to slow down the escalating momentum behind adaptive clinical trial designs to allow time to craft better working definitions of the new models and build a better base before moving forward. The FDA's efforts were primarily directed to issues current at the time, such as the impact of protocol amendments, challenges in by design adaptations, and obstacles of retrospective adaptations as described in the previous sections. Since 2010, there have been several successful regulatory submissions utilizing adaptive trial designs, following the draft guidance (Lee and Lin, 2016). As a result, the draft guidance has been revised and circulated for public comments in late 2018 (FDA, 2018).

In summary, from the clinical point of view, adaptive design methods reflect real clinical practice in clinical development. Adaptive design methods are very attractive due to their flexibility and are very useful especially in early clinical development. From the statistical point of view, the use of adaptive methods in clinical trials makes current good statistics practice even more complicated. The validity of the use of adaptive design methods is not well established and fully understood. The impact of statistical inference on treatment effect should be carefully evaluated under the framework of moving target patient population as the result of protocol amendments. In practice, regulatory agencies may not realize that the adaptive design methods for review and approval of regulatory submissions have been employed for years without any scientific basis. Guidelines regarding the use of adaptive design methods and Good Adaptive Design Practices (GADP) must be developed so that appropriate statistical methods and statistical software packages can be developed accordingly.

14

Criteria for Dose Selection

14.1 Introduction

In pharmaceutical/clinical development, a two-stage seamless adaptive trial design that combines two individual studies into a single study is commonly considered (FDA, 2010b, 2018; Chow and Chang, 2011; Bhatt and Mehta, 2016). Such designs include a two-stage phase II/III (or phase I/II) adaptive trial that combines one phase IIb study for dose finding or treatment selection and one phase III study for efficacy confirmation into a single study (Barnes et al., 2010; Lawrence et al., 2014; Chen et al., 2015; Chow and Lin, 2015). Thus, a two-stage seamless adaptive trial design consists of two stages: dose finding or treatment selection (stage 1) and efficacy confirmation (stage 2). The study objective at the first stage is dose finding or treatment selection, while the study objective at Stage 2 is to confirm the efficacy of the dose or treatment selected from the first stage as compared to a control. If the study endpoints and objectives at different stages are the same, the two-stage seamless adaptive trial design leads to a typical group sequential design.

For a two-stage seamless adaptive trial design, at the first stage, qualified subjects are usually randomly assigned to receive one of the dose or treatment at a 1:1 ratio. At the end of stage 1, promising dose(s) will be selected based on pre-specified selection criteria. In practice, since there is little power with limited subjects available at interim, commonly considered selection criteria for critical decision-making include (i) precision analysis, (ii) predictive probability of success, and (iii) probability of being the best dose or treatment (Chow and Lin, 2015; Lee and Lin, 2016). The selected promising dose(s) will then proceed to the next stage for efficacy confirmation. Based on the results observed at interim, some adaptations such as adaptive-randomization or sample size re-estimation are often applied for achieving the study objectives and/or increasing the probability of success.

In practice, for clinical studies utilizing two-stage seamless adaptive trial designs, the following questions are commonly asked by the regulatory agencies such as the FDA. First, it is a concern whether the overall type I error rate is well controlled at a pre-specified level of significance. Second, it is a concern that the selection criteria at Stage 1 may wrongly select inferior dose groups based on the limited number of subjects available at interim. Third, it is a concern that the interim data may deviate far from then expected (i.e., the assumptions made at the beginning of study). In this chapter, we attempt to address these concerns for two-stage seamless adaptive clinical trial (see also Zheng and Chow, 2019).

In Section 14.2, criteria that are commonly employed for dose selection at the first stage are briefly described. Section 14.3 illustrates the application of these selection criteria based on single primary study endpoint as well as co-primary study endpoints. The performances of these selection criteria in terms of the coverage probability of correctly selecting the promising dose(s) are evaluated through the conduct of extensive clinical trial simulation in Section 14.4. Concluding remarks are given in Section 14.5.

14.2 Dose Selection Criteria

Consider the dose finding stage in a two-stage seamless adaptive clinical trial. Assume there are k test doses and one control dose to investigate in this stage. Denote the measurement means corresponding to the k doses by $D_1, D_2,..., D_k$, and that of the control dose by D_0. We are interested in the difference $d_i = D_i - D_0, i = 1, ..., k$. Without loss of generality, assume the larger of this difference, the better effect of the corresponding dose. Define $X_{ij}, i = 0, 1,..., k, j = 1,..., N_i$, where N_i is the sample size of the ith dose, as the measurement of the jth subject with the ith dose. For each dose, i.e., each i, assume $X_{ij}, j = 1,..., N_i$ are independent and identical distributed, and follow a normal distribution with an unknown mean D_i and an unknown variance σ_i^2. Denote the sample mean by $\overline{X}_i$ and let S_i^2 denote the sample variance for dose i. Based on the total sample size of $N = \sum_{i=1}^{k} N_i$, we can construct the $1 - \alpha$ confidence interval of the mean d_i as follows:

$$(L_i, U_i) = \left(\overline{X}_i - \overline{X}_0 - q_{\alpha/2}\tilde{S}_i, \overline{X}_i - \overline{X}_0 + q_{\alpha/2}\tilde{S}_i\right),$$

where $q_{\alpha/2}$ is the upper $\alpha/2$ quantile of the standard normal distribution and the pooled sample standard deviation

$$\tilde{S}_i = \sqrt{\frac{(N_i - 1)S_i^2 / N_i + (N_0 - 1)S_0^2 / N_0}{N_i + N_0 - 2}}.$$

Assume that at interim, the data from $n = \sum_{i=1}^{k} n_i$ subjects are available, with n_i subjects available for each dose. Denote X_i' as the sample mean and $S_i'^2$ as the sample variance for dose i at interim. Then similarly, we have the $1-\alpha$ confidence interval of the mean d_i:

$$\left(L_i', U_i'\right) = \left(\overline{X_i'} - \overline{X_0'} - q_{\alpha/2}\widetilde{S}_i', \overline{X_i'} - \overline{X_0'} + q_{\alpha/2}\widetilde{S}_i'\right),$$

where $q_{\alpha/2}$ is the upper $\alpha/2$ quantile of the standard normal distribution and the pooled sample standard deviation

$$\widetilde{S}_i = \sqrt{\frac{(n_i-1)S_i'^2/n_i + (n_0-1)S_0'^2/n_0}{n_i + n_0 - 2}}.$$

In practice, commonly considered criteria for dose selection include conditional power, predictive probability of success, probability being the best, which are briefly describe below.

14.2.1 Conditional Power

At interim (dose finding stage), denote the observed effect size by

$$\widehat{\delta}'_i = \frac{\overline{X_i'} - \overline{X_0'}}{\sqrt{S_i'^2 + S_0'^2}} \text{ for dose } i.$$

Take $\widehat{\delta}'_i$ (or $(\overline{X_i'}, \overline{X_0'}, S_i'^2, S_0'^2)$) as the true value of the effect size (or means of dose i and control dose, and variances of dose i and control dose, respectively) and calculate the power with the pre-specified sample size, denoted as p_i^{power}, which can be expressed as

$$pr\left\{ \frac{\overline{X_i} - \overline{X_0}}{\sqrt{\dfrac{\dfrac{(N_i-1)S_i^2}{N_i} + \dfrac{(N_0-1)S_0^2}{N_0}}{N_i + N_0 - 2}}} > q_\alpha \,\middle|\, \left(D_i, D_0, \sigma_i^2, \sigma_0^2\right) = \left(\overline{X_i'}, \overline{X_0'}, S_i'^2, S_0'^2\right) \right\},$$

where q_α is the upper α quantile of the standard normal distribution. Noncentral t-distribution can be used to calculate this probability.

Select the dose with the largest p_i^{power}. If $N_i, i = 0, 1, \ldots, k$ are equal, then this criterion is equivalent to choosing the one having the largest observed effect size $\widehat{\delta}'_i$.

14.2.2 Precision Analysis Based on Confidence Interval

Denote (L_i^+, U_i^+) as the positive part of (L_i', U_i'). Define $p_i^+ = pr\{d_i \in (L_i^+, U_i^+)\}$ as the confidence level for achieving statistical significance. If $L_i' \geq 0$, $p_i^+ = 1 - \alpha$. If $U_i' < 0$, $p_i^+ = 0$. If (L_i', U_i') contains 0, $p_i^+ = pr\{d_i \in (0, U_i')\}$ and we derive the exact expression of p_i^+ in this case. Denote $w_i = \Phi\left(\frac{\overline{X_i'} - \overline{X_0'}}{\widetilde{S_i'}}\right)$. Then $p_i^+ = w_i - \alpha/2$. Based on $p_i^+, i = 1, \ldots, k$, select the dose having the highest p_i^+.

14.2.3 Predictive Probability of Success

Assume that the difference observed at interim, $\overline{X_i'}, i = 1, \ldots, k$, is preserved to the end of the study. We calculate the predictive probability of success $pr\{d_i > d\}$ at interim, where d is a pre-specified value. Denote $p_i^s = pr\{d_i > d\}$. We have the mean difference $\overline{X_i'} - \overline{X_0'}$ and the pooled sample standard deviation

$$\widetilde{S}_i^p = \sqrt{\frac{(N_i - 1)S_i'^2 / N_i + (N_0 - 1)S_0'^2 / N_0}{N_i + N_0 - 2}}$$

based on the data obtained at interim. Then

$$p_i^s = \Phi\left(\frac{\overline{X_i'} - \overline{X_0'} - d}{\widetilde{S}_i^p}\right),$$

where Φ is the cumulative distribution function of standard normal distribution. We may select the dose with the highest p_i^s.

14.2.4 Probability of Being the Best Dose

Calculate $p_i^b = pr\{\max_i(d_i) = d_i\}$ which is the probability of being the best dose, based on the interim data. Taking $X_{0j}, j = 1, \ldots, N_0$ as fixed values, then we can assume that $(\overline{X}_{i_1}, S_{i_1}^2)$ is statistically independent from $(\overline{X}_{i_2}, S_{i_2}^2)$ for any $i_1 > i_2$. It also holds for

$$\left(\overline{X_i'}, S_i'^2\right), i = 1, \ldots, k.$$

We have

$$p_i^b = \int \frac{1}{\widetilde{S_i'}} \phi\left(\frac{z - \left(\overline{X_i'} - \overline{X_0'}\right)}{\widetilde{S_i'}}\right) \prod_{g=1, i \neq g}^{k} \Phi\left(\frac{z - \left(\overline{X_g'} - \overline{X_0'}\right)}{\widetilde{S_g'}}\right) dz,$$

where ϕ and Φ are the probability density function and the cumulative distribution function of standard normal distribution, respectively. Similarly, as predictive probability of success, $\tilde{S}_i'$ and $\tilde{S}_g'$ in the above expression of p_i^b can be replaced with $\tilde{S}_i^p$ and $\tilde{S}_g^p$.

Thus, we may select the dose with the largest p_i^b.

14.3 Implementation and Example

In this section, we describe the practical implementation under the case of single primary endpoint and the case of co-primary endpoints. A numerical example is given to illustrate these methods.

14.3.1 Single Primary Endpoint

As introduced in the previous section, we summarize the criteria for dose selection under the case of single primary endpoint as follows.

Conditional power—Select the dose with the largest p_i^{power}, that is

$$p_i^{\text{power}} = pr\left\{ \frac{\overline{X}_i - \overline{X}_0}{\sqrt{\dfrac{(N_i-1)S_i^2}{N_i} + \dfrac{(N_0-1)S_0^2}{N_0}}{N_i+N_0-2}}} > q_\alpha \mid \left(D_i, D_0, \sigma_i^2, \sigma_0^2\right) = \left(\overline{X_i'}, \overline{X_0'}, S_i'^2, S_0'^2\right) \right\}$$

Precision analysis based on CI—Select the dose having the highest p_i^+.

$$\text{If } L_i' \geq 0,\ p_i^+ = 1-\alpha;$$

$$\text{If } U_i' < 0,\ p_i^+ = 0;$$

$$\text{If } \left(L_i', U_i'\right) \text{ contains } 0,\ p_i^+ = \Phi\left(\frac{\overline{X_i'} - \overline{X_0'}}{\tilde{S}_i'}\right) - \alpha/2,$$

where

$$\tilde{S}_i' = \sqrt{\frac{(n_i-1)S_i'^2/n_i + (n_0-1)S_0'^2/n_0}{n_i+n_0-2}}.$$

Predictive probability of success—Select the dose with the highest p_i^s. That is,

$$p_i^s = \Phi\left(\frac{\overline{X_i'} - \overline{X_0'} - d}{\tilde{S}_i^p}\right),$$

where

$$\tilde{S}_i^p = \sqrt{\frac{(N_i - 1)S_i'^2 / N_i + (N_0 - 1)S_0'^2 / N_0}{N_i + N_0 - 2}}$$

and d is a pre-specified value.

Probability of being the best dose—Select the dose with the largest p_i^b. That is,

$$p_i^b = \int \frac{1}{\tilde{S}_i'} \phi\left(\frac{z - \left(\overline{X_i'} - \overline{X_0'}\right)}{\tilde{S}_i'}\right) \prod_{g=1, i \neq g}^{k} \Phi\left(\frac{z - \left(\overline{X_g'} - \overline{X_0'}\right)}{\tilde{S}_g'}\right) dz.$$

14.3.2 Co-primary Endpoints

Assume under the case of co-primary endpoint, the type I error allocated to each endpoint is equal (i.e., $\alpha/2$). Denote the measurement means corresponding to the k doses by D_1, $D_2 \supset D_k$, and that of the control dose by D_0 for the first endpoint. Similarly, denote those for the second endpoint by E_1, $E_2 \supset E_k$ and E_0. We are interested in the differences $d_i = D_i - D_0$ and $g_i = E_i - E_0$, $i = 1, \ldots, k$. Without loss of generality, assume the larger of this difference, the better effect of the corresponding dose. Denote X_{ij}, $i = 0, 1, \ldots, k$, $j = 1, \ldots, N_i$, where N_i is the sample size of the ith dose, as the measurement of the jth subject with the ith dose for the first endpoint. Similarly, denote Y_{ij} as the measurement of the jth subject with the ith dose relative to the second endpoint. For each dose, assume $X_{ij}, j = 1, \ldots, N_i$ are independent and identical distributed, and follow a normal distribution with an unknown mean D_i and an unknown variance σ_{1i}^2. Similarly, assume $Y_{ij} = 1, \ldots, N_i$, are independent and identical distributed, and follow a normal distribution with an unknown mean E_i and an unknown variance σ_{2i}^2. Define $\overline{X}_i$ and $\overline{Y}_i$ as the sample means and S_i^2 and Q_i^2 as the sample variance for dose i. Based on the total sample size of $N = \sum_{i=1}^{k} N_i$, we can construct the $1 - \alpha$ confidence intervals of the mean d_i and g_i as follows:

$$\left(L_{1i}, U_{1i}\right) = \left(\overline{X}_i - \overline{X}_0 - q_{\alpha/2}\tilde{S}_i, \overline{X}_i - \overline{X}_0 + q_{\alpha/2}\tilde{S}_i\right),$$

$$\left(L_{2i}, U_{2i}\right) = \left(\overline{Y}_i - \overline{Y}_0 - q_{\alpha/2}\tilde{Q}_i, \overline{Y}_i - \overline{Y}_0 + q_{\alpha/2}\tilde{Q}_i\right),$$

where $q_{\alpha/2}$ is the upper $\alpha/2$ quantile of the standard normal distribution and the pooled sample standard deviations

$$\tilde{S}_i = \sqrt{\frac{(N_i-1)S_i^2 / N_i + (N_0-1)S_0^2 / N_0}{N_i + N_0 - 2}} \quad \text{and}$$

$$\tilde{Q}_i = \sqrt{\frac{(N_i-1)Q_i^2 / N_i + (N_0-1)Q_0^2 / N_0}{N_i + N_0 - 2}}.$$

Assume that at interim, the data from $N = \sum_{i=1}^{k} n_i$ subjects are available, with n_i subjects available for each dose. Denote X_i' and Y_i' as the sample means and $S_i'^2$ and $Q_i'^2$ as the sample variances for dose i at interim. Then similarly, we have the $1 - \alpha$ confidence interval of the means d_i and g_i:

$$\left(L_{1i}', U_{1i}'\right) = \left(\overline{X_i'} - \overline{X_0'} - q_{\alpha/2}\tilde{S}_i', \overline{X_i'} - \overline{X_0'} + q_{\alpha/2}\tilde{S}_i'\right),$$

$$\left(L_{2i}', U_{2i}'\right) = \left(\overline{Y_i'} - \overline{Y_0'} - q_{\alpha/2}\tilde{Q}_i', \overline{Y_i'} - \overline{Y_0'} + q_{\alpha/2}\tilde{Q}_i'\right),$$

where $q_{\alpha/2}$ is the upper $\alpha/2$ quantile of the standard normal distribution and the pooled sample standard deviations

$$\tilde{S}_i' = \sqrt{\frac{(n_i-1)S_i'^2 / n_i + (n_0-1)S_0'^2 / n_0}{n_i + n_0 - 2}} \quad \text{and}$$

$$\tilde{Q}_i' = \sqrt{\frac{(n_i-1)Q_i'^2 / n_i + (n_0-1)Q_0'^2 / n_0}{n_i + n_0 - 2}}.$$

Then, we have the criteria for dose selection under the case of co-primary endpoints as follows.

Conditional power—Select the dose with the largest p_i^{power}. That is

$$p_i^{power} = pr \left\{ \left. \dfrac{\overline{X}_i - \overline{X}_0}{\sqrt{\dfrac{(N_i-1)S_i^2}{N_i} + \dfrac{(N_0-1)S_0^2}{N_0}}{N_i + N_0 - 2}} > q_{\frac{\alpha}{2}}, \right. \right.$$

$$\left. \dfrac{\overline{Y}_i - \overline{Y}_0}{\sqrt{\dfrac{(N_i-1)Q_i^2}{N_i} + \dfrac{(N_0-1)Q_0^2}{N_0}}{N_i + N_0 - 2}} > q_{\frac{\alpha}{2}} \,\middle|\, \left(D_i, D_0, \sigma_{1i}^2, \sigma_{10}^2\right) \right.$$

$$\left. = \left(\overline{X}_i', \overline{X}_0', S_i'^2, S_0'^2\right), \left(E_i, E_0, \sigma_{2i}^2, \sigma_{20}^2\right) = \left(\overline{Y}_i', \overline{Y}_0', Q_i'^2, Q_0'^2\right) \right\}$$

Precision analysis based on CI—Select the dose having the highest p_i^+. That is,

$$p_i^+ = pr\left\{ d_i \in \left(L_{1i}^+, U_{1i}^+\right), g_i \in \left(L_{2i}^+, U_{2i}^+\right) \right\}$$

$$= pr\left\{ d_i \in \left(\max(0, L_{1i}^+), \max(0, U_{1i}^+)\right), \ g_i \in \left(\max\left(0, L_{2i}^+\right), \max(0, U_{2i}^+)\right) \right\},$$

where (L_{1i}^+, U_{1i}^+) and (L_{2i}^+, U_{2i}^+) as the positive parts of (L_{1i}', U_{1i}') and (L_{2i}', U_{2i}'), respectively. Under the assumption of the statistical independence between X_{ij} and Y_{ij}, we can express p_i^+ in detail as follows:

If $L_{2i}' \geq 0$, and $L_{1i}' \geq 0$, $p_i^+ = (1-\alpha)^2$;

If $L_{2i}' \geq 0$ and (L_{1i}', U_{1i}') contains 0, $p_i^+ = (1-\alpha)*\left(\Phi \frac{\overline{X}_i - \overline{X}_0}{\tilde{S}_i} - \alpha/2\right)$,

where

$$\tilde{S}_i = \sqrt{\dfrac{(n_i-1)S_i'^2/n_i + (n_0-1)S_0'^2/n_0}{n_i + n_0 - 2}};$$

If $L'_{1i} \geq 0$ and (L'_{2i}, U'_{2i}) contains 0, $p_i^+ = (1-\alpha)*\left(\Phi\left(\frac{Y'_i - Y'_0}{\tilde{Q}'_i}\right) - \alpha/2\right),$

where

$$\tilde{Q}'_i = \sqrt{\frac{(n_i - 1)Q'^2_i/n_i + (n_0 - 1)Q'^2_0/n_0}{n_i + n_0 - 2}};$$

If both (L'_{1i}, U'_{1i}) and (L'_{2i}, U'_{2i}) contain 0,

$$p_i^+ = \left(\Phi\left(\frac{X'_i - X'_0}{\tilde{S}'_i}\right) - \frac{\alpha}{2}\right)*\left(\Phi\left(\frac{X'_i - X'_0}{\tilde{Q}'_i}\right) - \frac{\alpha}{2}\right);$$

otherwise, $p_i^+ = 0$.

Predictive probability of success—Select the dose with the highest p_i^s. That is,

$$p_i^s = pr\{d_i > d, g_i > g\}.$$

Under the assumption of the statistical independence between X_{ij} and Y_{ij}, we have

$$p_i^s = \Phi\left(\frac{X'_i - X'_0 - d}{\tilde{S}^p_i}\right)\Phi\left(\frac{Y'_i - Y'_0 - g}{\tilde{Q}^p_i}\right),$$

where

$$\tilde{S}^p_i = \sqrt{\frac{(N_i - 1)S'^2_i/N_i + (N_0 - 1)S'^2_0/N_0}{N_i + N_0 - 2}},$$

$$\tilde{Q}^p_i = \sqrt{\frac{(N_i - 1)Q'^2_i/N_i + (N_0 - 1)Q'^2_0/N_0}{N_i + N_0 - 2}},$$

and d, g are pre-specified values.

Probability of being the best dose—Select the dose with the largest p_i^b. That is

$$p_i^b = pr\{\max_i(d_i) = d_i, \max_i(g_i) = g_i\}$$

$$= \int \frac{1}{\tilde{S}'_i} \phi\left(\frac{z - (X'_i - X'_0)}{\tilde{S}'_i}\right) \prod_{g=1, i \neq g}^k \Phi\left(\frac{z - (X'_g - X'_0)}{\tilde{S}'_g}\right) dz$$

$$\int \frac{1}{\widetilde{Q}'_i} \phi\left(\frac{w-\left(\overline{Y'_i}-\overline{Y'_0}\right)}{\widetilde{Q}'_i}\right) \prod_{h=1, i\neq h}^{k} \Phi\left(\frac{w-\left(\overline{Y'_h}-\overline{Y'_0}\right)}{\widetilde{Q}'_h}\right) dw,$$

under the assumption that X_{ij} and Y_{ij} are statistically independent.

14.3.3 A Numeric Example

A clinical study was intended to be designed as a two-stage phase II/III seamless, adaptive, randomized, double-blind, parallel group, placebo-controlled, multicenter study to evaluate the efficacy, safety, and dose-response of a compound in treating postmenopausal women and adjuvant breast cancer patients with vasomotor symptoms (VMS). Hot flashes, flushes, and night sweats, sometimes accompanied by shivering and a sense of cold, are well-known VMS. The primary study endpoints would be frequency and severity of VMS evaluated at Weeks 4 and 12 as co-primary endpoints. Since the real data for this study is unavailable, we provide this example with numerical simulated data to illustrate the use of the dose-selection methods, based on such studies.

Under this two-stage adaptive trial design, the first stage is a phase II trial for dose selection. The second stage is a phase III trial for confirming the efficacy and safety of the compound. An interim analysis will be conducted to evaluate the dose effect and to review the efficacy and safety effects when about 50% subjects at Stage 1 complete Week 4 of the Double-blind Treatment period. One or two optimal dose(s) will be selected and the sample size will be re-estimated. The second stage of the study starts after the interim analysis. Newly enrolled subjects will be randomized to receive the selected optimal dose(s) or placebo for 12 weeks. The second stage will serve as a pivotal study.

To establish the measure frequency of VMS, at week 4, each study subject reported the number of VMS each day and those numbers were averaged for each subject. For severity, each subject was asked to rate how much they were bothered by VMS at the end of week 4 using scores from 1 to 10 with higher value indicating more severity. Three doses and one placebo were compared. 800 subjects were recruited into the study with 200 in each group. At interim, each group had 100 subjects having completed week 4 of the treatment period. The logarithm-transformation of frequency and the original scale of severity were treated as following normal distributions and were used for interim analysis. Using the approach of probability of being the best dose, which was shown to be the best one in the simulation studies, we selected dose 3 which had the largest overall probability of being the best dose. The results were shown in Table 14.1 with means and standard errors of the two co-primary endpoints.

TABLE 14.1

Results of the Example

		Placebo	Dose 1	Dose 2	Dose 3
Endpoint 1	Sample mean	3.37	3.42	3.07	2.56
	Sample standard error	1.61	2.10	1.62	1.53
	Probability of being the best		0.0%	1.1%	98.9%
Endpoint 2	Sample mean	6.74	6.17	5.94	5.93
	Sample standard error	1.74	1.76	1.89	1.87
	Probability of being the best		7.0%	45.0%	47.9%
Overall probability of being the best			0.0%	0.5%	47.4%

14.4 Clinical Trial Simulation

In this section, we conducted simulation studies to investigate the performances of the four methods described in Section 14.2.

14.4.1 Single Primary Endpoint

Assume the mean effects of four doses (dose 1 being a placebo) are to be compared, and all effects follow normal distributions. Different parameter settings were given to investigate the performances of the methods. Dose 4 was always set to be the most effective dose among the four doses compared. We consider the following total sample sizes: 100, 150, and 200. The proportion of interim sample size was set to be 20%, resulting in interim sample sizes of 20, 30, and 40, respectively. For each case, 5,000 repetitions were implemented. We summarized the simulation results as the frequency of being selected as the most effective dose in Tables 14.2 through 14.4, according to difference sample sizes.

From the simulation results, we observed that the last method probability of being the best dose performed better in all cases than other methods, with higher probability of choosing the right best dose and lower probability of choosing the wrong dose.

14.4.2 Co-primary Endpoints

Assume the mean effects of the co-primary endpoints of four doses (dose 1 being a placebo) are to be compared, and all effects follow normal distributions. Different parameter settings were given to investigate the performances of the methods. In the first scenario, dose 4 was set to be the most effective dose among the four doses compared, with both co-primary endpoints being the most effective. In the second scenario, the first endpoint of

TABLE 14.2

Simulation Results for Single Primary Endpoint When the Total Sample Size Is 100

N = 100

	Dose 1	Dose 2	Dose 3	Dose 4	Dose 1	Dose 2	Dose 3	Dose 4	Dose 1	Dose 2	Dose 3	Dose 4
Mean	0.5	0.51	0.52	0.53	0.5	0.51	0.525	0.53	0.5	0.505	0.51	0.53
Criterion	SD	Probability of being selected			SD	Probability of being selected			SD	Probability of being selected		
C1	0.005	33.3%	33.3%	33.3%	0.005	33.3%	33.4%	33.4%	0.005	18.5%	40.7%	40.8%
C2		33.3%	33.3%	33.3%		33.3%	33.3%	33.3%		32.5%	33.8%	33.8%
C3		33.3%	33.3%	33.3%		33.3%	33.3%	33.3%		23.3%	38.4%	38.4%
C4		0.0%	0.0%	100.0%		0.0%	0.7%	99.3%		0.0%	0.0%	100.0%
C1	0.01	18.5%	40.7%	40.8%	0.01	18.8%	40.6%	40.6%	0.01	3.0%	26.7%	70.3%
C2		32.6%	33.7%	33.7%		32.5%	33.8%	33.8%		22.4%	37.7%	39.9%
C3		23.7%	38.1%	38.1%		23.5%	38.2%	38.2%		5.7%	32.6%	61.7%
C4		0.0%	0.1%	99.9%		0.0%	6.1%	93.9%		0.0%	0.0%	100.0%
C1	0.02	2.9%	27.7%	69.4%	0.02	2.9%	41.6%	55.6%	0.02	0.8%	4.4%	94.8%
C2		22.3%	38.0%	39.7%		22.3%	38.7%	39.0%		13.1%	28.2%	58.7%
C3		5.5%	33.7%	60.9%		5.5%	43.3%	51.2%		1.7%	8.0%	90.3%
C4		0.1%	5.5%	94.4%		0.0%	20.8%	79.2%		0.0%	0.2%	99.8%
C1	0.03	2.1%	18.2%	79.7%	0.03	1.7%	34.3%	64.0%	0.03	0.9%	3.1%	95.9%
C2		16.1%	34.6%	49.3%		15.5%	40.2%	44.3%		10.1%	20.7%	69.2%
C3		3.2%	21.7%	75.1%		2.9%	36.0%	61.1%		1.4%	4.6%	94.0%
C4		1.0%	13.6%	85.4%		0.7%	30.3%	69.0%		0.4%	1.8%	97.8%

(Continued)

TABLE 14.2 (Continued)

Simulation Results for Single Primary Endpoint When the Total Sample Size Is 100

N = 100

	Dose 1	Dose 2	Dose 3	Dose 4	SD	Dose 1	Dose 2	Dose 3	Dose 4	SD	Dose 1	Dose 2	Dose 3	Dose 4	SD
Mean	0.5	0.51	0.52	0.53	0.5	0.5	0.51	0.525	0.53	0.5	0.5	0.505	0.51	0.53	0.5
Criterion		Probability of being selected			SD		Probability of being selected			SD		Probability of being selected			SD
C1		6.9%	25.6%	67.6%	0.05		5.7%	36.5%	57.8%	0.05		4.7%	9.7%	85.6%	0.05
C2		13.3%	30.6%	56.1%			12.8%	39.0%	48.3%			10.6%	16.9%	72.4%	
C3		7.3%	25.9%	66.8%			6.0%	36.9%	57.1%			4.9%	10.1%	85.0%	
C4		6.4%	24.5%	69.1			5.4%	36.3%	58.3%			4.4%	9.6%	86.0%	
C1		16.9%	30.0%	53.1%	0.1		14.6%	37.6%	47.9%	0.1		15.0%	22.5%	62.5%	0.1
C2		19.1%	31.1%	49.8%			16.9%	37.6%	45.6%			16.8%	24.4%	58.8%	
C3		17.0%	30.1%	53.0%			14.6%	37.7%	47.7%			15.0%	22.5%	62.4%	
C4		16.8%	29.9%	53.3%			14.3%	37.6%	48.2%			14.9%	22.5%	62.6%	
C1		23.9%	33.0%	43.1%	0.2		22.5%	36.2%	41.3%	0.2		24.2%	28.4%	47.4%	0.2
C2		24.6%	33.4%	42.0%			23.3%	35.9%	40.8%			25.1%	28.8%	46.1%	
C3		23.9%	33.0%	43.1%			22.5%	36.2%	41.3%			24.2%	28.4%	47.4%	
C4		23.9%	33.0%	43.2%			22.4%	36.6%	41.0%			24.1%	28.2%	47.7%	

Note: The darker value, the higher the probability.
C1: conditional power.
C2: precision analysis based on confidence interval.
C3: Predictive probability of success.
C4: Probability of being the best dose.

TABLE 14.3

Simulation Results for Single Primary Endpoint When the Total Sample Size Is 150

N = 150

Mean	Dose 1	Dose 2	Dose 3	Dose 4	Dose 1	Dose 2	Dose 3	Dose 4	Dose 1	Dose 2	Dose 3	Dose 4
	0.5	0.51	0.52	0.53	0.5	0.51	0.525	0.53	0.5	0.505	0.51	0.53
Criterion	SD	Probability of being selected			SD	Probability of being selected			SD	Probability of being selected		
C1	0.005	33.3%	33.3%	33.3%	0.005	33.3%	33.3%	33.3%	0.005	26.5%	36.8%	36.8%
C2		33.3%	33.3%	33.3%		33.3%	33.3%	33.3%		33.2%	33.4%	33.4%
C3		33.3%	33.3%	33.3%		33.3%	33.3%	33.3%		29.7%	35.2%	35.2%
C4		0.0%	0.0%	100.0%		0.0%	0.1%	99.9%		0.0%	0.0%	100.0%
C1	0.01	26.4%	36.8%	36.8%	0.01	26.2%	36.9%	36.9%	0.01	4.9%	37.2%	57.9%
C2		33.2%	33.4%	33.4%		33.2%	33.4%	33.4%		26.0%	36.8%	37.2%
C3		29.6%	35.2%	35.2%		29.5%	35.2%	35.2%		8.5%	40.5%	50.9%
C4		0.0%	0.0%	100.0%		0.0%	2.8%	97.2%		0.0%	0.0%	100.0%
C1	0.02	4.9%	37.1%	58.1%	0.02	4.8%	45.9%	49.3%	0.02	1.0%	7.7%	91.3%
C2		26.1%	36.8%	37.1%		25.8%	37.1%	37.1%		15.4%	32.7%	51.9%
C3		8.6%	40.0%	51.4%		8.3%	45.2%	46.5%		2.2%	12.6%	85.2%
C4		0.0%	3.1%	96.9%		0.0%	16.3%	83.7%		0.0%	0.0%	100.0%
C1	0.03	1.9%	20.4%	77.7%	0.03	1.6%	36.1%	62.4%	0.03	0.4%	2.7%	96.9%
C2		18.6%	37.4%	44.0%		18.3%	40.1%	41.6%		10.9%	22.8%	66.3%
C3		3.7%	25.4%	70.9%		3.5%	39.3%	57.2%		1.2%	5.1%	93.8%
C4		0.4%	9.6%	90.0%		0.2%	26.3%	73.5%		0.0%	0.6%	99.4%

(Continued)

TABLE 14.3 (Continued)

Simulation Results for Single Primary Endpoint When the Total Sample Size Is 150

N = 150

Criterion	Dose 1 (0.5) SD	Dose 2 (0.51)	Dose 3 (0.52)	Dose 4 (0.53)	Dose 1 (0.5) SD	Dose 2 (0.51)	Dose 3 (0.525)	Dose 4 (0.53)	Dose 1 (0.5) SD	Dose 2 (0.505)	Dose 3 (0.51)	Dose 4 (0.53)
		Probability of being selected				Probability of being selected				Probability of being selected		
C1	0.05	4.4%	22.3%	73.3%	0.05	2.9%	34.5%	62.6%	0.05	2.2%	6.9%	90.9%
C2		13.6%	31.9%	54.5%		13.1%	38.5%	48.4%		9.7%	17.5%	72.8%
C3		4.8%	23.3%	71.9%		3.4%	34.9%	61.7%		2.5%	7.6%	89.9%
C4		3.8%	21.2%	75.0%		2.3%	33.9%	63.7%		2.1%	6.6%	91.3%
C1	0.1	15.0%	29.1%	55.9%	0.1	12.3%	37.8%	49.9%	0.1	11.9%	18.8%	69.3%
C2		17.5%	30.7%	51.9%		15.3%	37.7%	47.0%		14.6%	21.3%	64.1%
C3		15.0%	29.1%	55.9%		12.4%	37.8%	49.8%		12.0%	18.9%	69.0%
C4		14.7%	28.9%	56.4%		12.2%	37.3%	50.5%		11.5%	18.8%	69.7%
C1	0.2	23.0%	31.0%	45.9%	0.2	20.4%	37.9%	41.8%	0.2	22.4%	26.1%	51.5%
C2		24.0%	31.4%	44.6%		21.8%	37.2%	41.0%		23.5%	26.9%	49.6%
C3		23.0%	31.1%	45.9%		20.4%	37.9%	41.7%		22.5%	26.1%	51.4%
C4		23.0%	31.1%	45.9%		20.4%	37.8%	41.8%		22.4%	26.0%	51.6%

Note: The darker value, the higher the probability.

C1: conditional power.

C2: precision analysis based on confidence interval.

C3: Predictive probability of success.

C4: Probability of being the best dose.

TABLE 14.4

Simulation Results for Single Primary Endpoint When the Total Sample Size Is 200

	Dose 1	Dose 2	Dose 3	Dose 4	Dose 1	Dose 2	Dose 3	Dose 4	Dose 1	Dose 2	Dose 3	Dose 4
Mean	0.5	0.51	0.52	0.53	0.5	0.51	0.525	0.53	0.5	0.505	0.51	0.53
Criterion	SD	Probability of being selected			SD	Probability of being selected			SD	Probability of being selected		
C1	0.005	33.3%	33.3%	33.3%	0.005	33.3%	33.3%	33.3%	0.005	30.8%	34.6%	34.6%
C2		33.3%	33.3%	33.3%		33.3%	33.3%	33.3%		33.3%	33.3%	33.3%
C3		33.3%	33.3%	33.3%		33.3%	33.3%	33.3%		32.3%	33.9%	33.9%
C4		0.0%	0.0%	100.0%		0.0%	0.0%	100.0%		0.0%	0.0%	100.0%
C1	0.01	30.8%	34.6%	34.6%	0.01	30.7%	34.7%	34.7%	0.01	7.4%	42.5%	50.1%
C2		33.3%	33.3%	33.3%		33.3%	33.3%	33.3%		28.2%	35.9%	35.9%
C3		32.3%	33.8%	33.8%		32.2%	33.9%	33.9%		11.6%	42.5%	45.9%
C4		0.0%	0.0%	100.0%		0.0%	1.5%	98.5%		0.0%	0.0%	100.0%
C1	0.02	7.5%	42.3%	50.3%	0.02	7.6%	45.9%	46.5%	0.02	1.4%	10.1%	88.5%
C2		28.4%	35.8%	35.8%		28.5%	35.8%	35.8%		16.6%	35.0%	48.4%
C3		12.0%	42.2%	45.8%		11.9%	44.0%	44.1%		2.9%	16.3%	80.8%
C4		0.0%	1.2%	98.8%		0.0%	12.6%	87.4%		0.0%	0.0%	100.0%
C1	0.03	2.5%	25.1%	72.5%	0.03	2.1%	39.9%	58.0%	0.03	0.7%	3.4%	96.0%
C2		21.3%	37.7%	41.1%		20.7%	39.4%	39.9%		12.3%	26.5%	61.2%
C3		4.7%	31.2%	64.1%		4.2%	42.2%	53.6%		1.5%	6.6%	91.9%
C4		0.1%	6.7%	93.2%		0.0%	23.0%	76.9%		0.0%	0.2%	99.7%

N = 200

(Continued)

TABLE 14.4 (Continued)

Simulation Results for Single Primary Endpoint When the Total Sample Size Is 200

									N = 200						
Mean	Dose 1	Dose 2	Dose 3	Dose 4		Dose 1	Dose 2	Dose 3	Dose 4		Dose 1	Dose 2	Dose 3	Dose 4	
	0.5	0.51	0.52	0.53		0.5	0.51	0.525	0.53		0.5	0.505	0.51	0.53	
Criterion	SD	\	Probability of being selected			SD	Probability of being selected				SD	Probability of being selected			
	0.05					0.05					0.05				
C1		2.5%	20.2%	77.3%	0.05		1.8%	33.8%	64.4%	0.05		1.3%	4.2%	94.4%	0.05
C2		14.5%	32.8%	52.7%			13.6%	39.5%	46.9%			9.2%	17.5%	73.3%	
C3		3.1%	21.9%	75.0%			2.7%	35.2%	62.1%			1.7%	5.2%	93.1%	
C4		2.0%	18.1%	79.9%			1.2%	32.1%	66.6%			1.1%	3.5%	95.4%	
C1		12.2%	27.9%	59.9%	0.1		9.8%	38.5%	51.8%	0.1		10.0%	15.4%	74.6%	0.1
C2		16.5%	30.3%	53.2%			14.1%	38.5%	47.4%			13.8%	19.2%	67.0%	
C3		12.4%	27.8%	59.8%			9.9%	38.5%	51.6%			10.1%	15.5%	74.4%	
C4		12.1%	27.6%	60.2%			9.7%	38.5%	51.9%			9.9%	15.2%	74.9%	
C1		21.3%	32.0%	46.8%	0.2		19.0%	36.2%	44.9%	0.2		20.4%	24.5%	55.1%	0.2
C2		22.6%	31.8%	45.6%			20.0%	36.3%	43.7%			21.5%	25.2%	53.2%	
C3		21.3%	32.0%	46.7%			18.9%	36.2%	44.8%			20.4%	24.5%	55.1%	
C4		21.4%	31.8%	46.9%			18.7%	36.7%	44.6%			20.0%	24.7%	55.3%	

Note: The darker value, the higher the probability.

C1: conditional power.

C2: precision analysis based on confidence interval.

C3: Predictive probability of success.

C4: Probability of being the best dose.

dose 4 is most effective, while the second endpoint of dose 3 is most effective. We consider the following total sample sizes: 100 and 200. The proportion of interim sample size was set to be 20%, resulting in interim sample sizes of 20 and 40, respectively. For each case, 5,000 repetitions were implemented. We summarized the simulation results as the frequency of being selected as the most effective dose in Tables 14.5 and 14.6, according to difference sample sizes.

TABLE 14.5

Simulation Results for Co-primary Endpoints: Scenario 1

				Scenario 1						
	Dose 1	Dose 2	Dose 3	Dose 4		Dose 1	Dose 2	Dose 3	Dose 4	
Mean1	0.5	0.51	0.52	0.53	**Mean2**	0.5	0.51	0.52	0.53	
		N = 100						*N* = 200		
SD1	**SD2**	Probability of being selected			**SD1**	**SD2**	Probability of being selected			
0.03	0.03	0.5%	12.9%	86.6%	0.03	0.03	0.2%	12.1%	87.8%	
		5.2%	29.4%	65.4%			9.9%	39.9%	50.2%	
		0.7%	14.8%	84.6%			0.7%	21.4%	78.0%	
		0.2%	6.7%	93.1%			0.0%	1.9%	98.1%	
0.03	0.05	3.0%	20.2%	76.7%	0.03	0.05	0.9%	14.8%	84.3%	
		5.4%	26.2%	68.4%			6.6%	31.5%	61.9%	
		3.1%	20.5%	76.4%			1.1%	17.0%	81.9%	
		0.7%	11.2%	88.2%			0.0%	4.7%	95.3%	
0.03	0.1	9.9%	30.1%	60.0%	0.03	0.1	7.4%	28.4%	64.2%	
		10.4%	31.4%	58.2%			9.1%	31.4%	59.5%	
		10.3%	30.3%	59.4%			7.5%	28.6%	63.8%	
		1.6%	15.5%	82.8%			0.2%	9.8%	90.0%	
0.05	0.05	4.4%	22.1%	73.4%	0.05	0.05	1.3%	14.9%	83.8%	
		5.4%	24.1%	70.6%			4.4%	25.5%	70.1%	
		4.5%	22.3%	73.2%			1.4%	15.5%	83.1%	
		2.4%	18.7%	79.0%			0.4%	9.7%	89.9%	
0.05	0.1	9.8%	27.8%	62.5%	0.05	0.1	5.9%	26.3%	67.8%	
		10.0%	28.1%	61.9%			7.2%	27.7%	65.1%	
		10.3%	27.8%	61.9%			6.1%	26.4%	67.5%	
		5.5%	23.3%	71.2%			1.7%	18.1%	80.2%	
0.1	0.1	14.4%	29.2%	56.4%	0.1	0.1	8.2%	27.3%	64.5%	
		14.3%	29.4%	56.3%			8.4%	27.7%	63.9%	
		14.5%	29.6%	55.9%			8.4%	27.6%	64.0%	
		12.0%	29.2%	58.9%			6.3%	24.2%	69.6%	

Note: The darker value, the higher the probability.
Mean1: mean of 1st endpoint; Mean2: mean of 2nd endpoint.
SD1: standard deviation of 1st endpoint; SD2: standard deviation of 2nd endpoint.

From the simulation results, we observed that the last method probability of being the best dose performed better in all cases than other methods, with higher probability of choosing the right best dose and lower probability of choosing the wrong dose.

TABLE 14.6

Simulation Results for Co-Primary Endpoints: Scenario 2

					Scenario 2				
	Dose 1	Dose 2	Dose 3	Dose 4		Dose 1	Dose 2	Dose 3	dose 4
Mean1	0.5	0.51	0.52	0.53	Mean2	0.5	0.51	0.53	0.52
		N = 100					*N* = 200		
SD1	SD2	Probability of being selected			SD1	SD2	Probability of being selected		
0.03	0.03	1.1%	49.8%	49.2%	0.03	0.03	0.2%	50.1%	49.6%
		5.6%	47.4%	47.0%			9.9%	45.1%	45.0%
		1.2%	49.5%	49.3%			0.8%	49.9%	49.4%
		0.3%	50.0%	49.7%			0.0%	50.7%	49.3%
0.03	0.05	3.0%	56.8%	40.2%	0.03	0.05	1.4%	67.6%	31.1%
		5.2%	55.0%	39.9%			7.1%	57.8%	35.1%
		3.0%	57.0%	40.0%			1.6%	67.4%	31.0%
		0.8%	39.3%	59.8%			0.1%	34.7%	65.1%
0.03	0.1	11.1%	50.8%	38.2%	0.03	0.1	7.6%	60.0%	32.4%
		11.6%	50.5%	37.9%			9.9%	57.5%	32.7%
		11.5%	50.7%	37.8%			7.8%	60.0%	32.2%
		1.9%	30.3%	67.7%			0.4%	24.5%	75.1%
0.05	0.05	5.3%	46.6%	48.1%	0.05	0.05	1.8%	49.5%	48.8%
		6.4%	46.0%	47.6%			4.6%	48.2%	47.2%
		5.5%	46.6%	47.9%			1.9%	49.5%	48.6%
		2.9%	48.1%	49.0%			0.5%	49.8%	49.8%
0.05	0.1	10.3%	45.0%	44.7%	0.05	0.1	6.7%	51.2%	42.0%
		10.4%	44.8%	44.8%			7.8%	50.5%	41.8%
		10.6%	45.1%	44.3%			7.1%	51.2%	41.7%
		6.2%	38.7%	55.1%			2.3%	37.4%	60.4%
0.1	0.1	14.4%	42.5%	43.2%	0.1	0.1	9.9%	45.2%	44.9%
		14.2%	42.8%	43.1%			10.0%	45.0%	45.1%
		14.7%	42.3%	43.0%			10.3%	45.3%	44.4%
		11.8%	44.0%	44.2%			7.4%	45.3%	47.3%

Note: The darker value, the higher the probability.

Mean1: mean of 1st endpoint; Mean2: mean of 2nd endpoint.

SD1: standard deviation of 1st endpoint; SD2: standard deviation of 2nd endpoint.

14.5 Concluding Remarks

In this chapter, we discussed the problem of dose selection in two-stage adaptive clinical trials with the aim of selecting the most effective dose for further investigation in the next stage. Four criteria for dose selection were introduced and compared through extensive simulation studies. From the simulation results, the method of probability of being the best dose performed better than all other three methods. Thus, we recommend this method in practical application as appropriate. Besides, a numerical example concerning dose-response of compound in treating postmenopausal women and adjuvant breast cancer patients with vasomotor symptoms was given to illustrate the use of this method.

15

Generics and Biosimilars

15.1 Introduction

In the United States, for traditional chemical (small molecule) drug products, when an innovative (brand-name) drug product is going off patent, pharmaceutical and/or generic companies may file an abbreviated new drug application (ANDA) for approval of generic copies of the brand-name drug product. In 1984, the FDA was authorized to approve generic drug products under the Drug Price Competition and Patent Term Restoration Act, which is also known as the Hatch-Waxman Act. For approval of small molecule generic drug products, the FDA requires that evidence of *average* bioavailability, in terms of the rate and extent of drug absorption, be provided. The assessment of bioequivalence as a surrogate endpoint for quantitative evaluation of drug safety and efficacy is based on the *Fundamental Bioequivalence Assumption* that if two drug products are shown to be bioequivalent in average bioavailability, it is assumed that they will reach the same therapeutic effect or they are therapeutically equivalent and hence can be used interchangeably. Under the Fundamental Bioequivalence Assumption, regulatory requirements, study design, criteria, and statistical methods for assessment of bioequivalence have been well established (see, e.g., Schuirmann, 1987; EMEA, 2001; FDA 2001, 2003a, 2003b; WHO, 2005; Chow and Liu, 2008).

Unlike small molecule drug products, the *generic versions* of biologic products are viewed as similar biological drug products (SBDP). The SBDP are *not* generic drug products, which are drug products with *identical* active ingredient(s) as the innovative drug product. Thus, the concept for development of SBDP, which are made of living cells, is very different from that of the generic drug products for small molecule drug products. The SBDP are usually referred to as biosimilars by the European Medicines Agency (EMA) of the European Union, follow-on biologics (FOB) by the FDA, and subsequent entered biologics (SEB) by the Public Health Agency (PHA) of Canada. As a number of biologic products are due to expire in the next few years, the subsequent production of follow-on products has aroused interest within the pharmaceutical/biotechnology industry as biosimilar manufacturers strive to obtain part of an already large and rapidly growing market.

The potential opportunity for price reductions versus the original biologic products remains to be determined, as the advantage of a slightly cheaper price may be outweighed by the hypothetical increased risk of side effects from biosimilar molecules that are not exact copies of their originators.

In this chapter, the focus will not only be placed on the fundamental differences between small molecule drug products and biologic products, but also on issues surrounding quantitative evaluation of bioequivalence (for small molecule drug products) and biosimilarity (for biosimilars or follow-on biologics). In Section 15.2, fundamental differences between small molecule drug products and biologic drug products are briefly described. Sections 15.3 and 15.4 provide brief descriptions of current process for quantitative evaluation of bioequivalence and biosimilarity, respectively. A general approach using biosimilarity index for assessment of bioequivalence and biosimilarity, which was derived based on the concept of reproducibility probability is proposed and discussed in Section 15.5. Section 15.6 summarizes some current scientific factors and practical issues regarding the assessment of biosimilarity. Brief concluding remarks are given in Section 15.7.

15.2 Fundamental Differences

Biosimilars or follow-on biologics are fundamentally different from those of traditional chemical generic drugs. Unlike traditional chemical generic drug products, which contain identical active ingredient(s), the generic versions of biologic products are made of living cells. Unlike classical generics, biosimilars are not identical to their originator products and therefore should not be brought to market using the same procedure applied to generics. This is partly a reflection of the complexities of manufacturing and safety and efficacy controls of biosimilars when compared to their small molecule generic counterparts (see, e.g., Chirino and Mire-Sluis, 2004; Schellekens, 2004; Crommelin et al., 2005; Roger and Mikhail, 2007).

Some of the fundamental differences between biosimilars and generic chemical drugs are summarized in Table 15.1. For example, biosimilars are known to be variable and very sensitive to the environmental conditions such as light and temperature. A small variation may translate to a drastic change in clinical outcomes (e.g., safety and efficacy). In addition to differences in the size and complexity of the active substance, important differences also include the nature of the manufacturing process. Since biologic products are often recombinant protein molecules manufactured in living cells, manufacturing processes for biologic products are highly complex and require hundreds of specific isolation and purification steps. Thus, in practice, it is impossible to produce an *identical* copy of a biologic product, as changes to the structure of the molecule can occur with changes in the

TABLE 15.1

Fundamental Differences

Chemical Drugs	Biologic Drugs
Made by chemical synthesis	Made by living cells
Defined structure	Heterogeneous structure
	Mixtures of related molecules
Easy to characterize	Difficult to characterize
Relatively stable	Variable
	Sensitive to environmental conditions such as light and temperature
No issue of immunogenicity	Issue of immunogenicity
Usually taken orally	Usually injected
Often prescribed by a general practitioner	Usually prescribed by specialists

manufacturing process. Since a protein can be modified during the process (e.g., a side chain may be added, the structure may have changed due to protein misfolding, and so on), different manufacturing processes may lead to structural differences in the final product, which result in differences in efficacy and safety, and may have a negative impact on the immune responses of patients. It should be noted that these issues occur also during the post-approval changes of the innovator's biological products.

Thus, SBDP are not generic products. Hence, the standard generic approach is not applicable and acceptable due to the complexity of biological/biotechnology derived products. Instead, similar biological approach depending upon the state-of-art of analytical procedures should be applied.

15.3 Quantitative Evaluation of Generic Drugs

For approval of small molecule generic drug products, the FDA requires that evidence of average bioequivalence in drug absorption in terms of some pharmacokinetic (PK) parameters such as the area under the blood and/or plasma concentration-time curve (AUC) and peak concentration (Cmax) be provided through the conduct of bioequivalence studies. In practice, we may claim that a test drug product is bioequivalent to an innovative (reference) drug product if the 90% confidence interval for the ratio of geometric means of the primary PK parameter is completely within the bioequivalence limits of (80%, 125%). The confidence interval for the ratio of geometric means of the primary PK parameter is obtained based on log-transformed data. In what follows, study designs and statistical methods that are commonly considered in bioequivalence studies are briefly described.

15.3.1 Study Design

As indicated in the *Federal Register* [Vol. 42, No. 5, Sec. 320.26(b) and Sec. 320.27(b), 1977], a bioavailability study (single-dose or multi-dose) should be crossover in design, unless a parallel or other design is more appropriate for valid scientific reasons. Thus, in practice, a standard two-sequence, two-period (or 2 × 2) crossover design is often considered for a bioavailability or bioequivalence study. Denote by T and R the test product and the reference product, respectively. Thus, a 2 × 2 crossover design can be expressed as (TR, RT), where TR is the first sequence of treatments and RT denotes the second sequence of treatments. Under the (TR, RT) design, qualified subjects who are randomly assigned to sequence 1 (TR) will receive the test product T first and then crossed over to receive the reference product R after a sufficient length of wash-out period. Similarly, subjects who are randomly assigned to sequence 2 (RT) will receive the reference product (R) first and then receive the test product (T) after a sufficient length of wash-out period.

One of the limitations of the standard 2 × 2 crossover design is that it does not provide independent estimates of intra-subject variabilities since each subject will receive the same treatment only once. In the interest of assessing intra-subject variabilities, the following alternative higher-order crossover designs for comparing two drug products are often considered: (i) Balaam's design, i.e., (TT, RR, RT, TR); (ii) two-sequence, three-period dual design, e.g., (TRR, RTT); and (iii) four-sequence, four-period design, e.g., (TTRR, RRTT, TRTR, RTTR).

For comparing more than two drug products, a Williams' design is often considered. For example, for comparing three drug products, a six-sequence, three-period (6 × 3) Williams' design is usually considered, while a 4 × 4 Williams' design is employed for comparing four drug products. Williams' design is a variance stabilizing design. More information regarding the construction and good design characteristics of Williams' designs can be found in Chow and Liu (2008).

In addition to the assessment of average bioequivalence (ABE), there are other types of bioequivalence assessment such as population bioequivalence (PBE) that is intended for addressing drug prescribability and individual bioequivalence (IBE) that is intended for addressing drug switchability. For assessment IBE/PBE, the FDA recommends that a *replicated* design be considered for obtaining independent estimates of intra-subject and inter-subject variabilities and variability due to subject-by-drug product interaction. A commonly considered replicate crossover design is the replicate of a 2 × 2 crossover design is given by (TRTR, RTRT). In some cases, an incomplete block design or an extra-reference design such as (TRR, RTR) may be considered depending upon the study objectives of the bioavailability/bioequivalence studies (Chow et al., 2002b).

15.3.2 Statistical Methods

As indicated in Chapter 3, the FDA recommends using a two one-sided tests (TOST) procedure for testing interval hypotheses for average bioequivalence (ABE) assessment of generic drugs. However, ABE is evaluated based on confidence interval approach. That is, ABE is claimed if the ratio of average bioavailabilities between test and reference products is within the bioequivalence limit of (80%, 125%) with 90% assurance based on log-transformed data. In many cases, TOST at the 5% level of significance at each side is operationally equivalent to the 90% confidence interval approach although they are very different conceptually.

Along this line, commonly employed statistical methods for assessing bioequivalence for generic drugs are the TOST for interval hypotheses and the confidence interval approach. For the confidence interval approach, a 90% confidence interval for the ratio of means of the primary pharmacokinetic response such as AUC or Cmax is obtained under an analysis of variance model. We claim bioequivalence if the obtained 90% confidence interval is totally within the bioequivalence limit of (80%, 125%) inclusively. For the method of interval hypotheses testing, the interval hypotheses that

$$H_0 : \text{Bioinequivalence} \quad \text{vs.} \quad H_a : \text{Bioequivalence}. \tag{15.1}$$

Note that the above hypotheses are usually decomposed into two sets of one-sided hypotheses. For the first set of hypotheses is to verify that the average bioavailability of the test product is not too low, whereas the second set of hypotheses is to verify that average bioavailability of the test product is not too high. Under the two one-sided hypotheses, Schuirmann's two one-sided tests procedure is commonly employed for testing ABE (Schuirmann, 1987). Note that Schuirmann's two one-sided tests procedure is a size-α test (Chow and Shao, 2002b).

In practice, other statistical methods such as Westlake's symmetric confidence interval approach, confidence interval based on Fieller's theorem, Chow and Shao's joint confidence region approach, Bayesian methods, and non-parametric methods such as Wilcoxon-Mann-Whitney two one-sided tests procedure, distribution-free confidence interval based on the Hodges-Lehmann estimator, and bootstrap confidence interval are sometimes considered (Chow and Liu, 2008).

It, however, should be noted that the concept of interval hypotheses testing and the concept of confidence interval approach for bioequivalence assessment are very different, although they are operationally equivalent under certain conditions. In the case of binary responses, TOST is not operationally equivalent to that of the confidence interval approach. Thus, bioequivalence evaluation using statistical methods other than the official method TOST should be proceed with caution to avoid possible false positive and/or false negative rates.

15.3.3 Other Criteria for Bioequivalence Assessment

Although the assessment of ABE for generic approval has been in practice for years, it has the following limitations: (i) it focuses only on population average; (ii) it ignores the distribution of the metric; (iii) it does not provide independent estimates of intra-subject variabilities and ignores the subject-by-formulation interaction. In addition, the use of one-fits-all criterion for assessment of ABE has been criticized in the past decade. It is suggested that the one-fits-all criterion be flexible by adjusting intra-subject variability of the reference product and therapeutic window whenever possible. Many authors criticize that (i) the assessment of ABE does not address the question of drug interchangeability, and (ii) it may penalize drug products with lower variability.

15.3.3.1 Population Bioequivalence and Individual Bioequivalence (PBE/IBE)

To address the issue of drug interchangeability, between early 1990s and early 2000s, as more and more generic drug products become available, it became a point of concern as to whether generic drugs are safe and efficacious when they are used interchangeably. To address this issue, an aggregated criterion which take into consideration of both inter- and intra-subject variabilities of the test product and the reference product and the variability due to subject-by-drug interaction. The criteria were proposed to address interchangeability (in terms of prescribability and switchability) through the assessment of population bioequivalence (PBE) for prescribability and individual bioequivalence (IBE) for switchability. The proposed PBE/IBE criteria have some undesirable properties (see, e.g., Chow, 1999) due to possible masking and cancellation effects in the complicated aggregated criteria. As a result, as indicated in the 2003 FDA guidance, PBE/IBE are not required for BE assessment for generic approval.

15.3.3.2 Scaled Average Bioequivalence (SABE)

To address the issue that current ABE may penalize drug products with lower variability, Haidar et al. (2008) proposed a scaled average bioequivalence (SABE) criterion for assessment of bioequivalence for highly variable drug products. The SABE criterion is useful not only for assessment of highly variable drug products, but also for drug products with narrow therapeutic index (NTI). However, it should be noted that the SABE criterion is, in fact, an ABE criterion adjusted for the standard deviation of the reference product. Thus, it is a special case of the following criteria for IBE:

$$\frac{(\mu_T - \mu_R)^2 + \sigma_D^2 + (\sigma_{WT}^2 - \sigma_{WR}^2)}{\max(\sigma_{WR}^2, \sigma_{W0}^2)} \le \theta_I, \tag{15.2}$$

where σ_{WT}^2 and σ_{WR}^2 are the within-subject variances of the test drug product and the reference drug product, respectively, σ_D^2 is the variance component due to subject-by-drug interaction, σ_{W0}^2 is a constant that can be adjusted to control the probability of passing IBE, and θ_I is the bioequivalence limit for IBE.

15.3.3.3 Scaled Criterion for Drug Interchangeability (SCDI)

Chow et al. (2015) proposed a criterion based on the first two components of the criterion for IBE in (15.2), which consists of (i) criterion for ABE adjusted for intra-subject variability of the reference product (i.e., SABE), and (ii) correction for variability due to subject-by-product variability (i.e. σ_D^2). The proposed criterion for assessing interchangeability is briefly derived below.

Step 1: Unscaled ABE criterion

Let BEL be the BE limit which equals 1.25. Thus, bioequivalence requires that

$$\frac{1}{\text{BEL}} \leq \text{GMR} \leq \text{BEL}$$

This implies

$$-\log(\text{BEL}) \leq \log(\text{GMR}) \leq \log(\text{BEL})$$

or

$$-\log(\text{BEL}) \leq \mu_T - \mu_R \leq \log(\text{BEL}),$$

where μ_T and μ_R are logarithmic means.

Step 2: Scaled ABE (SABE) criterion

Difference in logarithmic means is adjust for intra-subject variability as follows:

$$-\log(\text{BELS}) \leq \frac{\mu_T - \mu_R}{\sigma_W} \leq \log(\text{BELS}),$$

or

$$-\log(\text{BELS})\sigma_W \leq \mu_T - \mu_R \leq \log(\text{BELS})\sigma_W$$

where σ_W^2 is a within-subject variation. In practice, σ_{WR}^2, within-subject variation of the reference product is often considered.

Step 3: Development of SCDI

Consider the first two components of the IBE in (15.2), we have the following relationship

$$\frac{(\mu_T - \mu_R)^2 + \sigma_D^2}{\sigma_W^2} = \frac{(\delta + \sigma_D)^2 - 2\delta\sigma_D}{\sigma_W^2},$$

where $\delta = \mu_T - \mu_R$. When $\delta = 0$ and σ_D approaches to 0, we have

$$\frac{(\delta + \sigma_D)^2}{\sigma_W^2} \approx \frac{2\delta\sigma_D}{\sigma_W^2}.$$

Thus, Chow et al. (2015) proposed that SCDI can be summarized as follows:

$$-\log(\text{BELS}) \leq \left(\frac{\mu_T - \mu_R}{\sigma_W}\right)\left(\frac{2\sigma_D}{\sigma_W}\right) \leq \log(\text{BELS})$$

Now, let $f = \sigma_W / (2\sigma_D)$, correction factor for drug interchangeability. Then, Chow et al. (2015) proposed SCDI criterion is given by

$$-\log(\text{BELS}) f \sigma_W \leq \mu_T - \mu_R \leq \log(\text{BELS}) f \sigma_W$$

Note that statistical properties and finite sample performance need further research.

15.3.3.4 Remarks

Following the concept of criterion for IBE and the idea of SABE, the Chow et al.'s (2015) proposed SCDI criterion is developed based on the one-size-fits-all criterion adjusted for both intra-subject variability of the reference product and the variability due to subject-by-product interaction. As compared to SABE, SCDI may result in a wider or narrower limit depending upon the correction factor f, which is a measurement of the relative magnitude between σ_{WR} and σ_D.

Chow et al.'s (2015) proposed SCDI criterion for drug interchangeability depends upon the selection of regulatory constants of σ_{WR} and σ_D. In practice, the observed variabilities may be deviated far off the regulatory constants. Thus, it is suggested that the following hypotheses be tested before the use of SCDI criterion:

$$H_{01}: \sigma_{WR} \leq \sigma_{W0} \quad \text{vs.} \quad H_{a1}: \sigma_{WR} \leq \sigma_{W0},$$

and

$$H_{02}: \sigma_D \leq \sigma_{D0} \quad \text{vs.} \quad H_{a2}: \sigma_D \leq \sigma_{D0}.$$

If we fail to reject the null hypotheses H_{01} or H_{02}, then we will stick with the individual regulatory suggested constants; otherwise, estimates of σ_{WR} and/or σ_D should be used in SCDI criterion. It, however, should be noted that statistical properties and/or the finite sample performance of SCDI with estimates of σ_{WR} and/or σ_D are not well established. Further research is needed.

15.4 Quantitative Evaluation of Biosimilars

As indicated earlier, the assessment of bioequivalence is possible under the Fundamental Bioequivalence Assumption. Due to the fundamental differences between the small molecule drug products and biological products, the Fundamental Bioequivalence Assumption and the well-established standard methods may not be appropriately applied directly for assessment of biosimilarity.

15.4.1 Regulatory Requirement

On March 23, 2010, the Biologics Price Competition and Innovation (BPCI) Act was written into law. This action gave the FDA the authority to approve similar biological drug products. Following the passage of the BPCI Act, in order to obtain input on specific issues and challenges associated with the implementation of the BPCI Act, the FDA conducted a two-day public hearing November 2–3, 2010, in Silver Spring, Maryland, on the approval pathway for biosimilar and interchangeability, biological products. Several scientific factors were suggested and discussed at this public hearing, include criteria for assessing biosimilarity, study design and analysis methods for assessment of biosimilarity, and tests for comparability in quality attributes of manufacturing process and/or immunogenicity (see, e.g., Chow et al., 2010). These issues primarily focus on the assessment of biosimilarity. The issue of interchangeability in terms of the concepts of alternating and switching were also mentioned and discussed. These discussions have led to the development of regulatory guidance. On February 9, 2012, the FDA circulated three draft guidances on the demonstration of biosimilarity for comments. These three draft guidance were (i) *Scientific Considerations in Demonstrating Biosimilarity to a Reference Product*, (ii) *Quality Considerations in Demonstrating Biosimilarity to a Reference Protein Product*, and (iii) *Biosimilars: Questions and Answers Regarding Implementation of the Biologics Price Competition and Innovation (BPCI) Act of 2009* (FDA, 2012a, 2012b, 2012c). Another FDA public hearing devoted to the discussion of these draft guidance was held at the FDA on May 11, 2012. These three guidances were finalized in 2015.

As indicated in the guidance on scientific considerations, the FDA recommends a stepwise approach for obtaining the totality-of-the-evidence for demonstrating biosimilarity between a proposed biosimilar (test) product and an innovative biological (reference) product. The stepwise approach starts with analytical studies for structural and functional characterization of critical quality attributes and followed by the assessment of pharmacokinetics/pharmacodynamics (PK/PD) similarity and the demonstration of clinical similarity including immunogenicity and safety/efficacy evaluation.

Based on the BPCI Act (part of the Affordable Care Act), quantitative evaluation of biosimilarity includes the concepts of biosimilarity and drug interchangeability, which will be briefly described below.

15.4.2 Biosimilarity

In the BPCI Act, a biosimilar product is defined as a product that is *highly similar* to the reference product notwithstanding minor differences in clinically inactive components if there are no clinically meaningful differences in terms of safety, purity, and potency. Based on this definition, a biological medicine is considered biosimilar to a reference biological medicine if it is highly similar to the reference in safety, purity (quality) and efficacy. However, little or no discussion regarding *"How similar is considered highly similar?"* is given in the BPCI Act.

15.4.2.1 Basic Principles

The BPCI Act seems to suggest that a biosimilar product should be highly similar to the reference drug product in all spectrums of good drug characteristics such as identity, strength, quality, purity, safety, and stability. In practice, however, it is almost impossible to demonstrate that a biosimilar product is highly similar to the reference product in all aspects of good drug characteristics in a single study. Thus, to ensure a biosimilar product is highly similar to the reference product in terms of these good drug characteristics, different biosimilar studies may be required. For example, if safety and efficacy is a concern, then a clinical trial must be conducted to demonstrate that there are no clinically meaningful differences in terms of safety and efficacy. On the other hand, to ensure highly similar in quality, assay development/validation, process control/validation, and product specification of the reference product are necessarily established. In addition, tests for comparability in manufacturing process between biosimilars and the reference must be performed. In some cases, if a surrogate endpoint such as pharmacokinetic (PK), pharmacodynamics (PD), or genomic marker is predictive of the primary efficacy/safety clinical endpoint, then a PK/PD or genomic study may be used to assess biosimilarity between biosimilars and the reference product.

It should be noted that current regulatory requirements are guided based on a case-by-case basis following basic principles that requirements reflect: (i) the extent of the physicochemical and biological characterization of the product; (ii) the nature or possible changes in the quality and structure of the biological product due the changes in the manufacturing process (and their unexpected outcomes); (iii) clinical/regulatory experiences with the particular class of the product in question; and (iv) several factors that need to be considered for biocomparability.

15.4.2.2 Criteria for Biosimilarity

For the comparison between drug products, some criteria for the assessment of bioequivalence, similarity (e.g., the comparison of dissolution profiles), and consistency (e.g., comparisons between manufacturing processes) are available in either regulatory guidelines/guidances or the literature. These criteria, however, can be classified into either (i) absolute change versus relative change, (ii) aggregated versus disaggregated, or (iii) moment-based versus probability-based.

In practice, we may consider assessing bioequivalence or biosimilarity by comparing average and variability separately or simultaneously. This leads to the so-called disaggregated criterion and aggregated criterion. A disaggregate criterion will provide different levels of biosimilarity. For example, the study that passes criteria of both average and variability of biosimilarity provides stronger evidence of biosimilarity as compared to those studies that pass only the average biosimilarity. On the other hand, it is not clear whether an aggregated criterion would provide a stronger evidence of biosimilarity due to potential offset (or masked) effect between the average and variability in the aggregated criterion. Further research for establishing the appropriate statistical testing procedures based on the aggregate criterion and comparing its performance with the disaggregate criterion may be needed.

Chow et al. (2010) compared the moment-based criterion with the probability-based criterion for assessment of bioequivalence or biosimilarity under a parallel group design. The results indicate that the probability-based criterion is not only a much more stringent criterion, but also has sensitivity to any small change in variability. This justifies the use of the probability-based criterion for assessment of biosimilarity between follow-on biologics if a certain level of precision and reliability of biosimilarity is desired.

15.4.2.3 Study Design

As indicated earlier, a crossover design is often employed for bioequivalence assessment. In a crossover study, each drug product is administered to each subject. Thus, estimated (approximate) within-subject variance can be made to serve to address switchability and interchangeability. For a parallel-group study, each drug product is administered to a different group of subjects. Thus, we can only estimate total variance (between and within subject variances) not individual variance components. For follow-on biologics with long half-lives, crossover study would be ineffective and unethical. In this case, we need to undertake study with parallel groups. However, a parallel-group study does not provide an estimate for within-subject variation (since there is no R vs. R).

15.4.2.4 Statistical Methods

Similar to the assessment of average bioequivalence, FDA recommends that Schuirmann's two one-sided tests (TOST) procedure be used for assessment of biosimilarity, although this method has been mix-used with the confidence interval approach. On the other hand, if similar criteria for assessment of population/individual bioequivalence are considered, the 95% confidence upper bound can be used for assessing biosimilarity based on linearized criteria of population/individual bioequivalence.

Note that the clarification of the confusion between when to use a 90% CI and when to use a 95% CI can be found in Chapter 3.

15.4.3 Interchangeability

As indicated in the Subsection (b)(3) amended to the Public Health Act Subsection 351(k)(3), the term *interchangeable* or *interchangeability* in reference to a biological product that is shown to meet the standards described in subsection (k)(4), means that the biological product may be substituted for the reference product without the intervention of the health care provider who prescribed the reference product. Along this line, in what follows, definitions and basic concepts of interchangeability (in terms of switching and alternating) are given.

15.4.3.1 Definition and Basic Concepts

As indicated in the Subsection (a)(2) amends the Public Health Act Subsection 351(k)(3), a biological product is considered to be interchangeable with the reference product if (i) the biological product is biosimilar to the reference product; and (ii) it can be expected to produce the same clinical result in *any given patient*. In addition, for a biological product that is administered more than once to an individual, the risk in terms of safety or diminished efficacy of alternating or switching between use of the biological product and the reference product is not greater than the risk of using the reference product without such alternation or switch.

Thus, there is a clear distinction between biosimilarity and interchangeability. In other words, biosimilarity does not imply interchangeability which is much more stringent. Intuitively, if a test product is judged to be interchangeable with the reference product then it may be substituted, even alternated, without a possible intervention, or even notification, of the health care provider. However, interchangeability implies that a test product is expected to produce the *same* clinical result in *any given patient*, which can be interpreted as that the same clinical result can be expected in *every single patient*. Conceivably, lawsuits could be filed if adverse effects are recorded in a patient after switching from one product to another, interchangeable product.

It should be noted that when the FDA declares the biosimilarity of two drug products, it may not be assumed that they are interchangeable. Therefore, labels ought to state whether for a follow-on biologic which is biosimilar to a reference product, interchangeability has or has not been established. However, payers and physicians may, in some cases, switch products even if interchangeability has not been established.

15.4.3.2 Switching and Alternating

Unlike drug interchangeability (in terms of prescribability and switchability (Chow and Liu, 2008), the FDA has slightly different perception of interchangeability for biosimilars. From the FDA's perspectives, interchangeability includes the concept of switching and alternating between an innovative biologic product (R) and its follow-on biologics (T). The concept of switching is referred to as a single switch including not only the switch from "R to T" or "T to R" (narrow sense of switchability), but also "T to T" and "R to R" (broader sense of switchability). As a result, in order to assess switching, biosimilarity for "R to T," "T to R," "T to T," and "R to R" need to be assessed based on some biosimilarity criteria under a valid switching design.

On the other hand, the concept of alternating is referred to as multiple switches including either the switch from T to R and then switch back to T (i.e., "T to R to T") or the switch from R to T and then switch back to R (i.e., "R to T to R." Thus, the difference between "the switch from T to R" or "the switch from R to T" and "the switch from R to T" or "the switch from T to R" needs to be assessed for addressing the concept of alternating.

15.4.3.3 Study Design

For assessment of bioequivalence for chemical drug products, a standard two-sequence, two-period (2 × 2) crossover design is often considered, except for drug products with relatively long half-lives. Since most biosimilar products have relatively long half-lives, it is suggested that a parallel group design should be considered. However, parallel group design does not provide independent estimates of variance components such as inter- and intra-subject variabilities and variability due to subject-by-product interaction. Thus, it is a major challenge for assessing biosimilars under parallel group designs.

In order to assess biosimilarity for "R to T," "T to R," "T to T," and "R to R," the Balaam's 4 × 2 crossover design, i.e., (TT, RR, TR, RT) may be useful. For addressing the concept of alternating, a two-sequence, three-period dual design, i.e., (TRT, RTR) may be useful. For addressing both concepts of switching and alternating for drug interchangeability of biosimilars, a modified Balaam's crossover design, i.e., (TT, RR, TRT, RTR) is then recommended.

For switching designs, FDA recommends (RT,RR) (single switch) and (RTR, RRR) and (RTRT, RRRR) (multiple switches) be used. However, Chow and Lee (2009) suggest a complete n-of-1 design be considered because the FDA recommended switching designs are partial designs of complete n-of-1 trial designs.

15.4.4 Remarks

With small molecule drug products, bioequivalence generally reflects therapeutic equivalence. Drug prescribability, switching, and alternating are generally considered reasonable. With biologic products, however, variations are often higher (factors other than pharmacokinetic factors may be sensitive to small changes in conditions). Thus, often only parallel-group design rather than crossover kinetic studies can be utilized. It should be noted that very often, with follow-on biologics, biosimilarity does *not* reflect therapeutic comparability. Therefore, switching and alternating should be pursued only with substantial caution.

15.5 General Approach for Assessment of Bioequivalence/Biosimilarity

As indicated earlier, the concept of biosimilarity and interchangeability for follow-on biologics is very different from that of bioequivalence and drug interchangeability for small molecule drug products. It is debatable whether standard methods for assessment of bioequivalence and drug interchangeability can be applied to assessing biosimilarity and interchangeability of follow-on biologics due to the fundamental differences as described in Section 2. While appropriate criteria or standards for assessment of biosimilarity and interchangeability are still under discussion within the regulatory agencies and among the pharmaceutical industry and academia, we would like to propose the a general approach for assessing biosimilarity and interchangeability by comparing the relative difference between *"a test product vs. a reference product"* and *"the reference vs. the reference"* based on the concept of reproducibility probability of claiming biosimilarity between a test product and a reference product in a future biosimilarity study provided that the biosimilarity between the test product and the reference product has been established in the current study.

15.5.1 Development of Bioequivalence/Biosimilarity Index

Shao and Chow (2002) proposed a reproducibility probability as an index for determining whether it is necessary to require a second trial when the result of the first clinical trial is strongly significant. Suppose that the null hypothesis H_0 is rejected if and only if $|T| > c$, where c is a positive known constant and T is a test statistic. Thus, the reproducibility probability of observing a significant clinical result when H_a is indeed true is given by

$$p = P\left(|T| > c \mid H_a\right) = P\left(|T| > c \mid \hat{\theta}\right), \tag{15.3}$$

where $\hat{\theta}$ is an estimate of θ, which is an unknown parameter or vector of parameters. Following the similar idea, a reproducibility probability can also be used to evaluate biosimilarity and interchangeability between a test product and a reference product based on any pre-specified criteria for biosimilarity and interchangeability. As an example, biosimilarity index proposed by Chow et al. (2011) is illustrated based on the well-established bioequivalence criterion by the following steps:

Step 1. Assess the average bioequivalence/biosimilarity between the test product and the reference product based on a given bioequivalence/biosimilarity criterion. For illustration purpose, consider bioequivalence criterion. That is, bioequivalence/biosimilarity is claimed if the 90% confidence interval of the ratio of means of a given study endpoint falls within the bioequivalence/biosimilarity limit of (80%, 125%) inclusive based on log-transformed data.

Step 2. Once the product passes the test for bioequivalence/biosimilarity in Step 1, calculate the reproducibility probability based on the observed ratio (or observed mean difference) and *variability*. We will refer to the calculated reproducibility probability as the *bioequivalence/biosimilarity index*.

Step 3. We then claim bioequivalence/biosimilarity if the following null hypothesis is rejected:

$$H_0 : P \le p_0 \text{ vs. } H_a : P > p_0. \tag{15.4}$$

A confidence interval approach can be similarly applied. In other words, we claim bioequivalence/biosimilarity if the lower 95% confidence bound of the reproducibility probability is larger than a pre-specified number p_0. In practice, p_0 can be obtained based on an estimated of reproducibility probability for a study comparing a reference product to itself (the reference product). We will refer to such a study as an R-R study.

In an R-R study, define

$$P_{TR} = P \left(\begin{array}{l} \text{concluding average biosimiliarity between the test and the} \\ \text{reference products in a future trial given that the average} \\ \text{biosimiliarity based on ABE criterion has been established} \\ \text{in first trial} \end{array} \right) \tag{15.5}$$

Alternatively, a reproducibility probability for evaluating the bioequivalence/biosimilarity of the two same reference products based on ABE criterion is defined as:

$$P_{RR} = P \left(\begin{array}{l} \text{concluding average biosimiliarity of the two same reference} \\ \text{products in a future trial given that the average biosimilarity} \\ \text{based on ABE criterion have been established in first trial} \end{array} \right) \quad (15.6)$$

Since the idea of the bioequivalence/biosimilarity index is to show that the reproducibility probability in a study for comparing generics/biosimilars with the innovative (reference) product is higher than a reference product with the reference product. The criterion of an acceptable reproducibility probability (p_0) for assessment of bioequivalence/biosimilarity can be obtained based on the R-R study. For example, if the R-R study suggests the reproducibility probability of 90%, i.e., $P_{RR} = 90\%$, the criterion of the reproducibility probability for bioequivalence/biosimilarity study could be chosen as 80% of the 90%, i.e., $p_0 = 80\% \times P_{RR} = 72\%$.

The above described bioequivalence/biosimilarity index has the advantages that (i) it is robust with respect to the selected study endpoint, bioequivalence/biosimilarity criteria, and study design; (ii) it takes variability into consideration (one of the major criticisms in the assessment of average bioequivalence); (iii) it allows the definition and assessment of the degree of similarity (in other words, it provides a partial answer to the question "How similar is considered similar?"); and (iv) the use of the bioequivalence/biosimilarity index will reflect the sensitivity of heterogeneity in variance.

Most importantly, the bioequivalence/biosimilarity index proposed by Chow et al. (2011) can be applied to different functional areas (domains) of biological products such as good drug characteristics; as safety (e.g., immunogenicity); purity, and potency (as described in BPCI Act); pharmacokinetics (PK) and pharmacodynamics (PD); biological activities; biomarkers (e.g., genomic markers); and manufacturing process, etc. for an assessment of *global* biosimilarity. An overall biosimilar index across domains can be obtained by the following steps:

Step 1. Obtain P_i, the probability of reproducibility for the ith domain, $i = 1, ..., K$.

Step 2. Define the global biosimilarity index $P = \sum_{i=1}^{K} w_i P_i$, where w_i is the weight for the ith domain.

Step 3. Claim global biosimilarity if the lower 95% confidence bound of the reproducibility probability (P) is larger than a pre-specified number p_0, where p_0 is a pre-specified acceptable reproducibility probability.

15.5.2 Remarks

Hsieh et al. (2010) studied the performance of the biosimilarity index under an R-R study for establishing a baseline for assessment of biosimilarity based on current criterion for average bioequivalence. The results indicate that biosimilarity index is sensitive to the variability associated with the reference product. The biosimilarity index decreases as the variability increases. As an example, Figure 15.1 gives reproducibility probability curves under a 2×2 crossover design with sample sizes $n_1 = n_2 = 10, 20, 30, 40, 50,$ and 60 at the 0.05 level of significance and $(\theta_L, \theta_U) = (80\%, 125\%)$ when $\sigma_d = 0.2$ and 0.3, where σ_d is the standard deviation of period difference within each subject.

In practice, alternative approaches for assessment of the proposed biosimilarity index are available (see, e.g., Hsieh et al., 2010, Yang et al., 2010). The methods include maximum likelihood approach and Bayesian approach. For the Bayesian approach, let $p(\theta)$ be the power function, where θ is an unknown parameter or vector of parameters. Under this Bayesian approach, θ is random with a prior distribution assumed to be known. The reproducibility probability can be viewed as the posterior mean of the power function for the future trial

$$\int p(\theta)\pi(\theta \mid x)d\theta, \tag{15.7}$$

where $\pi(\theta \mid x)$ is the posterior density of θ, given the data set x observed for the previous trial(s). However, there may exist no explicit form for the estimation of the biosimilarity index. As a result, statistical properties of the

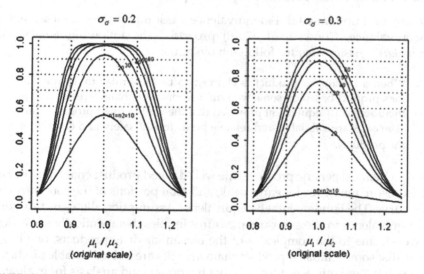

FIGURE 15.1
Impact of variability on reproducibility probability. (The reproducibility decreases when μ_1/μ_2 (original scale) moves away from 1 and σ_d (log scale) is larger.)

derived biosimilarity index may not be known. In this case, the finite sample size performance of the derived biosimilarity index may only be evaluated by clinical trial simulations.

As an alternative measure for assessment of global biosimilarity across domains, we may consider $rd = \sum_{i=1}^{K} w_i rd_i$, where $rd_i = \frac{P_{TRi}}{P_{RRi}}$, which is the relative measure of biosimilarity between T and R as compared to that of between R and R. Based on rd_i, $i = 1, ..., K$, we may conduct a *profile analysis* as described in the 2003 FDA guidance on *Bioavailability and Bioequivalence Studies for Nasal Aerosols and Nasal Sprays for Local Action* (FDA, 2003b). However, statistical properties of the profile analysis based on rd_i, $i = 1, ..., K$ are not fully studied. Thus, further research is required.

15.6 Scientific Factors and Practical Issues for Biosimilars

Following the passage of the BPCI Act, in order to obtain input on specific issues and challenges associated with the implementation of the BPCI Act, the FDA conducted a two-day public hearing on *Approval Pathway for Biosimilar and Interchangeability Biological Products* held on November 2–3, 2010, at the FDA in Silver Spring, Maryland, USA. In what follows, some of the scientific factors and practical issues are briefly described.

15.6.1 Fundamental Biosimilarity Assumption

Similar to Fundamental Bioequivalence Assumption for assessment of bioequivalence, Chow et al. (2010) proposed the following *Fundamental Biosimilarity Assumption* for follow-on biologics:

> When a biosimilar product is claimed to be biosimilar to an innovator's product based on some well-defined product characteristics and is therapeutically equivalent provided that the well-defined product characteristics are validated and reliable predictors of safety and efficacy of the products.

For the chemical generic products, the well-defined product characteristics are the exposure measures for early, peak, and total portions of the concentration-time curve. The Fundamental Bioequivalence Assumption allows us to assume that equivalence in the exposure measures implies therapeutically equivalent. However, due to the complexity of the biosimilar drug products, one has to verify that some validated product characteristics are indeed reliable predictors of the safety and efficacy. It follows that the design and analysis for evaluation of equivalence between the biosimilar drug product and innovator products are substantially different from those of the chemical generic products.

15.6.2 Endpoint Selection

For assessment of biosimilarity of follow-on biologics, the following questions are commonly asked. First, what endpoints should be used for assessment of biosimilarity? Second, should a clinical trial always be conducted?

To address these two questions, we may revisit the definition of biosimilarity as described in the BPCI Act. A biological product that is demonstrated to be *highly similar* to an FDA-licensed biological product may rely on certain existing scientific knowledge about safety, purity (quality), and potency (efficacy) of the reference product. Thus, if one would like to show that the safety and efficacy of a biosimilar product are highly similar to that of the reference product, then a clinical trial may be required. In some cases, clinical trials for assessment of biosimilarity may be waived if there exists substantial evidence that surrogate endpoints or biomarkers are predictive of the clinical outcomes. On the other hand, clinical trials are required for assessment of drug interchangeability in order to show that the safety and efficacy between a biosimilar product and a reference product are similar in any given patient of the patient population under study.

15.6.3 How Similar Is Similar?

Current criteria for assessment of bioequivalence/biosimilarity are useful for determining whether a biosimilar product is similar to a reference product. However, it does not provide additional information regarding the *degree* of similarity. As indicated in the BPCI Act, a biosimilar product is defined as a product that is *highly similar* to the reference product. However, little or no discussion regarding the degree of similarity implied by highly similar was provided. Besides, it is also of concern to the sponsor that "what if a biosimilar product turns out to be superior to the reference product?" A simple answer to the concern is that superiority is not biosimilarity.

15.6.4 Guidance on Analytical Similarity Assessment

On September 27, 2017, FDA circulated a draft guidance on analytical similarity assessment for comments (FDA, 2017b). In the draft guidance, the FDA recommended that equivalence tests be used for analytical similarity assessment between a proposed biosimilar (test) product and an innovative biological (reference) product for critical quality attributes (CQAs) with high risk ranking relevant to clinical outcomes. For CQAs with mild to moderate risk ranking relevant to clinical outcomes, FDA suggested the quality range (QR) approach be considered. The equivalence test for analytical similarity evaluation have been criticized by many authors due to its inflexibility in similarity margin selection of 1.5 σ_R, where σ_R is the standard deviation of the reference product (Chow et al., 2016), while the QR method is considered inadequate because it relies on the primary assumption that the proposed biosimilar product and the reference product have similar mean and standard deviation, which is never true in practice. The draft guidance was subsequently withdrawn in June 2018, although the equivalence test (for CQAs with high risk ranking relevant clinical outcomes) and the QR method (for CQAs with mild to moderate risk ranking relevant clinical outcomes) are still used for analytical similarity assessment in some biosimilar regulatory submissions.

Recently, the FDA circulated a new draft guidance for comparative analytical assessment (FDA, 2019). In the draft guidance, FDA suggested the use of QR method for comparative analytical assessment. As indicated in the draft guidance, the objectives of the QR method is to verify the assumption that the proposed biosimilar product and the reference product have similar mean and similar standard deviation. The guidance also indicated that analytical similarity for a quality attribute would generally be supported when a sufficient percentage of biosimilar lot values (e.g., 90%) fall within the quality range defined for that attribute. This statement, however, may have been misinterpreted by non-statistical reviewers that the concept of comparative analytical assessment is eventually the same as that of analytical similarity evaluation.

As indicated by Chow et al. (2016), the QR method is designed for the purpose of quality control/assurance in the sense that we would expect that there will be about 95% (99%) test results of the test lots will fall within the quality range developed based on 2 (3) standard deviations below and above the mean test results of the reference lots. The QR method is only valid under the assumption that the test product and the reference product have similar population mean and similar population standard deviation (i.e., they are highly similar and can be viewed as coming from the same population).

15.6.5 Practical Issues

Since there are many critical (quality) attributes of a potential patient's response in follow-on biologics, for a given critical attribute, valid statistical methods are necessarily developed under a valid study design and a given set of criteria for similarity, as described in the previous section. Several areas can be identified for developing appropriate statistical methodologies for the assessment of biosimilarity of follow-on biologics. These areas include, but are not limited to:

15.6.5.1 *Criteria for Biosimilarity (in Terms of Average, Variability, or Distribution)*

We suggest establishing disaggregated criteria for biosimilarity in terms of average, variability, and/or distribution address the question "How similar is similar?" We suggest establishing disaggregated criteria for biosimilarity in terms of average, variability, and/or distribution. In other words, we can establish generally similar by demonstrating similarity in average first. Then, we can establish highly similar by demonstrating similarity in variability or distribution.

15.6.5.2 Criteria for Interchangeability

In practice, it is recognized that drug interchangeability is related to the variability due to subject-by-drug interaction. However, it is not clear whether criterion for interchangeability should be based on the variability due to subject-by-drug interaction or the variability due to subject-by-drug interaction adjusted for intra-subject variability of the reference drug.

15.6.5.3 Reference Product Changes

In practice, it is not uncommon to observe a shift in mean response over time for the reference product. This reference product change over time may be due to (i) minor changes in manufacturing process, (ii) use of new and/or advanced technology, and/or (iii) some unknown factors. During the review process of a biosimilar regulatory submission, it is of great concern when there are reference products shifts because products before and after shift may not be similar. In this case, which lots (e.g., lots before shift, lots after shift, or all lots combined) should be used for biosimilarity assessment is a major challenge to the reviewers. On the second thought, should a significant reference product change be considered a major violation (e.g., 483 observation) and appropriate action should be taken for the purpose of quality control/assurance? Since the possible reference product change over time will have an impact on the biosimilarity assessment, *"How to detect potential reference product change?"* has become an important issue in the review and approval process of biosimilar regulatory submissions. FDA is currently working on a specific guidance for reference product change based on similar to SUPAC (scaled-up and post-approval change) guidance for generic drugs.

15.6.5.4 Extrapolation

For a given indication and a CQA, the validity of extrapolation depends upon that there is a well-established relationship (linear or non-linear) between the CQA and PK/clinical outcomes. Without such a well-established relationship, a notable difference in a given CQA (e.g., CQAs in Tier 1, which are considered most relevant to PK/clinical outcomes) may or may not be translated to a clinically meaningful difference in clinical outcomes. In practice, a notable difference in Tier 1 CQAs may vary from one indication to another even they have similar PK profile or mechanism of action (MOA). Thus, the validity of extrapolation across indications without collecting any clinical data is a great concern. In this case, statistical method for evaluation of sensitivity index proposed by Lu et al. (2017) may be helpful.

15.6.5.5 Non-medical Switch

Non-medical switch is referred to as the switch from the reference product (more expensive) to an approved biosimilar product (less expensive) based on factors unrelated to clinical/medical considerations. Typical approaches for assessment of non-medical switch include (i) observational studies and (ii) limited clinical studies. However, there are concerns regarding (i) validity, quality and integrity of the data collected, and (ii) scientific validity of design and analysis of studies conducted for assessment of safety and efficacy of non-medical switch (see also Chow, 2018).

In recent years, several observational studies and a national clinical study (NOR-SWITCH) were conducted to evaluate the risk of non-medical switch from a reference product to an approved biosimilar product (Løvik Goll, 2016). The conclusions from these studies, however, are biased and hence may be somewhat misleading due to some scientific and/or statistical deficiencies in design and analysis of the data collected. Chow (2018) recommended some valid study designs and appropriate statistical methods for a more accurate and reliable assessment of potential risk of medical/non-medical switch between a proposed biosimilar product and a reference product. The results can be easily extended for evaluation of the potential risk of medical/non-medical switch among multiple biosimilar products and a reference product.

15.6.5.6 Bridging Studies for Assessing Biosimilarity

As most biosimilars studies are conducted using a parallel design rather than a replicated crossover design, independent estimates of variance components such as the intra-subject and the variability due to subject-by-drug interaction are not possible. In this case, bridging studies may be considered.

Other practical issues include (i) the use of a percentile method for the assessment of variability; (ii) comparability in biologic activities; (iii) assessment of immunogenicity; (iv) consistency in manufacturing processes (see, e.g., ICH, 1996b, 1999, 2005b); (v) stability testing for multiple lots and/or multiple labs (see, e.g., ICH, 1996c); (vi) the potential use of sequential testing procedures and multiple testing procedures; and (vii) assessing biosimilarity using a surrogate endpoint or biomarker such as genomic data (see, e.g., Chow et al., 2004).

Further research is needed in order to address the above-mentioned scientific factors and practical issues recognized at the FDA Public Hearings, FDA Public Meeting, and FDA biosimilars review process.

15.7 Concluding Remarks

As indicated earlier, we claim that a test drug product is bioequivalent to a reference (innovative) drug product if the 90% confidence interval for the ratio of means of the primary PK parameter is totally within the bioequivalence limits of (80%, 125%). This one size-fits-all criterion only focuses on average bioavailability and ignores heterogeneity of variability. Thus, it is not scientifically/statistically justifiable for assessment of biosimilarity of follow-on biologics. In practice, it is then suggested that appropriate criteria, which can take the heterogeneity of variability into consideration be developed since biosimilars are known to be variable and sensitive to small variations in environmental conditions (Chow and Liu, 2010; Chow et al., 2010; Hsieh et al., 2010).

At the FDA public hearing, questions that are commonly asked are *"How similar is considered similar?"* and *"How the degree of similarity should be measured and translated to clinical outcomes (e.g., safety and efficacy)*[?]" These questions closely related to drug interchangeability of biosimilars or follow-on biologics which have been shown to be biosimilar to the innovative product (Roger, 2006; Roger and Mikhail, 2007).

For assessment of bioequivalence for chemical drug products, a crossover design is often considered, except for drug products with relatively long half-lives. Since most biosimilar products have relatively long half-lives, it is suggested that a parallel group design should be considered. However, parallel group design does not provide independent estimates of variance components such as inter- and intra-subject variabilities and variability due to subject-by-product interaction. Thus, it is a major challenge for assessing biosimilars under parallel group designs.

Although the EMA has published several product-specific guidance based on the concept papers (e.g., EMEA 2003a, 2003b, 2005a, 2005b, 2005c, 2005d, 2005e, 2005f, 2005g), it has been criticized that there are no objective *standards* for assessment of biosimilars because it depends upon the nature of the products. Product-specific standards seem to suggest that a *flexible* biosimilarity criterion should be considered and the flexible criterion should be adjusted for variability and/or the therapeutic index of the innovative (or reference) product.

As described above, there are many uncertainties for assessment of biosimilarity and interchangeability of biosimilars. As a result, it is a major challenge to both clinical scientists and biostatisticians to develop valid and robust clinical/statistical methodologies for assessment of biosimilarity and interchangeability under the uncertainties. In addition, how to address the issues of quality and comparability in manufacturing process is another challenge to both the pharmaceutical scientists and biostatisticians. The proposed general approach using the bioequivalence/biosimilarity index (derived based on the concept of reproducibility probability) may be useful. However, further research on the statistical properties of the proposed bioequivalence/biosimilarity index is required.

16

Precision Medicine

16.1 Introduction

In clinical trials, a typical approach for evaluation of safety and efficacy of a test treatment under investigation is to first test for the null hypothesis of no treatment difference in efficacy based on clinical data collected from adequate and well-controlled clinical studies. If significant, the investigator would reject the null hypothesis of no treatment difference and then conclude the alternative hypothesis that there is a difference in favor of the test treatment. If there is a sufficient power for correctly detecting a clinically meaningful difference (treatment effect) when such a difference truly exists, we claim that the test treatment is efficacious. The test treatment will then be reviewed and approved by the regulatory agency such as the FDA if the test treatment is well tolerated and there appear to be no safety concerns. We will refer to medicine developed based on this typical approach as *traditional medicine*.

In his State of the Union address on January 20, 2015, President Barack Obama announced that he was launching the *Precision Medicine Initiative* – a bold new research effort to revolutionize how we improve health and treat disease. As President Obama indicated, *precision medicine* is an innovative approach that takes into account individual differences in people's genes, environments, and lifestyles. Under the auspices of the traditional approach (traditional medicine), most medical treatments have been designed for the average patient. This *one-size-fits-all* approach, treatments can be very successful for some patients but not for others. Precision medicine, on the other hand, gives medical professionals the resources they need to target the specific treatments of the illnesses we encounter (News Release, 2015). In response to President Obama's Precision Medicine Initiative, the NIH initiated off cohort grants for precision medicine to develop treatments tailored to an individual based on their genetics and other personal characteristics subsequently (McCarthy, 2015). Seeking for precision medicine of cure has become the center of clinical research in pharmaceutical development since then.

The purpose of this chapter is to provide a comprehensive summarization regarding concept, design, and analysis of precision medicine in pharmaceutical research and development. In the next section, the concept of precision medicine will be described. Design and analysis of precision medicine will be reviewed and discussed in Section 16.3. Section 16.4 provides alternative enrichment designs for precision medicine. Some concluding remarks are given in the Section 16.5.

16.2 The Concept of Precision Medicine

16.2.1 Definition of Precision Medicine

Unlike traditional medicine, precision medicine (PM) is a medical model that proposes the customization of healthcare, with medical decisions, practices, and/or products being tailored to the individual patient (NRC, 2011). In this model, diagnostic testing is often employed for selecting appropriate and optimal therapies in the context of a patient's genetics or other molecular or cellular analysis. Tools employed in PM could include molecular diagnostics, imaging, and analytics/software. This has led to biomarker development in genomics studies for target clinical trial. A validated biomarker (diagnostic tool) is then used to identify patients who are most likely to respond to the test treatment under investigation in the enrichment process of the target clinical trials (FDA, 2005; FDA, 2007a, 2007b; Liu et al., 2009). As a result, precision medicine will benefit the subgroup of patients who are biomarker positive. In practice, however, there may exist no perfect diagnostic tool for determining whether a given patient is with or without molecular target in the enrichment process of the target clinical trials. Possible misclassification, which could cause significant bias in assessment of treatment effect in target clinical trials, is probably the most challenging issue in precision medicine.

16.2.2 Biomarker-Driven Clinical Trials

With the surge in advanced technology, especially in the "-omics" space (e.g., genomics, proteomics, etc.), the clinical trial designs that incorporate biomarker information for interim decisions have attracted much attention lately. The biomarker, which usually is a short-term endpoint that is indicative of the behavior of the primary endpoint, has the potential to provide substantial added value to interim study population selection (e.g., biomarker-enrichment designs) and interim treatment selection (e.g., biomarker-informed adaptive designs). As an example, for biomarker-enrichment designs, it is always of particular interest to clinicians to identify patients with disease targets under study, who are most likely to respond to the treatment under study. In practice, an enrichment process is often employed to identify such a target patient population. Clinical trials utilizing an enrichment design are referred to as targeted clinical trials. After the completion of a human genome project, the disease targets at a certain molecular level can be identified and should be utilized for treatment of diseases (Maitournam and Simon, 2005; Casciano and Woodcock, 2006). As a result, diagnostic devices for detection of diseases using biotechnology such microarray; polymerase chain reaction (PCR); mRNA transcript profiling; and others have become possible in practice (FDA, 2005, 2007). The treatments specific for specific molecular targets could then be developed for those patients who were most likely to benefit. Consequently, personalized medicine could become a reality. Clinical development of Herceptin (trastuzumab), which is targeted at patients suffering from metastatic breast cancer with an over-expression of HER2 (human epidermal growth factor receptor) protein, is a typical example (see Table 16.1).

As it can be seen from Table 16.1, Herceptin plus chemotherapy provides statistically significantly additional clinical benefit in terms of overall survival over chemotherapy alone for patients with a staining score of 3+, while Herceptin plus chemotherapy fails to provide additional survival benefit for patients with a FISH (fluorescence in situ hybridization) or CTA (clinical trial assay) score of 2+. Note that CTA is an investigational immunohistochemica (IHC) assay consists of four-point ordinal score system (0, 1+, 2+, 3+). However, as indicated in the Decision Summary of HercepTest (a commercial IHC assay for over-expression of HER2 protein), about 10% of samples have discrepancy results between 2+ and 3+ staining intensity. In other words, some patients tested with a score of 3+ may actually have a score of 2+ and vice versa.

TABLE 16.1

Treatment Effects as a Function of HER2 Over-Expression or Amplification

HER2 Assay Result	Number of Patients	Relative Risk for Mortality (95%)
CTA 2+ or 3+	469	0.80 (0.64, 1.00)
FISH (+)	325	0.70 (0.53, 0.91)
FISH (−)	126	1.06 (0.70, 1.63)
CTA 2+	120	1.26 (0.82, 1.94)
FISH (+)	32	1.31 (0.53, 3.27)
FISH (−)	83	1.11 (0.68, 1.82)
CTA 3+	349	0.70 (0.51, 0.89)
FISH (+)	293	0.67 (0.51, 0.89)
FISH (−)	43	0.88 (0.39, 1.98)

Source: From U.S. FDA Annotated Redlined Draft Package Insert for Herceptin, Rockville, Maryland, 2006.

We will refer to these treatments as targeted treatments or drugs. Development of targeted treatments involves translation from the accuracy and precision of diagnostic devices for the molecular targets to the effectiveness and safety of the treatment modality for the patient population with the targets. Therefore, evaluation of targeted treatments is much more complicated than that of the traditional drugs. To address the issues of development of the targeted drugs, in April 2005, the FDA published *The Drug-Diagnostic Co-development Concept Paper*. In clinical trials, subjects with and without disease targets may respond to the treatment differently with different effect sizes. In other words, patients with disease targets may show a much larger effect size, while patients without disease targets may exhibit a relatively small effect size. In practice, fewer subjects are required for detecting a bigger effect size. Thus, the traditional clinical trials may conclude that the test treatment is ineffective based on the detection of a combined effect size, while the test treatment is in fact effective for those patients with positive disease targets. Consequently, personalized medicine is possible if we can identify those subjects with positive disease targets.

16.2.3 Precision Medicine versus Personalized Medicine

The term precision medicine is often confused with the term personalized medicine. To distinguish the difference between precision medicine and personalized (or individualized) medicine, the National Research Council (NRC) indicates that precision medicine refers to the tailoring of medical treatment to the individual characteristics of each patient. It does not literally mean the creation of drugs or medical devices that are unique to a patient, but rather the ability to classify individuals into subpopulations that differ in their susceptibility to a particular disease, in the biology and/or prognosis of those diseases they may develop, or in their response to a specific treatment. In summary, precision medicine is to benefit subgroup of patients with the diseases under study, while personalized medicine is to benefit individual subjects with the diseases under investigation.

Statistically, the term precision is usually referred to the degree of closeness of the observed data to the truth. High degree of closeness is an indication of high precision. Thus, precision is related to the variability associated with the observed data. In practice, the variability associated with observed data include (i) intra-subject variability, (ii) inter-subject variability, and (iii) variability due to subject-by-treatment interaction. As a result, precision medicine can be viewed as the identification of subgroup population with larger effect size (i.e., smaller variability) assuming that the difference in mean response is fixed. Consequently, precision medicine focuses on minimizing inter-subject variability, while personalized medicine focuses on minimizing intra-subject variability. Table 16.2 provides a comparison between precision medicine and personalized medicine.

TABLE 16.2

Precision Medicine versus Personalized Medicine

Characteristic	Traditional Medicine	Precision Medicine	Personalized Medicine[a]
Active Ingredient	Single	Single	Multiple
Target population	Population	Population	Individuals
Primary focus	Mean	Inter-subject variability	Intra-subject variability
Dose/Regimen	Fixed	Fixed	Flexible
Beneficial	Average patient	Subgroup of patients	Individual patients
Statistical Method	Hypotheses testing Confidence interval	Hypotheses testing Confidence interval	Hypotheses testing Confidence interval
Use of biomarker	No	Yes	Yes
Blinding	Yes	Yes	May be difficult
Objective	Accuracy	Accuracy Precision	Accuracy Precision Reproducibility
Study design	Parallel/crossover	Parallel/crossover Adaptive design	Parallel/Crossover Adaptive design
Probability of success	Low	Mild-to-moderate	High

[a] Personalized medicine = individualized medicine

16.3 Design and Analysis of Precision Medicine

16.3.1 Study Designs

As indicated in the FDA's *Drug-Diagnostic Co-development Concept Paper*, one of the useful designs for evaluation of the targeted treatments is the enrichment design (see also Chow and Liu, 2003). Under the enrichment design, the targeted clinical trials consist of two phases. The first phase is the enrichment phase in which each patient is tested by a diagnostic device for detection of the pre-defined molecular targets. Then, patients with a positive result by the diagnostic device are randomized to receive either the targeted treatment or a concurrent control. However, in practice, no diagnostic test is perfect with 100% positive predicted value (PPV). As a result, some of the patients enrolled in targeted clinical trials under the enrichment design might not have the specific targets and hence the treatment effects of the drug for the molecular targets could be under-estimated due to misclassification (Liu and Chow, 2008). Under the enrichment design, following the idea as described in Liu and Chow (2008), Liu et al. (2009) proposed using the EM algorithm (Dempster et al., 1977; McLachlan and Krishnan, 1997) in conjunction with the bootstrap technique (Efron and

Tibshirani, 1993) for obtaining the inference of the treatment effects. Their method, however, depends upon the accuracy and reliability of the diagnostic device. A poor (i.e., less accurate and reliable) diagnostic device may result in a large proportion of misclassification that has an impact on the assessment of the true treatment effect. To overcome (correct) the problem of an inaccurate diagnostic device, we propose using Bayesian approach in conjunction with the EM algorithm and bootstrap technique for obtaining a more accurate and reliable estimate of treatment effect under various study designs recommended by FDA.

Under an enrichment design, one of the objectives of targeted clinical trials is to evaluate the treatment effects of the molecular targeted test treatment in the patient population with the molecular target. The diagram in the FDA concept paper (FDA, 2005) for demonstration of these designs are reproduced in Figure 16.1 (Design A) and Figure 16.2 (Design B), respectively.

Let Y_{ij} be the responses of the jth subject in the ith group, where $j = 1,\ldots,n_i; i = T, C$. Y_{ij} are assumed approximately normality distributed with homogeneous variances between the test and control treatments. Also, let $\mu_{T+}, \mu_{C+}, (\mu_{T-}, \mu_{C-})$ be the means of test and control groups for the patients with (without) the molecular target. Table 16.3 summarizes population means by treatment and diagnosis.

Under the enrichment design (Design A), Liu et al. (2009) considered a two-group parallel design in which patients with a *positive* result by the diagnostic device are randomized in a 1:1 ratio are randomized to receive the molecular targeted test treatment (T) or a control treatment (C) (see Figure 16.2). In other words, only patients with positive diagnosed results are included in

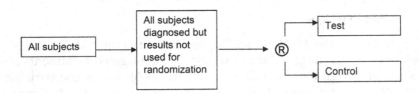

FIGURE 16.1
Design A—Targeted clinical trials under an enrichment design.

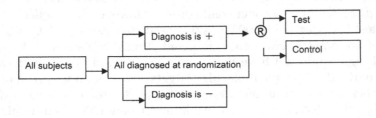

FIGURE 16.2
Design B—Enrichment design for patients with positive results.

TABLE 16.3

Population Means by Treatment and Diagnosis

Positive Diagnosis	True Target Condition	Indicator of Diagnostic	Test Group	Control Group	Difference
+	+	γ_1	μ_{T+}	μ_{C+}	$\mu_{T+} - \mu_{C+}$
	−	$1 - \gamma_1$	μ_{T-}	μ_{C-}	$\mu_{T-} - \mu_{C-}$

Note: γ is the positive predicted value.

the study. For simplicity, Liu et al. (2009) assumed that the primary efficacy endpoint is a continuous variable. The results can be easily extended to other date types such as binary response and time-to-event data.

16.3.2 Statistical Methods

Under Design B (Figure 16.2) for target clinical trials, it is of interest to estimate the treatment effect for the patients truly having the molecular target, i.e., $\theta = \mu_{T+} - \mu_{C+}$. However, this effect may be contaminated due to misclassification, i.e., for those subjects who do not have the molecular target but got positive diagnosed results and those subjects who have the molecular target but got negative diagnosed results. The following hypothesis for detecting a clinically meaningful treatment difference in the patient population truly with the molecular target is of interest:

$$H_0 : \mu_{T+} - \mu_{C+} = 0 \text{ vs. } H_a : \mu_{T+} - \mu_{C+} \neq 0 \tag{16.1}$$

Let $\bar{y}_T$ and $\bar{y}_C$ be the sample means of test and control treatments, respectively. Since no diagnostic test is perfect for diagnosis of the molecular target of interest without error, therefore, some patients with a positive diagnostic result may in fact do not have the molecular target. It follows that

$$E(\bar{y}_T - \bar{y}_C) = \gamma(\mu_{T+} - \mu_{C+}) + (1 - \gamma)(\mu_{T-} - \mu_{C-}), \tag{16.2}$$

where γ is the positive predicted value (PPV), which is often unknown. Thus, an accurate and reliable estimate of γ is the key to the success of target clinical trials (Liu et al., 2009) and hence precision medicine.

Liu and Chow (2008) indicated that the expected value of the difference in sample means consists of two parts. The first part is the treatment effects of the molecular target drug in patients with a positive diagnosis who truly have the molecular target of interest. The second part is the treatment effects of the patients with a positive diagnosis but in fact they do not have the molecular target. The reason for developing the targeted treatment is based on the assumption that the efficacy of the targeted treatment is greater than in the patients truly with the molecular target than those without the target. In addition, the targeted treatment is also expected to be more efficacious than the untargeted control in the patient population truly with the molecular targets. It follows that $\mu_{T+} - \mu_{C+} > \mu_{T-} - \mu_{C-}$. As a result, the difference in sample means obtained under the enrichment design for targeted clinical trials actually under-estimated the true treatment effects of the molecular target test drug in the patient population truly with the molecular target of interest. As it can be seen from (16.2), the bias of the difference in sample means decreases as the positive predicted value increases. On the other hand, the positive predicted value of a diagnostic test increases as the prevalence of the disease increases (Fleiss et al., 2003). For a disease that is highly prevalent, say greater than 10%, even with a high diagnostic accuracy of 95% sensitivity and specificity for the diagnostic device, the positive predicted value is only about 67.86%. It follows that the downward bias of the traditional difference in sample means could be substantial for estimation of treatment effects of the molecular target drug in patients who do have the target of interest.

The traditional unpaired two-sample t-test approach is to reject the null hypothesis in (16.1) at the at the α level of significance level if

$$t = \left| \left(\bar{y}_T - \bar{y}_C \right) / \sqrt{s_p^2 \left(1/n_T + 1/n_C \right)} \right| \geq t_{\alpha/2, n_T + n_C - 2},$$

where s_p^2 is the pooled sample variance, and $t_{\alpha, n_T + n_C - 2}$, is the αth upper percentile of a central t distribution with $n_T + n_C - 2$ degrees of freedom. Since $\bar{y}_T - \bar{y}_C$ under-estimates $\mu_{T+} - \mu_{C+}$, the planned sample size may not be sufficient for achieving the desired power for detecting the true treatment effects in the patients truly with molecular target of interest. Based on the above t-statistic, the corresponding $(1-\alpha) \times 100\%$ confidence interval can be obtained as follows

$$\left(\bar{y}_T - \bar{y}_C \right) \pm t_{\alpha/2, n_T + n_C - 2} \sqrt{s_p^2 \left(\frac{1}{n_T} + \frac{1}{n_C} \right)}.$$

Although all patients randomized under the enrichment design have a positive diagnosis, the true status of the molecular target for individual patients in the targeted clinical trials is in fact unknown. It follows that under the assumption of homogeneity of variance, Y_{ij} are independently distributed as a mixture of two normal distributions with mean μ_{i+} and μ_{i-} respectively, and common variance σ^2 (McLachlan and Peel, 2000):

$$\varphi \left(y_{ij} \mid \mu_{i+}, \sigma^2 \right)^\gamma \varphi \left(y_{ij} \mid \mu_{i-}, \sigma^2 \right)^{1-\gamma} \quad i = T, C \, ; j = 1, \ldots, n_i \tag{16.3}$$

where $\varphi \left(\cdot \mid \cdot \right)$ denotes the density of a normal variable.

However, γ is an unknown positive predicted value, which is usually estimated from the data. Therefore, the data obtained from the targeted clinical trials are incomplete because the true status of the molecular target of the patients is missing. The EM algorithm is a one of the methods for obtaining the maximum likelihood estimators of the parameters for an underlying distribution from a given data set when the data is incomplete or has missing values. On the other hand, the diagnostic device for detection of molecular targets has been validated in diagnostic effectiveness trials for its diagnostic accuracy. Therefore, the estimates of the positive predictive value for the diagnostic device can be obtained from the previously conducted diagnostic effectiveness trials. As a result, we can apply the EM algorithm to estimate the treatment effect for the patients truly with the molecular target by incorporating the estimates of the positive predictive value of the device obtained from the diagnostic effectiveness trials as the initial values.

For each patient, we have a pair of variables (Y_{ij}, X_{ij}), where Y_{ij} is the observed primary efficacy endpoint of patient j in treatment i and X_{ij} is the latent variable indicting the true status of the molecular target of patient j in treatment $i; j = 1, \ldots, n_i, I = T, C$. In other words, X_{ij} is an indicator variable with value of 1 for the patients truly with the molecular target and with a value of 0 for the patients truly without the target. In addition, X_{ij} are assumed i.i.d. Bernoulli random variables with probability γ for the molecular target. Let $\Psi = (\gamma, \mu_{T+}, \mu_{T-}, \mu_{C+}, \mu_{C-}, \sigma^2)'$ be the vector containing all unknown parameters and $y_{obs} = (y_{T1}, \ldots, y_{Tn_T}, y_{C1}, \ldots, y_{Cn_C})'$ be the vector of the observed primary efficacy endpoints from the targeted clinical trials. It follows that the complete-data log-likelihood function is given by

$$
\log L_c(\Psi) = \sum_{j=1}^{n_T} X_{Tj} \left[\log \gamma + \log \varphi(y_{Tj} \mid \mu_{T+}, 0^2) \right]
$$

$$
+ \sum_{j=1}^{n_T} (1 - x_{Tj}) \left[\log(1 - \gamma) + \log \varphi(y_{Tj} \mid \mu_{T-}, 0^2) \right]
$$

$$
+ \sum_{j=1}^{n_C} X_{Cj} \left[\log \gamma + \log \varphi(y_{Cj} \mid \mu_{C+}, 0^2) \right] \tag{16.4}
$$

$$
+ \sum_{j=1}^{n_C} (1 - x_{Cj}) \left[\log(1 - \gamma) + \log \varphi(y_{Cj} \mid \mu_{c-}, 0^2) \right].
$$

Furthermore, from the previous diagnostic effectiveness trials, an estimate of the positive predictive value of the device is known. Therefore, at the initial step of the EM algorithm for estimation the treatment effects in the patients with the molecular target, the observed latent variable X_{ij} are generated as i.i.d. Bernoulli random variables with the positive predicted value γ estimated by that obtained from the diagnostic effectiveness trial. The procedures for implementation of the EM algorithm in conjunction with the bootstrap procedure for inference of θ in the patient population truly with the molecular target are briefly described below.

At the $(k+1)$st iteration, the E-step requires the calculation of the conditional expectation of the complete-data log-likelihood $L_c(\Psi)$, given the observed data y_{obs}, using currently fitting $\widehat{\Psi}^{(k)}$ for Ψ.

$$Q\left(\Psi;\widehat{\Psi}^{(k)}\right)=E_{\Psi(k)}\left\{\log L_c(\Psi)\mid y_{obs}\right\}$$

Since $\log L_c(\Psi)$ is a linear function of the unobservable component labeled variables x_{ij}, the E-step is calculated by replacing x_{ij}, by its conditional expectation given y_{ij}, using $\widehat{\Psi}^{(k)}$ for Ψ. That is, x_{ij} is replaced by

$$\widehat{X}_{ij}^{(k)}=E_{\Psi(k)}\left\{x_{ij}\mid y_{ij}\right\}=\frac{\gamma_i^{(k)}\varphi\left(y_{ij}\mid\widehat{\mu}_{i+}^{(k)},\left(\widehat{\sigma}_i^2\right)^{(k)}\right)}{\gamma_i^{(k)}\varphi\left(y_{ij}\mid\widehat{\mu}_{i+}^{(k)},\left(\widehat{\sigma}_i^2\right)^{(k)}\right)+\left(1-\gamma_i^{(k)}\right)\varphi\left(y_{ij}\mid\widehat{\mu}_{i-}^{(k)},\left(\widehat{\sigma}_i^2\right)^{(k)}\right)},i=T,C.$$

which is the estimate of the posterior probability of the observation y_{ij} with molecular target after the kth iteration. The M-step requires the computation of

$$\gamma_i^{(k+1)},\widehat{\mu}_{i+}^{(k+1)},\widehat{\mu}_{i-}^{(k+1)},\text{ and }(\widehat{\sigma}_i^2)^{(k+1)},\ i=T,C,$$

by maximizing $\log L_c(\Psi)$. It is equivalent to computing the sample proportion, the weighted sample mean and sample variance with the weight x_{ij}. Since $\log L_c(\Psi)$ is linear in the x_{ij}, it follows that x_{ij} are replaced by their conditional expectations $\widehat{x}_{ij}^{(k)}$. On the $(k+1)th$ iteration, the intent is to choose the value of Ψ, say $\widehat{\Psi}^{(k+1)}$, that maximizes $Q(\Psi;\widehat{\Psi}^{(k)})$. It follows that on the M-step of the $(k+1)st$ iteration, the current fit for the positive predicted value of test drug group and control group is given by

$$\gamma_i^{(k+1)}=\frac{\sum_{j=1}^{n_i}\widehat{X}_{ij}^{(k)}}{n_i},\quad i=T,C.$$

Under the assumption that $n_T=n_c$, it follows that the overall positive predicted value is estimated by

$$\gamma^{(k+1)}=\left(\gamma_T^{(k+1)}+\gamma_C^{(k+1)}\right)/2.$$

The means of the molecularly target test drug and control can then be estimated respectively as

$$
\hat{\mu}_{T+}^{(k+1)} = \sum_{j=1}^{n_T} \hat{X}_{Cj}^{(k)} y_{Tj} \bigg/ \sum_{j=1}^{n_T} \hat{X}_{Tj}^{(k)}, \quad \hat{\mu}_{T-}^{(k+1)} = \sum_{j=1}^{n_T} \left(1 - \hat{X}_{Tj}^{(k)}\right) y_{Tj} \bigg/ \sum_{j=1}^{n_T} \left(1 - \hat{X}_{Tj}^{(k)}\right),
$$

$$
\hat{\mu}_{C+}^{(k+1)} = \sum_{j=1}^{n_C} \hat{X}_{Cj}^{(k)} y_{Cj} \bigg/ \sum_{j=1}^{n_C} \hat{X}_{Cj}^{(k)}, \text{ and } \hat{\mu}_{C-}^{(k+1)} = \sum_{j=1}^{n_C} \left(1 - \hat{X}_{Cj}^{(k)}\right) y_{Cj} \bigg/ \sum_{j=1}^{n_C} \left(1 - \hat{X}_{Cj}^{(k)}\right),
$$

with unbiased estimators for the variances of molecularly targeted drug and control given respectively by

$$
(\hat{\sigma}_T^2)^{(k+1)} = \left(\sum_{j=1}^{n} \hat{X}_{Tj}^{(k)} \left(y_{Tj} - \hat{\mu}_{T+}^{(k)} \right)^2 + \sum_{j=1}^{n} \left(1 - \hat{X}_{Tj}^{(k)}\right) \left(y_{Tj} - \hat{\mu}_{T-}^{(k)} \right)^2 \right) \bigg/ (n_T - 2),
$$

and

$$
(\hat{\sigma}_C^2)^{(k+1)} = \left(\sum_{j=1}^{n} \hat{X}_{Cj}^{(k)} \left(y_{Cj} - \hat{\mu}_{C+}^{(k)} \right)^2 + \sum_{j=1}^{n} \left(1 - \hat{X}_{Cj}^{(k)}\right) \left(y_{Cj} - \hat{\mu}_{C-}^{(k)2} \right) \right) \bigg/ (n_c - 2).
$$

It follows that an unbiased estimated for the pooled variance is given as

$$
(\hat{\sigma}^2)^{(k+1)} = \frac{\left[(n_T - 2) \times (\hat{\sigma}_T^2)^{(k+1)} + (n_c - 2) \times \left(\hat{\sigma}_C^2\right)^{(k+1)} \right]}{(n_T + n_c - 4)}.
$$

Therefore, the estimator for the treatment effects in the patients with the molecular target θ obtained from the EM algorithm is given as $\hat{\theta} = \hat{\mu}_{T+} - \hat{\mu}_{C+}$. Liu et al. (2009) proposed to apply the parametric bootstrap method to estimate the standard error of $\hat{\theta}$.

Step 1: Choose a large bootstrap sample size, say $B = 1{,}000$. For $1 \le b \le B$, generate the bootstrap sample y_{obs}^b according to the probability model in (16.3). The parameters in (6.3) for generating bootstrap samples y_{obs}^b are substituted by the estimators obtained from the EM algorithm based on the original observations of primary efficacy endpoints from the targeted clinical trials.

Step 2: The EM algorithm is applied to the bootstrap sample y_{obs}^b to obtain estimates θ_b^*, $b = 1, \ldots, B$.

Step 3: An estimator for the variance of $\hat{\theta}$ by the parametric bootstrap

procedure is given as $S_B^2 = \sum_{b=1}^{B}(\hat{\theta}_b^* - \overline{\hat{\theta}^*})^2 / (B-1)$, where $\overline{\hat{\theta}^*} = \sum_{b=1}^{B} \hat{\theta}_b^* / B$.

Let $\hat{\theta}$ be the estimator for the treatment effects in the patients truly with the molecular target obtained from the EM algorithm. Nityasuddhi and Böhning (2003) show that the estimator obtained under the EM algorithm is asymptotic unbiased. Let S_B^2 denote the estimator of the variance of $\hat{\theta}$ obtained by the bootstrap procedure. It follows that the null hypothesis is rejected and the efficacy of the molecular targeted test drug is different from that of the control in the patient population truly with the molecular target at the α level if

$$t = \left| \hat{\theta} / \sqrt{S_B^2} \right| \geq z_{\alpha/2}, \tag{16.5}$$

where $z_{\alpha/2}$ is the $\alpha/2$ upper percentile of a standard normal distribution. Thus, the corresponding $(1-\alpha) \times 100\%$ asymptotic confidence interval for $\theta = \mu_{T+} - \mu_{C+}$ can be constructed as

$$\hat{\theta} \pm z_{1-\alpha/2} \sqrt{S_B^2}$$

(see, e.g., Basford et al., 1997). It should be noted that although the assumption that

$$\mu_{T+} - \mu_{C+} > \mu_{T-} - \mu_{C-}$$

is one of the reasons for developing the targeted treatment, this assumption is not used in the EM algorithm for estimation of θ. Hence, the inference for θ by the proposed procedure is not biased in favor of the targeted treatment.

16.3.3 Simulation Results

Liu et al. (2009) conducted a simulation study to evaluate finite sample performance of the proposed method of EM algorithm. In the simulation, μ_{T-}, μ_{C+}, and μ_{C-} are assumed equal and set to be a generic value of 100. To investigate the impact of the positive predictive value, sample size, difference in means, and variability, Liu et al. (2009) considered the following specifications of parameters: (1) the positive predicted value is set to be 0.5, 0.7, 0.8, and 0.9 which reflect a range of low, median, and high positive predicted value, and (2) the range of the standard deviation σ is set as 20, 40, or 60. To investigate the finite sample properties, the sample sizes are set as 50, 100, and 200 per group. The mean differences are chosen a fraction of the standard deviation, from 10% to 60% by 10%; and 75% and 100%. In addition, the size of the proposed testing procedure was investigated at $\mu_{T+} = 100$. For each of 288 combinations, 5,000 random samples were generated and the number of the bootstrap samples was set to be 1000. The simulation results indicate that the absolute relative bias of the estimator for θ by the current method ranges from 10% to more than 50% and increases as the positive predictive value decreases. On the other hand, most of absolute relative bias measurements of the estimator for θ obtained by the EM algorithm are smaller than 0.05% although it can as high as 10% for few combinations when the difference in means is 2. The variability has little impact on the bias of both methods. However, for the EM procedure, the relative bias tends to decrease as the sample size increases. The bias of the current method with consideration of the true status of molecular target can be as high as 50% when the positive predictive value is low. Consequently, the empirical coverage probabilities of the corresponding 95% confidence interval can be as low as only 0.28% when the positive predictive value is 50%, mean difference is 20, standard deviation is 20, and n is 200. The coverage probability of the 95% confidence interval by the current method is an increasing function of the positive predictive value. On the other hand, only 36 of the 288 coverage probabilities (12.5%) of the 95% confidence intervals by the current method exceed 0.9449 and 24 of them occurred when the positive predictive value is 0.9. On the contrary, only 14.6% of the 288 coverage probabilities of the 95% confidence intervals by the EM method are below 0.9449. However, 277 of the 288 coverage probability of the 95% confidence interval constructed by the EM algorithm are above 0.94. No coverage probability of the EM method is below 0.91. Therefore, the proposed procedures for estimation of the treatment effects in the patient population with the molecular target by the EM algorithm is not only unbiased but also provide sufficient coverage probability.

16.4 Alternative Enrichment Designs

16.4.1 Alternative Designs with/without Molecular Targets

As indicated above, Liu et al. (2009) proposed statistical methods for assessment of the treatment effect for patients with positive diagnosed results under the enrichment design described in Figure 16.2. Their methods suffer from the lack information regarding the proportion of subjects who truly have molecule targets in the patient population and the unknown positive predicted value. Consequently, the conclusion drawn from the collected data may be biased and misleading. In addition to the study designs as given in Figures 16.1 and 16.2, the 2005 FDA concept paper also recommended the following two study designs for different study objectives (see Figure 16.3 for Design C and Figure 16.4 for Design D).

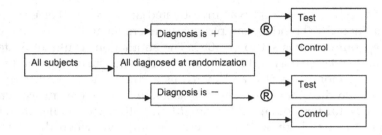

FIGURE 16.3
Design C—Enrichment design for patients with and without molecular targets.

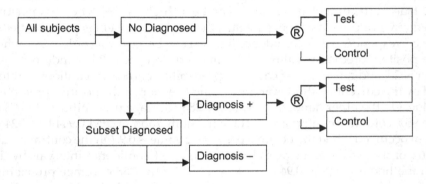

FIGURE 16.4
Design D—Alternative enrichment design for targeted clinical trials.

This study design allows the evaluation of the treatment effect within subpopulations, i.e., the subpopulation of patients with positive or negative results. Similar to Table 16.2 for the study design given in Figure 16.1, the expected values of Y_{ij} by treatment and diagnostic result of the molecular targets are summarized in Tables 16.3 and 16.4.

As a result, it may of interest in estimating the following treatment effects:

$$\theta_1 = \gamma_1\left(\mu_{T++} - \mu_{C++}\right) + \left(1-\gamma_1\right)\left(\mu_{T+-} - \mu_{C+-}\right);$$

$$\theta_2 = \gamma_2\left(\mu_{T-+} - \mu_{C-+}\right) + \left(1-\gamma_2\right)\left(\mu_{T-} - \mu_{C}\right);$$

$$\theta_3 = \delta\gamma_1\left(\mu_{T++} - \mu_{C++}\right) + \left(1-\delta\right)\gamma_2\left(\mu_{T-+} - \mu_{C-+}\right);$$

$$\theta_4 = \delta\gamma_1\left(\mu_{T+-} - \mu_{C+-}\right) + \left(1-\delta\right)\gamma_1\left(\mu_{T-} - \mu_{C--}\right);$$

$$\theta_5 = \delta\left[\gamma_1\left(\mu_{T+-} - \mu_{C+-}\right) + \left(1-\gamma_1\right)\left(\mu_{T+-} - \mu_{C+-}\right)\right]$$
$$+ \left(1-\delta\right)\left[\gamma_2\left(\mu_{T-+} - \mu_{C-+}\right) + \left(1-\gamma_2\right)\left(\mu_{T-} - \mu_{C--}\right)\right],$$

where δ is the proportion of subjects with positive molecular targets. Following similar ideas described in the previous section, estimates

TABLE 16.4

Population Means by Treatment and Diagnosis

Positive Diagnosis	True Target Condition	Indicator of Diagnostic	Test Group	Control Group	Difference
+	+	γ_1	μ_{T++}	μ_{C++}	$\mu_{T++} - \mu_{C++}$
	−	$1-\gamma_1$	μ_{T+-}	μ_{C+-}	$\mu_{T+-} - \mu_{C+-}$
−	+	γ_2	μ_{T-+}	μ_{C-+}	$\mu_{T-+} - \mu_{C-+}$
	−	$1-\gamma_2$	μ_{T--}	μ_{C--}	$\mu_{T--} - \mu_{C--}$

Notes: γ_i is the positive predicted value, $i = 1$ (positive diagnosis) and $i = 2$ (negative diagnosis).

μ_{ijk} is the mean for subjects in the ith group with kth true target status but with jth diagnosed result.

of $\theta_1 - \theta_5$ can be obtained. In other words, estimates of θ_1 and θ_2 can be obtained based on data collected from the subpopulations of subjects with and without positive diagnoses who truly have the molecular target of interest. Similarly, the combined treatment effect θ_5 can be assessed. These estimates, however, depend upon both $\gamma_i, i = 1,2$, and δ. To obtain some information regarding $\gamma_i, i = 1,2$, and δ, the FDA recommends the following alternative enrichment design that includes a group of subjects without any diagnoses and a subset of subjects who will be diagnosed at the screening stage.

16.4.2 Statistical Methods

As indicated earlier, the method proposed by Liu et al. (2009) suffers from the lack of information regarding the uncertainty in accuracy of the diagnostic device. As an alternative, we propose considering a Bayesian approach to incorporate the uncertainty in accuracy and reliability of the diagnostic device for the molecular target into the inference of treatment effects of the targeted drug. For each patient, we have a pair of variables $\left(y_{ij}, x_{ij}\right)$, where y_{ij} is the observed primary efficacy endpoint of patient j in treatment i and x_{ij} is the latent variable indicting the true status of the molecular target of patient j in treatment i, $j = 1,\ldots,n_i, i = T,C$. In other words, x_{ij} is an indicator variable with value of 1 for patients with the molecular target and with a value of 0 for patients without the target. The x_{ij} are assumed i.i.d. Bernoulli random variables with probability of the molecular target being γ. Thus,

$$x_{ij} = 1 \text{ if } y_{ij} \sim N\left(\mu_{i+}, \sigma^2\right)$$

and

$$x_{ij} = 0 \text{ if } y_{ij} \sim N\left(\mu_{i-}, \sigma^2\right), i = T, C; j = 1,\ldots,n_i.$$

The likelihood function is given by

$$L(\Psi \mid Y_{obs}, x_{ij}) = \prod_{j,x_{Tj}=1} \gamma \varphi(y_{Tj} \mid \mu_{T+}, \sigma^2) \times \prod_{j,x_{Tj}=0} (1-\gamma)\varphi(y_{Tj} \mid \mu_{T-}, \sigma^2)$$

$$\times \prod_{j,x_{Cj}=1} \gamma \varphi(y_{Cj} \mid \mu_{C+}, \sigma^2) \times \prod_{j,x_{Cj}=0} (1-\gamma)\varphi(y_{Cj} \mid \mu_{c-}, \sigma^2),$$

where $i = T, C; j = 1, \ldots, n_i$ and $\varphi(.\mid.)$ denotes the density of a normal variable. For Bayesian approach, a beta distribution can be employed as the prior distribution for γ, while normal prior distributions can be used for μ_{i+} and μ_{i-}. In addition, a gamma distribution can be used as a prior for σ^{-2}. Under the assumptions of these prior distributions, the conditional posterior distributions of $\gamma, \mu_{i+}, \mu_{i-}, \sigma^{-2}$ can be derived. In other words, assuming that

$$f(\gamma) \sim \beta(\alpha_r, \beta_r), f(\mu_{i+}) \sim N\left(\lambda_{i+}, \sigma_0^2\right), f(\mu_{i-}) \sim N\left(\lambda_{i-}, \sigma_0^2\right),$$

and

$$f(\sigma^{-2}) \sim \gamma(\alpha_g, \beta_g),$$

where μ_{i+}, μ_{i-} and γ are assumed to be independent and $\alpha_r, \beta_r, \alpha_g, \beta_g, \lambda_{i+}, \lambda_{i-}$ and σ_0^2 are assumed to be known. Thus, the conditional posterior distribution of $x_{ij}i$ is given by

$$f(x_{ij} \mid \gamma, \mu_{i+}, \mu_{i-}, Y_{obs}) \sim \text{Bernoulli}\left(\frac{\gamma \varphi(y_{ij} \mid \mu_{i+}, \sigma_0^2)}{\gamma \varphi(y_{ij} \mid \mu_{i+}, \sigma_0^2) + (1-\gamma)\varphi(y_{ij} \mid \mu_{i-}, \sigma_0^2)}\right),$$

where

$$E_\Psi[x_{ij} \mid \gamma, \mu_{i+}, \mu_{i-}, Y_{obs}] = \frac{\gamma \varphi(y_{ij} \mid \mu_{i+}, \sigma^2)}{\gamma \varphi(y_{ij} \mid \mu_{i+}, \sigma^2) + (1-\gamma)\varphi(y_{ij} \mid \mu_{i-}, \sigma^2)}, i = T, C; j = 1, \ldots, n_i$$

in the EM algorithm. The joint distribution of $\gamma, \mu_{i+}, \mu_{i-}$ and σ^2 is given by

$$f(\gamma, \mu_{i+}, \mu_{i-}, \sigma^2 \mid Y_{obs}, x_{ij})$$

$$= \prod_{j, x_{Tj}=1} \varphi(y_{Tj} \mid \mu_{T+}, \sigma^2) \times \prod_{j, x_{Tj}=0} \varphi(y_{Tj} \mid \mu_{T-}, \sigma^2)$$

$$\times \prod_{j, x_{Cj}=1} \varphi(y_{Cj} \mid \mu_{C+}, \sigma^2) \times \prod_{j, x_{Cj}=0} \varphi(y_{Cj} \mid \mu_{C-}, \sigma^2)$$

$$\times \varphi(\mu_{T+} \mid \lambda_{7+}, \sigma_0^2) \times \varphi(\mu_{T-} \mid \lambda_{T-}, \sigma_0^2)$$

$$\times \varphi(\mu_{C+} \mid \lambda_{C+}, \sigma_0^2) \times \varphi(\mu_{C-} \mid \lambda_{c-}, \sigma_0^2)$$

$$\times \frac{\Gamma(\alpha_r + \beta_r)}{\Gamma(\alpha_r)\Gamma(\beta_r)} (\gamma)^{\sum_{j=1}^{n_T} x_{Tj} + \sum_{j=1}^{n_C} x_{Cj} + \alpha_\gamma - 1} (1-\gamma)^{\sum_{j=1}^{n_T}(1-x_{Tj}) + \sum_{j=1}^{n_C}(1-x_{Cj}) + \beta_\gamma - 1} .$$

Thus, the conditional posterior distribution of $\gamma, \mu_{i+}, \mu_{i-}$ and σ^{-2} can be obtained as follows:

$$f(\gamma \mid \mu_{i+}, \mu_{i-}, \sigma^{-2}, Y_{obs}, x_{ij}) \sim \beta \left(\begin{array}{c} \sum_{j=1}^{n_T} x_{Tj} + \sum_{j=1}^{n_C} x_{Cj} \\ \\ +\alpha_\gamma, \sum_{j=1}^{n_T}(1-x_{Tj}) + \sum_{j=1}^{n_C}(1-x_{Cj}) + \beta_\gamma, \end{array} \right),$$

$$f(\mu_{i+} \mid \gamma, \mu_{i-}, \sigma^{-2}, Y_{obs}, x_{ij}) \sim N \left[\frac{\sigma^{-2} \sum_{j=1}^{n_i} x_{ij} y_{ij} + \sigma_0^{-2} \lambda_{i+}}{\sigma^{-2} \sum_{j=1}^{n_i} x_{ij} + \sigma_0^{-2}}, \frac{1}{\sigma^{-2} \sum_{j=1}^{n_i} x_{ij} + \sigma_0^{-2}} \right],$$

$$f(\mu_{i-} \mid \gamma, \mu_{i+}, \sigma^{-2}, Y_{obs}, x_{ij}) \sim N \left(\begin{array}{c} \frac{\sigma^{-2} \sum_{j=1}^{n_i}(1-x_{ij}) y_{ij} + \sigma_0^{-2} \lambda_{i-}}{\sigma^{-2} \sum_{j=1}^{n_i}(1-x_{ij}) + \sigma_0^{-2}}, \\ \\ \frac{1}{\sigma^{-2} \sum_{j=1}^{n_i}(1-x_{ij}) + \sigma_0^{-2}} \end{array} \right),$$

$$f(\sigma^{-2} \mid \gamma, \mu_{i+}, \mu_{i-}, Y_{obs}, x_{ij})$$

$$\sim \gamma \left(\begin{array}{c} \dfrac{n_T + n_c}{2} \\[2em] +\sigma_g, \dfrac{1}{2} \displaystyle\sum_{i=T,C} \left[\begin{array}{c} \displaystyle\sum_{j=1}^{n_i} x_{ij}\left(y_{ij} - \mu_{i+}\right)^2 \\[1.5em] + \displaystyle\sum_{j=1}^{n_i} \left(1 - x_{ij}\right)\left(y_{ij} - \mu_{i-}\right)^2 \end{array} \right] \\[3em] +\beta_g \end{array} \right),$$

respectively. Consequently, the conditional posterior distribution of $\theta = \mu_{T+} - \mu_{C+}$ can be obtained as follows:

$$f\left(\hat{\theta} \mid \gamma, \mu_{i+}, \mu_{i-}, \sigma^2, Y_{obs}, x_{ij}\right)$$

$$\sim N \left(\begin{array}{c} \dfrac{\sigma^{-2} \displaystyle\sum_{j=1}^{n_T} x_{Tj}y_{Tj} + \sigma_0^{-2}\lambda_{T+}}{\sigma^{-2} \displaystyle\sum_{j=1}^{n_T} x_{Tj} + \sigma_0^{-2}} \\[3em] + \dfrac{\sigma^{-2} \displaystyle\sum_{j=1}^{n_C} x_{Cj}y_{Cj} + \sigma_0^{-2}\lambda_{C+}}{\sigma^{-2} \displaystyle\sum_{j=1}^{n_C} x_{Cj} + \sigma_0^{-2}} \end{array} , \quad \dfrac{1}{\sigma^{-2} \displaystyle\sum_{j=1}^{n_T} x_{Tj} + \sigma_0^{-2}} + \dfrac{1}{\sigma^{-2} \displaystyle\sum_{j=1}^{n_C} x_{Cj} + \sigma_0^{-2}} \right).$$

As a result, statistical inference for $\theta = \mu_{T+} - \mu_{C+}$ can be obtained. Following similar ideas, statistical inferences for the treatment effects can be derived. Note that different priors for $\gamma, \mu_{i+}, \mu_{i-}$ and σ^{-2} may be applied depending upon disease targets across different therapeutic areas. However, different prior assumptions will result in different statistical inference for assessment of treatment effect under study.

16.4.3 Remarks

In addition to the study designs proposed by FDA, Freidlin and Simon (2005) proposed an adaptive signature design for randomized clinical trials of targeted agents in settings where an assay or signature that identified sensitive patients was not available at the outset of the study. The design combined prospective development of a gene expression-based classifier to select sensitive patients with a properly powered test for overall effect. Jiang et al. (2007) proposed a biomarker-adaptive threshold design for settings in which a putative biomarker to identify patients who were sensitive to the new agent was measured on a continuous or graded scale. The design combined a test for overall treatment effect in all randomly assigned patients with the establishment and validation of a cut point for a prespecified biomarker of the sensitive subpopulation. Freidlin et al. (2010) proposed a cross-validation extension of the adaptive signature design that optimizes the efficiency of both the classifier development and the validation components of the design. Zhou et al. (2008) and Lee et al. (2010) proposed Bayesian adaptive randomization enrichment designs for targeted agent development. On the other hand, Todd and Stallard (2005) presented a group sequential design, which incorporates interim treatment selection based upon a biomarker, followed by a comparison of the selected treatment with control in terms of the primary endpoint. A statistical approach that controls type I error rate for the design was proposed. Stallard (2010) later proposed a method for group sequential trials that use both the available biomarker and primary endpoint information for treatment selections. The proposed method controlled the type I error rate in the strong sense. Shun et al. (2008) studied a biomarker informed two-stage winner design with normal endpoints. Di Scala and Glimm (2011) studied the case with correlated time-to-event biomarker and primary endpoint where Bayesian predictive power combining evidence from both endpoints is used for interim selection, they investigated the precise conditions under which type I error control is attained. Friede et al. (2011) considered an adaptive seamless phase II/III design with treatment selection based on early outcome data ("biomarker informed drop-the-losers design"). Bringing together combination tests for adaptive designs and the closure principle for multiple testing, control of the familywise type I error rate in the strong sense was achieved.

Focused clinical trials using a biomarker strategy have been shown the potential to: result in shorter trial duration; allow smaller study sizes,; provide higher probability of trial success; result in enhancement of the benefit-risk relationship; and potentially mitigating ever-escalating development costs. In the planning of a study that uses biomarker informed adaptive procedures, it is desirable to perform statistical simulations in order to understand the operating characteristics of the design, including sample size required for a target power. It is therefore necessary to specify a model for simulation of trial data. Friede et al. (2011) proposed a simulation model based on standardized test statistics that allows the generation of virtual trials for a variety of outcomes. The test statistics of the trial were simulated directly instead of trial data. To simulate individual patient data for the trial, on the other hand, a model that describes the relationship between biomarker and primary endpoint needs to be specified.

If both endpoints follow normal distribution, Shun et al. (2008) used a bivariate normal distribution for modeling the two endpoints. Wang et al. (2014) showed that the bivariate normal model that only considers the individual level correlation between biomarker and primary endpoint is inappropriate when little is known about how the means of the two endpoints are related. Wang et al. (2014) further proposed a two-level correlation (individual level correlation and mean level correlation) model to describe the relationship between biomarker and primary endpoint. The two-level correlation model incorporates a new variable that describes the mean level correlation between the two endpoints. The new variable, together with its distribution, reflects the uncertainty about the mean-level relationship between the two endpoints due to a small sample size of historical data. It was shown that the two-level correlation model is a better choice for modeling the two endpoints.

16.5 Concluding Remarks

As discussed in the previous sections, traditional medicine can only benefit average patients with the diseases under study, while precision medicine can further benefit a specific group (subgroup) of patients who have certain characteristics (e.g., specific genotype or molecular target). President Obama's Precision Medicine Initiative is an important step moving away from the traditional medicine and toward personalized medicine so that the specific group of patients could be beneficial. A typical example is the development of Herceptin for treatment of female patients with breast cancer. As indicated in Section 16.2.2, Herceptin plus chemotherapy provides statistically significantly additional clinical benefit in terms of overall survival over chemotherapy alone for patients with a staining score of 3+ (see also Table 16.1).

Under a valid design (e.g., Designs B, C, and D) for assessment of precision medicine, the accuracy and reliability of the analysis results depend upon an accurate and reliable estimate of positive predicted value (PPV). The method of EM algorithm in conjunction with Bayesian approach may be useful to resolve the issue. Alternative methods for estimation of treatment effects as described in Section 16.4.1, especially under FDA recommended Design C and Design D, however, are necessary developed.

In addition to precision medicine, the ultimate goal of personalized (individualized) medicine is seeking cure for individual patients if it is not impossible. President Obama's Precision Medicine Initiative is an important step moving toward personalized (or individualized) medicine so that the individual patients could be beneficial from the treatment for treating specific diseases under investigation. For this purpose, traditional Chinese medicine, which often consists of multiple components and focuses on global dynamic harmony (or balance) among specific organs within individual patients is expected to be the center of attention in the next century (Chow, 2015). Personalized (individualized) medicine development (e.g., traditional Chinese medicine) moving toward the next century, however, regulatory requirement and quantitative/statistical methods for assessment of treatment effect for drug products with multiple components are necessarily developed. More details regarding traditional Chinese medicine (TCM) development are given in Chapter 12.

17

Big Data Analytics

17.1 Introduction

In healthcare-related biomedical research, big data analytics refers to the analysis of large data sets that contain a variety of data sets (with similar or different data types) from various data structured, semi-structured, or unstructured sources such as registry; randomized or non-randomized studies; published or unpublished studies; and healthcare databases. The purpose of big data analytics is to detect any possible hidden signals, patterns, and/or trends of certain test treatments under study as that pertain to safety and efficacy. In addition, big data analytics may uncover possible unknown associations and/or correlations between potential risk factors and clinical outcomes, and other useful biomedical information such as risk/benefit ratio of certain clinical endpoints/outcomes. The finding of big data analytics could lead to more efficient assessment of treatments under study and/or identification of new intervention opportunities; better disease management; other clinical benefits; and improvement of operational efficiency for planning of future biomedical studies.

As indicated in an Request for proposals (RFP) at the website of the United States National Institutes of Health (NIH), biomedical research is rapidly becoming data-intensive as investigators are generating and using increasingly large, complex, multidimensional, and diverse data sets. However, the ability to release data; to locate, integrate, and analyze data generated by others; and to utilize the data is often limited by the lack of tools, accessibility, and training. Thus, the NIH has developed the Big Data to Knowledge (BD2K) initiative to solicit development of software tools and statistical methods for data analysis in the four topic areas of data compression and reduction, data visualization, data provenance, and data wrangling as part of the overall BD2K initiative.

Big data analytics is promising and provides the opportunities of uncover hidden medical information, determining possible associations or correlations between potential risk factors and clinical outcomes; predictive model building; validation and generalization; data mining for biomarker development; and critical information for future clinical trials planning (see, e.g., Bollier, 2010; Ohlhorst, 2012; Raghupathi and Raghupathi, 2014). Although big data analytics is promising and has great potential in biomedical research, there are some limitations due to possible selection bias and heterogeneity across data sets from different data sources that may affect the representativeness and validity of big data analytics. In biomedical research, the most commonly considered big data analytics is probably a meta-analysis by combining several independent studies. In meta-analysis, the most commonly employed method is probably the application of a random effects model or a mixed effects model (see, e.g., DerSimonian and Laird, 1986; Chow and Liu, 1997).

Chow and Kong (2015) pointed out that the findings from the big data analytics may be biased due to selection bias that is commonly encountered in big data analytics. This is because of the diverse sources of data sets that may be accepted into the big data center. In practice, it is likely that published and/or positive studies would be included in the big data center. The published or positive studies tend to report much larger treatment effect size, which will over-estimate the true treatment effect of the target patient population. As a result, the findings obtained from the big data analytics could be biased and misleading. The bias could be substantial especially when there is a large portion of unpublished and/or negative studies which are not included in the big data analytics. To overcome these problems, in this chapter, we attempt to propose a method for estimation of the true treatment effect of the target patient population by taking possible selection bias into consideration.

The remaining of this chapter is organized as follows. In Section 17.2, in addition to the challenges outlined by Raghupathi and Raghupathi (2014), we will focus on some basic considerations for assuring the quality, integrity, and validity of big data analytics in biomedical research. These basic considerations include, but are not limited to, representativeness; quality and integrity of big data; validity of big data analytics; FDA Part 11 compliance for electronic records; and statistical methodology and software development. Section 17.3 will briefly outline types of big data analytics in clinical research. Selection bias of big data analytics is explored in Section 17.4. Section 17.5 discusses statistical methods for estimation of bias; bias adjustment; treatment effect in big data analytics; and issues that are commonly encountered bias in big data analytics. Section 17.6 presents a simulation study that was conducted to evaluate the performances of the proposed statistical methods for estimation of bias, bias adjustment, and treatment effect under various scenarios. Some concluding remarks are given in Section 17.7.

17.2 Basic Considerations

17.2.1 Representativeness of Big Data

In biomedical research, a big data often contains a variety of data sets (with different data types) from various data sources including registry; randomized or non-randomized clinical studies; published or unpublished data and health care databases. As a result, it is a concern whether the big data is a truly representative of the target patient population with the diseases under study because possible selection bias may have occurred when accepting individual data sets into the big data. In addition, heterogeneity is expected within and across individual data sets (studies). The issues of selection bias, heterogeneity, and consequently reproducibility and generalizability are briefly discussed below.

17.2.2 Selection Bias

In practice, it is likely that most data sets with positive results will enter the big data, in which case selection bias may have occurred. Let μ and μ_b be the true means of the target patient population and big data, respectively; let μ_P and μ_N be the true means of data sets with positive and negative results, respectively; and suppose that r is the true proportion of data with positive results. In this case, $\mu = r\mu_P + (1-r)\mu_N$, where r is often unknown. Thus, selection bias for accepting individual data sets could have a significant impact on the finding of big data analytics. In other words, the assessment of μ through the big data analytics $\hat{\mu}_b$ could be biased. If the big data only contains data sets with positive results, then $\mu_b = \mu_P$. Consequently, the bias be substantial if μ_P is far away from μ_N. As a result, the findings of big data analytics could be biased and hence misleading due to selection bias.

17.2.3 Heterogeneity

In addition to the representativeness and selection bias of data sets, heterogeneity within and between individual data sets from different sources is also a great concern. In practice, although individual data sets may come from clinical studies conducted with the same patient population, data from these studies may be collected under similar but different study protocols with similar but different doses or dose regimens at different study sites with local laboratories. These differences will cause heterogeneity within and between individual data sets. In other words, these data sets may follow similar distributions with different means and different variances. The heterogeneity could decrease the reliability of the assessment of the treatment effect.

17.2.4 Reproducibility and Generalizability

As indicated above, the heterogeneity within and across individual data sets (studies) in the big data center could have an impact on the reliability of the assessment of the treatment effect. In addition, as the big data continues growing, it is a concern whether the findings from the big data analytics is reproducible and generalizable from one big data center (database) to another big data center (database) of similar patient population with the same diseases or conditions under study. For evaluation of reproducibility and generalizability, the concept using a sensitivity index proposed by Shao and Chow (2002) is useful. Let (μ_0, σ_0) and (μ_1, σ_1) denote the population of the original database (big data center) and another database (another big data center), respectively. Thus, since the two databases are for similar patient populations with the same diseases and/or conditions, it is reasonable to assume that $\mu_1 = \mu_0 + \varepsilon$, and $\sigma_1 = C\sigma$, where ε and C are shift parameters in location and scale, respectively. After some algebra, it can be verified that

$$\left|\frac{\mu_1}{\sigma_1}\right| = \left|\frac{\mu_0 + \varepsilon}{C\sigma_0}\right| = |\Delta|\left|\frac{\mu_0}{\sigma_0}\right|,$$

where $\Delta = \left(1 + \varepsilon / (\mu_0) / C\right)$ is the sensitivity index for generalizability. In other words, if $|1 - \Delta| \leq \delta$, where δ is a pre-specified small number, we then claim that the results from the original big data center are generalizable to another big data center with data obtained from similar patient population with the same diseases and/or conditions. In practice, since ε and C are random, statistical methodology for assessment of Δ is necessarily developed.

17.2.5 Data Quality, Integrity, and Validity

In biomedical research, data management, that ensures the quality, integrity and validity of data collected from trial subjects to a database system. Proper data management delivers a clean and high-quality database for statistical analysis and consequently enables clinical scientists to draw conclusions regarding the effectiveness, safety, and clinical benefit/ risk of the test treatment under investigation. An invalid and/or poor quality database may result in wrong and/or misleading conclusions regarding the drug product under investigation. Thus, the objective of the data management process in clinical trials is not only to capture the information that the intended clinical trials are designed to capture, but also to ensure the quality, integrity and validity of the collected data. These data sets are then aggregated into a big data through a database system. Since the big data center contains electronic data records from a variety of sources, some regulatory requirements must be met for assurance of data quality, integrity and validity of the electronic data in the big data center.

17.2.6 FDA Part 11 Compliance

The FDA Part 11 compliance is referred to as requirements or criteria as described in 21 Codes of Federal Registration (CFR) Part 11 under which the FDA will consider electronic records and signatures to be generally equivalent to paper records and handwritten signatures. It applies to any records required by the FDA or submitted to the FDA under agency regulations. To reinforce Part 11 compliance, the FDA has published a compliance policy guide–CPG 7153.17, Enforcement Policy: 21 CFR Part 11 Electronic Records, Electronic Signatures. In addition, the FDA also published numerous draft guidance documents to assist the sponsors for Part 11 compliance. FDA Part 11 compliance has a significant impact on the process of clinical data management and consequently the big data management, which has recently become the focus for Good Data Management Practice (GDMP) in compliance with Good Statistics Practice (GSP) and Good Clinical Practice (GCP) for data quality, integrity, and validity. For example, 21 CFR Part 11 requires that procedures regarding creation, modification, maintenance, and transmission of records must be in place to ensure the authenticity and integrity of the records. In addition, the adopted systems must ensure that electronic records are accurately and reliably retained. 21 CFR Part 11 has specific requirements for audit trail systems to discern invalid or altered records. For electronic signatures, they must be linked to their respective electronic records to ensure that signatures cannot be transferred to falsify an electronic record. The FDA requires that systems must have the ability to generate documentation suitable for FDA inspection to verify that the requirements set forth by the 21 CFR Part 11 are met.

In practice, data management of big data is the top priority in the plan for 21 CFR Part 11 compliance for assurance of data quality, integrity and validity. A typical plan for Part 11 compliance for data management process usually includes (1) gap assessment; (2) user requirements specification; (3) validation master plan; and (4) tactical implementation plan. The task is implemented through a team consisting of senior experienced personnel from multiple disciplinary areas such as information technology (IT), programming, and data managers.

17.2.7 Missing Data

Missing values or incomplete data are commonly encountered in biomedical research and hence have become a major issue for big data analytics. One of the primary causes of missing data is the dropouts. Reasons for dropouts include, but are limited to: refusal to continue in the study (e.g., withdrawal of informed consent); perceived lack of efficacy; relocation; adverse events; unpleasant study procedures; worsening of disease; unrelated disease; non-compliance with the study; need to use prohibited medication; and death. How to handle incomplete data is always a challenge to statisticians in practice. Imputation is a very popular methodology to compensate for the missing data and is widely used in biomedical research. Despite to its popularity, however, its theoretical properties are far from well understood. Addressing missing data in clinical trials also involves missing data prevention and missing data analysis. Missing data prevention is usually done through the enforcement of GCP during protocol development and clinical operations and personnel training for data collection. This will lead to reduced biases, increased efficiency, less reliance on modeling assumption and less need for sensitivity analysis. However, in practice, missing data cannot be totally avoided. Missing data often occur due to factors beyond the control of patients, investigators, and clinical project team.

17.3 Types of Big Data Analytics

17.3.1 Case-Control Studies

In clinical research, studies utilizing big data commonly include, but are not limited to: incorporating retrospective cohort and/or case-control studies; meta-analysis by combining several independent studies; and data mining in genomic studies for biomarker development. For illustration purpose, in this chapter, we will focus on case-control studies only. The ideas explored here can be similarly applied to meta-analysis and data mining.

The primary objective of case-control studies in clinical research is not only to study possible risk factors and to develop a medical predictive model based on the identified risk factors, but also to examine the generalizability of the established predictive model (e.g., from one patient population such as adults to another similar but different patient population such as pediatrics or from one medical center to another). For this purpose, multivariate (logistic) regression process is probably the most commonly used method. The use of logistic regression analysis for identifying potential risk factors (or predictors) in order to build a predictive model has become very popular in clinical research (Hosmer and Lemeshow, 2000). Typically, logistic process for model building in case-control studies starts with propensity score matching between the case and control groups and followed by the steps of (i) descriptive analysis to have better understanding of the data; (ii) univariate analysis to test associations between the variables and the outcomes; (iii) collinearity analysis to test associations/correlations between explanatory variables; (iv) multivariate analysis to test association of a variable after adjusting for other variables or confounders; and (v) model diagnostic/ validation to assess whether the final model fulfils the assumptions it was based on. The first three steps can be achieved by the use of univariate logistic regression and the remaining two by the use of multiple logistic regression analysis, both of which are described below.

17.3.1.1 Propensity Score Matching

In clinical research, one of the major concerns of a case-control study is selection bias, which often caused by significant difference or imbalance between the case and control groups, especially for large observational studies (Rosenbaum and Rubin, 1983, 1984; Austin, 2011). In this case, the target patient population under study in the control group may not be comparable to that of the case group. This selection bias could alter the conclusion of the treatment effect due to possible confounding effect. Consequently, the conclusion may be biased and hence misleading. To overcome this problem, Rosenbaum and Rubin (1983) proposed the concept of *propensity score* as a method to reduce selection bias in observational studies. Propensity score is a conditional probability (or score) of the subject being in a particular group when given chosen characteristics. That is, consider the case group as those who received a certain treatment ($T = 1$), and the control group as those who did not receive this treatment ($T = 0$). Let X be a vector represents baseline demographics and/or patient characteristics that are important (e.g., possible confounding factors) for matching the case and control populations for reducing the selection bias. The propensity score is then given by

$$p(X) = \Pr[T = 1 | X] = E(T | X),$$

where $0 < p(X) < 1$. Propensity score matching is a powerful tool reducing the bias due to possible confounding effects. Sometimes, propensity score matching is also considered as *post-study* randomization as compared to a randomized clinical trial for reducing bias. For propensity score matching, it should be noted that data available for matching will decrease as the number of matching factors (potential confounding factors) increase.

17.3.1.2 Model Building

Let Y be the outcome variable, which could be a discrete response variable, e.g., a binary response variable where $Y = 1$ if it is a success and $Y = 0$ if it is a failure or a categorical variable where $Y = 1$ if there is an improvement, $Y = 0$ where there is no change, or $Y = -1$ when the symptom has worsen. Also, let X be any type of covariate (e.g., continuous or dichotomous). For the study of six-month survival of cirrhotic patients, X could be TCR functional status, TCR marked muscle wasting and serum creatinine, or demographics characteristics which could be potential risk factors (predictors) such as gender, age, or weight. Consider univariate logistic regression, the general model with on covariate is given by

$$logit(\pi) = \log\left(\frac{\pi}{1-\pi}\right) = \alpha + \beta x,$$

where π is the probability of success at covariate level x. The logistic regression model can be rewritten as:

$$\frac{\pi}{1-\pi} = e^{\alpha + \beta x} = e^{\alpha}\left\{e^{\beta}\right\}^{x},$$

where e^{β} represents the change in the odds of the outcome by increasing x by one unit. In other words, every one-unit increase in x increases the odds by a factor of e^{β}. More specifically, if $\beta = 0$ (i.e., $e^{\beta} = 1$), the probability of success is the same at each level of x. When $\beta > 0$ (i.e., $e^{\beta} > 1$), the probability of success increases as x increases. Similarly, when $\beta < 0$ (i.e., $e^{\beta} < 1$), the probability of success decreases as x increases. Similarly, the general logistic regression model with multiple covariates can be written as follows:

$$\log\left(\frac{\pi}{1-\pi}\right) = \alpha + \beta_1 x_1 + \beta_2 x_2 + \ldots + \beta_K x_K,$$

where log odds are a linear function of the covariates. If two explanatory variables are highly correlated with each other, it is suggested examining associations/correlation between explanatory variables and exclude one of a pair of highly correlated variables before conducting multivariable analysis. For quantitative variables, correlation is tested by Pearson correlation coefficient while for ordinal variables by Spearman rank correlation coefficient. Association between nominal (binominal or multinomial) variables can be tested by a chi-square test (and p-value). Association between a categorical and a continuous variable can be assessed by t-test (if the categorical variable has two categories) or analysis of variance (ANOVA) if there are more than two categories.

For model building, selection of an efficient subset of the explanatory variables or predictors that account for maximum variability in the outcome is the ultimate goal of the multiple logistic analysis. The following is a summary of commonly considered logistic selection process. First, one may start with a null model (without any variables) and then adding one variable at a time to the model, retaining a variable if it is significant. This selection process is usually referred to as the *forward selection process*. Second, one may start with a full model (model with all the variables) and then dropping one variable at a time, excluding a variable from the model if it is non-significant. This process is known as the *backward selection process*. Alternatively, one may consider the stepwise selection approach, where both forward and backward selection is done at each step. In practice, two selection criteria are commonly employed to assist in variable selection during logistic regression multivariable analyses. They are either based on likelihood and Wald Chi-square tests or using information criteria such as Akaike's information criteria (AIC), Bayesian information criteria (BIC), or Schwartz criteria (SC).

17.3.1.3 Model Diagnosis and Validation

The developed model can then be tested for goodness-of-fit, which can serve as the diagnostic tool for model selection. Commonly considered goodness-of-fit tests include (i) Pearson goodness-of-fit test; (ii) deviance goodness-of-fit test; (iii) Hosmer and Lemeshow goodness-of-fit test (only for binomial outcome); (iv) pseudo R-square; (v) pseudo adjusted R-square; and (vi) score test for assumption of proportional odds (only for ordinal outcome). Detailed information regarding these tests can be found in Hosmer and Lemeshow (2000).

For the validation of the final model, an internal validation of the developed model is usually performed, while an external validation for assessment of its generalizability is often also performed. In practice, a commonly considered procedure for model validation is to randomly split the data set into two sub data sets: one contains 90% of the data and the other one contains 10% of data. The data set contains 90% of the data is usually referred to as *training* data set which is used to build up the predictive model based on the identified possible risk factors (predictors). As indicated earlier, commonly considered criterion for model selection is the so-called information criterion such as AIC, BIC, or SC. The second data set that contains 10% of the data is known as the *validation* data set, which is used to validate the selected final model. If the closeness of the predictive values and the observed responses is within a pre-specified range, we claim that the model is validated.

17.3.1.4 Model Generalizability

In clinical research, it is often of interest to generalize clinical results obtained from a given target patient population at a medical center to a similar patient population at another medical center. Denote the original target patient population by (μ_0, σ_0) and the similar but different patient population by (μ_1, σ_1). Since the two populations are similar but different, it is reasonable to assume that $\mu_1 = \mu_0 + \varepsilon$ and $\sigma_1 = C\sigma_0$ $(C > 0)$, where ε is referred to as the shift in location parameter (population mean) and C is the inflation factor of the scale parameter (population standard deviation). Thus, the (treatment) effect size adjusted for standard deviation of population (μ_1, σ_1) can be expressed as follows:

$$E_1 = \left|\frac{\mu_1}{\sigma_1}\right| = \left|\frac{\mu_0 + \varepsilon}{C\sigma_0}\right| = |\Delta|\left|\frac{\mu_0}{\sigma_0}\right| = |\Delta|E_0,$$

where $\Delta = (1 + \varepsilon / (\mu_0)/C)$, and E_0 and E_1 are the effect size (of clinically meaningful importance) of the original target patient population and the similar but different patient population, respectively. Chow et al. (2002) and Chow and Chang (2006) refer to Δ as a sensitivity index measuring the change in effect size between patient populations. As it can be seen from the above, if $\varepsilon = 0$ and $C = 1$ (there is no shift in target patient population; the two patient populations are identical), $E_1 = E_0$. That is, the effect sizes of the two populations are identical. In this case, we claim that the results observed from the original target patient population (e.g., adults) can be generalized to the similar but different patient populations. Thus, if we can show that the sensitivity index Δ is within an acceptable range say 80% and 120%, we may claim that the results observed at the original medical center can be generalized to another medical center with similar but different patient population.

As indicated in Chow et al. (2005), the effect sizes of the two population could be linked by baseline demographics or patient characteristics if there is a relationship between the effect sizes and the baseline demographics and/or patient characteristics (a covariate vector). However, such covariates may not exist or may not be observable in practice. In this case, Chow et al. (2005) suggested assessing the sensitivity index by simply replacing ε and C with their corresponding estimates. Intuitively, ε and C can be estimated by

$$\hat{\varepsilon} = \hat{\mu}_1 - \hat{\mu}_0 \quad \text{and} \quad C = \hat{\sigma}_1 / \hat{\sigma}_0,$$

where $(\hat{\mu}_0, \hat{\sigma}_0)$ and $(\hat{\mu}_1, \hat{\sigma}_1)$ are some estimates of (μ_0, σ_0) and (μ_1, σ_1), respectively. Thus, the sensitivity index can be estimated by

$$\hat{\Delta} = \frac{1 + \hat{\varepsilon} / \hat{\mu}_0}{\hat{C}}.$$

Note that in practice, the shift in location parameter (i.e., ε) and/or the change in scale parameter (i.e., C) could be random. If both ε and C are fixed, the sensitivity can be assessed based on the sample means and sample variances obtained from the two populations. In real world problems, however, ε and C could be either a fixed or a random variable. In other words, there are three possible scenarios: (1) the case where ε is random and C is fixed, (2) the case where ε is fixed and C is random, and (3) the case where both ε and C are random. These possible scenarios have been studied by Lu et al. (2017).

17.3.2 Meta-analysis

L'Abbe et al. (1987) defined a meta-analysis as a systematic reviewing strategy for addressing research questions that is especially useful when (1) results from real studies disagree with regard to direction of effect, (2) sample sizes are individually too small to detect an effect, or (3) a large trial is too costly and time-consuming to perform. The primary objectives of a meta-analysis are not only to reduce bias (e.g., have a much narrower confidence interval for the true treatment effect) but also to increase statistical power (e.g., increase the probability of correctly detecting the true treatment effect if it does exist). In addition, it is to provide a more accurate and reliable statistical inference on the true treatment effect of a given therapy.

17.3.2.1 Issues in Meta-analysis

There are, however, several critical issues such as representativeness, selection bias, heterogeneity (similarities and dis-similarities), and poolability that may affect the validity of a meta-analysis. These issues are inevitably encountered because a meta-analysis often combines a number of studies with similar but different protocols, target patient population, doses or dose regimens, sample sizes, study endpoints, laboratories (local or central laboratories), equipment/analyst/times, and so on.

In practice, it is often a concern that only positive clinical studies are included in the meta-analysis, which may have introduced selection bias for a fair and unbiased comparison. To avoid possible selection bias, criteria for selection of studies to be included in the meta-analysis need to be clearly stated in the study protocol. Also, the time period (e.g., in the past 5 years or the most recent 10 studies) should also be specified. If there is a potential trend over time, a statement need to be provided for scientific justification for inclusion of the studies selected for the meta-analysis.

Before the studies are combined for a meta-analysis, similarity and/or dis-similarity among studies need to be assessed in order to reduce variability for a more accurate and reliable assessment of the test treatment under investigation. This is critical because different studies may be conducted with similar but different: (1) study protocols; (2) drug products and doses; (3) patient populations; (4) sample sizes; and (5) evaluability criteria.

Thus, the FDA suggested that before the data sets obtained from different studies can be combined for a meta-analysis, it is suggested that a statistical test for poolability be performed to determine whether there is significant treatment-by-study interaction. If a significant qualitative treatment-by-study interaction is observed, the studies should be combined for a meta-analysis, while if a significant quantitative treatment-by-study interaction is detected, the studies could be combined for a meta-analysis.

In practice, several methods are commonly employed for meta-analysis. These methods include, but are not limited to: (1) simple lumping or collapsing (which may be misleading); (2) graphical presentation (which provides a visual impression of inter-study consistency but does not provide any statistical inference and estimation of the treatment effect); (3) averaging p-values (which cannot show effects in different direction and does not reflect sample size of each study; (4) averaging test statistics (which provides adjustment by sample size but cannot test for interaction and heterogeneity); (5) blocking the data (which allows test for interaction and heterogeneity and provides estimates of inter-study and intra-study variations; and (6) random effects model for continuous variable and log-linear model for categorical data (see, e.g., DerSimonian and Laird, 1986). Most recently, it is suggested that a mixed effects model (by treating *study* as a random effect) be considered. The mixed effects model can utilize all data to (1) test for poolability, (2) compare the test treatment with a given control, and (3) handle missing data in conjunction with the GEE method.

17.4 Bias of Big Data Analytics

As indicated, big data usually consists of data sets from randomized and/or non-randomized (published or unpublished) studies. Thus, imbalances between the treatment group and the control group are likely to occur. In addition, studies with positive results are most likely to be published and accepted into the big data. In this case, even though the use of propensity score matching can help to reduce the bias due to some selection bias, the bias due to the fact that majority of data sets accepted to the big data are most likely those studies with positive results could be substantial and hence cannot be ignored. As a result, there is a bias of big data analytics regardless the big data analytics are case-control studies, meta-analyses, or data mining in genomics studies. In this section, we attempt to assess the bias due to the selection bias of accepting positive data sets into the big data.

Let μ and μ_B be the true mean of the target patient population with the disease under study and the true mean of the big data, respectively. Let $\varepsilon = \mu_B - \mu$, which depends upon the unknown percentage of data sets with positive results in the big data. Now, let μ_P and μ_N be the true means of data sets of positive studies and non-positive studies conducted in the target patient population, respectively. Also, let r be the proportion of positive studies conducted on the target patient population, which is usually unknown. For illustration and simplicity purposes, we assume that there is no treatment-by-center interaction for those multicenter studies and there is no treatment-by-study interaction. In this case, we have

$$\mu_B = r\mu_P + (1-r)\mu_N, \tag{17.1}$$

where $\mu_P > \delta > \mu_N$, in which δ is the effect of clinical importance. In other words, a study with an estimated effect greater than δ is considered a *positive* study. As it can be seen from (17.1), if the big data only contain data sets from positive studies, i.e., $r = 1$, (17.1) reduces to

$$\mu_B = \mu_P.$$

In other words, in this extreme case, the big data do not contain any studies with non-positive results. In practice, we would expect $\frac{1}{2} < r \leq 1$.

For a given big data, r can be estimated based on the number of positive studies in the big data (i.e., $\hat{r}$). In practice, $\hat{r}$ usually under-estimates the true r because the big data tends to accept data sets from published or positive studies. Thus, we have

$$E\left(\hat{r}\right) + \Delta = r.$$

Now, for simplicity, assume that all positive studies and non-positive studies are of the same size n_P and n_N, respectively. Let x_{ij} be the response of the ith subject in the jth positive study, $i = 1,\ldots,n_P$ and $j = 1,\ldots,rn$, where n is the total number of studies in the big data. Also, let y_{ij} be the response of the ith subject in the jth non-positive study, $i = 1,\ldots,n_N$ and $j = 1,\ldots,(1-r)n$. Thus, the bias of $\hat{\mu}_B$ is given by

$$\begin{aligned}\text{Bias}\left(\hat{\mu}_B\right) &= E\left(\hat{\mu}_B\right) - \mu = E\left[\hat{r}\hat{\mu}_P + \left(1-\hat{r}\right)\hat{\mu}_N\right] - \mu \\ &\approx \left(r - \Delta\right)\mu_P + \left(1 - r + \Delta\right)\mu_N - \mu \\ &= \varepsilon - \Delta\left(\mu_P - \mu_N\right),\end{aligned} \tag{17.2}$$

where $\hat{\mu}_P = \bar{x} = \frac{1}{rnn_P}\sum_{j=1}^{rn}\sum_{i=1}^{n_P} x_{ij}$ and $\hat{\mu}_N = \bar{y} = \frac{1}{rnn_N}\sum_{j=1}^{(1-r)n}\sum_{i=1}^{n_N} y_{ij}$. Thus, we have

$$\varepsilon = \Delta\left(\mu_P - \mu_N\right),$$

where $\mu_P > \delta > \mu_N$.

As an example, suppose there are 50% of positive studies and non-positive studies in the target patient population (i.e., $r = 0.5$) but 90% of positive studies were included in the big data (i.e., $\hat{r} = 0.9$). In this case, $\Delta = 0.9 - 0.5 = 0.4$. If we further assume that $\mu_P = 0.45$ and $\mu_N = 0.2$, then the bias of the big data analytics $\hat{\mu}_B$ could be as high as $\varepsilon = \Delta\left(\mu_P - \mu_N\right) = (0.4)(0.25) = 0.1$ or 10%.

To provide a better understanding, Table 17.1 summarizes potential biases that could occur due to the selection bias of accepting more positive studies into the big data when assessing the true treatment effect.

Regarding the power, by (1), the variance of $\hat{\mu}_B$ is given by

$$\begin{aligned}\text{Var}\left(\hat{\mu}_B\right) &= \frac{1}{n}\sigma_B^2 = r^2\left(\frac{\sigma_P^2}{nn_P}\right) + (1-r)^2\left(\frac{\sigma_N^2}{nn_N}\right) \\ &= \frac{1}{n}\left[r\sigma_P^2 / n_P + (1-r)\sigma_N^2 / n_N\right] \geq 0,\end{aligned}$$

where n_P and n_N are the size of positive and non-positive studies, respectively, and n is the total number of studies accepted into the big data. In addition, if we take derivative of the above, it leads to

TABLE 17.1

Potential Biases of Big Data Analytics

Δ (%)	$\mu_P - \mu_N$ (%)	Bias (ε)
10	20	2
	30	3
	40	4
20	20	4
	30	6
	40	8
30	20	6
	30	9
	40	12
40	20	8
	30	12
	40	16
50	20	10
	30	15
	40	20

$$\frac{\partial}{\partial r}\left[\mathrm{Var}\left(\hat{\mu}_B\right)\right] = \frac{1}{n}\left(\sigma_P^2 / n_P - \sigma_N^2 / n_N\right).$$

Thus, if $\sigma_P^2 / n_P > \sigma_N^2 / n_N$, σ_B^2 is an increasing function of r. In this case, it is expected that the power of big data analytics may decrease as r increases. The above discussion suggests that the power of the big data analytics can be studied through the evaluation of the following probability

$$P\left\{\hat{\sigma}_P^2 / n_P > \hat{\sigma}_N^2 / n_N \mid \mu_P, \mu_N, \sigma_P^2, \sigma_N^2, \text{and } r\right\}$$

based on data available in the big data.

17.5 Statistical Methods for Estimation of Δ and $\mu_P - \mu_N$

17.5.1 Estimation of Δ

As indicated in the previous section, $\hat{r}$ (the proportion of number of positive studies in the big data) is always an over-estimate of the true r (the proportion of the number of positive studies conducted with the target patient population), which is often unknown. However, following the concept of empirical power or reproducibility probability (Shao and Chow, 2002), we can estimate r by the reproducibility probability of observing positive result of a future study given that observed mean response and the corresponding sample variance as follows

$$p = P\left\{\text{future study is positive} \middle| \mu \equiv \hat{\mu}_B \text{ and } \sigma \equiv \hat{\sigma}_B \right\}. \tag{17.3}$$

The above expression can be interpreted as given the observed mean response ($\hat{\mu}_B$) and the corresponding sample standard deviation ($\hat{\sigma}_B$). We expect to see $p \times 100$ studies with positive results if we shall conduct the clinical trial under similar experimental conditions 100 times. Thus, intuitively, p is a reasonable estimate for the unknown r.

For simplicity and illustration purposes, suppose the investigator is interested in detecting a clinically meaningful difference (or an effect size that is of clinical importance). A typical approach is to test the following hypotheses of equality

$$H_0 : \mu_1 = \mu_0 \text{ versus } H_a : \mu_1 \neq \mu_0.$$

Under the null hypothesis, the test statistic is given by

$$T = \sqrt{\frac{n_1 n_0}{n_1 + n_0}} \frac{\left(\hat{\mu}_1 - \hat{\mu}_0\right)}{\hat{\sigma}}.$$

We reject the null hypothesis if $|T| > t_{\alpha/2, n_1 + n_0 - 2}$, where $t_{\alpha/2, n_1 + n_0 - 2}$ is the $(\alpha/2)$th upper quantile of a standard t distribution with $n_1 + n_0 - 2$ degrees of freedom. A typical approach is then to evaluate the power at under the alternative hypothesis that $\mu_1 - \mu_0 = \delta$. If there is at least 80% power for detecting the clinically meaningful difference of δ at the pre-specified α (say $\alpha = 5\%$) level of significance, we claim that the result is positive if the null hypothesis is rejected. In big data analytics, we have

$$\hat{\mu}_B = \hat{\mu}_1 - \hat{\mu}_0 \text{ and } \hat{\sigma}_B = \sqrt{\frac{n_1 \hat{\sigma}_1^2 + n_2 \hat{\sigma}_0^2}{n_1 + n_0}}.$$

Thus, (17.3) becomes

$$p = P\left\{T > t_{\alpha/2, n_1 + n_0 - 2} \middle| \mu \equiv \hat{\mu}_B \text{ and } \sigma \equiv \hat{\sigma}_B \right\}.$$

We then propose r be estimated by p, i.e., $\hat{r} = p$.

17.5.2 Estimation of $\mu_P - \mu_N$

Let (L_P, U_P) and (L_N, U_N) denote the $(1-\alpha) \times 100\%$ confidence interval for μ_P and μ_N, respectively. Under normality assumption and the assumption that $\sigma_P = \sigma_N = \sigma_B$, we have

$$(L_P, U_P) = \widehat{\mu}_P \pm z_{1-\alpha/2} \frac{\widehat{\sigma}_P}{\sqrt{n_P}} \text{ and } (L_N, U_N) = \widehat{\mu}_N \pm z_{1-\alpha/2} \frac{\widehat{\sigma}_N}{\sqrt{n_N}},$$

where $n_P = rn$, $n_N = (1-r)n$, and n is the sample size used to estimate μ_B.

Since $\mu_p > \delta > \mu_N$, the confidence intervals of μ_P and μ_N, i.e., (L_P, U_P) and (L_N, U_N) would not overlap each other. At extreme case, U_N is close to L_P. Thus, we have

$$\widehat{\mu}_P - z_{1-\frac{\alpha}{2}} \frac{\widehat{\sigma}_P}{\sqrt{n_P}} \approx \widehat{\mu}_N + z_{1-\frac{\alpha}{2}} \frac{\widehat{\sigma}_N}{\sqrt{n_N}}.$$

This leads to

$$\widehat{\mu}_P - \widehat{\mu}_N \approx z_{1-\frac{\alpha}{2}} \frac{\widehat{\sigma}_P}{\sqrt{n_P}} + z_{1-\frac{\alpha}{2}} \frac{\widehat{\sigma}_N}{\sqrt{n_N}}$$

$$= z_{1-\frac{\alpha}{2}} \left(\frac{\widehat{\sigma}_P}{\sqrt{n_P}} + \frac{\widehat{\sigma}_N}{\sqrt{n_N}} \right) \tag{17.4}$$

In some extreme cases, there are only data from positive studies achievable. The study with the least effect size would be used to estimate L_N and U_N.

17.5.3 Assumptions and Application

The two parameters, Δ and $(\mu_P - \mu_N)$, are corresponding to two assumptions: (i) the positive study is more likely to be published, and (ii) positive study and negative study have different distributions, which are reasonable considerations for selection bias in big data analytics. Based on the assumptions of the proposed approach, we suggest a two-step procedure to determine whether it is proper to apply the approach.

Step 1: Calculate the proportion of positive studies in big data center, $\widehat{r}$ and compare it with the designed power of each study included in the historical data set. If the proportion of positive studies is larger than power of most studies (mostly larger than all the studies in practice), then conduct step 2.

Step 2: Calculate the mean difference of positive studies and non-positive studies, $\hat{\mu}_P - \hat{\mu}_N$, and compare it with the theoretical distance given in (17.4). The adjustment could be conducted when

$$\hat{\mu}_P - \hat{\mu}_N < z_{1-\frac{\alpha}{2}}\left(\frac{\hat{\sigma}_P}{\sqrt{n_P}} + \frac{\hat{\sigma}_N}{\sqrt{n_N}}\right).$$

Through this two-step procedure, the estimated bias would be able to reduce bias that over-estimate M_1. The estimated power (EP) approach is considered as a conservative adjustment, which is reasonable when the bias is not extreme. When $\hat{r}$ is extremely larger than most power of studies included, the confidence bound (CB) approach would be more suitable.

17.6 Simulation Study

In this section, a simulation study was conducted to examine the performance of the proposed bias correction procedure by comparing the corrected treatment effect with the true effect size of the target patient population. A total of 1,000 simulation runs was considered.

In the simulation study, it is assumed that each trial contains two groups (treatment and control) of patients and the responses of two groups follow normal distribution $N(2,5)$ and $N(0,5)$ separately. Thus, 100 subjects per group would have a power of 80% to detect a clinically meaningful difference of 2. Sample size of 49, 63, 79, 100, and 133 per group were simulated for achieving a 50%, 60%, 70%, 80%, and 90% power. Positive studies and non-positive studies are divided through a t-test procedure.

To assess the performance of the proposed bias correction procedure, bias is defined as the relative difference between the estimated effect size $\hat{\mu}_B$ from pooled data and true effect size μ, which would be 2 in this simulation study, as follows:

$$\text{Bias} = \left(\hat{\mu}_B - \mu\right) / \mu$$

Adjusted bias is defined as $\left(\hat{\mu}_B - \mu - \varepsilon\right) / \mu$, in which ε is estimated bias correction factor by the proposed method.

To assess the performance of the proposed approach, the following scenarios were considered:

Scenario 1: performance of proposed bias adjustment approach using the EP or the CB approach.

Scenario 2: performance of the proposed bias adjustment approach when negative studies are absent.

Scenario 3: performance of the proposed bias adjustment approach for a small number of historical studies.

For each combination of parameters in these scenarios, 100 repetitions were conducted. The simulation study and data analysis were performed by SAS 9.4.

In scenario 1, the goal of the simulation study is to assess the performance of the proposed bias adjustment approach and compare the performance of the EP approach or the CB approach. Table 17.2 and Figure 17.1 summarize the amount of bias and adjustment. As expected, the selection bias increases with the observed proportion of positive study in the big data center. The EP approach can reduce the bias slightly when r is larger than the designed power by 15%. When the proportion r is close to the designed power ($\pm 10\%$), the EP approach is not suitable as it may increase the bias. The CB approach yields similar performance. The amount of adjustment made by the confidence approach is larger than the EP approach. When the proportion of positive study, r, is nearly 15% larger than the designed power, the adjusted bias is shown to be very close to 0. Both the proposed approach can achieve a conservative estimated of M_1 by adjusting the bias toward the right direction, and the adjustment of the EP approach is shown to be more moderate than the CB approach.

Table 17.3 summarizes the performance of the EP approach and the CB approach when the big data contains only positive studies. In this situation, with a fixed $\hat{r} = 100\%$, bias becomes smaller when designed power increase. By considering the study with least effect size as the negative study, the EP approach showed a slightly reduction of bias. Compared with Table 17.2, the performance of the EP approach shows no different, bias is also slightly reduced from 1.6% to 6.4%, and the reduction is increased with the amount of bias. The CB approach also yield similar outcomes in Table 17.2. When the designed power is around 80%, the adjusted bias is distributed around 0. Among all tested designed power, the simulation shows that both approaches are still reliable when the historical data is constituted by only positive studies.

Table 17.4 shows the result when the historical data is constituted by a small number of positive studies, which is a usual condition in practice. In this scenario, we assume historical data sets are constituted by 2 to 10

TABLE 17.2

Performance of Proposed Bias Adjustment Approach at Different Power

Power	r	Bias	Adjusted Bias	
			EP Approach	CB Approach
0.5080	0.70	0.2597	0.2571	0.0473
	0.75	0.3340	0.3304	0.1012
	0.80	0.3945	0.3872	0.1388
	0.85	0.4617	0.4481	0.1739
	0.90	0.5082	0.4812	0.1664
	0.95	0.5888	0.5449	0.1483
0.6122	0.70	0.1324	0.1355	−0.0529
	0.75	0.1904	0.1910	−0.0109
	0.80	0.2573	0.2560	0.0359
	0.85	0.3128	0.3053	0.0639
	0.90	0.3726	0.3565	0.0803
	0.95	0.4256	0.3918	0.0464
0.7102	0.70	0.0386	0.0477	−0.1230
	0.75	0.0934	0.1002	−0.0825
	0.80	0.1543	0.1589	−0.0383
	0.85	0.2111	0.2129	−0.0013
	0.90	0.2625	0.2553	0.0139
	0.95	0.2975	0.2725	−0.0349
0.8074	0.70	−0.0351	−0.0190	−0.1740
	0.75	0.0149	0.0290	−0.1343
	0.80	0.0640	0.0758	−0.0978
	0.85	0.1156	0.1229	−0.0651
	0.90	0.1510	0.1498	−0.0620
	0.95	0.2112	0.1995	−0.0620
0.9035	0.70	−0.1024	−0.0770	−0.2128
	0.75	−0.0680	−0.0465	−0.1881
	0.80	−0.0256	−0.0076	−0.1542
	0.85	0.0129	0.0261	−0.1314
	0.90	0.0715	0.0806	−0.0896
	0.95	0.1121	0.1111	−0.0974

Abbreviations: EP: estimated power; CB: confidence bound.

positive studies. The designed power of each study is set to be 80%. Both approaches are shown to be capable of bias reduction. However, as shown in Figure 17.2, the bias of historical data that constituted by a small number of studies is more uncertainty, in other words, small historical data is more likely to produce extreme outcomes. In this case, the CB approach may

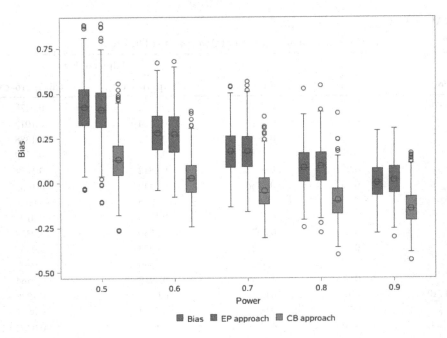

FIGURE 17.1
Performance of the proposed bias adjustment approach at different power.

TABLE 17.3

Performance of the Proposed Bias Adjustment Approach When Negative Studies
Are Absent

| | | Adjusted Bias ||
Power	Bias	EP Approach	CB Approach
0.5080	0.6384	0.5747	0.1704
0.6122	0.5047	0.4597	0.1184
0.7102	0.3602	0.3228	0.0242
0.8074	0.2751	0.2519	0.0064
0.9035	0.1444	0.1283	−0.0702

increase the variability of bias. With the number of studies included in the
historical data accumulated, the EP approach shows better performances.
This is because the bias of historical data becomes larger when more positive
studies are included.

In summary, both the EP approach and the CB approach are capable of bias
reduction when the bias indeed exists. The EP approach is more conservative
which is appropriate when the bias is not extreme. The CB approach would

TABLE 17.4

Performance of Proposed Bias Adjustment Approach for Different Number of Studies

		Adjusted Bias	
Number of Studies	Bias	EP Approach	CB Approach
2	0.2660	0.1988	−0.1577
3	0.2751	0.2310	−0.0957
4	0.2663	0.2255	−0.0850
5	0.2500	0.2109	−0.0898
6	0.2636	0.2290	−0.0553
7	0.2625	0.2297	−0.0503
8	0.2283	0.1911	−0.0942
9	0.2544	0.2228	−0.0529
10	0.2483	0.2158	−0.0594

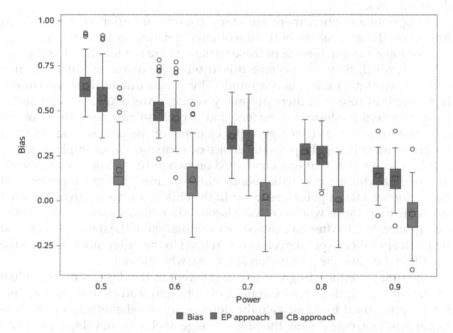

FIGURE 17.2

Performance of the proposed bias adjustment approach when negative studies are absent.

significantly reduce the bias when the estimated power of each study that included in the historical data is significantly lower than the proportion of positive studies observed. Either approach is not suitable when the number of studies is too small or the situation that the proportion of positive studies is close to the designed power.

17.7 Concluding Remarks

As big data include data sets from a variety of sources including registries, randomized or non-randomized clinical studies, published or unpublished data, positive or negative clinical results (data), and healthcare database, heterogeneity within and across these data sets will have an impact on the assessment of treatment effects of interest. Big data analytics provides opportunities for uncovering hidden important medical information, determining possible associations or correlations between possible risk factors and clinical outcomes, predictive model building, validation, and generalization, critical information for planning of future studies. Statistical methodology and software development are necessary for achieving these ultimate goals. Although there are benefits for big data analytics, statistical issues regarding representativeness of the big data and its quality, integrity, and validity as described in this chapter must be addressed to ensure the success of the big data analytics.

In big data analytics, there are many sources of variation that have an impact on the assessment of treatment effect relating to a certain new regimen or intervention. If some of these variations are not identified and properly controlled, they can become mixed with the treatment effect. In this case, the treatment effect is confounded by effects due to these variations. In biomedical research, there are many subtle, unrecognizable, and seemingly innocent confounding factors that can cause ruinous results of big data analytics. Moses (1985) gave the example of the devastating result in the confounder being the personal choice of a patient. The example concerns a polio vaccine trial that was conducted on two million children worldwide to investigate the effect of Salk poliomyelitis vaccine. This trial reported that the incidence rate of polio was lower in the children whose parents refused injection than those who received placebo after their parent gave permission (Meier, 1972). After an exhaustive examination of the data, it was found that susceptibility to poliomyelitis was related to the differences between the families who gave the permission and those who did not.

Sometimes, confounding factors are inherent in the designs of individual studies in the big data. For example, dose titration studies in escalating levels are often used to investigate the dose-response relationship of the antihypertensive agents during the phase 2 stage of clinical development. For a typical dose titration study, after a washout period during which previous medication stops and the placebo is prescribed, N subjects start at the lowest dose for a pre-specified time interval. At the end of the interval, each patient is evaluated as a responder to the treatment or a non-responder according to some criteria pre-specified in the protocol. In a titration study, a subject will continue to receive the next higher dose if he or she fails, at the current level, to meet some objective physiological criteria such as reduction of diastolic blood pressure by a pre-specified amount and has not experienced any unacceptable adverse experience. Dose titration studies are quite popular among clinicians because they mimic real clinical practice in the care of patients (Ohlhorst, 2012). The major problem with this typical design for a dose titration study is that the dose-response relationship is often confounded with time course and the unavoidable carryover effects from the previous dose levels which cannot be estimated and eliminated. Thus, in big data analytics, appropriate statistical methodology must be developed in order to address the issue of possible confounding factors for a valid assessment of the treatment effect under investigation.

As indicated earlier, NIH has launched the bd2K initiative to focus on the following areas: data compression/reduction, data visualization, data provenance, and data wrangling, which require innovative analytical methods and software tools with the objective of addressing critical current and emerging needs of the biomedical research community for using, managing, and analyzing the larger and more complex data sets inherent to biomedical big data. Data compression is referred to as the algorithm-based conversion of large data sets into alternative representations that require less space in memory, while data reduction is the reduction of data volume via the systematic removal of unnecessary data bulk. Data visualization refers to human-centric data representation that aids information presentation, exploration, and manipulation. Data provenance, on the other hand, is referred to as the chronology or record of transfer, use, and alteration of data that document the reverse path from a particular set of data back to the initial creation of a source data set. Finally, data wrangling is a term that is applied to activities that make data more usable by changing their form but not their meaning, which may involve reformatting data, mapping data from one data model to another, and/or converting data into more consumable forms.

One of the most controversial issues in big data analytics occurs when the finding of the big data analytics (with a large scale) is inconsistent with that from a relatively small scale of adequate well controlled randomized clinical trial which was conducted under the similar target patient population. In this case, the representativeness of the big data is questionable which may be due to the possible selection bias of accepting poor data sets into the big data. The inconsistency may indicate that there are major dissimilarities among individual data sets (studies) in the big data. Thus, it is suggested that similarities/ dissimilarities, possible interactions, and poolability be carefully assessed for identifying the possible causes of inconsistencies. The other controversial issue that is commonly seen is related to reproducibility of an established predictive model from big data analytics using similar but slightly different statistical methods. For example, in a case-control study utilizing the technique of propensity score matching with respect to some selected variables for matching, the use of (logistic) regression analysis with forward or backward stepwise approach often arrive similar but different predictive models with different sets of risk factors (predictors). Another controversial issue in big data analytics is related to the possible time effect.

In practice, it is likely that the findings from big data analytics at different time periods are different. This may be due to the availability of advanced technology, genetic changes in patient population, and health care over time. As a result, it is suggested that these factors be taken into consideration for a more accurate and reliable assessment of treatment effect (or clinically meaningful difference) under study.

In practice, the bias of meta-analysis in big data analytics inevitably occurs due to the ignorance of data sets (studies) with non-positive results. The proposed method provides a way to correct the bias based on empirical power by treating the observed treatment effect and the corresponding variability as the true values. Although the performance of the proposed method is satisfactory, the primary assumptions that the variabilities and sizes of studies with positive results and studies with non-positive results are the same, i.e., σ_P^2 and n_P and σ_N^2 and n_N, respectively, are rather strong. In practice, $\sigma_{Pi}^2 \neq \sigma_{Pj}^2$ for $i \neq j, i, j = 1, \ldots, rn$ and $\sigma_{Ni}^2 \neq \sigma_{Nj}^2$ for $i \neq j, i, j = 1, \ldots, (1-r)n$. It is then suggested that (1) treatment-by-center interaction within individual studies and (2) the poolability (i.e., treatment-by-study interaction) be tested before the big data analytics for statistical validity.

18

Rare Diseases Drug Development

18.1 Introduction

A rare disease is defined by the Orphan Drug Act of 1983 as a disorder or condition that affects less than 200,000 persons in the United States. As indicated in an Food and Drug Administration (FDA) draft guidance, most rare diseases are genetic, and thus are present throughout the person's entire life, even if symptoms do not immediately appear (FDA, 2019). Many rare diseases appear early in life, and about 30% of children with rare diseases will die before reaching their fifth birthday. FDA is to advance the evaluation and development of products including drugs, biologics, and devices that demonstrate promise for the diagnosis and/or treatment of rare diseases or conditions. Along this line, FDA evaluates scientific and clinical data submissions from sponsors to identify and designate products as promising for rare diseases and to further advance scientific development of such promising medical products. Following the Orphan Drug Act, FDA also provides incentives program for sponsors to develop products for rare diseases. The program has successfully enabled the development and marketing of over 600 drugs and biologic products for rare diseases since 1983.

FDA's incentives program include (i) fast track designation (ii) breakthrough therapy designation (iii) priority review designation and (iv) accelerated approval for approval of rare disease drug products. In its recent guidance, however, the FDA emphasizes that the FDA will not create a statutory standard for approval of orphan drugs that is different from the standard for approval of other, more typical drugs (FDA, 2019). For approval of drug products, the FDA requires that substantial evidence regarding effectiveness and safety of the drug products be provided. Substantial evidence is based on the results of adequate and well-controlled investigations (21 CFR 314.126(a)).

As rare diseases may affect far fewer persons, one of the major concerns of rare disease clinical trials is that often there are only a small number of potential subjects available. With the limited number of subjects available, it is feared that there may not be sufficient power for detecting a clinically meaningful difference (treatment effect). In rare disease drug development, power calculation for required sample size may not be feasible for rare disease clinical trials. In this case, alternative methods such as precision analysis, reproducibility analysis, or probability monitoring approach may be considered for providing substantial evidence with certain statistical assurance. In practice, however, small patient population is a challenge to rare disease clinical trials for obtaining substantial evidence of safety and effectiveness of the drug under investigation.

For development of drug products for rare disease development, data collection is essential for obtaining substantial evidence for approval of the drug products. Proper data collection is a key to the success of the intended trial. Thus, utilizing innovative trial designs and statistical methods in a rare disease setting is necessary important for the success of rare diseases drug development.

In Section 18.2, some basic considerations in rare disease clinical trials are outlined. Section 18.3 introduces several innovative trial designs such as a complete *n*-of-1 trial design, an adaptive design, master protocols (e.g., platform trial design), and a Bayesian design. Statistical methods for data analysis under these innovative trial designs are derived and discussed in Section 18.4. Section 18.5 proposes several criteria for evaluation of rare disease clinical trials. Some concluding remarks are given in Section 18.6.

18.2 Basic Considerations

For approval of drug products, the FDA requires that substantial evidence regarding the effectiveness and safety of the drug products under development be provided. This, however, is a major challenge for rare disease drug product development because there are only a limited number of subjects available. Thus, some basic principles are necessarily considered. These basic considerations are described below.

18.2.1 Historical Data

In its draft guidance, the FDA encourages product sponsors to evaluate existing natural history knowledge in the course of formulating drug development programs. Natural history studies help in defining the disease population, selecting clinical endpoints (sensitive and specific measures), and developing new or optimized biomarkers in early rare disease drug development. Most importantly, natural history studies provide an external control group for clinical trials. This is important especially for rare disease clinical trials with small patient population.

In practice, natural history studies could be conducted either prospectively or retrospectively (e.g., based on existing medical records such as patient charts).

Data could be collected from cohorts of patients under either a cross-sectional study or a longitudinal study. As indicated by the FDA, for a prospective design, a cross-sectional study may be conducted more quickly than a longitudinal study. However, cross-sectional studies are unable to provide a comprehensive description of the course of progressive or recurrent diseases (FDA, 2019).

18.2.2 Ethical Consideration

As indicated by Grady (2017), the principles that contribute to ethical research include collaborative partnership with stakeholders, social value, scientific validity, fair subject evaluation, a balance of risk and benefit, informed consent, respect for participants, and independent review (see also Coors et al., 2017). Independent review is usually conducted by an institutional review board (IRB). An IRB ensures that ethical requirements are fulfilled including biases are eliminated, ethical issues between investigators and participants are balanced, and the research does not exploit individuals or groups. In practice, several critical and emerging ethical issues present challenges to rare disease clinical research. These challenges include relatively few participants; the need for multi-site studies; innovative designs; and the need to protect privacy in a contained research environment.

As an example, as is the case in most clinical trials, it may not be ethical to conduct placebo-controlled trials for rare diseases when effective treatments exist. Placebo-controlled trials, however, are generally considered acceptable when risks are minimized and there is no increased risk of serious harm to those in the placebo arm. In case of placebo-controlled trials, an unequal treatment allocation, e.g., 2 (test treatment): 1 (placebo) may be considered to minimize possible ethical concern. For another example, since there are a relatively small number of potential subjects to enroll, an innovative trial design such as n-of-1 crossover design and adaptive trial design may be useful for obtaining substantial evidence regarding the effectiveness and safety of the test treatment under investigation.

18.2.3 The Use of Biomarkers

In clinical trials, a biomarker is usually used to select *right* patient population (i.e., patient population who are most likely to respond to the test treatment). This process is usually referred to as an enrichment process assuming that the biomarker is predictive of clinical outcomes. A biomarker is also often used in the development of diagnostic procedure for early detection of disease. It can also help in achieving personalized medicine (individualized medicine or precision medicine).

As compared to a hard (or gold standard) endpoint such as survival, a biomarker often has the following characteristics: (i) it can be measured earlier, easier, and more frequently; (ii) it is less subjected to competing risks; (iii) it is less affected by other treatment modalities; (iv) it can detect a larger effect size (i.e., a smaller sample size is required); and (v) it is predictive of clinical endpoints. The use of biomarker has the following advantages: (i) it can lead to better target population, (ii) it can detect a larger effect size (clinical benefit) with a smaller sample size, and (iii) it allows an early and faster decision-making.

Thus, under that assumption that there is a well-established relationship between a biomarker and clinical outcomes, the use of biomarker in rare disease clinical trials cannot only allow screening for possible responders at enrichment phase but also provide the opportunity to detect signal of potential safety concerns early and provides supportive evidence of efficacy with small number of patients available.

18.2.4 Generalizability

As noted in the FDA draft guidance, about half of the people affected by rare diseases are children. Thus, conducting clinical trials with pediatric patients is critical in rare disease drug development so that the drug can be properly labeled for pediatric use. Note that a pediatric population is defined as the group of subjects whose ages are between birth to 17 years. The FDA encourages sponsors to develop formulations for pediatric population.

In clinical development, after the drug product has been shown to be effective and safe with respect to the targeted patient population, it is often of interest to determine how likely the clinical results can be reproducible to a *different but similar* patient population with the same disease. For example, if an approved drug product is intended for the adult patient population with a certain rare disease, it is often of interest to study the effect of the drug product on a different but similar patient population, such as pediatric or elderly patient population with the same rare disease. In addition, it is also of interest to determine whether the clinical results can be generalized from one patient population to other patient population with/without ethnic differences. Shao and Chow (2002) proposed the concept of the *probability of generalizability*, which is the measurement of reproducibility probability of a future trial with the population slightly deviated from the targeted patient population of the current trials. The assessment of generalizability probability can be used to determine whether the clinical results can be generalized from the targeted patient population to a different but similar patient population with the same rare disease.

18.2.5 Sample Size

For rare disease drug development, the Orphan Drug Act provides incentives associated with orphan drug designation to make the development of rare disease drug products with small number of patients financially viable (FDA, 2019). However, FDA does not have intention to create a statutory standard for approval of orphan drugs that is different from the standard for approval of drugs in common conditions. Thus, sample size requirement has become one of the most challenging issues in rare disease clinical trials.

In rare disease clinical trials, power calculation for required sample size may be irrelevant for rare disease clinical trials due to limited number of subjects available, especially when the anticipated treatment effect is relatively small. In this case, alternative methods such as precision analysis (or confidence interval approach), reproducibility analysis, or probability monitoring approach, and Bayesian approach may be considered for providing substantial evidence with certain statistical assurance (Chow et al., 2017). It, however, should be noted that the resultant sample sizes from these different analyses could be very different with different levels of statistical assurance achieved. Thus, for rare disease clinical trials, it is suggested that an appropriate sample size should be selected for achieving certain statistical assurance under a valid trial design.

In practice, it is suggested that statistical methods used for data analysis should be consistent with statistical methods used for sample size estimation for scientific validity of the intended clinical trial. The concepts of power analysis, precision analysis, reproducibility analysis, probability monitoring approach, and Bayesian approach should not be mixed up with, while the statistical methods for data analysis should reflect the desired statistical assurance under the trial design.

18.3 Innovative Trial Designs

As indicated earlier, small patient population is a challenge to rare disease clinical trials. Thus, there is a need for innovative trial designs in order to obtain substantial evidence with small number of subjects available for achieving the same standard for regulatory approval. In this section, several innovative trial designs including *n*-of-1 trial design, an adaptive trial design, master protocols, and a Bayesian design are discussed.

18.3.1 *n*-of-1 Trial Design

One of the major dilemmas for rare diseases clinical trials is the in-availability of patients with the rare diseases under study. In addition, it is unethical to consider a placebo control in the intended clinical trial. Thus, it is suggested that an *n*-of-1 crossover design be considered. An *n*-of-1 trial design is to apply *n* treatments (including placebo) in an individual at different dosing periods with sufficient washout in between doing periods. A complete *n*-of-1 trial design is a crossover design that consists of all possible combinations of treatment assignment at different dosing periods.

18.3.1.1 Complete n-of-1 Trial Design

Suppose there are p dosing periods and two test treatments, e.g., a test (T) treatment and a reference (R) product, to be compared. A complete n-of-1 trial design for comparing two treatments consists of $\prod_{i=1}^{p} 2$, where $p \geq 2$, sequences of p treatments (either T or R at each dosing period). In this case, $n = p$. If $p = 2$, then the n-of-1 trial design is a 4×2 crossover design, i.e., (RR, RT, TT, TR), which is a typical Balaam's design. When $p = 3$, the n-of-1 trial design becomes an 8×3 crossover design, while the complete n-of-1 trial design with $p = 4$ is a 16×4 crossover design, which is illustrated in Table 18.1.

As indicated in a recent FDA draft guidance, a two-sequence dual design, i.e., (RTR, TRT), and a 4×2 crossover designs, i.e., (RTRT, RRRR), are commonly considered switching designs for assessing interchangeability in biosimilar product development (FDA, 2017a). These two switching designs, however, are limited for fully characterizing relative risk (i.e., reduction in efficacy or increase in incidence rate of adverse event rate). On the other hand, these two trial designs are special cases of the complete n-of-1 trial design with 3 or 4 dosing periods, respectively. Under the complete n-of-1 crossover design with 4 dosing periods, all possible switching and alternations can be assessed, and the results can be compared within the same group of patients and between different groups of patients.

TABLE 18.1

Examples of Complete n-of-1 Designs with $p = 4$

Group	Period 1	Period 2	Period 3	Period 4
1	R	R	R	R
2	R	T	R	R
3	T	T	R	R
4	T	R	R	R
5	R	R	T	R
6	R	T	T	T
7	T	R	T	R
8	T	T	T	T
9	R	R	R	T
10	R	R	T	T
11	R	T	R	T
12	R	T	T	R
13	T	R	R	T
14	T	R	T	T
15	T	T	R	T
16	T	T	T	R

Note: The first block (a 4×2 crossover design) is a complete n-of-1 design with 2 periods, while the second block is a complete n-of-1 design with 3 periods.

18.3.1.2 Merits and Limitations

A complete *n*-of-1 trial design has the following advantages: (i) each subject is at his/her own control; (ii) it allows a comparison between the test
product and the placebo if the intended trial is a placebo-controlled study
(this has lifted the ethical issue of using placebo on the patients with
critical conditions) (iii) it allows estimates of intra-subject variability;
(iv) it provides estimates for treatment effect in the presence of possible
carryover effect, and most importantly; (v) it requires less subjects for
achieving the study objectives of the intended trial design. However, the
n-of-1 trial design suffers from the drawbacks of (i) possible dropouts or
missing data and (ii) patients' disease status may change at each dosing
period prior to dosing.

18.3.2 Adaptive Trial Design

Another useful innovative trial design for rare disease clinical trials is an
adaptive trial design. In its draft guidance on adaptive clinical trial design,
the FDA defines an adaptive design as a study that includes a *prospectively*
planned opportunity for modification of one or more specified aspects of
the study design and hypotheses based on analysis of (usually interim) data
from subjects in the study (FDA, 2010b, 2018). The FDA draft guidance has
been served as an official document describing the potential use of adaptive
designs in clinical trials since it was published in 2010. It, however, should
be noted that the FDA draft guidance on adaptive clinical trial design is currently being revised in order to reflect pharmaceutical practice and FDA's
current thinking.

In practice, a two-stage seamless adaptive trial design is probably
the most commonly considered adaptive trial design in clinical trials.
A seamless trial design is referred to as a program that addresses study
objectives within a single trial that are normally achieved through separate trials in clinical development. An adaptive seamless design is a
seamless trial design that would use data from patients enrolled before
and after the adaptation in the final analysis. Thus, a two-stage seamless adaptive design consists of two phases (stages), namely a learning
(or exploratory) phase (Stage 1) and a confirmatory phase (Stage 2).
The learning phase provides that opportunity for adaptations such as
stop the trial early due to safety and/or futility/efficacy based on accrued
data at the end of the learning phase. A two-stage seamless adaptive trial
design reduces lead time between the learning (i.e., the first study for
the traditional approach) and confirmatory (i.e., the second study for the
traditional approach) phases. Data collected at the learning phase are
combined with those data obtained at the confirmatory phase for final
analysis. The specific design features of a two-stage seamless adaptive
trial design can overcome the limitations and dilemma of rare disease

TABLE 18.2

Types of Two-Stage Seamless Adaptive Designs

Study Objectives	Study Endpoint	
	Same (S)	Different (D)
Same (S)	I = SS	II = SD
Different (D)	III = DS	IV = DD

clinical trials. Thus, a two-stage seamless adaptive trial design is not only feasible but also useful for rare disease clinical trials (Table 18.2).

In practice, two-stage seamless adaptive trial designs can be classified into the following four categories depending upon study objectives and study endpoints used at different stages. In other words, we have Category I (SS)—same study objectives and same study endpoints, Category II (SD)—same study objectives but different study endpoints, Category III (DS)—different study objectives but same study endpoints, and Category IV (DD)—different study objectives and different study endpoints. Note that different study objectives are usually referred to dose finding (selection) at the first stage and efficacy confirmation at the second stage, while different study endpoints are directed to biomarker versus clinical endpoint or the same clinical endpoint with different treatment durations.

Category I trial design is often viewed as a similar design to a group sequential design with one interim analysis despite that there are differences between a group sequential design and a two-stage seamless design. In this article, our emphasis will be placed on Category II designs. The results obtained can be similarly applied to Category III and Category IV designs with some modification for controlling the overall type I error rate at a prespecified level. In practice, typical examples for a two-stage adaptive seamless design include a two-stage adaptive seamless phase I/II design and a two-stage adaptive seamless phase II/III design. For the two-stage adaptive seamless phase I/II design, the objective at the first stage is for biomarker development and the study objective for the second stage is to establish early efficacy. For a two-stage adaptive seamless phase II/III design, the study objective is for treatment selection (or dose finding), while the study objective at the second stage is for efficacy confirmation.

18.3.3 Other Designs

18.3.3.1 Master Protocol

Woodcock and LaVange (2017) introduced the concept of master protocol for studying multiple therapies, multiple diseases, or both in order to answer more questions in a more efficient and timely fashion (see also Redman and Allegra, 2015). Master protocols include the following types of trials: umbrella, basket, and platform. The umbrella trial is designed to study multiple targeted therapies in the context of a single disease, while the type of basket trial is meant to study a single therapy in the context of multiple diseases or disease subtypes. The platform trial is intended to study multiple targeted therapies in the context of a single disease in an ongoing manner, with therapies allowed to enter or leave the platform on the basis of decision algorithm. As indicated by Woodcock and LaVange (2017), if designed correctly, master protocols offer a number of benefits that include streamlined logistics; improved data quality, collection and sharing; as well as the potential to use innovative statistical approaches to study design and analysis. Master protocols may be a collection of sub-studies or a complex statistical design or platform for rapid learning and decision-making.

In practice, master protocol is intended for the addition or removal of drugs, arms, and study hypotheses. Thus, in practice, master protocols may or may not be adaptive, umbrella, or basket studies. Since master protocol has the ability to combine a variety of logistical, innovative, and correlative elements, it allows learning more from smaller patient populations. Thus, the concept of master protocols in conjunction with adaptive trial design described in the previous section may be useful for rare diseases clinical investigation although it has been most frequently implemented in oncology research. For rare diseases drug development, a platform trial with master protocol is often considered. A platform trial is an exploratory multi-arm clinical trial evaluating one or more treatments on one or more cohorts (or populations) with an objective to screen and identify promising treatments in connection with some cohorts for further investigation (see also Table 18.3). Note that a platform trial is typically followed by confirmative studies further investigating potential arms identified by the screening outcomes.

TABLE 18.3

Types of Platform Trials for Rare Diseases Drug Development

Design Type	Description
T1 / P1 P2 P3	This design is to evaluate one new treatment for multiple cohorts or populations.
T1 T2 T3 / P1	This design is to evaluate multiple new treatments for the same cohort or population.
T1/SOC T2/SOC T3/SOC / P1	This design is to evaluate one or more new treatments in combination with different standard of cares (SOCs) for the same cohort or population.
T1 T2 T3 / P1 P2 P3	This design is to evaluate multiple new treatments for multiple cohorts or populations.

18.3.3.2 Bayesian Approach

Under the assumption that historical data (e.g., previous studies or experience) are available, Bayesian methods for borrowing information from different data sources may be useful. These data sources could include, but are not limited to, natural history studies and expert's opinion regarding prior distribution about the relationship between endpoints and clinical outcomes. The impact of borrowing on results can be assessed through the conduct of sensitivity analysis. One of the key questions of particular interest to the investigator and regulatory reviewer is that how much to borrow in order to (i) achieve desired statistical assurance for substantial evidence and (ii) maintain the quality, validity, and integrity of the study.

Although Bayesian approach provides a formal framework for borrowing historical information, which is useful in rare disease clinical trials, borrowing can only be done under the assumption that there is a well-established relationship between patient populations (e.g., from previous studies to current study). In practice, it is suggested not to borrow any data from previous studies whenever possible. The primary analysis should rely on the data collected from the current study. When borrowing, the associated risk to scientific/statistical validity of the final conclusion should be carefully evaluated. It should be noted that Bayesian approach may not be feasible if there are no prior experience or study available. The determination of prior in Bayesian is always debatable because the primary assumption of the selected prior is often difficult, if not impossible, to verify.

18.4 Statistical Methods for Data Analysis

Statistical methods used for data analysis should be consistent with statistical methods used for sample size calculation. Moreover, statistical methods should be able to overcome the issue of small sample size and achieve certain statistical assurance. Statistical methods for data analysis should be innovative in order to confirm the observed treatment effect has provided substantial evidence to support the safety and efficacy of the test treatment under investigation.

18.4.1 Analysis under a Complete *n*-of-1 Trial Design

18.4.1.1 Statistical Model

As it can be seen from Table 18.1, the complete *n*-of-1 trial design with 4 dosing periods can be generally described as a K-sequence, J-period (i.e., KxJ) crossover design. In this section, consider the following model for comparing two treatments, i.e., a test (T) drug and a reference (R) drug:

$$Y_{ijk} = \mu + G_k + S_{ik} + P_j + D_{d_{(j,k)}} + C_{d_{(j-1,k)}} + e_{ijk}$$

$$i = 1,2,\cdots,n_k; j = 1,2,\cdots,J; k = 1,2,\cdots,K; d = T \text{ or } R,$$

where μ is the overall mean, G_k is the fixed kth sequence effect, S_{ik} is the random effect for the ith subject within the kth sequence with mean 0 and variance σ_s^2, P_j is the fixed effect for the jth period, $D_{d_{(j,k)}}$ is the drug effect for the drug at the kth sequence in the jth period, $C_{d_{(j-1,k)}}$ is the carryover effect, and e_{ijk} is the random error with mean 0 and variance σ_e^2. Under the model, it is assumed that S_{ik} and e_{ijk} are mutually independent.

Define P as a parameter vector, referring to $(\mu, G_1, G_2, P_1, P_2, P_3, P_4, D_T, D_R, C_T, C_R)'$, which contains all unknown parameters in the model. The aim is to construct method of moment estimator in linear form of the observed cell means $\tilde{Y} = \beta'\bar{Y}$, where $\bar{Y}$ is the vector of observed mean vector. Then $E(\tilde{Y}) = E(\beta'\bar{Y}) = L'P$ which is the parameter of interest based on the linear contrast L. Let ω_{jk} is the expected value at the kth sequence and jth period, then $\omega = X'P$ with X as the design matrix. Setting $E(\beta'\bar{Y}) = L'P$ implies that $\beta'\omega = L'P \Rightarrow \beta'X'P = L'P$. Thus, in order to have $E(\beta'\bar{Y}) = L'P$, we should set $\beta'X' = L' \Rightarrow L = X\beta$. Therefore, the method of moment estimates of $L'P$ is $\beta'\bar{Y}$ with $\beta = (X'X)^- X'L$.

18.4.1.2 Statistical Analysis

To evaluate the biosimilarity in the complete n-of-1 design, it is necessary to derive the coefficients for estimation of the difference in drug effect (Table 18.4). Based on coefficients given in Table 18.4, the expected value of $\tilde{D}$ is given by

$$E(\tilde{D}) = D_T - D_R, \text{Var}(\tilde{D}) = \frac{\sigma_e^2}{11n},$$

where n is the number of subjects enrolled in each sequence (assuming equally allocated).

The null hypothesis will be rejected, and the bioequivalence will be demonstrated when the statistic

$$T_D = \frac{\tilde{D} - \theta}{\hat{\sigma}_e^2 \sqrt{\frac{1}{11n}}} > t\left[\frac{\alpha}{2}, 16n - 5\right].$$

The corresponding confidence interval at α significance level is

$$\tilde{D} \pm t\left[\frac{\alpha}{2}, 16n - 5\right] \hat{\sigma}_e^2 \sqrt{\frac{1}{11n}}.$$

TABLE 18.4

Coefficients for Estimates of Drug in Complete n-of-1 Design

| Sequence | 132βs to estimate $D = D_T - D_R$ | | | |
	I	II	III	IV
1	3	−1	−1	−1
2	−3	−7	−7	17
3	−5	−9	15	−1
4	−11	−15	9	17
5	−5	15	−1	−9
6	−11	9	−7	9
7	−13	7	15	−9
8	−19	1	9	9
9	19	−1	−9	−9
10	13	−7	−15	9
11	11	−9	7	−9
12	5	−15	1	9
13	11	15	−9	−17
14	5	9	−15	1
15	3	7	7	−17
16	−3	1	1	1

Note: Adjusting for Carryover Effect.

Similarly, the carryover effect coefficient estimates are derived, and the corresponding statistic for testing bioequivalence is shown in Table 18.5. Based on Table 18.5, the unbiased estimator $\tilde{C}$ can be constructed, and we have

$$E\left(\tilde{C}\right) = C_T - C_R, \operatorname{Var}\left(\tilde{C}\right) = \frac{4\sigma_e^2}{33n}.$$

Under the model without the first-order carryover effect, we can derive a different unbiased estimator $\tilde{D}$ such that

$$E\left(\tilde{D}\right) = D_T - D_R, \operatorname{Var}\left(\tilde{D}\right) = \frac{\sigma_e^2}{12n}.$$

Thus, the corresponding confidence interval at α significance level can be similarly constructed.

18.4.1.3 Sample Size Requirement

The sample size determination under the fixed power and significance level is derived based on the following hypothesis testing.

TABLE 18.5

Coefficients for Estimates of Carryover Effect
in Complete n-of-1 Design

	132βs to estimate $C = C_T - C_R$			
Sequence	I	II	III	IV
1	12	−4	−4	−4
2	10	−6	−6	2
3	2	−14	−6	18
4	0	−16	−8	24
5	2	−6	18	−14
6	0	−8	16	−8
7	−8	−16	16	8
8	−10	−18	14	14
9	10	18	−14	−14
10	8	16	−16	−8
11	0	8	−16	8
12	−2	6	−18	14
13	0	16	8	−24
14	−2	14	6	−18
15	−10	6	6	−2
16	−12	4	4	4

$$H_0 : |D_T - D_R| > \theta \quad \text{versus} \quad H_1 : |D_T - D_R| \le \theta$$

According to the ±20% rule, the bioequivalence is concluded if the average bioavailability of the test drug effect is within ±20% of that of the reference drug effect with a certain assurance. Therefore, θ is usually represented by $\nabla \mu_R$ where $\nabla = 20\%$ and the hypothesis testing can be rewritten as follows.

$$H_0 : \mu_T - \mu_R < -\nabla \mu_R \ \text{or} \ \mu_T - \mu_R > \nabla \mu_R \ \text{versus} \ H_a : -\nabla \mu_R \le \mu_T - \mu_R \le \nabla \mu_R$$

The power function can be written as

$$P(\theta) = F_v\left(\left[\frac{\nabla - R}{CV\sqrt{b/n}}\right] - t(\alpha, \upsilon)\right) - F_v\left(t(\alpha, \upsilon) - \left[\frac{\nabla + R}{CV\sqrt{b/n}}\right]\right),$$

where $R = \frac{\mu_T - \mu_R}{\mu_R}$ is the relative change; $CV = \frac{S}{\mu_R}$; μ_T and μ_R are the average bioavailability of the test and reference formulations, respectively; S is the squared root of the mean square error from the analysis of variance table for each crossover design; $[-\nabla \mu_R, \nabla \mu_R]$ is the bioequivalence limit interval; $t(\alpha, \upsilon)$ is the upper αth quantile of a t-distribution with υ degrees of freedom; F_v is the cumulative distribution function of the t-distribution; and b is the constant value for the variance of drug effect.

Accordingly, the exact sample size formula when $R = 0$ is given by

$$n \geq b\left[t(\alpha, \upsilon) + t\left(\frac{\beta}{2}, \upsilon\right)\right]^2 [CV/\nabla]^2 ;$$

the approximate sample size formula when $R > 0$ can be obtained as

$$n \geq b\left[t(\alpha, \upsilon) + t(\beta, \upsilon)\right]^2 \left[CV/(\nabla - R)\right]^2 .$$

Another way to determine the sample size is based on testing the ratio of drug effect between the biosimilar product and the reference product. Consider $\delta \in (0.8, 1.25)$ is the bioequivalence range of μ_T / μ_R, the hypothesis changes to

$$H_0 : \frac{\mu_T}{\mu_R} < 0.8 \text{ or } \frac{\mu_T}{\mu_R} > 1.25 \text{ versus } H_a : 0.8 \leq \frac{\mu_T}{\mu_R} \leq 1.25$$

In case of the skewed distribution, the hypotheses are transformed to logarithmic scale,

$$H_0 : \log\mu_T - \log\mu_R < \log(0.8) \text{ or } \log\mu_T - \log\mu_R > \log(1.25)$$

$$\text{versus } H_a : \log(0.8) \leq \log\mu_T - \log\mu_R \leq \log(1.25)$$

Then, the sample size formulas for different δ are given below (See details in Appendix),

$$n \geq b\left[t(\alpha, \upsilon) + t\left(\frac{\beta}{2}, \upsilon\right)\right]^2 [CV/\ln 1.25]^2 \quad \text{if } \delta = 1$$

$$n \geq b\left[t(\alpha, \upsilon) + t(\beta, \upsilon)\right]^2 \left[CV/(\ln 1.25 - \ln \delta)\right]^2 \quad \text{if } 1 < \delta < 1.25$$

$$n \geq b\left[t(\alpha, \upsilon) + t(\beta, \upsilon)\right]^2 \left[CV/(\ln 0.8 - \ln \delta)\right]^2 \quad \text{if } 0.8 < \delta < 1$$

18.4.2 Analysis under an Adaptive Trial Design

Statistical analysis for Category 1 (SS) two-stage seamless designs is similar to that of a group sequential design with one interim analysis. Thus, standard statistical methods for a group sequential design can be applied. For other kinds of two-stage seamless trial designs, standard statistical methods for a group sequential design are not appropriate and hence should not be applied directly. In this section, statistical methods for other types of two-stage adaptive seamless design are described. Without loss of generality, in this section, we will discuss n-stage adaptive design based on individual p-values from each stage (Chow and Chang, 2011).

Consider a clinical trial with K interim analyses. The final analysis is treated as the Kth interim analysis. Suppose that at each interim analysis, a hypothesis test is performed followed by some actions that are dependent on the analysis results. Such actions could be an early stopping due to futility/ efficacy or safety, sample size re-estimation, modification of randomization, or other adaptations. In this setting, the objective of the trial can be formulated using a global hypothesis test, which is an intersection of the individual hypothesis tests from the interim analyses

$$H_0 : H_{01} \cap \cdots \cap H_{0K},$$

where $H_{0i}, i = 1,...,K$ is the null hypothesis to be tested at the ith interim analysis. Note that there are some restrictions on H_{0i}, that is, rejection of any $H_{0i}, i = 1,...,K$ will lead to the same clinical implication (e.g., drug is efficacious); hence all $H_{0i}, i = 1,...,K$ are constructed for testing the *same* endpoint within a trial. Otherwise the global hypothesis cannot be interpreted.

In practice, H_{0i} is tested based on a sub-sample from each stage, and without loss of generality, assume H_{0i} is a test for the efficacy of a test treatment under investigation, which can be written as

$$H_{0i} : \eta_{i1} \geq \eta_{i2} \quad \text{versus} \quad H_{ai} : \eta_{i1} < \eta_{i2},$$

where η_{i1} and η_{i2} are the responses of the two treatment groups at the ith stage. It is often the case that when $\eta_{i1} = \eta_{i2}$, the p-value p_i for the sub-sample at the ith stage is uniformly distributed on $[0, 1]$ under H_0 (Bauer and Kohne, 1994). This desirable property can be used to construct a test statistic for multiple-stage seamless adaptive designs. As an example, Bauer and Kohne (1994) used Fisher's combination of the p-values. Similarly, Chang (2007) considered a linear combination of the p-values as follows

$$T_k = \sum_{i=1}^{K} w_{ki} p_i, i = 1,...,K, \tag{18.1}$$

where $w_{ki} > 0$, and K is the number of analyses planned in the trial. For simplicity, consider the case where $w_{ki} = 1$. This leads to

$$T_k = \sum_{i=1}^{K} p_i, i = 1,...,K. \tag{18.2}$$

The test statistic T_k can be viewed as cumulative evidence against H_0. The smaller the T_k is, the stronger the evidence is. Equivalently, we can define the test statistic as $T_k = \sum_{i=1}^{K} p_i / K$, which can be viewed as an average of the evidence against H_0. The stopping rules are given by

$$\begin{cases} \text{Stop for efficacy} & \text{if } T_k \leq \alpha_k \\ \text{Stop for futility} & \text{if } T_k \geq \beta_k \\ \text{Continue} & \text{otherwise} \end{cases} \tag{18.3}$$

where $T_k, \alpha_k,$ and β_k are monotonic increasing functions of k, $\alpha_k < \beta_k$, $k = 1,...,K-1$, and $\alpha_K = \beta_K$. Note that α_k and β_k are referred to as the efficacy and futility boundaries, respectively. To reach the kth stage, a trial has to pass 1 to $(k-1)$th stages. Therefore, a so-called proceeding probability can be defined as the following unconditional probability:

$$\psi_k(t) = P\left(T_k < t, \alpha_1 < T_1 < \beta_1,..., \alpha_{k-1} < T_{k-1} < \beta_{k-1}\right)$$

$$= \int_{\alpha_1}^{\beta_1} \cdots \int_{\alpha_{k-1}}^{\beta_{k-1}} \int_{-\infty}^{t} f_{T_1 \cdots T_k}(t_1,...,t_k) dt_k dt_{k-1} \cdots dt_1, \tag{18.4}$$

where $t \geq 0$; $t_i, i = 1,...,k$ is the test statistic at the ith stage; and $f_{T_1 \cdots T_k}$ is the joint probability density function. The error rate at the kth stage is given by

$$\pi_k = \psi_k(\alpha_k). \tag{18.5}$$

When efficacy is claimed at a certain stage, the trial is stopped. Therefore, the type I error rates at different stages are mutually exclusive. Hence, the experiment-wise type I error rate can be written as follows:

$$\alpha = \sum_{k=1}^{K} \pi_k. \tag{18.6}$$

Note that (18.4–18.6) are the keys to determine the stopping boundaries, which will be illustrated in the next sub-section, with two-stage seamless adaptive designs. The adjusted p-value calculation is the same as the one in a classic group sequential design. The key idea is that when the test statistic at the kth stage $T_k = t = \alpha_k$ (i.e., just on the efficacy stopping boundary), the p-value is equal to alpha spent $\sum_{i=1}^{k} \pi_i$. This is true regardless of which error spending function is used and consistent with the p-value definition of the classic design. The adjusted p-value corresponding to an observed test statistic $T_k = t$ at the kth stage can be defined as

$$p(t;k) = \sum_{i=1}^{k-1} \pi_i + \psi_k(t), k = 1,...,K. \tag{18.7}$$

This adjusted p-value indicates weak evidence against H_0, if the H_0 is rejected at a late stage because one has spent some alpha at previous stages. On the other hand, if the H_0 was rejected at an early stage, it indicates strong evidence against H_0 because there is a large portion of overall alpha that has not been spent yet. Note that p_i in (18.1) is the stage-wise naive (unadjusted) p-value from a sub-sample at the ith stage, while $p(t;k)$ are adjusted p-values calculated from the test statistic, which are based on the cumulative sample up to the kth stage where the trial stops; Equations (18.6) and (18.7) are valid regardless how the p_i values are calculated.

18.4.2.1 Two-Stage Adaptive Design

In this sub-section, we will apply the general framework to the two-stage designs. Chang (2007) derived the stopping boundaries and p-value formula for three different types of adaptive designs that allow (i) early efficacy stopping, (ii) early stopping for both efficacy and futility, and (iii) early futility stopping. The formulation can be applied to both superiority and non-inferiority trials with or without sample size adjustment.

18.4.2.1.1 Early Efficacy Stopping

For a two-stage design $(K = 2)$ allowing for early efficacy stopping $(\beta_1 = 1)$, the type I error rates to spend at Stages 1 and 2 are

$$\pi_1 = \psi_1(\alpha_1) = \int_0^{\alpha_1} dt_1 = \alpha_1, \tag{18.8}$$

and

$$\pi_2 = \psi_2(\alpha_2) = \int_{\alpha_1}^{\alpha_2} \int_t^{\alpha_1} dt_2 dt_1 = \frac{1}{2}(\alpha_2 - \alpha_1)^2, \tag{18.9}$$

respectively. Using (18.8) and (18.9), (18.6) becomes

$$\alpha = \alpha_1 + \frac{1}{2}(\alpha_2 - \alpha_1)^2. \tag{18.10}$$

Solving for α_2, we obtain

$$\alpha_2 = \sqrt{2(\alpha - \alpha_1)} + \alpha_1. \tag{18.11}$$

Note that when the test statistic $t_1 = p_1 > \alpha_2$, it is certain that $t_2 = p_1 + p_2 > \alpha_2$. Therefore, the trial should stop when $p_1 > \alpha_2$ for futility. The clarity of the method in this respect is unique, and the futility stopping boundary is often hidden in other methods. Furthermore, α_1 is the stopping probability (error spent) at the first stage under the null hypothesis condition, and $\alpha - \alpha_1$ is the

TABLE 18.6

Stopping Boundaries for Two-Stage Efficacy Designs

One-sided α	α_1	0.005	0.010	0.015	0.020	0.025	0.030
0.025	α_2	0.2050	0.1832	0.1564	0.1200	0.0250	—
0.05	α_2	0.3050	0.2928	0.2796	0.2649	0.2486	0.2300

Source: Chang, M., *Stat. Med.*, 26, 2772–2784.

error spent at the second stage. Table 18.6 provides some examples of the stopping boundaries from (18.11).

The adjusted *p*-value is given by

$$p(t;k) = \begin{cases} t & \text{if } k = 1 \\ \alpha_1 + \dfrac{1}{2}(t - \alpha_1)^2 & \text{if } k = 2 \end{cases}, \tag{18.12}$$

where $t = p_1$ if the trial stops at Stage 1, and $t = p_1 + p_2$ if the trial stops at Stage 2.

18.4.2.1.2 Early Efficacy or Futility Stopping

It is obvious that if $\beta_1 \geq \alpha_2$, the stopping boundary is the same as it is for the design with early efficacy stopping. However, futility boundary β_1 when $\beta_1 \geq \alpha_2$ is expected to affect the power of the hypothesis testing. Therefore,

$$\pi_1 = \int_0^{\alpha_1} dt_1 = \alpha_1, \tag{18.13}$$

and

$$\pi_2 = \begin{cases} \displaystyle\int_{\alpha_1}^{\beta_1}\int_{t_1}^{\alpha_2} dt_2 dt_1 & \text{for } \beta_1 \leq \alpha_2 \\ \displaystyle\int_{\alpha_1}^{\alpha_2}\int_{t_1}^{\alpha_2} dt_2 dt_1 & \text{for } \beta_1 > \alpha_2 \end{cases} \tag{18.14}$$

Carrying out the integrations in (18.13) and substituting the results into (18.6), we have

$$\alpha = \begin{cases} \alpha_1 + \alpha_2(\beta_1 - \alpha_1) - \dfrac{1}{2}(\beta_1^2 - \alpha_1^2) & \text{for } \beta_1 < \alpha_2 \\ \alpha_1 + \dfrac{1}{2}(\alpha_2 - \alpha_1)^2 & \text{for } \beta_1 \geq \alpha_2 \end{cases} \tag{18.15}$$

Various stopping boundaries can be chosen from (18.15). See Table 18.7 for examples of the stopping boundaries.

TABLE 18.7

Stopping Boundaries for Two-Stage Efficacy and Futility Designs

One-sided α		$\beta_1 = 0.15$				
0.025	α_1	0.005	0.010	0.015	0.020	0.025
	α_2	0.2154	0.1871	0.1566	0.1200	0.0250
		$\beta_1 = 0.2$				
0.05	α_1	0.005	0.010	0.015	0.020	0.025
	α_2	0.3333	0.3155	0.2967	0.2767	0.2554

Source: Chang, M., *Stat. Med.*, 26, 2772–2784.

The adjusted *p*-value is given by

$$p(t;k) = \begin{cases} t & \text{if } k = 1 \\ \alpha_1 + t(\beta_1 - \alpha_1) - \dfrac{1}{2}(\beta_1^2 - \alpha_1^2) & \text{if } k = 2 \text{ and } \beta_1 < \alpha_2 \\ \alpha_1 + \dfrac{1}{2}(t - \alpha_1)^2 & \text{if } k = 2 \ \beta_1 \geq \alpha_2 \end{cases} \quad (18.16)$$

where $t = p_1$ if the trial stops at Stage 1 and $t = p_1 + p_2$ if the trial stops at Stage 2.

18.4.2.1.3 Early Futility Stopping

A trial featuring early futility stopping is a special case of the previous design, where $\alpha_1 = 0$ in (18.15). Hence, we have

$$\alpha = \begin{cases} \alpha_2\beta_1 - \dfrac{1}{2}\beta_1^2 & \text{for } \beta_1 < \alpha_2 \\ \dfrac{1}{2}\alpha_2^2 & \text{for } \beta_1 \geq \alpha_2 \end{cases} \quad (18.17)$$

Solving for α_2, it can be obtained that

$$\alpha_2 = \begin{cases} \dfrac{\alpha}{\beta_1} + \dfrac{1}{2}\beta_1 & \text{for } \beta_1 < \sqrt{2\alpha} \\ \sqrt{2\alpha} & \text{for } \beta_1 \geq \alpha_2 \end{cases} \quad (18.18)$$

Examples of the stopping boundaries generated using (18.18) are presented in Table 18.8.

TABLE 18.8

Stopping Boundaries for Two-Stage Futility Design

One-sided α	β_1	0.1	0.2	0.3	≥ 0.4
0.025	α_2	0.3000	0.2250	0.2236	0.2236
0.05	α_2	0.5500	0.3500	0.3167	0.3162

Source: Chang, M., *Stat. Med.*, 26, 2772–2784.

The adjusted p-value can be obtained from (18.16), where $\alpha_1 = 0$, that is,

$$p(t;k) = \begin{cases} t & \text{if } k = 1 \\ \alpha_1 + t\beta_1 - \dfrac{1}{2}\beta_1^2 & \text{if } k = 2 \text{ and } \beta_1 < \alpha_2 \\ \alpha_1 + \dfrac{1}{2}t^2 & \text{if } k = 2 \; \beta_1 \geq \alpha_2 \end{cases}$$

18.4.2.2 Remarks

In practice, one of the questions that are commonly asked when applying a two-stage adaptive seamless design in clinical trials is about sample size calculation/allocation. For the first kind of two-stage seamless designs, the methods based on individual p-values as described in Chow and Chang (2006) can be applied. However, these methods are not appropriate for Category IV (DD) trial designs with different study objectives and endpoints at different stages. For Category IV (DD) trial designs, the following issues are challenging to the investigator and the biostatistician. First, how do we control the overall type I error rate at a pre-specified level of significance? Second, is the typical O'Brien-Fleming type of boundaries feasible? Third, how to perform a valid final analysis that combines data collected from different stages?

18.5 Evaluation of Rare Disease Clinical Trials

Due the small sample size in rare disease clinical trials, the conclusion drawn may not achieve a desired level of statistical inference (e.g., power or confidence interval). In this case, it is suggested that the following methods can be considered for evaluation of the rare disease clinical trial to determine whether substantial evidence of safety and efficacy has been achieved. Let n_1, n_2, and N be the sample size of the intended trial at interim, sample size of the data borrowed from previous studies, and sample size required for achieving a desired power (say 80%), respectively.

18.5.1 Predictive Confidence Interval (PCI)

Let $\bar{T}_i$ and $\bar{R}_i$ be the sample mean of the ith sample for the test product and the reference product, respectively. Also, let $\hat{\sigma}_1$, $\hat{\sigma}_2$, and $\hat{\sigma}_*$ be the pooled sample standard deviation of difference in sample mean between the test product and the reference product based on the first sample (n_1), the second sample $(n_1 + n_2)$, and the third sample (N). Under a parallel design, the usual confidence interval of the treatment effect can be obtained based on the ith sample and jth sample as follows:

$$CI_i = \bar{T}_i - \bar{R}_i \pm z_{1-\alpha} \hat{\sigma}_i,$$

where $i = 1, 2,$ and N. In practice, for rare disease clinical trials, we can compare these confidence intervals in terms of their relative efficiency for a complete clinical picture. Relative efficiency of CI_i as compared to CI_j is defined as

$$R_{ij} = \hat{\sigma}_i / \hat{\sigma}_j,$$

where i and j represent the ith sample and jth sample, respectively.

18.5.2 Probability of Reproducibility

Although there will not always be sufficient power due to the small sample size available in rare diseases clinical trials, alternatively, we may consider empirical power based on the observed treatment effect and the variability associated with the observed difference adjusted for the sample size required for achieving the desired power. The empirical power is also known as reproducibility probability of the clinical results for future studies if the studies shall be conducted under similar experimental conditions. Shao and Chow (2002) studied how to evaluate the reproducibility probability using this approach under several study designs for comparing means with both equal and unequal variances.

When the reproducibility probability is used to provide substantial evidence of the effectiveness of a drug product, the estimated power approach may produce an optimistic result. Alternatively, Shao and Chow (2002) suggested that the reproducibility probability be defined as a lower confidence bound of the power of the second trial. The reproducibility probability can be used to determine whether the observed clinical results from previous studies is reproducible in future studies for evaluation of safety and efficacy of the test treatment under investigation.

In addition, Shao and Chow also suggested a more sensible definition of reproducibility probability using the Bayesian approach. Under the Bayesian approach, the unknown parameter θ is a random vector with a prior distribution, say, $\pi(\theta)$, which is assumed known. Thus, the reproducibility probability can be defined as the conditional probability of $|T| > C$ in the future trial, given the data set x observed from the previous trial(s), that is,

$$P\{|T| > C \mid x\} = \int P(|T| > C \mid \theta)\pi(\theta \mid x)d\theta$$

where $T = T(y)$ is based on the data set y from the future trial, and $\pi(\theta \mid x)$ is the posterior density of θ, given x.

Moreover, similar idea can be applied to assess generalizability from one patient population (adults) to another (e.g., pediatrics or elderly). The generalizability can be derived as follows.

18.6 Some Proposals for Regulatory Consideration

For rare disease drug development, the Orphan Drug Act provides incentives associated with orphan drug designation to make the development of rare disease drug products financially viable with small number of patients (FDA, 2019). However, the FDA does not intend to create a statutory standard for approval of orphan drugs that is different from the standard for approval of more typical drugs. Thus, the level of substantial evidence and sample size required for approval of rare diseases drug products are probably the most challenging issues in rare disease drug product development. Given these facts, we suggest the following proposals for regulatory considerations.

18.6.1 Demonstrating *Effectiveness* or Demonstrating *Not Ineffectiveness*

For approval of a new drug product, the sponsor is required to provide substantial evidence regarding safety and efficacy of the drug product under investigation. In practice, a typical approach is to conduct adequate and well-controlled clinical studies and test the following point hypotheses:

$$H_0 : \text{Ineffectiveness versus } H_a : \text{Effectiveness.} \tag{18.19}$$

The rejection of the null hypothesis of *in*effectiveness is in favor of the alternative hypothesis of effectiveness. Most researchers interpret that the rejection of the null hypothesis is the demonstration of the alternative hypothesis of effectiveness. It, however, should be noted that "in favor of effectiveness" does not imply "the demonstration of effectiveness." In practice, hypotheses (18.19) should be

$$H_0 : \text{Ineffectiveness versus } H_a : \text{Not ineffectiveness.} \qquad (18.20)$$

In other words, the rejection of H_0 would lead to the conclusion of *"not H_0,"* which is H_a given in (18.20). As it can be seen from H_a in (18.19) and (18.20), the concept of *effectiveness* (18.19) and the concept of *not ineffectiveness* (18.20) are not the same. Not ineffectiveness does not imply effectiveness in general. Thus, the traditional approach for clinical evaluation of the drug product under investigation can only demonstrate *"not ineffectiveness"* but not *"effectiveness"*. The relationship between demonstrating "effectiveness" (18.19) and demonstrating "not ineffectiveness" (18.20) is illustrated in Figure 18.1. As it can be seen from Figure 18.1, "not ineffectiveness" consists of two parts, namely, the portion of "inconclusiveness" and the portion of "effectiveness."

For a placebo-control clinical trial comparing a test treatment (T) and a placebo control (P), let $\theta = \mu_T - \mu_R$ be the treatment effect of the test treatment as compared to the placebo, where μ_T and μ_R are the mean response of the test treatment and the placebo, respectively. For a given sample, e.g., test results from a previous or pilot study, let (θ_L, θ_U) be a $(1-\alpha) \times 100\%$ confidence interval of θ. In this case, hypothesis (18.19) becomes

$$H_0 : \theta \le \theta_L \text{ versus } H_a : \theta > \theta_U, \qquad (18.21)$$

while hypothesis (18.20) is given by

$$H_0 : \theta \le \theta_L \text{ versus } H_a : \theta > \theta_L. \qquad (18.22)$$

Hypotheses (18.21) is similar to the hypotheses set in Simon's two-stage optimal design for cancer research. At the first stage, Simon suggested testing whether the response rate has exceeded a pre-specified undesirable response rate. If yes, then proceed to test whether the response rate has achieved a pre-specified desirable response rate. Note that Simon's hypothesis testing actually is an interval hypothesis testing. On the other hand, hypotheses (18.22) is a typical one-sided test for non-inferiority of the test treatment as compared

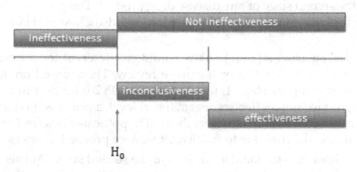

FIGURE 18.1
Demonstrating *"effectiveness"* or *"not ineffectiveness."*

to the placebo. Thus, the rejection of inferiority leads to the conclusion of non-inferiority that consists of equivalence (the area of inconclusiveness, i.e., $\theta_L < \theta < \theta_U$) and superiority (i.e., effectiveness). For a given sample size, the traditional approach for clinical evaluation of the drug product under investigation can only demonstrate that the drug product is *not ineffective* when the null hypothesis is rejected. For demonstrating the drug product is truly effective, we need to perform another test to rule out the possibility of inconclusiveness (i.e., to reduce the probability of inconclusiveness).

In practice, however, we typically test point hypotheses of equality at the $\alpha = 5\%$ level of significance. The rejection of the null hypothesis leads to the conclusion that there is a treatment effect. An adequate sample size is then selected to have a desired power (say 80%) to determine whether the observed treatment effect is clinically meaningful and hence the claim that the effectiveness is demonstrated. For testing point hypotheses of no treatment effect, many researchers prefer testing the null hypothesis at the $\alpha = 1\%$ rather than $\alpha = 5\%$ in order to account for the possibility of inconclusiveness. In other words, if the observed p-value falls between 1% and 5%, we claim the test result is *inconclusive*. It should be noted that the concept of point hypotheses testing for no treatment effect is very different from interval hypotheses testing (18.19) and one-sided hypotheses testing for non-inferiority (18.20). In practice, however, point hypotheses testing, interval hypotheses testing, and one-sided hypotheses for testing non-inferiority have been mixed up and used in pharmaceutical research and development.

18.6.2 Two-Stage Adaptive Trial Design for Rare Disease Product Development

As discussed in the previous sections, in-availability of patients in clinical trials due to small population size of rare disease and how to achieve the same standard for regulatory review and approval are probably the most obstacles and challenges in rare disease product development. In this section, to address these dilemmas, we propose a two-stage adaptive trial design for demonstrating "not ineffectiveness" at the first stage and then demonstrating "effectiveness" at the second stage of rare disease drug product. The proposed two-stage adaptive trial design is briefly outlined below (see also Chow and Huang, 2019):

Stage 1. Construct a $(1-\alpha) \times 100\%$ confidence interval for θ based on previous/pilot studies or literature review. Then, based on n_1 subjects available at Stage 1, test hypotheses (18.20) for non-inferiority (i.e., test for not ineffectiveness at the α_1 level, a pre-specified level of significance). If it fails to reject the null hypothesis of ineffectiveness, then stop the trial due to futility. Otherwise proceed to next stage.

Stage 2. Recruit additional n_2 subjects at the second stage. At this stage, sample size re-estimation may be performed for achieving the desirable statistical assurance (say 80%) for establishment of effectiveness of the test treatment under investigation. At the second stage, a statistical test is performed to assure that the probability of the area of inclusiveness is within an acceptable range at the α_2 level, a pre-specified level of significance.

Under the proposed two-stage adaptive trial design, it can be showed that the overall type I error rate is a function of α_1 and α_2 (see also Chow and Huang, 2019). Thus, with appropriate choice of α_1, we may reduce the sample size required for demonstrating "not ineffectiveness." However, it is suggested that the selection of α_1 and α_2 be specified in the study protocol. The post-study adjustment is not encouraged.

For review and approval of rare diseases drug products, thus, we propose to first demonstrate *not ineffectiveness* with limited information available at a pre-specified level of significance and then collect additional information to rule out *inconclusiveness* for demonstration of effectiveness at a pre-specified level of significance under the proposed two-stage adaptive trial design.

18.6.3 Probability Monitoring Procedure for Sample Size

In rare disease clinical trials, power calculation for required sample size may not be feasible for rare disease clinical trials due to limited number of subjects available, especially when the anticipated treatment effect is relatively small. In this case, alternative methods such as precision analysis (or confidence interval approach), reproducibility analysis, or probability monitoring approach, and Bayesian approach may be considered for providing substantial evidence with certain statistical assurance (Chow et al., 2017). It, however, should be noted that the resultant sample sizes from these different analyses could be very different with different levels of statistical assurance achieved. Thus, for rare disease clinical trials, it is suggested that an appropriate sample size should be selected for achieving certain statistical assurance under a valid trial design.

As an example, an appropriate sample size may be selected based on a probability monitoring approach such that the probability of crossing safety boundary is controlled at a pre-specified level of significance. Suppose an investigator plans to monitor the safety of a rare disease clinical trial sequentially at several times, $t_i, i = 1,...,K$. Let n_i and P_i be the sample size and the probability of observing an event at time t_i. Thus, an appropriate sample size can be selected such that the following probability of crossing safety stopping boundary is less than a pre-specified level of significance

$$p_k = P\{\text{across safety stopping boundary} \mid n_k, P_k\} < \alpha, k = 1,...,K. \quad (18.23)$$

In practice, it is suggested that statistical methods used for data analysis should be consistent with statistical methods used for sample size estimation for scientific validity of the intended clinical trial. The concepts of power analysis, precision analysis, reproducibility analysis, probability monitoring approach, and Bayesian approach should not be mixed up with, while the statistical methods for data analysis should reflect the desired statistical assurance under the trial design.

18.7 Concluding Remarks

As discussed, for rare disease drug development, power analysis for sample size calculation may not be feasible due to the fact that there is small patient population. The FDA draft guidance emphasizes that the same standards for regulatory approval will be applied to rare diseases drug development despite of small patient population. Thus, often there is insufficient power for rare disease drug clinical investigation. In this case, it is suggested that sample size calculation or justification should be performed based on precision analysis, reproducibility analysis, or probability monitoring approach for achieving certain statistical assurance.

In practice, it is a dilemma for having same standard with fewer subjects in rare disease drug development. Thus, it is suggested that innovative design and statistical methods should be considered and implemented for obtaining substantial evidence regarding effectiveness and safety in support of regulatory approval of rare disease drug products. In this chapter, several innovative trial designs such as complete *n*-of-1 trial design, adaptive seamless trial design, trial design utilizing the concept of master protocols, and Bayesian trial design are introduced. The corresponding statistical methods and sample size requirement under respective study designs are derived. These study designs are useful in speeding up rare disease development process and identifying any signal, pattern or trend, and/or optimal clinical benefits of the rare disease drug products under investigation.

Due to the small patient population in rare disease clinical development, the concept of generalizability probability can be used to determine whether the clinical results can be generalized from the targeted patient population (e.g., adults) to a different but similar patient population (e.g., pediatrics or elderly) with the same rare disease. In practice, the generalizability probability can be evaluated through the assessment of sensitivity index between the targeted patient population and the different patient population (Lu et al., 2017). The degree of generalizability probability can then be used to judge whether the intended trial has provided substantial evidence regarding the effectiveness and safety for the different patient population (e.g., pediatrics or elderly).

In practice, although an innovative and yet complex trial design may be useful in rare disease drug development, it may introduce operational bias to the trial and consequently increase the probability of making errors. It is then suggested that quality, validity, and integrity of the intended trial utilizing an innovative trial design should be maintained.

Bibliography

Afonja, B. (1972). The moments of the maximum of correlated normal and t-variates. *Journal of the Royal Statistical Society Series B*, 34, 251–262.

Agin, M.A., Aronstein, W.S., Ferber, G., Geraldes, M.C., Locke, C., and Sager, P. (2008). QT/QTc prolongation in placebo-treated subjects: A PhRMA collaborative data analysis. *Journal of Biopharmaceutical Statistics*, 18, 408–426.

Agresti, A. and Min, Y. (2005). Simple improved confidence intervals for comparing matched proportions. *Statistics in Medicine*, 24(5), 729–740.

Akaike, H. (1974). A new look at statistical model identification. *IEEE Transactions on Automatic Control*, 19, 716–723.

Alosh, M. (2009). The impact of missing data in a generalized integer-valued autoregression model for count data. *Journal of Biopharmaceutical Statistics*, 19, 1039–1054.

Anderson, S. and Hauck, W.W. (1990). Consideration of individual bioequivalence. *Journal of Pharmacokinetics and Biopharmaceutics*, 18, 259–273.

Atkinson, A.C. and Donev, A.N. (1992). *Optimum Experimental Designs*, Oxford University Press, New York.

Austin, P.C. (2011). An Introduction to propensity score methods for reducing the effects of confounding in observational studies. *Multivariate Behavioral Research*, 46, 399–424.

Babb, J., Rogatko, A., and Zacks, S. (1998). Cancer phase I clinical trials: Efficient dose escalation with overdose control. *Statistics in Medicine*, 17, 1103–1120.

Babb, J.S. and Rogatko, A. (2004). Bayesian methods for cancer phase I clinical trials. In *Advances in Clinical Trial Biostatistics*, Geller, N.L. (Ed.), Marcel Dekker, New York.

Bailar, J.C. (1992). Some use of statistical thinking. In *Medical Use of Statistics*, Bailar, J.C. and Mosteller, F. (Eds.), New England Journal of Medicine Books, Boston, MA, pp. 5–26.

Barnes, P.J., Pocock, S.J., Magnussen, H., Iqbal, A., Kramer, B., Higgins, M., and Lawrence, D. (2010). Integrating indacaterol dose selection in a clinical study in COPD using an adaptive seamless design. *Pulmonary Pharmacology & Therapeutics*, 23(3), 165–171.

Barrentine, L.B. (1991). *Concepts for R&R Studies*, ASQC Quality Press, Milwaukee, WI.

Barry, M.J., Fowler, F.J. Jr., O'Leary, M.P., Bruskewitz, R.C., Holtgrewe, H.L., Mebust, W.K., and Cockett, A.T. (1992). The American Urological Association Symptom Index for benign prostatic hyperplasia. *Journal of Urology*, 148, 1549–1557.

Basford, K.E., Greenway, D.R., McLachlan, G.J., and Peel, D. (1997). Standard errors of fitted component means of normal mixtures, *Computational Statistics*, 12, 1–17.

Bauer, P. and Kieser, M. (1999). Combining different phases in the development of medical treatments within a single trial. *Statistics in Medicine*, 14, 1595–1607.

Bauer, P. and Kohne, K. (1994). Evaluation of experiments with adaptive interim analysis. *Biometrics*, 1029–1041.

Bauer, P. and Rohmel, J. (1995). An adaptive method for establishing a dose-response relationship. *Statistics in Medicine*, 14, 1595–1607.

Benjamini, Y. and Hochberg, Y. (1995). Controlling the false discovery rate: A practical and powerful approach to multiple testing. *Journal of Royal Statistical Society, Series B*, 57, 289–300.

Bensoussan, A., Talley, N.J., Hing, M., Menzies, R., Guo, A., and Ngu, M. (1998). Treatment of irritable bowel syndrome with Chinese herbal medicine. *Journal of American Medical Association*, 280, 1585–1589.

Berger, J.O. (1985). *Statistical Decision Theory and Bayesian Analysis*, 2nd ed., Springer-Verlag, New York.

Bergner, M., Bobbitt, R.A., Carter, W.B., and Gilson, B.S. (1981). The sickness impact profile: Development and final revision of a health status measure. *Medical Care*, 19, 787–805.

Berger, R.L. and Hsu, J.C. (1996). Bioequivalence trials, intersection-union tests and equivalence confidence sets (with discussion). *Statistical Science*, 11, 283–319.

Bergum, J.S. (1988). Constructing acceptance limits for multiple stage USP tests. *Proceedings of the Biopharmaceutical Section of the American Statistical Association*, pp. 197–201.

BHAT (1983). Beta-Blocker Heart Attack Trial Research Group. A randomized trial of propranolol in patients with acute myocardial infarction: Morbidity. *JAMA*, 250, 2814–2819.

Bhatt, D.L. and Mehta, C. (2016). Adaptive designs for clinical trials. *New England Journal of Medicine*, 375(1), 65–74.

Blackwell, D. and Hodges, J.L. Jr. (1957). Design for the control of selection bias. *The Annals of Mathematical Statistics*, 28, 449–460.

Blair, R.C. and Cole, S.R. (2002). Two-sided equivalence testing of the difference between two means. *Journal of Modern Applied Statistical Methods*, 1, 139–142.

Bofinger, E. (1985). Expanded confidence intervals. *Communications in Statistics, Theory and Methods*, 14, 1849–1864.

Bofinger, E. (1992). Expanded confidence intervals, one-sided tests and equivalence testing. *Journal of Biopharmaceutical Statistics*, 2, 181–188.

Bollier, D. (2010). *The Promise and Peril of Big Data*, The Aspen Institute, Washington, DC.

Brannath, W., Koening, F., and Bauer, P. (2003). Improved repeated confidence bounds in trials with a maximal goal. *Biometrical Journal*, 45, 311–324.

Branson, M. and Whitehead, W. (2002). Estimating a treatment effect in survival studies in which patients switch treatment. *Statistics in Medicine*, 21, 2449–2463.

Breunig, R. (2001). An almost unbiased estimator of the coefficient of variation. *Economics Letters*, 70, 15–19.

Brookmeyer, R. and Crowley, J. (1982). A confidence interval for the median survival time. *Biometrics*, 38, 29–41.

Brown, B.W. (1980). The crossover experiment for clinical trials. *Biometrics*, 36, 69–79.

Brown, L.D., Hwang, J.T.G., and Munk, A. (1997). An unbiased test for the bioequivalence problem. *The Annals of Statistics*, 25, 2345–2367.

Brownell, K.D. and Stunkard, A.J. (1982). The double-blind in danger untoward consequences of informed consent. *The American Journal of Psychiatry*, 139, 1487–1489.

Canales, R.D., Luo, Y., Willey, J.C., Austermiller, B., Barbacioru, C.C., Boysen, C., Hunkapiller, K. et al. (2006). Evaluation of DNA microarray results with quantitative gene expression platforms. *Nature Biotechnology*, 24, 1115–1122.

Caraco, Y. (2004). Genes and the response to drugs. *The New England Journal of Medicine*, 351, 2867–2869.

Cardoso, F., Piccart-Gebhart, Van't Veer L., Rutgers, E. on behalf of the TRANSBIG (2007). The MINDACT trial: the first prospective clinical validtion of a genomic tool. *Molecular Oncology*, 1, 246–251.

Carrasco, J.L. and Jover, L. (2003). Assessing individual bioequivalence using structural equation model. *Statistics in Medicine*, 22, 901–912.

Casciano, D.A. and Woodcock, J. (2006). Empowering microarrays in the regulatory setting. *Nature Biotechnology*, 24, 1103.

Casella, G. and Berger, R.L. (2002). *Statistical Inference*, 2nd ed., Duxbury Advanced Series, Duxbury, Pacific Grove, CA.

CAST (1989). Cardiac Arrhythmia Supression Trial. Preliminary report: Effect of encainide and flecainide on mortality in a randomized trial of arrhythmia supresssion after myocardial infarction. *The New England Journal of Medicine*, 321, 406–412.

CBER/FDA. (1999). CBER/FDA Memorandum. Summary of CBER considerations on selected aspects of active controlled trial design and analysis for the evaluation of thrombolytics in acute MI, June 1999.

Chang, M. (2005a). Bayesian adaptive design with biomarkers. Invited presentation at the IBC. *Second Annual Conference: Implementing Adaptive Designs for Drug Development*, November 7–8, 2005, Princeton, NJ.

Chang, M. (2005b). Adaptive clinical trial design. *Presented at International Conference for Stochastic Process and Data Analysis*, Brest, France, May, 2005.

Chang, M. (2007). Adaptive design method based on sum of p-values. *Statistics in Medicine*, 26, 2772–2784.

Chang, M. (2008). *Adaptive Design Theory and Implementation Using SAS and R*, Chapman and Hall/CRC Press, Taylor & Francis Group, New York.

Chang, M. and Chow, S.C. (2005). A Hybrid Bayesian adaptive design for dose response trials. *Journal of Biopharmaceutical Statistics*, 15, 677–691.

Chang, M. and Chow, S.C. (2006). Power and sample size for dose response studies. In *Dose Finding in Drug Development*, Ting, N. (Ed.), Springer, New York.

Charkravarty, A., Burzykowski, M., Molenberghs, G., and Buyse, T. (2005). Regulatory aspects in using surrogate markers in clinical trials. In *The Evaluation of Surrogate Endpoint*, Burzykoski, T., Molenberghs, G., and Buyse, M. (Eds.), Springer, New York.

Chen, J. and Chen, C. (2003). Microarray gene expression. In *Encyclopedia of Biopharmaceutical Statistics*, Chow, S.C. (Ed.), Marcel Dekker, New York, pp. 599–613.

Chen, M.L., Patnaik, R., Hauck, W.W., Schuirmann, D.F., Hyslop, T., and Williams, R. (2000). An individual bioequivalence criterion: Regulatory considerations. *Statistics in Medicine*, 19, 2821–2842.

Chen, X., Luo, X., and Capizzi, T. (2005). The application of enhanced parallel gatekeeping strategies. *Statistics in Medicine*, 24, 1385–1397.

Chen, Y.J., Gesser, R., and Luxembourg, A. (2015). A seamless phase IIB/III adaptive outcome trial: Design rationale and implementation challenges. *Clinical Trials*, 12(1), 84–90.

Cheng, B., Chow, S.C., Burt, D., and Cosmatos, D. (2008). Statistical assessment of QT/QTc prolongation based on maximum of correlated normal random variables. *Journal of Biopharmaceutical Statistics*, 18, 494–501.

Cheng, B. and Shao, J. (2007). Exact tests for negligible interaction in two-way linear models. *Statistica Sinica*, 17, 1441–1455.

Cheng, B., Zhang, B., and Chow, S.C. (2019). Unified approaches to assessing treatment effect of traditional Chinese medicine based on health profiles. *Journal of Biopharmaceutical Statistics*, To appear.

Chirino, A.J. and Mire-Sluis, A. (2004). Characterizing biological products and assessing comparability following manufacturing changes. *Nature Biotechnology*, 22, 1383–1391.

Chung, W.H., Hung, S.I., Hong, H.S., Hsih, M.S., Yang, L.C., Ho, H.C., Wu, J.Y., and Chen, Y.T. (2004). Medical genetics: A marker for Stevens-Johnson syndrome. *Nature*, 428 (6982), 486.

Chow, S.C. (1997). Good statistics practice in the drug development and regulatory approval process. *Drug Information Journal*, 31, 1157–1166.

Chow, S.C. (1999). Individual bioequivalence—a review of FDA draft guidance. *Drug Information Journal*, 33, 435–444.

Chow, S.C. (2007). Statistics in translational medicine. *Presented at Current Advances in Evaluation of Research & Development of Translational Medicine*, National Health Research Institutes, Taipei, Taiwan, October 19, 2007.

Chow, S.C. (2010). Generalizability probability of clinical results. In *Encyclopedia of Biopharmaceutical Statistics*, Chow, S.C. (Ed.), Informa Healthcare, Taylor & Francis Group, London, UK, pp. 534–536.

Chow, S.C. (2011). *Controversial Issues in Clinical Trials*, Chapman and Hall/CRC Press, Taylor & Francis Group, New York.

Chow, S.C. (2013). *Biosimilars: Design and Analysis of Follow-on Biologics*, Chapman and Hall/CRC Press, Taylor & Francis Group, New York.

Chow, S.C. (2015). *Quantitative Methods for Traditional Chinese Medicine Development*, Chapman and Hall/CRC Press, Taylor & Francis Group, New York.

Chow, S.C. (2018). Non-medical switch in biosimilar product development. *Enliven: Biosimilars Bioavailability*, 2(1): e001.

Chow, S.C. and Chang, M. (2005). Statistical consideration of adaptive methods in clinical development. *Journal of Biopharmaceutical Statistics*, 15, 575–591.

Chow, S.C. and Chang, M. (2006). *Adaptive Design Methods in Clinical Trials*, Chapman and Hall/CRC Press, Taylor & Francis Group, New York.

Chow, S.C. and Chang, M. (2008). Adaptive design methods in clinical trials–a review. *The Orphanet Journal of Rare Diseases*, 3, 1–13.

Chow, S.C. and Chang, M. (2011). *Adaptive Design Methods in Clinical Trials*, 2nd ed., Chapman and Hall/CRC Press, Taylor & Francis Group, New York, NY.

Chow, S.C., Chang, M., and Pong, A. (2005). Statistical consideration of adaptive methods in clinical development. *Journal of Biopharmaceutical Statistics*, 15, 575–591.

Chow, S.C., Cheng, B., and Cosmatos, D. (2008). On power and sample size calculation for QT studies with recording replicates at given time point. *Journal of Biopharmaceutical Statistics*, 18, 483–493.

Chow, S.C. and Corey, R. (2011). Benefits, challenges and obstacles of adaptive designs in clinical trials. *The Orphanet Journal of Rare Diseases*, 6, 79. doi:10.1186/1750-1172-6-79.

Chow, S.C., Corey, R., and Lin, M. (2012). On independence of data monitoring committee in adaptive clinical trial. *Journal of Biopharmaceutical Statistics*, 22, 853–867.

Chow, S.C., Endrenyi, L., Lachenbruch, P.A., Yang, L.Y., and Chi, E. (2011). Scientific factors for assessing biosimilarity and drug Interchangeability of follow-on Biologics. *Biosimilars*, 1, 13–26.

Chow, S.C. and Hsiao, C.F. (2010). Bridging diversity: Extrapolating foreign data to a new region. *Pharmaceutical Medicine*, 24, 349–362.

Chow, S.C., Hsieh, T.C., Chi, E., and Yang, J. (2010). A comparison of moment-based and probability-based criteria for assessment of follow-on biologics. *Journal of Biopharmaceutical Statistics*, 20, 31–45.

Chow, S.C. and Huang, Z.P. (2019). Demonstrating effectiveness or demonstrating not ineffectiveness – A potential solution for rare disease drug product development? *Journal of Biopharmaceutical Statistics*, In press.

Chow, S.C. and Ki, F. (1994). On statistical characteristics of quality of life assessment. *Journal of Biopharmaceutical Statistics*, 4, 1–17.

Chow, S.C. and Ki, F. (1996). Statistical issues in quality of life assessment. *Journal of Biopharmaceutical Statistics*, 6, 37–48.

Chow, S.C. and Kong, Y.Y. (2015). On big data analytics in biomedical research. *Journal of Biometrics and Biostatistics*, 6, 236. doi:10.4172/2155-6180.1000236.

Chow, S.C. and Lin, M. (2015). Analysis of two-stage adaptive seamless trial design. *Pharmaceutica Analytica Acta*, 6, 3. doi:10.4172/2153-2435.1000341.

Chow, S.C. and Liu, J.P. (1992a). *Design and Analysis of Bioavailability and Bioequivalence Studies*, Marcel Dekker, New York.

Chow, S.C. and Liu, J.P. (1992b). On assessment of bioequivalence under a higher-order crossover design. *Journal Biopharmaceutical Statistics*, 2, 239–256.

Chow, S.C. and Liu, J.P. (1995). *Statistical Design and Analysis in Pharmaceutical Science: Validation, Process Control, and Stability*, Marcel Dekker, New York.

Chow, S.C. and Liu, J.P. (1997). Meta-analysis for bioequivalence review. *Journal of Biopharmaceutical Statistics*, 7, 97–111.

Chow, S.C. and Liu, J.P. (1998a). *Design and Analysis of Animal Studies in Pharmaceutical Development*, Marcel Dekker, New York.

Chow, S.C. and Liu, J.P. (1998b). *Design and Analysis of Clinical Trials*, John Wiley & Sons, New York.

Chow, S.C. and Liu, J.P. (2000). *Design and Analysis of Bioavailability and Bioequivalence Studies, Revised and expanded*, 2nd ed., Marcel Dekker, New York.

Chow, S.C. and Liu, J.P. (2003). *Design and Analysis of Clinical Trials*, 2nd ed., John Wiley & Sons, New York.

Chow, S.C. and Liu, J.P. (2008). *Design and Analysis of Bioavailability and Bioequivalence Studies*, 3rd ed., Chapman Hall/CRC Press, Taylor & Francis Group, New York.

Chow, S.C. and Liu, J.P. (2010). Statistical assessment of biosimilar products. *Journal of Biopharmaceutical Statistics*, 20, 10–30.

Chow, S.C. and Liu, J.P. (2013). *Design and Analysis of Clinical Trials, Revised and Expanded*, 3rd ed., John Wiley & Sons, New York.

Chow, S.C., Lu, Q., and Tse, S.K. (2007). Statistical analysis for two-stage adaptive design with different study endpoints. *Journal of Biopharmaceutical Statistics*, 17, 1163–1176.

Chow, S.C., Pong, A., and Chang, Y.W. (2006). On traditional Chinese medicine clinical trials. *Drug Information Journal*, 40, 395–406.

Chow, S.C. and Shao, J. (1997). Statistical methods for two-sequence dual crossover designs with incomplete data. *Statistics in Medicine*, 16, 1031–1039.

Chow, S.C. and Shao, J. (2002a). *Statistics in Drug Research*, Marcel Dekker, New York.

Chow, S.C. and Shao, J. (2002b). A note on statistical methods for assessing therapeutic equivalence. *Controlled Clinical Trials*, 23, 515–520.

Chow, S.C. and Shao, J. (2004). Analysis of clinical data with breached blindness. *Statistics in Medicine*, 23, 1185–1193.

Chow, S.C. and Shao, J. (2005). Inference for clinical trials with some protocol amendments. *Journal of Biopharmaceutical Statistics*, 15, 659–666.

Chow, S.C. and Shao, J. (2006). On non-inferiority margin and statistical tests in active control trials. *Statistics in Medicine*, 25, 1101–1113.

Chow, S.C. and Shao, J. (2007). Stability analysis for drugs with multiple ingredients. *Statistics in Medicine*, 26, 1512–1517.

Chow, S.C., Shao, J., and Hu, Y.P. (2002). Assessing sensitivity and similarity in bridging studies. *Journal of Biopharmaceutical Statistics*, 12, 385–400.

Chow, S.C., Shao, J., and Li, L. (2004). Assessing bioequivalence using genomic data. *Journal of Biopharmaceutical Statistics*, 14, 869–880.

Chow, S.C., Shao, J., and Wang, H. (2002a). A note on sample size calculation for mean comparisons based on non-central t-statistics. *Journal of Biopharmaceutical Statistics*, 12, 441–456.

Chow, S.C., Shao, J., and Wang, H. (2002b). Individual bioequivalence testing under 2×3 crossover designs. *Statistics in Medicine*, 21, 629–648.

Chow, S.C., Shao, J., and Wang, H. (2003). Statistical tests for population bioequivalence. *Statistica Sinica*, 13, 539–554.

Chow, S.C., Shao, J., and Wang, H. (2008). *Sample Size Calculation in Clinical Research*, Chapman and Hall/CRC Press, Taylor & Francis Group, New York.

Chow, S.C., Shao, J., Wang, H., and Lokhnygina, Y. (2017). *Sample Size Calculations in Clinical Research*, 3rd ed., Taylor & Francis Group, New York.

Chow, S.C. and Tse, S.K. (1991). On the estimation of total variability in assay validation. *Statistics in Medicine*, 10, 1543–1553.

Chow, S.C. and Tu, Y.H. (2009). On two-stage seamless adaptive design in clinical trials. *Journal of Formosan Medical Association*, 107(12), S51–S59.

Chow, S.C. and Wang, H. (2001). On sample size calculation in bioequivalence trials. *Journal of Pharmacokinetics and Pharmacodynamics*, 28, 155–169.

Chow, S.C., Song, F.Y., and Bai, H. (2016). Analytical similarity assessment in biosimilar studies. AAPS Journal, 18(3), 670–677.

Chow, S.C., Xu, H., Endrenyi, L., and Song, F.Y. (2015). A new scaled criterion for drug interchangeability. *Chinese Journal of Pharmaceutical Analysis*, 35(5), 844–848.

Christensen, R. (1996). Exact tests for variance components. *Biometrics*, 52, 309–314.

Chuang, C. (1987). The analysis of a titration study. *Statistics in Medicine*, 6, 583–590.

Chung-Stein, C. (1996). Summarizing laboratory data with different reference ranges in multi-center clinical trials. *Drug Information Journal*, 26, 77–84.

Chung-Stein, C., Anderson, K., Gallo, P., and Collins, S. (2006). Sample size reestimation: A review and recommendations. *Drug Information Journal*, 40, 475–484.

Church, J.D. and Harris, B. (1970). The estimation of reliability from stress-strength relationships. *Technometrics*, 12, 49–54.

Cochran, W.G. (1977). *Sampling Techniques*, 3rd ed., Wiley, New York.

Cochran, W.G. and Cox, G.M. (1957). *Experimental Designs*, 2nd ed., Wiley, New York. P.18.

Coors, M., Bauer, L., Edwards, K., Erickson, K., Goldenberg, A., Goodale, J., Goodman, K. et al. (2017). Ethical issues related to clinical research and rare diseases. *Translational Science of Rare Diseases*, 2, 175–194.

Cosmatos, D. and Chow, S.C. (2008). *Translational Medicine*, Chapman and Hall/CRC Press, Taylor & Francis Group, New York.

CPMP. (1990). The Committee for Proprietary Medicinal Products Working Party on Efficacy of Medicinal Products. Note for Guidance; Good Clinical Practice for Trials on Medicinal Products in the European Community; Commission of European Communities: Brussels, Belgium 1990—111/396/88-EN Final.

CPMP. (1997). Points to consider: The assessment of the potential for QT interval prolongation by non-cardiovascular products. Available at: www.coresearch. biz/regulations/cpmp.pdf.

Crommelin, D., Bermejo, T., Bissig, M., Damianns, J., Kramer, I., Rambourg, P., Scroccaro, G., Strukelj, B., Tredree, R., and Ronco, C. (2005). Biosimilars, generic versions of the first generation of therapeutic proteins: Do they exist? *Cardiovascular Disorders in Hemodialysis*, 149, 287–294.

Crowley, J. (2001). *Handbook of Statistics in Clinical Oncology*, Marcel Dekker, New York.

CTriSoft Intl. (2002). *Clinical Trial Design with ExpDesign Studio*, www.ctrisoft.net., CTriSoft International, Lexington, MA.

Cui, C. and Chow, S.C. (2018). Clinical trial: n-of-1 design analysis. In *Encyclopedia of Biopharmaceutical Statistics*, 4th ed., Chow, S.C. (Ed.), CRC Press, Taylor & Francis Group, New York, pp. 564–571.

Cui, L., Hung, H.M.J., and Wang, S.J. (1999). Modification of sample size in group sequential trials. *Biometrics*, 55, 853–857.

Dalton, W.S. and Friend, S.H. (2006). Cancer biomarkers–an invitation to the table. *Science*, 312, 1165–1168.

D'Agostino, R.B., Massaro, J.M., and Sullivan, L.M. (2003). Non-inferiority trials: Design concepts and issues–the encounters of academic consultants in statistics. *Statistics in Medicine*, 22, 169–186.

Davison, A.C. (2003). *Statistical Models*, Cambridge University Press, New York, pp. 33–35.

DerSimonian, R. and Laird, N. (1986). Meta-analysis in clinical trials. *Controlled Clinical Trials*, 7, 177–188.

DeMets, D.L., Furberg, C.D., and Friedman, L.M. (2006). *Data Monitoring in Clinical Trials: A Case Studies Approach*, Springer, New York.

Dempster, A.P., Laird, N.M., and Rubin, D.B. (1977). Maximum likelihood from incomplete data via the EM algorithm. *Journal of the Royal Statistical Society*, 39, 1–38.

Dent, S.F. and Eisenhauer, E.A. (1996). Phase I trial design: Are new methodologies being put into practice? *Annals of Oncology*, 7, 561–566.

DerSimonian, R. and Laird, N. (1986). Meta-analysis in clinical trials. *Control Clin Trials*. 7, 177–188.

DeSouza, C.M., Legedza, T.R., and Sankoh, A.J. (2009). An overview of practical approaches for handling missing data in clinical trials. *Journal of Biopharmaceutical Statistics*, 19, 1055–1073.

Deuflhard, P. (2004). *Newton Methods for Nonlinear Problems. Affine Invariance and Adaptive Algorithms*, Springer Series in Computational Mathematics, Vol. 35, Springer, Berlin, Germany.

Di Scala, L. and Glimm, E. (2011). Time-to-event analysis with treatment arm selection at interim. *Statistics in Medicine*, 30, 3067–3081.

Diggle, P. and Kenward, M.G. (1994). Informative dropout in longitudinal data analysis (with discussion). *Applied Statistics*, 43, 49–94.

Dixon, D.O., Freedman, R.S., Herson, J., Hughes, M., Kim, K., Silerman, M.H., and Tangen, C.M. (2006). Guidelines for data and safety monitoring for clinical trials not requiring traditional data monitoring committees. *Clinical Trials*, 3, 314–319.

Dmitrienko, A., Molenberghs, G., Chung-Stein, C., and Offen, W. (2005). *Analysis of Clinical Trials Using SAS: A Practical Guide*, SAS Press, Gary, NC.

Dmitrienko, A., Offen, W., Wang, O., and Xiao D. (2006). Gatekeeping procedures in dose-response clinical trials based on the Dunnett test. *Pharmaceutical Statistics*, 5, 19–28.

Dmitrienko, A., Offen, W., and Westfall, P.H. (2003). Gatekeeping strategies for clinical trials that do not require all primary effects to be significant. *Statistics in Medicine*, 22, 2387–2400.

Dobbin, K.K., Beer, D.G., Meyerson, M., Yeatman, T.J., Gerald, W.L., Jacobson, J.W., Conley, B. et al. (2005). Interlaboratory comparability study of cancer gene expression analysis using oligonucleotide microarrays. *Clinical Cancer Research*, 11, 565–573.

DOH. (2004a). Draft Guidance for IND of Traditional Chinese Medicine. The Department of Health, Taipei, Taiwan.

DOH. (2004b). Draft Guidance for NDA of Traditional Chinese Medicine. The Department of Health, Taipei, Taiwan.

Dubey, S.D. (1991). Some thought on the one-sided and two-sided tests. *Journal of Biopharmaceutical Statistics*, 1, 139–150.

Dudoit, S., Yang, Y.H., Callow, M.J., and Speed, T.P. (2002). Statistical methods for identifying differentially expressed genes in replicated cDNA microarray experiments. *Statistica Sinica*, 12, 111–139.

Dunnett, C.W. (1955). Multivariate normal probability integrals with product correlation structure, Algorithm AS251. *Journal of American Statistical Association*, 50, 1096–1121.

Eaton, M.L., Muirhead, R.J., Mancuso, J.Y., and Kolluri, S. (2006). A confidence interval for the maximal mean QT interval change caused by drug effect. *Dug Information Journal*, 40, 267–271.

Efron, B. (1971). Forcing a sequential experiment to be balanced. *Biometrika*, 58, 403–417.

Efron, B. (1983). Estimating the error rate of a prediction rule: Improvement on cross-validation. *Journal of American Statistical Association*, 78, 316–331.

Efron, B. (1986). How biased is the apparent error rate of a prediction rule? *Journal of American Statistical Association*, 81, 461–470.

Efron, B. and Tibshirani, R.J. (1993). *An Introduction to the Bootstrap*, Chapman and Hall, New York.

Eisenhauer, E.A., O'Dwyer, P.J., Christian, M., and Humphrey, J.S. (2000). Phase I clinical trial design in cancer drug development. *Journal of Clinical Oncology*, 18, 684–692.

Ellenberg, J.H. (1990). Biostatistical collaboration in medical research. *Biometrics*, 46, 1–32.

Ellenberg, S.S., Fleming, T.R., and DeMets, D.L. (2002). *Data Monitoring Committees in Clinical Trials: A Practical Perspective*, John Wiley & Sons, New York.

EMEA. (2001). Note for guidance on the investigation of bioavailability and bioequivalence. The European Medicines Agency Evaluation of Medicines for Human Use. EMEA/EWP/QWP/1401/98, London, UK.

EMEA. (2002). Point to consider on methodological issues in confirmatory clinical trials with flexible design and analysis plan. The European Agency for the Evaluation of Medicinal Products Evaluation of Medicines for Human Use. CPMP/EWP/2459/02, London, UK.

EMEA. (2003a). Note for guidance on comparability of medicinal products containing biotechnology-derived proteins as drug substance: Non clinical and clinical issues. The European Medicines Agency Evaluation of Medicines for Human Use. EMEA/CHMP/3097/02, London, UK.

EMEA. (2003b). Rev. 1 Guideline on comparability of medicinal products containing biotechnology-derived proteins as drug substance: Quality issues. The European Medicines Agency Evaluation of Medicines for Human Use. EMEA/CHMP/BWP/3207/00/Rev 1, London, UK.

EMEA. (2005a). Guideline on similar biological medicinal products. The European Medicines Agency Evaluation of Medicines for Human Use. EMEA/CHMP/437/04, London, UK.

EMEA. (2005b). Draft guideline on similar biological medicinal products containing biotechnology-derived proteins as drug substance: Quality issues. The European Medicines Agency Evaluation of Medicines for Human Use. EMEA/CHMP/49348/05, London, UK.

EMEA. (2005c). Draft annex guideline on similar biological medicinal products containing biotechnology-derived proteins as drug substance: Non clinical and clinical issues–Guidance on biosimilar medicinal products containing recombinant erythropoietins. The European Medicines Agency Evaluation of Medicines for Human Use. EMEA/CHMP/94526/05, London, UK.

EMEA. (2005d). Draft annex guideline on similar biological medicinal products containing biotechnology-derived proteins as drug substance: Non clinical and clinical issues–Guidance on biosimilar medicinal products containing Recombinant Granulocyte-Colony Stimulating Factor. The European Medicines Agency Evaluation of Medicines for Human Use. EMEA/CHMP /31329 /05, London, UK.

EMEA. (2005e). Draft annex guideline on similar biological medicinal products containing biotechnology-derived proteins as drug substance: Non-clinical and clinical issues – Guidance on biosimilar medicinal products containing Somatropin. The European Medicines Agency Evaluation of Medicines for Human Use. EMEA/CHMP/94528/05, London, UK.

EMEA. (2005f). Draft annex guideline on similar biological medicinal products containing biotechnology-derived proteins as drug substance: Non clinical and clinical issues–Guidance on biosimilar medicinal products containing recombinant human insulin. The European Medicines Agency Evaluation of Medicines for Human Use. EMEA/CHMP/32775/05, London, UK.

EMEA. (2005g). Guideline on the clinical investigating of the pharmacokinetics of therapeutic proteins. The European Medicines Agency Evaluation of Medicines for Human Use. EMEA/CHMP/89249/04, London, UK.

EMEA. (2006). Reflection paper on methodological issues in confirmatory clinical trials with flexible design and analysis plan. The European Agency for the Evaluation of Medicinal Products Evaluation of Medicines for Human Use. CPMP/EWP/2459/02, London, UK.

EMEA. (2007). Reflection paper on methodological issues in confirmatory clinical trials planned with an adaptive design. EMEA Doc. Ref. CHMP/EWP/2459/02, October 20 Available at http://www.emea.europa.eu/pdfs/human/ewp/245902enadopted.pdf.

Emerson, J.D. (1982). Nonparametric confidence intervals for the median in the presence of right censoring. *Biometrics*, 38, 17–27.

Endrenyi, L., Declerck, P., and Chow, S.C. (2017). *Biosimilar Drug Product Development*, CRC Press, Taylor & Francis Group, New York.

Enis, P. and Geisser, S. (1971). Estimation of the probability that Y<X. *Journal of American Statistical Association*, 66, 162–168.

Fairweather, W.R. (1994). Statisticians, the FDA and a time of transition. *Presented at Pharmaceutical Manufacturers Association Education and Research Institute Training Course in Non-Clinical Statistics*, Georgetown University Conference Center, Washington, DC, February 6–8, 1994.

FDA. (1987a). Guideline for submitting documentation for the stability of human drugs and biologics. Center for Drugs and Biologics, Office of Drug Research and Review, Food and Drug Administration, Rockville, MD.

FDA. (1987b). Guideline on general principles of process validation. Center for Drug and Biologics and Center for Devices and Radiological Health, Food and Drug Administration, Rockville, MD.

FDA. (1988). Guideline for format and content of the clinical and statistical sections of new drug applications. Center for Drug Evaluation and Research, Food and Drug Administration, Rockville, MD.

FDA. (1989). Invited session on Meta Analysis, organized by the FDA at the 148th Annual Meeting of the American Statistical Association, New Orleans, LA.

FDA. (1991). Guidance for in vivo bioequivalence and in vitro drug release. Center for Drug Evaluation and Research, Food and Drug Administration, Rockville, MD.

FDA. (1992). Guidance on statistical procedures for bioequivalence using a standard two treatment crossover design. Division of Bioequivalence, Office of Generic Drugs, Center for Drug Evaluation and Research, U.S. Food and Drug Administration, Rockville, MD.

FDA. (1997). Guidance for industry: Dissolution testing of immediate release solid oral dosage forms. The United States Food and Drug Administration, Rockville, MD.

FDA. (2001). Guidance on statistical approaches to establishing bioequivalence. Center for Drug Evaluation and Research, the US Food and Drug Administration, Rockville, MD.

FDA. (2003a). Draft guidance for industry: Multiplex tests for heritable DNA markers, mutations and expression patterns. The United States Food and Drug Administration, Rockville, MD.

FDA. (2003b). Guidance on bioavailability and bioequivalence studies for orally administrated drug products—General considerations. Center for Drug Evaluation and Research, the US Food and Drug Administration, Rockville, MD.

FDA. (2004). Guidance for industry – Botanical drug products. The United States Food and Drug Administration, Rockville, MD.

FDA. (2005). Draft concept paper on drug-diagnostic co-development. The United States Food and Drug Administration, Rockville, MD.

FDA. (2006a). Draft guidance on in vitro diagnostic multivariate index assays. The United States Food and Drug Administration, Rockville, MD.

FDA. (2006b). Guidance for clinical trial sponsors: Establishment and operation of clinical trial data monitoring committees. CBER/CDER/CDRH, The United States Food and Drug Administration, Rockville, MD. http://www.fda.gov/cber/gdlns/clintrialdmc.pdf.

FDA. (2007a). Guidance on pharmacogenetic tests and genetic tests for heritable marker. The US Food and Drug Administration, Rockville, MD.

FDA. (2007b). Draft guidance on *In Vitro Diagnostic Multivariate Index Assays*. The US Food and Drug Administration: Rockville, MD.

FDA. (2010a). Guidance for industry – Non-inferiority clinical trials. The United States Food and Drug Administration, Rockville, MD.

FDA. (2010b). Draft guidance for industry – Adaptive design clinical trials for drugs and biologics. The United States Food and Drug Administration, Rockville, MD.

FDA (2012a). Scientific considerations in demonstrating biosimilarity to a reference product. The United States Food and Drug Administration, Silver Spring, MD.

FDA (2012b). Quality considerations in demonstrating biosimilarity to a reference protein product. The United States Food and Drug Administration, Silver Spring, MD.

FDA (2012c). Biosimilars: Questions and Answers Regarding Implementation of the Biologics Price Competition and Innovation Act of 2009.The United States Food and Drug Administration, Silver Spring, MD.

FDA. (2015a). Guidance for industry – Scientific considerations in demonstrating biosimilarity to a reference product. Center for Drug Evaluation and Research (CDER) and Center for Biologics Evaluation and Research (CBER), the United States Food and Drug Administration, Silver Spring, MD.

FDA. (2015b). Guidance for industry – Rare diseases: Common issues in drug development. United States Food and Drug Administration, Silver Spring, MD, August, 2015.

FDA. (2017a). Guidance for industry – Considerations in demonstrating interchangeability with a reference product. United States Food and Drug Administration, Silver Spring, MD.

FDA. (2017b). Guidance for industry – Statistical approaches to evaluate analytical similarity. Center for Drug Evaluation and Research (CDER) and Center for Biologics Evaluation and Research (CBER), the United States Food and Drug Administration, Silver Spring, MD, September 2017.

FDA. (2018). Guidance for industry – Adaptive design clinical trials for drugs and biologics. The United States Food and Drug Administration, Silver Spring, MD. FDA: https://www.fda.gov/downloads/drugs/guidances/ucm201790.pdf.

FDA. (2019a). Development of Therapeutic Protein Biosimilars: Comparative Analytical Assessment and Other Quality-Related Considerations. The United States Food and Drug Administration, Silver Spring, MD.

FDA. (2019b). Guidance for Industry – Rare Diseases: Common Issues in Drug Development. Center for Drug Evaluation and Research, US Food and Drug Administration, Silver Spring, MD.

FDA/TPD. (2003). Preliminary concept paper: The clinical evaluation of QT/QTc interval prolongation and proarrythmic potential for non-arrythmic drug products. Released on November 15, 2002. Revised on February 6, 2003.

Feeny, D.H. and Torrance, G.W. (1989). Incorporating utility-based quality-of-life assessment measures in clinical trials. *Medical Care*, 27, S198–S204.

Feng, H., Shao, J., and Chow, S.C. (2007). Group sequential test for clinical trials with moving patient population. *Journal of Biopharmaceutical Statistics*, 17, 1227–1238.

Filozof, C., Chow, S.C., Dimick-Santos, L. Chen, Y.F., Williams, R.N., Goldstein, B.J., and Sanyal, A. (2017). Clinical endpoints and adaptive clinical trials in precirrhotic nonalcoholic steatohepatitis: Facilitating development approaches for an emerging epidemic. *Hepatology Communications*, 1(7), 577–585.

Finney, D.J. (1979). *Statistical Method in Biological Assay*, 3rd ed., Oxford University Press, New York.

Fisher, L.D. (1991). The use of one-sided tests in during trials: An advisory committee member's perspective. *Journal of Biopharmaceutical Statistics*, 1, 151–156.

Fleiss, J. (1987). Some thoughts on two-tailed tests. (Letter to the Editor). *Controlled Clinical Trials*, 8, 394.

Fleiss, J.L., Levin, B., and Paik, M.C. (2003) Statistical Methods for Rates and Proportions. 3rd Edition, Wiley: New York.

Fontanarosa, P.B., Flanagin, A., and DeAngelis, C.D. (2005). Reporting conflicts of interest, financial aspects of research and role of sponsors in funded studies. *Journal of the American Medical Association*, 294, 110–111.

Freidlin, B. and Simon, R. (2005). Adaptive signature design: An adaptive clinical trial design for generating and prospectively testing a gene expression signature for sensitive patients. *Clinical Cancer Research*, 11, 7872–7878.

Freidlin, B., Jiang, W., and Simon, R. (2010). The cross-validated adaptive signature design. *Clinical Cancer Research*, 16, 692–698.

Friede, T. and Kieser, M. (2004). Sample size recalculation for binary data in internal pilot study designs. *Pharmaceutical Statistics*, 3, 269–279.

Friede, T., Parsons, N., Stallard, N., Todd, S., Marquez, E.V., et al. (2011). Designing a seamless phase II/III clinical trial using early outcomes for treatment selection: An application in multiple sclerosis. *Statistics in Medicine*, 30, 1528–1540.

Fritsch, K. (2012). Multiplicity issues in FDA-reviewed clinical trials. EMA Workshop on Multiplicity Issues, London, UK.

Frueh, F.W. (2006). Impact of microarray data quality on genomic data submissions to the FDA. *Nature Biotechnology*, 24, 1105–1107.

Fukuoka, M., Yano, S., Giaccone, G., Tamura, T., Nakagawa, K., Douillard, J.Y., Nishiwaki, Y. et al. (2003). Multi-institutional randomized phase II trial of gefitinib for previously treated patients with advanced non-small-cell lung cancer. *Journal of Clinical Oncology*, 21(12): 2237–2246.

Gail, M.H. and Simon, R. (1985). Testing for qualitative interactions between treatment effects and patient subsets. *Biometrics*, 41, 361–372.

Gallo, P., Chuang-Stein, C., Dragalin, V., Gaydos, B., Krams, M., and Pinheiro, J. (2006). Adaptive design in clinical drug development—an executive summary of the PhRMA Working Group (with discussions). *Journal of Biopharmaceutical Statistics*, 16(3), 75–283.

Gallo, J. and Khuri, A.I. (1990). Exact tests for the random and mixed effects in an unbalanced mixed two-way cross-classification model. *Biometrics*, 46, 1087–1095.

Gelfand, A.E. and Smith, A.F.M. (1990). Sampling-based approaches to calculating densities. *Journal of the American Statistical Association*, 85, 398–409.

Genevois, E., Lelouer, V., Vercken, J.-B., and Caillon, R. (1996). Study design, methodology and statistical analyses in the clinical development of sparfloxacin. *Journal Antimicrobial Chemotherapy*, 37, 65–72.

Gentle, J.E. (1998). *Random Number Generator and Monte Carlo Methods*, Springer-Verlag, New York.

Genz, A. and Bretz, F. (2002). Methods for the computation of multivariate t-probabilities. *Journal of Computational and Graphical Statistics*, 11, 950–971.

Girman, C.J., Ibia, E., Menjoge, S., Mak, C., Chen, J., Agarwal, A., Binkowitz, B. (2011). Impact of different regulatory requirements for trial endpoints in multiregional clinical trials. *Drug Information Journal*, 45, 587–594.

Global orphan drug market to reach US$ 120 billion by 2018 (press release). New Delhi, India: Kuick Research, February 7, 2014, retrieved March 20, 2014.

Goldberg, J.D. and Kury, K.J. (1990). Design and Analysis of Multicenter Trials. In *Statistical Methodology in the Pharmaceutical Industry*, Berry, D. (Ed.), Marcel Dekker, New York, pp. 201–237.

Goodman, S.N. (1992). A comment on replication, p-values and evidence. *Statistics in Medicine*, 11, 875–879.

Goodman, S.N., Zahurak, M.L., and Piantadosi, S. (1995). Some practical improvements in the continual reassessment method for phase I studies. *Statistics in Medicine*, 14, 1149–1161.

Gormley, G.J., Stoner, E., Bruskewitz, R.C., Imperato-McGinley, J., Walsh, P.C., McConnell, J.D., Andriole, G.L., Geller, J., Bracken, B.R., Tenover, J.S., Vaughan, E.D., Pappas, F., Taylor, A., Binkowitz, B., Ng, J., for the Finasteride Study Group (1992). The effect of finasteride in men with benign prostatic hyperplasia. *New Engl. J. Med.*, 327, 1185–1191.

Grady, C. (2017). Ethics ad IRBs, proposed changes to the common rule, and rare disease research. *Translational Science of Rare Diseases*, 2, 176–178.

Graybill, F.A., and Wang, C.M. (1980) Confidence intervals on nonnegative linear combinations of variances. *Journal of the American Statistical Association*, 75, 869–873.

Greene, J.G. and Hart, D.M. (1987). Evaluation of a psychological treatment programme for climacteric women. *Maturitas*, 9, 41–48.

Griggs, R.C., Batshaw, M., Dunkle, M., Gopal-Srivastava, R., Kaye, E., Krischer, J., Nguyen, T., Paulus, K., and Merkel, P.A. (2009). Clinical research for rare disease: Opportunities, challenges, and solutions. *Molecular Genetics and Metabolism*, 96(1), 20–26.

Guilford, J.P. (1954). *Psychometric Methods*, 2nd ed., McGraw-Hill, New York.

Gunst, G.F. and Mason, R.L. (1980). *Regression Analysis and Its Application*, Marcel Dekker, New York.

Guyatt, G.H., Veldhuyen Van Zanten S.J.O., Feeny, D.H., Patric, D.L. (1989). Measuring quality of life in clinical trials: A taxonomy and review. *Canadian Medical Association Journal*, 140, 1441–1448.

Haidar, S.H., Davit, B., Chen, M.L., Conner, D., Lee, L., Li, Q.H., Lionberger, R. et al. (2008). Bioequivalence approaches for highly variable drugs and drug products. *Pharmaceutical Research*, 25, 237–241.

Hall, P. (1992). *The Bootstrap and Edgeworth Expansion*, Springer-Verlag, New York.

Hardwick, J.P. and Stout, Q.F. (2002). Optimal few-stage designs. *Journal of Statistical Planning and Inference*, 104, 121–145.

Hauck, W.W. and Anderson, S. (1984). A new statistical procedure for testing equivalence in two-group comparative bioavailability trials. *Journal of Pharmacokinetics and Biopharmaceutics*, 12, 83–91.

Hemmings, R. and Day, S. (2004). Regulatory perspectives on data safety monitoring boards: Protecting the integrity of data. *Drug Safety*, 27, 1–6.

Herson, J. (2009). *Data and Safety Monitoring Committees in Clinical Trials*, Chapman and Hall/CRC Press, Taylor & Francis Group, New York.

Heyd, J.M. and Carlin, B.P. (1999). Adaptive design improvements in the continual reassessment method for phase I studies. *Statistics in Medicine*, 18, 1307–1321.

Hochberg, Y. (1988). A sharper Bonferroni procedure for multiple tests of significance. *Biometrika*, 75, 800–803.

Hochberg, Y. and Benjamini, Y. (1990). More powerful procedures for multiple significance testing. *Statistics in Medicine*, 9, 811–818.

Hochberg, Y. and Tamhane, A.C. (1987). *Multiple Comparison Procedures*, Wiley, New York.

Hollenberg, N.K., Testa, M., and Williams, G.H. (1991). Quality of life as a therapeutic end-point: An analysis of therapeutic trials in hypertension. *Drug Safety*, 6, 83–93.

Holm, S. (1979). A simple sequentially rejective multiple test procedure. *Scandinavian Journal of Statistics*, 6, 65–70.

Holmgren, E.B. (1999). Establishing equivalence by showing that a prespecified percentage of the effect of the active control over placebo is maintained. *Journal of Biopharmaceutical Statistics*, 9, 651–659.

Holy, D.C., Rattray, M., Jupp, R., and Brass, A. (2002). Making sense of microarray data distributions. *Bioinformatics*, 18, 576–584.

Hommel, G. (1988). A stagewise rejective multiple test procedure based on a modified Bonferroni test. *Biometrika*, 75, 383–386.

Hommel, G. (2001). Adaptive modifications of hypotheses after an interim analysis. *Biometrical Journal*, 43, 581–589.

Hommel, G., Lindig, V., and Faldum, A. (2005). Two stage adaptive designs with correlated test statistics. *Journal of Biopharmaceutical Statistics*, 15, 613–623.

Hosmane, B. and Locke, C. (2005). A simulation study of power in thorough QT/QTc studies and a normal approximation for planning purposes. *Dug Information Journal*, 39, 447–455.

Hosmer, D.W. and Lemeshow, S.L. (2000). *Applied Logistic Regression*, 2nd ed., Wiley-Interscience, New York.

Howe, W.G. (1974). Approximate confidence limit on mean of X+Y where X and Y are two tabled independent random variables. *Journal of the American Statistical Association*, 69, 789–794.

Hsiao, C.F., Hsu, Y.Y., Tsou, H.H., and Liu, J.P. (2007). Use of prior information for Bayesian evaluation of bridging studies. *Journal of Biopharmaceutical Statistics*, 17, 109–121.

Hsiao, C.F., Tsou, H.H., Pong, A., Liu, J.P., Lin, C.H., Chang, Y.J., and Chow, S.C. (2009). Statistical validation of traditional Chinese diagnostic procedure. *Drug Information Journal*, 43, 83–95.

Hsieh, T.C., Chow, S.C., Liu, J.P., Hsiao, C.F., and Chi, E. (2010). Statistical test for evaluation of biosimilarity of follow-on biologics. *Journal of Biopharmaceutical Statistics* 20, 75–89.

Hsieh, T.C., Chow, S.C., Yang, L.Y., and Chi, E. (2013). The evaluation of biosimilarity index based on reproducibility probability for assessing follow-on biologics. *Statistics in Medicine*, 32, 406–414.

Hsu, J.C. (1984). Constrained two-sided simultaneous confidence intervals for multiple comparisons with the best. *Annals of Statistics*, 12, 1136–1144.

Hsu, J.C. (1996). *Multiple Comparisons: Theory and Methods*, Chapman & Hall, London, UK.

Hung, H.M.J. (2003). Statistical issues with design and analysis of bridging clinical trial. *Presented at the 2003 Symposium on Statistical Methodology for Evaluation of Bridging Evidence*, Taipei, Taiwan.

Hung, H.M.J. and Wang, S.J. (2009). Some controversial multiple testing problems in regulatory applications. *Journal of Biopharmaceutical Statistics*, 19, 1–11.

Hung, H.M.J., Wang, S.J., and O'Neil, R. (2007). Issues with statistical risks for testing methods in noninferiority trial without a placebo arm. *Journal of Biopharmaceutical Statistics*, 17, 201–213.

Hung, H.M., Wang, S.J., and O'Neill, R.T. (2010). Consideration of regional difference in design and analysis of multi-regional trials. *Pharmaceutical Statistics*, 9, 173–178.

Hung, H.M., Wang, S.J., Tsong, Y., Lawrence, J., and O'Neill, R.T. (2003). Some fundamental issues with non-inferiority testing in active controlled trials. *Statistics in Medicine*, 22, 213–225.

Huque, M. and Dubey, S.D. (1990). A three arm design and analysis for clinical trials in establishing therapeutic equivalence with clinical endpoints. *Proceedings of the Biopharmaceutical Section of the American Statistical Association*, pp. 91–98.

Hwang, I.K. and Morikawa, T. (1999). Design issues in noninferiority/equivalence trials. *Drug Information Journal*, 33, 1205–1218.

ICH. (1993). Q1A. Stability Testing of New Drug Substances and Products. Tripartite International Conference on Harmonization Guideline Q1A, Geneva, Switzerland.

ICH. (1995). Guideline for structure and content of clinical study report. International Conference on Harmonization, Yokohama, Japan.

ICH. (1996a). E6 good clinical practice. Tripartite International Conference on Harmonization Guideline, http://www/ich.org/LOB/media/MEDIA482.pdf.

ICH. (1996b). Validation of analytical procedures: Methodology. Tripartite International Conference on Harmonization Guideline.

ICH. (1996c). Q5C guideline on quality of biotechnological products: Stability testing of biotechnological/biological products. Center for Drug Evaluation and Research, Center for Biologics Evaluation and Research, the US Food and Drug Administration, Rockville, MD.

ICH. (1997). E5 guideline on ethnic factors in the acceptability of foreign data. *The U.S. Federal Register*, 83, 31790–31796.

ICH. (1998). E9 Guideline on *Statistical Principles for Clinical Trials*. Tripartite International Conference on Harmonization, Geneva, Switzerland.

ICH. (1999). E10 Guideline on *Choice of Control Group in Clinical Trials*. Tripartite International Conference on Harmonization, Geneva, Switzerland.

ICH. (1999). Q6B guideline on test procedures and acceptance criteria for biotechnological/biological products. Center for Drug Evaluation and Research, Center for Biologics Evaluation and Research, the US Food and Drug Administration, Rockville, MD.

ICH. (2000). E10 International conference on harmonization guideline: Guidance on choice of control group and related design and conduct issues in clinical trials. Food and Drug Administration, DHHS, July 2000.

ICH. (2005a). E14 The clinical evaluation of QT/QTc interval prolongation and proarrythmic potential for non-antiarrythmic drugs. Tripartite International Conference on Harmonization Guideline, Geneva, Switzerland, May 2005.

ICH. (2005b). Q5E guideline on comparability of biotechnological/biological products subject to changes in their manufacturing process. Center for Drug Evaluation and Research, Center for Biologics Evaluation and Research, the US Food and Drug Administration, Rockville, MD.

ICH. (2017). ICH E9(R1) Estimands and sensitivity analysis in clinical trials; 2017. Available from: http://www.ich.org/products/ guidelines/efficacy/article/efficacy-guidelines.html. Accessed August 20, 2018.

ICH. (2018). E9 guideline for statistical principles for clinical trials. Tripartite International Conference on Harmonization Guideline, Geneva, Switzerland.

Ideker, T., Thorsson, V., Siegel, A.F., and Hood, L.E. (2000). Testing for differentially expressed genes by maximum-likelihood analysis of microarray data. *Journal of Computational Biology*, 7, 805–817.

IHTT. (2013). Transforming Health Care Through Big Data: Strategies for leveraging big data in the health care industry. Institute for Health Technology Transformation (IHTT), New York.

Irizarry, R.A., Warren, D., Spencer, F., Kim, I.F., Biswal, S., Frank, B.C., Gabrielson, E. et al. (2005). Multi-laboratory comparison of microarray platforms. *Nature Methods*, 2, 345–349.

ISIS-2 Group. (1988). Randomized trial of intravenous streptokinase, oral aspirin, both, or neither among 17,187 cases of suspected acute myocardial infarction. *Lancet*, 13, 349–360.

Jachuck, S.J., Brierley, H., and Wilcox, P.M. (1982). The effect of hypotensive drugs on the quality of life. *Journal of the Royal College General Practitioners*, 32, 103–105.

Janerich, D.J., Piper, J.M., and Glebatis, D.M. (1980). Oral contraceptives and birth defects. *American Journal of Epidemiology*, 112, 73–79.

Jennison, C. and Turnbull, B.W. (2000). *Group Sequential Tests with Applications to Clinical Trials*, Chapman & Hall, London/Boca Raton, FL.

Jennison, C. and Turnbull, B.W. (2005). Meta-analysis and adaptive group sequential design in the clinical development process. *Journal of Biopharmaceutical Statistics*, 15, 537–558.

Ji, H. and Davis, R.W. (2006). Data quality in genomics and microarray. *Nature Biotechnology*, 24, 1112–1113.

Jiang, W., Freidlin, B., and Simon, R. (2007). Biomarker-adaptive threshold design: A procedure for evaluating treatment with possible biomarker-defined subset effect. *Journal of National Cancer Institute*, 99, 1036–1043.

JMP. (2012). Quality and Reliability Methods. JMP Version 10.1, A Business Unit of SAS, SAS Campus Drive, Cary, NC 27513.

Johnson, N.L. and Kotz, S. (1970). *Distributions in Statistics: Continuous Univariate Distributions*, John Wiley & Sons, New York.

Johnson, J., Williams, G., and Pazdur, R. (2003). End points and United States Food and Drug Administration approval of oncology drugs. *Journal of Clinical Oncology*, 21, 1404–1411.

Johnson, R.A. and Wichern, D.W. (1992). *Applied Multivariate Statistical Analysis*, 5th edition, Prentice Hall, Upper Saddle River, NJ.

Jones, B., Jarvis, P., Lewis, J.A., and Ebbutt, A.F. (1996). Trials to assess equivalence: The importance of rigorous methods. *British Medical Journal*, 313, 36–39.

Jones, B. and Kenward, M.G. (1989). *Design and Analysis of Crossover Trials*, Chapman-Hall, London, UK.

Julious, S.A., Tan, S.B., and Machin, D. (2009). *An Introduction to Statistics in Early Phase Clinical Trials*, Wiley-Blackwell, Chichester, UK.

Jung, S.H., Chow, S.C., and Chi, E.M. (2007). A note on sample size calculation based on propensity analysis in non-randomized trials. *Journal of Biopharmaceutical Statistics*, 17, 35–41.

Kalton, G. and Kasprzyk, D. (1986). The treatment of missing data. *Survey Methodology*, 12, 1–16.

Kamp, B., Bretz, F., Dmitrienko, A., Enas, G., Gaydos, B., Hsu, C.H., Konig, F. et al. (2007). Innovative approaches for designing and analyzing adaptive dose-ranging trials. *Journal of Biopharmaceutical Statistics*, 17, 965–995.

Kaplan, R.M., Bush, J.W., and Berry, C.C. (1976). Health status: Types of validity and index of well-being. *Health Services Research*, 4, 478–507.

Karlowski, T.R., Chalmers, T.C., Frenkel, L.D., Kapikian, A.Z., Lewis, T.L., and Lynch, J.M. (1975). Ascorbic acid for the common cold: A prophylactic and therapeutic trial. *Journal of American Medical Association*, 231, 1038–1042.

Kawai, N., Stein, C., Komiyama, O., and Li, Y. (2008). An approach to rationalize partitioning sample size into individual regions in a multiregional trial. *Drug Information Journal*, 42, 139–147.

Kelly, P.J., Sooriyarachchi, M.R., Stallard, N., and Todd, S. (2005a). A practical comparison of group-sequential and adaptive designs. *Journal of Biopharmaceutical Statistics*, 15, 719–738.

Kelly, P.J., Stallard, N., and Todd, S. (2005b). An adaptive group sequential design for phase II/III clinical trials that select a single treatment from several. *Journal of Biopharmaceutical Statistics*, 15, 641–658.

Kessler, D.A. (1989). The regulation of investigational drugs. *New England Journal of Medicine*, 320, 281–288.

Kessler, D.A. and Feiden, K.L. (1995). Faster evaluation of vital drugs. *Scientific American*, 272, 48–54.

Khongphatthanayothin, A., Lane, J., Thomas, D., Yen, L., Chang, D., and Bubolz, B. (1998). Effects of cisapride on QT interval in children. *Journal of Pediatrics*, 133, 51–56.

Khuri, A.I., Mathew, T., and Sinha, B.K. (1998). *Statistical Tests for Mixed Linear Models*, John Wiley & Sons, New York.

Ki, F.Y.C. and Chow, S.C. (1994). Analysis of quality of life with parallel questionnaires. *Drug Information Journal*, 28, 69–80.

Ki, F.Y.C. and Chow, S.C. (1995). Statistical justification for the use of composite score in quality assessment. *Drug Information Journal*, 29, 715–727.

Kimko, H.C. and Duffull, S.B. (Eds.) (2003). *Simulation for Designing Clinical Trials*, Marcel Dekker, New York.

Ko, F.S., Tsou, H.H., Liu, J.P., and Hsiao, C.F. (2010). Sample size determination for a specific region in a multi-regional trial. *Journal of Biopharmaceutical Statistics*, 20, 870–885.

Koch, G.C. (1991). One-sided and two-sided tests and p-values. *Journal of Biopharmaceutical Statistics*, 1, 161–170.

Kong, F., Chen, Y.F., and Jin, K. (2009). A bias correction in testing treatment effect under informative dropout in clinical trials. *Journal of Biopharmaceutical Statistics*, 19, 980–1000.

Korn, E.L., Albert, P.S., and McShane, L.M. (2005). Assessing surrogated as trial endpoints using mixed models. *Statistics in Medicine*, 24, 163–182.

Korteweg, M. (2002). Benchmarking of GRP – quality management system in the framework of PERF. *The Regulatory Affairs Journals Ltd*, 109–113.

Koti, K.M. (2007a). Use of the Fieller-Hinkley distribution of the ratio of random variables in testing for noninferiority. *Journal of Biopharmaceutical Statistics*, 17, 215–228.

Koti, K.M. (2007b). New tests for null hypothesis of non unity ratio of proportions. *Journal of Biopharmaceutical Statistics*, 17, 229–245.

Koyfman, S.A., Agrawal, M., Garrett-Mayer, E., Krohmal, B., Wolf, E., Emanuel, E.J., and Gross, C.P. (2007). Risks and benefits associated with novel phase 1 oncology trial designs. *Cancer*, 110, 1115–1124.

Kozlowski, S. (2007). FDA Policy on follow on biologics. Presented at Biosimilars 2007, George Washington University, Washington, DC.

Krams, M., Burman, C.F., Dragalin, V., Gaydos, B., Grieve, A.P., Pinheiro, J., and Maurer, W. (2007). Adaptive designs in clinical drug development: Opportunities challenges, and dcope reflections following PhRMA's November 2006 Workshop. *Journal of Biopharmaceutical Statistics*, 17, 957–964.

L'Abbe, K.A., Detsky, A.S., and O'Rourke, K. (1987). Meta-analysis in clinical research. *Annals of Internal Medicine*, 107, 224–233.

Lachin, J.M. (1988). Statistical properties of randomization in clinical trials. *Controlled Clinical Trials*, 9, 289–311.

Lachin, J.M. and Foulkes, M.A. (1986). Evaluation of sample size and power for analysis of survival with allowance for nonuniform patient entry, losses to follow-up, noncompliance, and stratification. *Biometrics*, 42, 507–519.

Lakatos, E. (1986). Sample size determination in clinical trials with time dependent rates of losses and noncompliance. *Controlled Clinical Trials*, 7, 189–199.

Lakshminarayanan, M.Y. (2010). Multiple comparisons. In *Encyclopedia of Biopharmaceutical Statistics*, Chow, S.C. (Ed.), Taylor & Francis Group, New York.

Lan, K.K.G. and DeMets, D.L. (1987). Group sequential procedures: Calendar versus information time. *Statistics in Medicine*, 8, 1191–1198.

Lange, K.L., Little, R.J.A., and Taylor, J.M.G. (1989). Robust statistical modeling using the t distribution. *Journal of the American Statistical Association*, 84, 881–896.

Larkin, J.E., Frank, B.C., Gavras, H., Sultana, R., and Quackenbush, J. (2005). Independence and reproducibility across microarray platforms. *Nature Methods*, 2, 337–343.

Lasser, K.E., Allen, P.D., Woolhandler, S.J., Himmelstein, D.U., Wolfe, S.M., and Bor, D.H. (2002). Timing of new black box warnings and withdrawals for prescription medications. *The Journal of American Medical Association*, 287, 2215–2220.

Laster L.L. and Johnson, M.F. (2003). Non-inferiority trials: The 'at least as good as' criterion. *Statistics in Medicine*, 22, 187–200.

Lawrence, D., Bretz, F., and Pocock, S.J. (2014). INHANCE: An adaptive confirmatory study with dose selection at interim. In *Indacaterol*, Trifilieff, A. (Ed.), Springer, Basel, Switzerland, pp. 77–92.

Le Tourneau, C., Lee, J.J., and Siu, L.L. (2009). Dose escalation methods in phase I cancer clinical trials. *Journal of the National Cancer Institute*, 101, 708–720.

Leber, P.D. (1989). Hazards of inference: the active control interpretation. *Epilepsia*, 30, S57–S63.

Lee, J.J., Gu, X., and Liu, S. (2010). Bayesian adaptive randomization designs for targeted agent development. *Clinical Trials*, 7, 584–596.

Lee, M.S. and Chen, M.C. (2004). Predicting antigenic variants of influenza A/H3N2 viruses. *Emergent Infectious Diseases*, 10, 1385–1390.

Lee, M.S., Chen, M.C., Liao, Y.C., and Hsiung, A.G. (2007). Identifying potential Immunodominant positions and predicting Antigenic variants of influenza A/H3N2 viruses. *Vaccine*, 25, 8133–8139.

Lee, S.J. and Lin, M. (2016). Adaptive trial design–case studies. Presented at the 2016 Duke-Industry Statistics Symposium (DISS), Durham, NC, September 14, 2016.

Lee, Y., Shao, J., Chow, S.C., and Wang, H. (2002a). Test for inter-subject and total variabilities under crossover design. *Journal of Biopharmaceutical Statistics*, 12, 503–534.

Lee, Y., Wang, H., and Chow, S.C. (2002b). Comparing variabilities in clinical research. In *Encyclopedia of Biopharmaceutical Statistics*, Chow, S.C. (Ed.), Marcel Dekker, New York.

Leeson, L.J. (1995). In Vitro/in vivo correlation. *Drug Information Journal*, 29, 903–915.

Lehmacher, W. and Wassmer, G. (1999). Adaptive sample size calculations in group sequential trials. *Biometrics*, 55, 1286–1290.

Lehmann, E.L. (1952). Testing multiparameter hypotheses. *Ann. Math. Stat.*, 23, 541–552.

Lehmann, E.L. (1986). *Testing Statistical Hypotheses*, 2nd ed. Wiley, New York.

Lepor, H., Williford, W.O., Barry, M.J., Brawer, M.K., Dixon, C.M., Gormley, G., Haakenson, C., Machi., M., Narayan, P., and Padley, R.J. (1996). The efficacy of terazosin, finasteride, or both in benign prostatic hyperplasia. *New Engl. J. Med.*, 335, 533–539.

Lewis, J.A. (1995). Statistical issues in the regulation of medicine. *Statistics in Medicine*, 14, 127–136.

Li, C.R., Liao, C.T., and Liu, J.P. (2007). On the exact interval estimation for the difference in paired areas under the ROC curves, *Statistics in Medicine*, Published on line on 10/29/2007.

Li, N. (2006). Adaptive trial design – FDA statistical reviewer's view. Presented at the CRT2006 Workshop with the FDA, Arlington, VA, April 4, 2006.

Li, W.J., Shih, W.J., and Wang, Y. (2005). Two-stage adaptive design for clinical trials with survival data. *Journal of Biopharmaceutical Statistics*, 15, 707–718.

Liang, B.A. (2007). Regulating follow-on biologics. *Harvard Journal on Legislation*, 44, 363–373.

Liao, C.T., Lin, C.Y., and Liu, J.P. (2007). Noninferiority tests based on concordance correlation coefficient for assessment of the agreement for gene expression data from microarray experiments. *Journal of Biopharmaceutical Statistics*, 17, 309–327.

Liao, Y.C., Lee, M.S., Ko, C.Y., and Hsiung, C.A. (2008). Bioinformatics models for predicting variants of influenza A/H3N2 viruses. *Bioinformatics*, 24, 505–512.

Lin, L.I. (1989). A concordance correlation coefficient to evaluate reproducibility. *Biometrics*, 45, 255–268.

Lin, L.I. (1992). Assay validation using the concordance correlation coefficient. *Biometrics*, 48, 599–604.

Lin, L.I. Hedayat, A.S., Sinha, B., and Yang, M. (2002). Statistical methods in assessing agreement: Models, issues, and tools. *Journal of the American Statistical Association*, 97, 257–270.

Lin, M., Chu, C.C., Chang, S.L., Lee, H.L., Loo, J.H., Akaza, T., Juji, T., Ohashi, J., and Tokunaga, K. (2001). The origin of Minnan and Hakka, the so-called "Taiwanese", inferred by HLA study. *Tissue Antigen*, 57, 192–199.

Lin, Y. and Shih, W.J. (2001). Statistical properties of the traditional algorithm-based designs for phase I cancer clinical trials. *Biostatistics*, 2, 203–215.

Little, R.J. (1994). A class of pattern-mixture models for normal missing data. *Biometrika*, 81, 471–483.

Little, R.J. and Rubin, D.B. (1987). *Statistical Analysis with Missing Data*, Wiley, New York.

Little, R.J. and Rubin, D.B. (2002). *Statistical Analysis with Missing Data*, 2nd ed., Wiley, New York.

Liu, C.H. and Rubin, D.B. (1995). ML estimation of the t distribution using EM and its extension ECM and ECME. *Statistica Sinica*, 5, 19–39.

Liu, H.K. (1990). Confidence intervals in bioequivalence. *Proceedings of the Biopharmaceutical Section of the American Statistical Association*, American Statistical Association, Alexandria, VA. pp. 51–54.

Liu, J.P. (1995). Use of the repeated crossover designs in assessing bioequivalence. *Statistics in Medicine*, 14, 1067–1078.

Liu, J.P. (1998). Statistical evaluation of individual bioequivalence. *Communications in Statistics, Theory and Methods*, 27, 1433–1451.

Liu, J.P. and Chow, S.C. (1996). Statistical issues on FDA conjugated estrogen tablets guideline. *Drug Information Journal*, 30, 881–889.

Liu, J.P. and Chow, S.C. (2008). Statistical issues on the diagnostic multivariate index assay and targeted clinical trials. *Journal of Biopharmaceutical Statistics*, 18, 167–182.

Liu, J.P., Chow, S.C., and Hsiao, C.F. (2013). *Design and Analysis of Bridging Studies*, Chapman & Hall/CRC Press, Taylor & Francis Group, New York.

Liu, J.P. and Lin, J.R. (2008). Statistical methods for targeted clinical trials under enrichment design. *Journal of the Formosan Medical Association*, 107, S34–S41.

Liu, J.P., Lin, J.R., and Chow, S.C. (2009). Inference on treatment effects for targeted clinical trials under enrichment design. *Pharmaceutical Statistics*, 8, 356–370.

Liu, J.P., Hsueh, H.M., and Hsiao, C.F. (2002a). Bayesian approach to evaluation of the bridging studies. *Journal of Biopharmaceutical Statistics*, 12, 401–408.

Liu, J.P., Hsueh, H.M., Hsieh, E., and Chen, J.J. (2002b). Tests for equivalence or non-inferiority for paired binary data. *Statistics in Medicine*, 21, 231–245.

Liu, J.P., Ma, M.C., Wu, C.Y., and Tai, J.Y. (2006). Tests of equivalence and non-inferiority for diagnostic accuracy based on the paired areas under ROC curves. *Statistics in Medicine*, 25, 1219–1238.

Liu, J.P., Dai, J.Y., Lee, T.C., and Liao, C.T. (2007). A new hypothesis to test minimal fold changes of gene expression levels. In: *The 5th International Conference on Multiple Comparison Procedures*, Vienna, Austria, July 9–11.

Liu, Q. and Chi, G.Y.H. (2001). On sample size and inference for two-stage adaptive designs. *Biometrics*, 57, 172–177.

Liu, Q., Proschan, M.A., and Pledger, G.W. (2002). A unified theory of two-stage adaptive designs. *Journal of American Statistical Association*, 97, 1034–1041.

Lohr, S.L. (1999). *Sampling Design and Analysis*, Duxbury Press, Pacific Grove, CA.

Loke, Y.C., Tan, S.B., Cai, Y., and Machin, D. (2006). A Bayesian dose finding design for dual endpoint Phase I trials. *Statistics in Medicine*, 25, 3–22.

Longford, N.T. (1993). *Random Coefficient Models*, Oxford University Press, New York.

Louis, T.A. (1982). Finding the observed information matrix when using the EM algorithm. *Journal of the Royal Statistical Society*, 44, 226–233.

Løvik Goll, G. (2016). NOR-SWITCH study: A randomized, doubleblind, parallel-group study to evaluate the safety and efficacy of switching from innovator infliximab to biosimilar infliximab compared with maintained treatment with innovator infliximab in patients with rheumatoid arthritis, spondyloarthritis, psoriatic arthritis, ulcerative colitis, Crohn's disease and chronic plaque psoriasis. http://www.lisnorway.no/.

Lu, Q., Chow, S.C., and Tse, S.K. (2007). Statistical quality control process for traditional Chinese medicine with multiple correlative components. *Journal of Biopharmaceutical Statistics*, 17, 791–808.

Lu, Y., Kong, Y., and Chow, S.C. (2017). Analysis of sensitivity index for assessing generalizability in clinical research. *Jacobs Journal of Biostatistics*, 2(1), 9–19.

Lu, Q., Tse, S.K., and Chow, S.C. (2010). Analysis of time-to-event data under a two-stage survival adaptive design in clinical trials. *Journal of Biopharmaceutical Statistics*, 20, 705–719.

Lu, Q., Tse, S.K., Chow, S.C., Chi, Y., and Yang, L.Y. (2009). Sample size estimation based on event data for a two-stage survival adaptive trial with different durations. *Journal of Biopharmaceutical Statistics*, 19, 311–323.

Lu, Q.S., Tse, S.K., Chow, S.C., and Lin, M. (2012). Analysis of time-to-event data with non-uniform patient entry and loss to follow-up under a two-stage seamless adaptive design with Weibull distribution. *Journal of Biopharmaceutical Statistics*, 22, 773–784.

Lyle, R.M., Melby, C.L., Hyner, G.C., Edmondson, J.W., Miller, J.Z., and Weinberger, M.H. (1987). Blood pressure and metabolic effects of calcium supplementation in normotensive white and black men. *Journal of American Medical Association*, 257, 1772–1776.

Ma, H., Smith, B., and Dmitrienko, A. (2008). Statistical analysis methods for QT/QTc prolongation. *Journal of Biopharmaceutical Statistics*, 18, 553–563.

Maca, J., Bhattacharya, S., Dragalin, V., Gallo, P., and Krams, M. (2006). Adaptive seamless phase II/III designs—background, operational aspects, and examples. *Drug Information Journal*, 40, 463–474.

Machin, D., Campbell, M.J., Tan, S.B., and Tan, S.H. (2008). *Sample Size Tables for Clinical Studies*, 3rd ed., Wiley-Blackwell, Chichester, UK.

Maitournam, A. and Simon, R. (2005). On the efficiency of targeted clinical trials. *Statistics in Medicine*, 24, 329–339.

Malik, M. and Camm, A.J. (2001). Evaluation of drug-induced QT interval prolongation. *Drug Safety*, 24, 323–351.

Mallows, C.L. (1973). Some comments on Cp. *Techniometrics*, 15, 661–675.

Mankoff, S.P., Brander, C., Ferrone, S., and Marincola, F.M. (2004). Lost in translation: Obstacles to translational medicine. *Journal of Translational Medicine*, 2, 14.

MAQC Consortium. (2006). The MAQC project shows inter- and intraplatform reproducibility of gene expression measurements. *Nature Biotechnology*, 24, 1151–1161.

Marcus, R., Peritz, E., and Gabriel, K.B. (1976). On closed testing procedures with special reference to ordered analysis of variance. *Biometrika*, 63, 655–660.

Margolies, M.E. (1994). Regulations of combination products. *Applied Clinical Trials*, 3, 50–65.

Mathew, T. and Sinha, B.K. (1992). Exact and optimum tests in unbalanced split-plot designs under mixed and random models, *Journal of the American Statistical Association*, 87, 192–200.

Maxwell, C., Domenet, J.G., and Joyce, C.R.R. (1971). Instant experience in clinical trials: A novel aid to teaching by simulation. *Journal of Clinical Pharmacology*, 11, 323–331.

McCarthy, J. (2015). NIH kicks off cohort grants for precision medicine. www.gov-healthit.com/news. November 23, 2015.

McLachlan, G.J. and Krishnan, T. (1997). *The EM Algorithm and Extensions*, Wiley, New York.

McLachlan, G.J. and Peel, D. (2000). *Finite Mixture Models*, Wiley, New York.

Meier, P. (1972). The biggest public health experiment ever, the 1954 field trial of the salk poliomyeitis vaccine. In Statistics A Guide to Unknown. Mosteller et al. eds. Holden-Day, Wadsworth, Belmont, CA. pp. 2–13.

Meier, P. (1989). The biggest public health experiment ever, the 1954 field trial of the Salk poliomyelitis vaccine. In *Statistics: A Guide to the Unknown*, 3rd ed., Tanur, J.M., Mosteller, F. and Kruskal, W.H. (Eds.), Wadsworth, Belmont, CA, pp. 3–14.

Members of the Toxicogenomic Research Consortium. (2005). Standardization of global gene expression analysis between laboratories and across platforms. *Nature Methods*, 2, 351–356.

MHLW. (2007). *Basic Principles on Global Clinical Trials*. Notification No. 0928010, September 28, 2007, Ministry of Helath, Labour and Welfare of Japan. Available at http://www.pmda.go.jp/eglish/service/pdf/notification/0928010-e.pdf.

Milliken, G.A. and Johnson, D.E. (1988). *Analysis of Messy Data: Designed Experiments*, John Wiley & Sons, New York.

Moore, J.W. and Flanner, H.H. (1996). Mathematical comparison of curves with an emphasis on dissolution profiles. *Pharmaceutical Technology*, 20, 64–74.

Moore, K.L. and van der Laan, M.J. (2009). Increasing power in randomized trials with right censored outcomes through covariate adjustment. *Journal of Biopharmaceutical Statistics*, 19, 1099–1131.

MOPH. (2002). Guidance for Drug Registration. Ministry of Public Health, Beijing, China.

Moses, L.E. (1985). Statistical concepts fundamental to investigations. *New England Journal of Medicine*, 312, 890–897.

Moss, A.J. (1993). Measurement of the QT interval and the risk associated with QT interval prolongation. *American Journal of Cardiology*, 72, 23B–25B.

Mosteller, F. and Kruskal, W.H. (1972). *Statistics A Guide to Unknown*, 3rd Edition, Holden-Day, Wadsworth, Belmont, CA.

Mosteller, F. and Tukey, J.W. (1987). *Data Analysis, including Statistics*, CRC Press, New York, pp. 601–720.

Motzer R., Penkov, K., Haanen, J., Rini, B., Albiges, L., Campbell, M.T., Venugopal, B., Kollmannsberger, B., Negrier, S., Uemura, M., Lee, J.L. Vasiliev, A., Miller, W.H., Gurney, H., Schmidinger, M., Larkin, J., Atkins, M.B., Bedke, J., Alekseev, B., Wang, J., Mariani, M., Robbins, P.B., Chudnovsky, A., Fowst, C., Hariharan, S., Huang, B., di Pietro, A., and Choueiri, T.K. et al. (2019). Avelumab plus Axitinib versus Sunitinib for Advanced Renal Cell Carcinoma. *The New England Journal of Medicine*, 380, 1103–1115.

Moye, L.A. (2003). *Multiple Analyses in Clinical Trials*, Springer-Verlag, New York.

Mugno, R., Zhus, W., and Rosenberger, W.F. (2004). Adaptive urn designs for estimating several percentiles of a dose response curve. *Statistics in Medicine*, 23, 2137–2150.

Muller, H.H. and Schafer, H. (2001). Adaptive group sequential designs for clinical trials: Combining the advantages of adaptive and classical group sequential approaches. *Biometrics*, 57, 886–891.

Myrand, S.P., Sekiguchi, K., Man, M.Z., Lin, X., Tzeng, R.Y., Teng, C.H., Hee, B. et al. (2008). Pharmacokinetics/genotype associations for major cytochrome P450 enzymes in native and first- and third-generation Japanese populations: Comparison with Korean, Chinese, and Caucasian populations. *Clinical Pharmacology & Therapeutics*, 84(3), 347–361.

Natarajan, J. and Tian, H. (2008). Effect of baseline measurement on the change from baseline in QTc intervals. *Journal of Biopharmaceutical Statistics*, 18, 542–552.

NRC. (2010). *The Prevention and Treatment of Missing Data in Clinical Trials*. Panel on Handling Missing Data in Clinical Trials. Committee on National Statistics, Division of Behavioral and Social Sciences and Education, The National Academies Press, Washington, DC.

NCCLS. (2001). *User Demonstration of Performance for Precision and Accuracy*. Approved Guidance, NCCLS document EP15-A, National Committee for Clinical Laboratory Standards, Wayne, PA.

Nevius, S.E. (1988). Assessment of evidence from a single multicenter trial. *Proceedings of Biopharmaceutical Section of the American Statistical Association*, Alexandria, VA, pp. 43–45.

News Release. (2015). FACT SHEET: President Obama's Precision Medicine Initiative. The White House, Office of the Press Secretary, January 30, 2015.

Ng, T.H. (2007). Simulatneous testing of noninferiority and superiority increases the false discover rate. *Journal of Biopharmaceutical Statistics*, 17, 259–264.

Nie, L., Chu, H., Cheng, Y., Spurney, C., Nagaraju, K., and Chen, J. (2009). Marginal and conditional approaches to multivariate variables subject to limit of detection. *Journal of Biopharmaceutical Statistics*, 19, 1151–1161.

Nie, L., Niu, Y., Yuan, M., Gwise, T., Leven, G. and Chow, S.C. (2019). Strategy for similarity margin selection in comparative biosimilar clinical studies. *Journal of Biopharmaceutical Statistics*, To appear.

NIH. (1998). NIH policy for data and safety monitoring. The United States National Institutes of Health. June, 1998. http://grants1.nih.gov/grants/guide/notice-files/not98-084.html.

NIH. (2000). Further guidance on data and safety monitoring for phase I and II trials. The United States National Institutes of Health, OD-00-038, June 2000.

Nityasuddhi, D. and Bohning, D. (2003). Asymptotic properties of the EM algorithm estimate for normal mixture models with component specific variances. *Computational Statistics and Data Analysis*, 41, 591–601.

NRC. (2011). *Toward Precision Medicine: Building a Knowledge Network for Biomedical Research and a New Taxonomy of Disease*. Committee on A Framework for Developing a New Taxonomy of Disease, National Research Council, The National Academies Press, Washington DC.

Oehlert, G.W. (1992). A note on the delta method. *The American Statistician*, 46, 27–29.

Ohlhorst, F.J. (2012). *Big Data Analytic Turning Big Data into Big Money*, John Wiley & Sons, New York.

Olschewski, M. and Schumacher, M. (1990). Statistical analysis of quality of life data in cancer clinical trials. *Statistics in Medicine*, 9, 749–763.

O'Neill, R.T. (2003). The ICH E5 guidance: An update on experiences with its implementation. *Presented at the 2003 Symposium on Statistical Methodology for Evaluation of Bridging Evidence*, Taipei, Taiwan.

O'Quigley, J. (2001). Dose-finding designs using continual reassessment method. In: *Handbook of Statistics in Clinical Oncology*, Crowley, J. (Ed.), Marcel Dekker, New York, pp. 35–72.

O'Quigley, J. and Chevret, S. (1991). Methods for dose finding studies in cancer clinical trials: A review and results of a Monte Carlo study. *Statistics in Medicine*, 10, 1647–1664.

O'Quigley, J., Hughes, M.D., and Fenton, T. (2001). Dose finding designs for HIV studies. *Biometrics*, 57, 1018–1029.

O'Quigley, J., Pepe, M., and Fisher, L. (1990). Continual reassessment method: A practical design for phase I clinical trials in cancer. *Biometrics*, 46, 33–48.

O'Quigley, J. and Shen, L. (1996). Continual reassessment method: A likelihood approach. *Biometrics*, 52, 673–684.

Owen, D.B. (1965). A special case of a noncentral t distribution. *Biometrika*, 52, 437–446.

Paez, J.G., Janne, P.A., Lee, J.C., Tracy, S., Greulich, H., Gabriel, S., Herman, P. et al. (2004). EGFR mutations in lung cancer: Correlation with clinical response to gefitinib therapy. *Science*, 304(5676), 1497–1500.

Paik, S., Shak, S., Tang, G., Kim, C., Baker, J., Cronin, M., Baehner, F.L. et al. (2004). A multigene assay to predict recurrence of tamoxifen-treated, node-negative breast cancer. *New England Journal of Medicine*, 351, 2817–2826.

Paik, S., Tang, G., Shak, S., Kim, C., Baker, J., Kim, W., Cronin, M. et al. (2006). Gene expression and benefit of chemotherapy in women with node-negative, estrogen receptor-positive breast cancer. *Journal of Clinical Oncology*, 24, 1–12.

Pariser, A. (2014). Rare disease and clinical trials. Office of Translational Sciences Center for Drug Evaluation and Research, Food and Drug Administration, November 4, 2014.

Paoletti, X. and Kramar, A. (2009). A comparison of model choices for the Continual Reassessment Method in phase I cancer trials. *Statistics in Medicine*, 28, 3012–3028.

Parmigiani, G. (2002). *Modeling in Medical Decision Making*, John Wiley & Sons, Chichester, UK.

Patterson, S., Agin, M., Anziano, R., Burgess, T., Chuang-Stein, C., Demitriwnko, A., et al. (2005a). Investigating drug-induced QT and QTc prolongation in the clinic: A review of statistical design and analysis considerations: Report from the Pharmaceutical Research and Manufacturers of America QT Statistics Expert Team. *Dug Information Journal*, 39, 243–266.

Patterson, S.D., Jones, B., and Zariffa, N. (2005b). Modelling and interpreting QTc prolongation in clinical pharmacology studies. *Dug Information Journal*, 39, 437–445.

Patterson, T.A., Lobenhofer, E.K., Fulmer-Smentek, S.B., Collins, P.J., Chu, T.M., Bao, W., Fang, H. et al. (2006). Performance comparison of one-color and two-color platforms with the MAQC project. *Nature Biotechnology*, 24, 1140–1150.

PDR. (1998). *Physicians' Desk Reference for Herbal Medicines*, Medical Economics Company, Montvale, NJ.

Peabody, F. (1927). The care of the patient. *JAMA*, 88, 877.

Petricciani, J.C. (1981). An overview of FDA, IRBs and regulations. *IRB*, 3, 1–3.

Pfanzagl, J., with the assistance of R. Hamböker (1994). *Parametric Statistical Theory*, Walter de Gruyter, Berlin, Germany, pp. 207–208.

Philipp, E. and Weihrauch, T.R. (1993). Multinational drug development and clinical research: A bird.s eye view of principles and practice. *Drug Information Journal*, 27, 1121–1132.

Phillips, K.F. (1990). Power of the two one-sided tests procedure in bioequivalence. *Journal of Pharmacokinetics and Biopharmaceutics*, 18, 137–144.

Phillips, K.F. (2003). A new test of non-inferiority for anti-infective trials. *Statistics in Medicine*, 22, 201–212.

PhRMA. (2003). Investigating drug-induced QT and QTc prolongation in the clinic: Statistical design and analysis considerations. Report from the Pharmaceutical Research an Manufacturers of America QT Statistics Expert Team, August 14, 2003.

Piantadosi, S. and Liu, G. (1996). Improved designs for dose escalation studies using pharmacokinetic measurements. *Statistics in Medicine*, 15, 1605–1618.

Pizzo, P.A. (2006). *The Dean's Newsletter*, Stanford University School of Medicine, Stanford, CA.

Pong, A. and Raghavarao, D. (2002). Comparing distributions of drug shelf lives for two components in stability analysis for different designs. *Journal of Biopharmaceutical Statistics*, 12, 277–293.

Posch, M. and Bauer, P. (1999). Adaptive two stage designs and the conditional error function. *Biometrical Journal*, 41, 689–696.

Posch, M. and Bauer, P. (2000). Interim analysis and sample size reassessment. *Biometrics*, 56, 1170–1176.

Posch, M., Konig, F., Brannath, W., Dunger-Baldauf, C., and Bauer, P. (2005). Testing and estimation in flexible group sequential designs with adaptive treatment selection. *Statistics in Medicine*, 24, 3697–3714.

Pratt, C.M., Hertz, R.P. Ellis, B.E., Crowell, S.P., Louv, W., and Moye, L. (1994). Risk of developing life-threatening ventricular arrhythmia associated with tefenadine in comparison with over-the-counter antihistamines, ibuprofen and clemastine. *American Journal of Cadiology*, 73, 346–352.

Press, W.H., Teukolsky, S.A., Vetterling, W.T., and Flannery, B.P. (2002). *Numerical Recipes in C++*, 2nd ed., Cambridge University Press, Cambridge, UK.

Proschan, M.A. and Hunsberger, S.A. (1995). Designed extension of studies based on conditional power. *Biometrics*, 51, 1315–1324.

Proschan, M.A. and Wittes, J. (2000). An improved double sampling procedure based on the variance. *Biometrics*, 56, 1183–1187.

Quan and Shih (1996). Assessing reproducibility by the within-subject coefficient of variation with random effects models. *Biometrics*, 52, 1195–1203.

Quan, H., Zhao, P.L., Zhang, J., Roessner, M., and Aizawa, K. (2010). Sample size considerations for Japanese patients based on MHLW guidance. *Pharmaceutical Statistics*, 9, 100–112.

Quinlan, J.A., Gallo, P., and Krams, M. (2006). Implementing adaptive designs: Logistical and operational consideration. *Drug Information Journal*, 40, 437–444.

Raghupathi, W. and Raghupathi, V. (2014). Big data analytics in healthcare promise and potential. *Health Information Science and Systems*, 2, 2–3.

Rao, J.N.K. and Scott, A.J. (1981). The analysis of categorical data from complex sample surveys: Chi-square tests for goodness-of-fit and independence in two-way tables. *Journal of American Statistical Association*, 76(374), 221–230.

Rao, J.N.K. and Scott, A.J. (1987). On simple adjustments to chi-square tests with sample survey data. *The Annals of Statistics*, 15, 1–12.

Rao, J.N.K. and Shao, J. (1992). Jackknife variance estimation with survey data under hot deck imputation. *Biometrika*, 79, 811–822.

Redman, M.W. and Allegra, C.J. (2015). The master protocol concept. *Seminars in Oncology*, 42(5), 724–730.

Reiser, B. and Faraggi, D. (1997). Confidence intervals for the general ROC criterion. *Biometrics*, 53, 644–652.

Roger, S.D. (2006). Biosimilars: How similar or dissimilar are they? *Nephrology*, 11, 341–346.

Roger, S.D. and Mikhail, A. (2007). Biosimilars: Opportunity or cause for concern? *Journal of Pharmaceutical Science*, 10, 405–410.

Rom, D.M. (1990). Asequentially rejective test procedure based on a modified Bonferroni inequality. *Biometrika*, 77, 663–665.

Rosenbaum, P.R. and Rubin, D.B. (1983). The central role of the propensity score in observational studies for causal effects. *Biometrika*, 70(1), 41–55.

Rosenbaum, P.R. and Rubin, D.B. (1984). Reducing bias in observational studies using subclassification on the propensity score. *Journal of American Statistical Association*, 95, 749–759.

Rosenberger, W.F. and Lachin, J.M. (2003). *Randomization in Clinical Trials*, John Wiley & Sons, New York.

Rosenberger, W.F., Stallard, N., Ivanova, A., Harper, C.N., and Rick, M.L. (2001). Optimal adaptive designs for binary response trials. *Biometrics*, 57, 909–913.

Rothmann, M.D., Koti, K., Lee, K.Y., Lu, H.L., and Shen, Y.L. (2009). Missing data in biologic oncology products. *Journal of Biopharmaceutical Statistics*, 19, 1074–1084.

Rotnitzky, A., Robins, J.M., and Scharfstein, D.O. (1998). Semiparametric regression for repeated measures outcomes with non-ignorable non-response. *Journal of the American Statistical Association*, 93, 1321–1339.

Rutgeerts, P., Sandborn, W.J., Feagan, B.G., Reinisch, W., Olson, A., Johanns, J., Travers, S. et al. (2005). Infliximab for induction and maintenance therapy for ulcerative colitis. *New England Journal of Medicine*, 353, 2462–2476.

Salah, S., Chow, S.C., and Song, F.Y. (2017). On evaluation of reliability, repeatability, and reproducibility in laboratory testing. *Journal of Biopharmaceutical Statistics*, 27, 331–337.

Sampson, A. and Sill, M.W. (2005). Drop-the-losers design: Normal case. *Biometrical Journal*, 47(3), 257–268.

Sargent, D.F., Wieand, S., Haller, D.G., Gray, R., Benedetti, J.K., Buyse, M., Labianca, R. et al. (2005). Disease-free survival versus overall survival as a primary end point for adjuvant colon cancer studies. *Journal of Clinical Oncology*, 23, 8864–8670.

Sarkar, S. and Chang, C.K. (1997). Simes method for multiple hypothesis testing with positiviely dependent test statistics. *Journal of American Statistical Association*, 91, 1601–1608.

Schafer, J.L. (1997). *Analysis of Incomplete Multivariate Data*, Chapman and Hall, London, UK.

Scheffé, H. (1959). *The Analysis of Variance*, Wiley, New York.

Schellekens, H. (2004). How similar do 'biosimilar' need to be? *Nature Biotechnology*, 22, 1357–1359.

Schuirmann, D.J. (1987). A comparison of the two one-sided tests procedure and the power approach for assessing the equivalence of average bioavailability. *Journal of Pharmacokinetics and Biopharmaceutics*, 15, 657–680.

Searle, S.R., Casella, G., and McCulloch, C.E. (1992). *Variance Components*. Wiley, New York.

Serfling, R.J. (1980). *Approximation Theorems of Mathematical Statistics*, Wiley, New York.

Shao, J. (1993). Linear model selection by cross-validation. *Journal of American Statistical Association*, 88, 486–494.

Shao, J. (1999). *Mathematical Statistics*, Springer-Verlag, New York.

Shao, J., Chang, M., and Chow, S.C. (2005). Statistical inference for cancer trials with treatment switching. *Statistics in Medicine*, 24, 1783–1790.

Shao, J. and Chow, S.C. (1993). Two-stage sampling with pharmaceutical applications. *Statistics in Medicine*, 12, 1999–2008.

Shao, J. and Chow, S.C. (2002). Reproducibility probability in clinical trials. *Statistics in Medicine*, 21, 1727–1742.

Shao, J. and Chow, S.C. (2007). Variable screening in predicting clinical outcome with high-dimensional microarrays. *Journal of Multivariate Analysis*, 98, 1529–1538.

Shao, J. and Tu, D. (1995). *The Jackknife and Bootstrap*, Springer, New York.

Shao, J. and Wang, H. (2002). Sample correlation coefficients based on survey data under regression imputation. *Journal of American Statistical Association*, 97, 544–552.

Shao, J. and Zhong, B. (2003). Last observation carry-forward and last observation analysis. *Statistics in Medicine*, 22, 2429–2441.

Shardell, M. and El-Kamary, S. (2009). Sensitivity analysis of informatively coarsened data using pattern mixture models. *Journal of Biopharmaceutical Statistics*, 19, 1018–1038.

Shibata, R. (1981). An optimal selection of regression variables. *Biometrika*, 68, 45–54.

Shih, W.J. (2001a). Sample size re-estimation–a journey for a decade. *Statistics in Medicine*, 20, 515–518.

Shih, W.J. (2001b). Clinical trials for drug registrations in Asian Pacific countries: Proposal for a new paradiam from a statistical perspective. *Controlled Clinical Trials*, 22, 357–366.

Shih, W.J., Gould, A.L., and Hwang, I.K. (1989). The analysis of titration studies in phase III clinical trials. *Statistics in Medicine*, 8, 583–591.

Shippy, R., Fulmer-Smentek, S., and Jensen, R.V., Jones, W.D., Wolber, P.K., Johnson, C.D., Pine, P.S. et al. (2006). Using RNA sample titrations to assess microarray platform performance and normalization techniques. *Nature Biotechnology*, 24, 1123–1131.

Shun, Z., Lan, K.K., and Soo, Y. (2008). Interim treatment selection using the normal approximation approach in clinical trials. *Statistics in Medicine*, 27(4), 597–618.

Sidak, Z. (1967). Rectangular confidence regions for the means of multivariate normal distributions. *Journal of the American Statistical Association*, 62, 626–633.

Simon, R. (1999). Bayesian design and analysis of active control clinical trials. *Biometrics*, 55, 484–487.

Simon, R. (2006). Validation of pharmacogenomics biomarker classifier for treatment selection. *Cancer Biomarkers*, 2, 89–96.

Simon, R. (2008). Development and validation of biomarker classifier for treatment selection. *Journal of Statistical Planning and Inference*, 138, 308–320.

Simon, R. and Maitournam, A. (2004). Evaluating the efficiency of targeted designs for randomized clinical trials. *Clinical Cancer Research*, 10, 6759–6763.

Simon, R.M., Korn, E.L., McShane, L.M., Radmacher, M.D., Wright, G.W., and Zhao, Y. (2003). *Design and Analysis of DNA Microarray Investigations*, Springer, New York.

Simes, R.J. (1986). An improved Bonferroni procedure for multiple test procedures. *Journal of American Statistical Association*, 81, 826–831.

Smith, N. (1992). FDA perspectives on quality of life studies. *Presented at DIA Workshop*, Hilton Head, SC.

Snapinn, S. (2017). Some remaining challenges regarding multiple endpoints in clinical trials. *Presented at 2017 Duke-Industry Statistics Symposium*, Durham, NC, September, 2017.

Sommer, A. and Zeger, S.L. (1991). On estimating efficacy from clinical trials. *Statistics in Medicine*, 10, 45–52.

Soon, G. (2009). Editorial: Missing data—prevention and analysis. *Journal of Biopharmaceutical Statistics*, 19, 941–944.

Sprarano, J., Hayes, D, Dees, E. et al. (2006). Phase III randomized study of adjuvant combination chemotherapy and hormonal therapy versus adjuvant hormonal therapy alone in women with previously resected axillary node-negative breast cancer with various levels of risk for recurrence (TAILORX Trial). http://www.cancer.gov/clinicaltrials/ECOG- PACCT-1. Accessed on June 5 2006.

Spriet, A. and Dupin-Spriet, T. (1992). Good biometrics practice proposals for a set of procedures. *Drug Information Journal*, 26, 405–409.

Sprung, C.L., Finch, R.G., Thijs, L.G., and Glauser, M.P. (1996). International sepsis trial (INTERSEPT): Role and impact of a clinical evaluation committee. *Critical Care Medicine*, 24, 1441–1447.

Srivastava, M.S. and Carter, E.M. (1986). The maximum likelihood method for nonresponse in sample surveys. *Survey Methodology*, 12, 61–72.

Stallard, N. (2010). A confirmatory seamless phase II/III clinical trial design incorporating short-term endpoint information. *Statistics in Medicine*. 29, 959–971.

Stampfer, M.J., Willett, W.C., and Colditz, G.A. (1985). A prospective study of postmenopause estrogen therapy and coronary heart disease. *New England Journal of Medicine*, 313, 1014–1049.

Storer, B.E. (1989). Design and analysis of phase I trials. *Biometrics*, 45, 925–937.

Storer, B.E. (1993). Small-sample confidence sets for the MTD in a phase I clinical trial. *Biometrics*, 49, 1117–1125.

Storer, B.E. (2001). An evaluation of phase I clinical trial designs in the continuous dose-response setting. *Statistics in Medicine*, 20, 2399–2408.

Strieter, D., Wu, W., and Agin, M. (2003). Assessing the effects of replicate ECGs on QT variability in healthy subjects. *Presented at Midwest Biopharmaceutical Workshop*, May 21, 2003, Muncie, Indiana.

Su, J.Q. and Liu, J.S. (1993). Linear combination of multiple diagnostic markers. *Journal of the American Statistical Association*, 88, 1350–1355.

SUPAC-IR. (1995). The United States Food and Drug Administration guideline immediate release solid oral dosage forms. Scale-Up and Postapproval Changes: Chemistry, Manufacturing, and Controls, In Vitro Dissolution Testing, and In Vivo Bioequivalence Documentation, Rockville, MD.

Suwelack, D. and Weihrauch, T.R. (1992). Practical issues in design and management of multinational trials. *Drug Information Journal*, 26, 371–378.

Swain, S.M. (2006). A step in the right direction. *Journal of Clinical Oncology*, 24(23), 1–2.

Tandon, P.K. (1990). Applications of global statistics in analyzing quality of life. *Statistics in Medicine*, 9, 819–827.

Tango, T. (1998). Equivalence test and confidence interval for the difference in proportions for the paired-sample design. *Statistics in Medicine*, 17(8), 891–908.

Temple, R. (1982). Government viewpoint of clinical trials. *Drug Information Journal*, 16, 10–17.

Temple, R. (1993). Trends in pharmaceutical development. *Drug Information Journal*, 27, 355–366.

Temple, R. (2003). Overview of the concept paper, history of the QT/TdP concern; Regulatory implications of QT prolongation. *Presentations at FDA/Industry Statistics Workshop*. September 18-19, 2003, Bethesda, MD.

Temple, R. and Ellenberg, S.S. (2000). Placebo-controlled trials and active-control trials in the evaluation of new treatments. Part 1: Ethical and scientific issues. *Annals of Internal Medicine*, 133(6), 455–463.

Tessman, D.K., Gipson, B., and Levins, M. (1994). Cooperative fast-track development: The fludara story. *Applied Clinical Trials*, 3, 55–62.

Testa, M.A. (1987). Interpreting quality of life clinical trial data for use in clinical practices of antihypertensive therapy. *Journal of Hypertension*, 5, S9–S13.

Testa, M.A., Anderson, R.B., Nackley, J.F., and Hollenberg, N.K. (1993). Quality of life And antihypertensive therapy in men: A comparison of Captopril with Enalapril. *New England Journal of Medicine*, 328, 907–913.

Thall, P.F. and Russel, K.E. (1998). A strategy for dose-finding and safety monitoring based on efficacy and adverse outcomes in phase I/II clinical trials. *Biometrics*, 54, 251–264.

Thomas, A.L. (1982). Finding the observed information matrix when using the EM algorithm. *Journal of the Royal Statistical Society*, 44, 226–233.

Tibshirani, R. (1996). Regression shrinkage and selection via the Lasso. *Journal of Royal Statistical Society: Series B*, 58, 267–288.

Todd, S. (2003). An adaptive approach to implementing bivariate group sequential clinical trial designs. *Journal of Biopharmaceutical Statistics*, 13, 605–619.

Todd, S. and Stallard, N. (2005). A new clinical trial design combining Phases 2 and 3: Sequential designs with treatment selection and a change of endpoint. *Drug Information Journal*, 39, 109–118.

Tong, W., Lucas, A.B., Shippy, R., Fan, X., Fang, H., Hong, H., Orr, M.S. et al. (2006). Evaluation of external RNA controls for the assessment of microarray performance. *Nature Technology*, 24, 1132–1139.

Torrance, G.W. (1976). Toward a utility theory foundation for health status index models. *Health Services Research*, 4, 349–369.

Torrance, G.W. (1987). Utility approach to measuring health-related quality of life. *Journal of Chronic Diseases*, 40, 593–600.

Torrance, G.W. and Feeny, D.H. (1989). Utilities and quality-adjusted life years. *Journal of Technology Assessment in Health Care*, 5, 559–575.

Tse, S.K., Chang, J.Y., Su, W.L., Chow, S.C., Hsiung, C., and Lu, Q.S. (2006). Statistical quality control process for traditional Chinese medicine. *Journal of Biopharmaceutical Statistics*, 16, 861–874.

Tsiatis, A.A. and Mehta, C. (2003). On the inefficiency of the adaptive design for monitoring clinical trials. *Biometrika*, 90, 367–378.

Tsong, Y. (2007). Special issue on active controlled clinical trials. *Journal of Biopharmaceutical Statistics*, 17, 197–199.

Tsong, Y., Higgins, K., Wang, S.J., and Hung, H.M.J. (1999). An overview of equivalence testing-CDER reviewers' perspective. *Proceedings of the Biopharmaceutical Section of American Statistical Association*, pp. 214–219.

Tsong, Y. and Shen, M. (2007). An alternative approach to assess exchangeability of a test treatment and the standard treatment with normally distributed response. *Journal of Biopharmaceutical Statistics*, 17, 329–338.

Tsong, Y., Shen, M., Zhong, J., and Zhang, J. (2008). Statistical issues of QT prolongation assessment based on linear concentration modeling. *Journal of Biopharmaceutical Statistics*, 18, 564–584.

Tsong, Y. and Zhang, J. (2005). Testing superiority and noninferiority hypotheses in active controlled clinical trials. *Biometrical Journal*, 47, 62–74.

Tsong, Y. and Zhang, J. (2007). Simultaneous test for superiority and noninferiority hypotheses in active-controlled clinical trials. *Journal of Biopharmaceutical Statistics*, 17, 247–257.

Tsong, Y. and Zhang, J. (2008). Guest editors' notes on statistical issues in design and analysis of thorough QTc studies. *Journal of Biopharmaceutical Statistics*, 18, 405–407.

Tsong, Y., Zhong, J., and Chen, W.J. (2008). Validation testing in thorough QT/QTc clinical trials. *Journal of Biopharmaceutical Statistics*, 18, 529–541.

Tsong, Y., Zhang, J., and Levenson, M. (2007). Choice of δ noninferiority margin and dependency of the noninferiority trials. *Journal of Biopharmaceutical Statistics*, 17, 279–288.

Tsou, H.H., Hsiao, C.F., Chow, S.C., Yue, L., Xu, Y., and Lee, S. (2007). Mixed noninferiority margin and statistical tests in active controlled trials. *Journal of Biopharmaceutical Statistics*, 17, 339–357.

Tsou, H.H., Tsong, Y., Wang, W.J., Dong, X., and Hsiao, C.F. (2012). Design and analysis issues of multi-regional clinical trials with different regional endpoints. *Journal of Biopharmaceutical Statistics*, 22, 1051–1059.

Tusher, V.G., Tibshirani, R., and Chu, G. (2001). Significance analysis of microarrays applied to ionizing radiation response, *Proceedings of National Academy of Sciences*, 98, pp. 5116–5121.

Uesaka, H. (2009). Sample size allocation to regions in multiregional trial. *Journal of Biopharmaceutical Statistics*, 19, 580–594.

Ueta, M., Sotozono, C., Tokunaga, K., Yabe, T., and Kinoshita, S. (2007). Strong association between HLA-A*0206 and Stevens-Johnson syndrome in the Japanese. *American Journal of Ophthalmology,* 143(2), 367–368.

USP/NF. (2000). United States Pharmacopeia 24 and National Formulary 19, United States Pharmacopeial Convention, Rockville, MD.

Van't Veer, L.J., Dai, H., van de Vijver, M.J., He, Y.D., Hart, A.A., Mao, M., Peterse, H.L. et al. (2002). Gene expression profiling predicts clinical outcome of breast cancer. *Nature,* 415, 530–536.

Van de Vijver, M.J., He, Y.D., van't Veer, L.J., Dai, H., Hart, A.A., Voskuil, D.W., Schreiber, G.J. et al. (2002). A gene-expression signature as a predictor of survival in breast cancer. *New England Journal of Medicine,* 347, 1999–2009.

Varmus, H. (2006). The new era in cancer research. *Science,* 312, 1162–1165.

Wang, H. (2001). Two-way contingency tables with marginally and conditionally imputed nonrespondents. Ph.D. Thesis, Department of Statistics, University of Wisconsin- Madison, WI.

Wang, H. and Chow, S.C. (2002a). On statistical power for average bioequivalence testing under replicated crossover design. *Journal of Biopharmaceutical Statistics,* 12, 295–309.

Wang, H. and Chow, S.C. (2002b). A practical approach for parallel trials without equal variance assumptions. *Statistics in Medicine,* 21, 3137–3151.

Wang, H., Chow, S.C., and Li, G. (2002). On sample size calculation based on odds ratio in clinical trials. *Journal of Biopharmaceutical Statistics,* 12, 471–483.

Wang, J., Chang, M., and Menon, S. (2014). Biomarker-informed adaptive design. In: Carini C, Menon S, Chang M (Eds), *Clinical and Statistical Considerations in Personalized Medicine.* CRC Press, Boca Raton, FL, pp. 129–148.

Wang, S. and Ethier, S. (2004). A generalized likelihood ratio test to identify differentially expressed genes from microarray data. *Bioinformatics,* 20, 100–104.

Wang, S.J., Hung, H.M.J., and Tsong, Y. (2002). Utility and pitfall of some statistical methods in active controlled clinical trials. *Controlled Clinical Trials,* 23, 15–28.

Wang, S.J., O'Neill, R.T., and Hung, H.M.J. (2007). Approaches to evaluation of treatment effect in randomized clinical trials with genomic subset. *Pharmaceutical Statistics,* 6, 227–244.

Wang, S.K. and Tsiatis, A.A. (1987). Approximately optimal one-parameter boundaries for a sequential trial. *Biometrics,* 43, 193–200.

Wang, X., Wu, Y., and Zhou, H. (2009). Outcome- and auxiliary-dependent subsampling and its statistical inference. *Journal of Biopharmaceutical Statistics,* 19, 1132–1150.

Wang, Y., Pan, G., and Balch, A. (2008). Bias and variance evaluation of QT interval correction methods. *Journal of Biopharmaceutical Statistics,* 18, 427–450.

Ware, J.E. (1987). Standards for validating health measures definition and content. *Journal of Chronic Diseases,* 40, 473–480.

Webber, K.O. (2007). Biosimilars: Are we there yet? Presented at Biosimilars 2007, George Washington University, Washington, DC.

Weerakkody, G.J. and Johnson, D.E. (1992). Estimation of within model parameters in regression models with a nested error structure. *Journal of the American Statistical Association,* 87, 708–713.

Wei, G.C.G. and Tanner, M.A. (1990). A Monte Carlo implementation of the EM algorithm and the poor man's data augmentation algorithm. *Journal of the American Statistical Association,* 85, 699–704.

Wei, L.J. (1978). The adaptive biased-coin design for sequential experiments. *The Annals of Statistics*, 9, 92–100.

Wei, X. and Chappel, R. (2005). A test for non-inferiority with a mixed multiplicative/additive null hypothesis. *Presentation in 2005 ENAR Spring Meeting*, March 20-23, 2005, Austin, Texas.

Westfall, P. and Bretz, F. (2010). Multiplicity in clinical trials. In *Encyclopedia of Biopharmaceutical Statistics*, 3rd ed., Chow, S.C. (Ed.), Taylor & Francis Group, New York.

Westfall, P.H., Tobias, R.D., Rom, D., Wolfinger, R.D., and Hochberg, Y. (1999). *Multiple Comparisons and Multiple Tests Using the SAS System*. SAS Institute, Gary, NC.

Westlake, W.J. (1976). Symmetrical confidence intervals for bioequivalence trials. *Biometrics*, 32, 741–744.

Whitehead, J. (1997). Bayesian decision procedures with application to dose-finding studies. *International Journal of Pharmaceutical Medicine*, 11, 201–208.

Whitehead, J. and Williamson, D. (1998). An evaluation of Bayesian decision procedures for dose-finding studies. *Journal of Biopharmaceutical Medicine*, 8, 445–467.

WHO. (2005). World Health Organization Draft Revision on Multisource (Generic) Pharmaceutical Products: Guidelines on Registration Requirements to Establish Interchangeability, Geneva, Switzerland.

Wikipedia. (2010). List of withdrawn drugs. http://en.wikipedia.org/wiki/List_of_withdrawn_drugs.

Wiles, A., Atkinson, G., Huson, L., Morse, P., and Struthers, L. (1994). Good statistical practices in clinical research: Guideline standard operating procedures. *Drug Information Journal*, 28, 615–627.

Williams, G.H. (1987). Quality of life and its impact on hypertensive patients. *American Journal of Medicine*, 82, 98–105.

Williams, G., Pazdur, R., and Temple, R. (2004). Assessing tumor-related signs and symptoms to support cancer drug approval. *Journal of Biopharmaceutical Statistics*, 14, 5–21.

Wilson, P.W.F., Garrison, R.J., and Castelli, W.P. (1985). Post-menopause estrogen use, cigarette smoking and cardiovascular morbidity in women over 50: The Framingham study. *New England Journal of Medicine*, 313, 1038–1043.

Woodcock, J. (2004). FDA's critical path initiative. FDA: http://www.fda.gov/oc/initiatives/criticalpath/woodcock0602/woodcock0602.html.

Woodcock, J. (2005). FDA introduction comments: Clinical studies design and evaluation issues. *Clinical Trials*, 2, 273–275.

Woodcock, J., Griffin, J., Behrman, R., Cherney, B., Crescenzi, T., Fraser, B., Hixon, D. et al. (2007). The FDA's assessment of follow-on protein products: A historical perspective. *Nature Reviews Drug Discovery*, 6, 437–442.

Woodcock, J. and LaVange, L.M. (2017). Master protocols to study multiple therapies, multiple diseases, or both. *New England Journal of Medicine*, 377, 62–70.

Wu, Y.J., Tan, T.S., Chow, S.C., and Hsiao, C.F. (2014). Sample size estimation of multiregional clinical trials with heterogeneous variability across regions. *Journal of Biopharmaceutical Statistics*, 24, 254–271.

Wysowski, D.K., Corken, A., Gallo-Torres, H., Talarico, L., and Rodriguez, E.M. (2001). Postmarketing reports of QT prolongation and ventricular arrhythmia in association with cisapride and Food and Drug Administration regulatory actions. *American Journal of Gastroenterology*, 96, 1698–1703.

Yan, X., Lee, S., and Li, N. (2009). Missing data handling methods in medical device clinical trials. *Journal of Biopharmaceutical Statistics*, 19, 1085–1098.

Yan, X., Wang, M.C., and Su, X. (2007). Test for the consistency of noninferiority from multiple clinical trials. *Journal of Biopharmaceutical Statistics*, 17, 265–278.

Ying, L., Song, F.Y., Chow, S.C., Zheng, J., Li, X., Henry, D., and Sethuraman, V. (2018). On evaluation of consistency in multi-regional clinical trials. *Journal of Biopharmaceutical Statistics*, 28(5), 840–856. doi:10.1080/10543406.2017.1397 008.

Ypma, T.J. (1995). Historical development of the Newton-Raphson method. *SIAM Review*, 37(4), 531–551.

Yue, L. (2006). Statistical and regulatory issues with the application of propensity score analysis to nonrandomized medical device clinical studies. *Journal of Biopharmaceutical Statistics*, 17(1), 1–13.

Zhang, H. and Paik, M.C. (2009). Handling missing responses in generalized linear mixed model without specifying missing mechanism. *Journal of Biopharmaceutical Statistics*, 19, 1001–1017.

Zhang, J. (2008). Testing for positive control activity in a thorough QTc study. *Journal of Biopharmaceutical Statistics*, 18, 517–528.

Zheng, J. and Chow, S.C. (2019). Criteria for dose-finding in two-stage seamless adaptive design. *Journal of Biopharmaceutical Statistics*, 1–12.

Zhang, J. and Machado, S.G. (2008). Statistical issues including design and sample size calculation in thorough QT/QTc studies. *Journal of Biopharmaceutical Statistics*, 18, 451–467.

Zhang, L., Dmitrienko, A., and Luta, G. (2008). Sample size calculations in thorough QT studies. *Journal of Biopharmaceutical Statistics*, 18, 468–482.

Zhang, W., Sargent, D.J., and Mandrekar, S. (2006). An adaptive dose-finding design incorporating both toxicity and efficacy. *Statistics in Medicine*, 25, 2365–2383.

Zheng, J., Yin, D., Yuan, M., and Chow, S.C. (2019). Simultaneous confidence interval approach for analytical similarity assessment. *Journal of Biopharmaceutical Statistics*, 29, In press.

Zhou, J. Vallejo, J., Kluetz, P., Pazdur, R., Kim, T., Keegan, P., Farrell, A., Beaver, J.A., and Sridhara, R. (2019). Overview of oncology and hematology drug approvals at US Food and Drug Administration between 2008 and 2016. *Journal of the National Cancer Institute*, 111(5), 449–458. doi:10.1093/jnci/djy130.

Zhou, X.H., Li, S.L., Tian, F., Cai, B.J., Xie, Y.M., Pei, Y., Kang, S., Fan, M. and Li, J.P. (2012). Building a disease risk model of osteoporosis based on traditional Chinese medicine symptoms and western medicine risk factors. *Statistics in Medicine*, 31, 643–652.

Zhou, X., Liu, S., Kim, E.S., Herbst, R.S., and Lee, J.J. (2008). Bayesian adaptive design for targeted therapy development in lung cancer-a step toward ersonalized medicine. *Clinical Trials*, 5, 181–193.

Zhou, Y. and Whitehead, J. (2003). Practical implementation of Bayesian dose-escalation procedures. *Drug Information Journal*, 37, 45–59.

Zhou, Y., Whitehead, J., Bonvini, E., and Stevens, J.W. (2006). Bayesian decision procedures for binary and continuous bivariate dose-escalation studies. *Pharmaceutical Statistics*, 5, 125–133.

Zwinderman, A.H. (1990). The measurement of change of quality of life in clinical trials. *Statistics in Medicine*, 9, 931–942.

Index

Note: Page numbers in italic and bold refer to figures and tables, respectively.

A

abbreviated new drug application (ANDA), 387
ABE (average bioequivalence), 390–392
ABP215 (Avastin biosimilar), 52–54
absolute change, 31, 93–94, **94**, 200, 202
absolute *versus* relative difference, 103, **104**
acceptance criteria, 308
active control trials, 31, 123
actual patient population, 209, 352
adaptations, 23–24, 344
adaptive design methods, 23
adaptive dose escalation design, 355–356
adaptive dose finding design, 347–348
adaptive group sequential design, 345, 353
adaptive-hypotheses design, 349–350
adaptive randomization design, 344–345
adaptive seamless design, 350, 463
adaptive treatment-switching design, 349
adaptive trial design, 22–24
 adaptations, 344
 challenges, 352–354
 clinical development, 363–364
 concept, 341–342
 definition, 343–344
 obstacles, retrospective adaptations, 354
 pharmaceutical/clinical development, 341
 potential use, 342
 protocol amendments, impact, 352
 regulatory/statistical perspectives, 351–352
 types, 344–351
adequate and well-controlled study, 48–49, **49**

advisory committees, 40–41
algorithm-based design, 355
alternative enrichment designs, PM
 clinical trials, 428–430
 statistical methods, 425–428
 with/without molecular targets, 423–425, *424*
alternative hypotheses, 8–9, 18
analysis of covariance (ANCOVA), 164
analysis of variance (ANOVA), 159, 183, 440
analysis with covariate adjustment, 259–260
analytical similarity assessment, 54–57
 biosimilars, 405–406
 in CQAs, 27, 51
 evaluation, 58–59
ANCOVA (analysis of covariance), 164
ANDA (abbreviated new drug application), 387
animal model/studies, 333
ANOVA (analysis of variance), 159, 183, 440
antidepressant agents, 15
antimicrobial agents, 15
Asian region, 213–214
aspirin, effect, 6–7, *7*
assay sensitivity, 16
asthma trial, statistics, **258**
auscultation, 297, 324
Avastin biosimilar (ABP215), 52–54
average bioequivalence (ABE), 390–392

B

backward selection process, 440
Barthel index, 302–303

Bayesian approach, 273–274, 416,
        425–426
    bioequivalence/biosimilarity
        index, 403
    CID, 26–27
    innovative trial designs, 466
    rare diseases drug development, 466
    reproducibility probability, 230–232,
        233, 234
bias, 418, 434
    big data analytics, *see* big data
        analytics
    performance, proposed, *452*, **452**,
        453, **453**
    relative, 423
    selection, 434–436
    -variance trade-off, 156
big data analytics
    bias, 444–446, **446**
    data quality/integrity/validity, 436
    diagnosis and validation, 441
    dose titration studies, 454–455
    FDA Part 11 compliance, 437
    generalizability model, 441–442
    heterogeneity, 435
    logistic regression analysis, 438
    meta-analysis, 442–444
    missing data, 437–438
    model building, 439–440
    propensity score matching, 439
    representativeness, 435
    reproducibility/generalizability, 435
    risk factors and clinical outcomes,
        433–434
    selection bias, 435
    simulation study, 449–453
    sources, variation, 454
big data center, 436
Big Data to Knowledge (BD2K)
        initiative, 433, 455
binary responses, 86, **87–88**
BIO (Biotechnology Industry
        Organization), 342
bioavailability, 387, 390
bioequivalence assessment, 68, 81, 243,
        387, 392
    approach for, 400
    for chemical drug products, 409
    development, 400–402

PBE/IBE, 392
    SABE, 392–393
    types, 390
bioequivalence/biosimilarity index, 57,
        59–60
    Bayesian approach, 403
    development, 400–402
    reproducibility probability curves,
        403, 403–404
bioequivalence limit, 68, 391
biological activity, *60*, 402
biological products, 60, 398, 402, 405
Biologics Price Competition and
        Innovation Act of 2009 (BPCI
        Act), 49–50, 395–396
biologic *versus* chemical drugs,
        388–389, **389**
biomarker-adaptive design, 348–349
    targeted clinical trials, 360–363, *362*
biomarker-driven clinical trials, 412–414
biomarkers, 459–460
biosimilar product, 27–28, 47–48, 404;
        see also generic drugs
    assessment, 67
    BPCI Act, 49–50, 396, 405
    development, 50, *50*
    proposed, 58
    versus reference product, 136
biosimilars/biosimilarity, 137; *see also*
        biosimilar product
    analytical similarity assessment,
        405–406
    bridging studies, 408
    criteria, 397, 406–407
    endpoint selection, 405
    extrapolation, 407
    Fundamental Biosimilarity
        Assumption, 404
    interchangeability, 398–399, 407
    non-medical switch, 408
    principles, 396–397
    reference product changes, 407
    regulatory requirement, 395–396
    small molecule drug products, 400
    statistical methods, 398
    study design, 397
Biotechnology Industry Organization
        (BIO), 342
bivariate, 62, 430

blinding, 49, 346
Bonferroni method, 183, 357
bootstrap, 216, 416, 421–423
    confidence interval, 391
    -median approach, 36
    sample size, 421
bridging studies, **239**, 408

C

calibration in clinical development,
    323–326
    CDP, 324–325
    reliability, 328–329
    ruggedness, 329–331
    validity, 326–328
CB approach, *see* confidence bound (CB)
    approach
CDP, *see* Chinese diagnostic
    procedures (CDP)
central laboratory, 265
chemical drugs *versus* biologic drugs,
    388–389, **389**
Chinese diagnostic procedures (CDP),
    294, 297–298, 323–325
Chinese Pharmacopoeia (CP), 333
Chow and Shao's method, 131–132
CI approach, *see* confidence interval (CI)
    approach
CID, *see* complex innovative
    design (CID)
clinical development, 19, 236, 303,
    363–364, 460
    calibration in, 323–331
    Herceptin (trastuzumab), 413
    strategies for, 363–364
clinical difference, 14
clinical endpoints, 93, 302
    selection, 94, 200
    translations, 96–99, 200–202
clinical evaluation, 67, 479
clinically meaningful difference, 50, 70,
    93–94, 179, 197, 294, 303
clinical significance and equivalence,
    14–17
clinical similarity, 28, 52, 54–57
clinical strategy
    comparison, 99–109, 202–205
    endpoint selection, 95–96

clinical trial simulation, 218, 365
co-primary endpoints, 377, **384**,
    384–385, **385**
    single primary endpoint, 377,
    378–379, 380–381, 382–383
closed testing procedure, 185–186
21 Code of Federal Regulation (CFR), 1,
    27, 47–48, 437
Committee for Proprietary Medicinal
    Products (CPMPs), 180
comparison-wise error rate (CWE), 180
complete n-of-1 trial design, 24–25, **25**,
    462, **462**
    coefficients for estimation, carryover
    effect, 468, **469**
    coefficients for estimation, drug
    effect, 467, **468**
    merits and limitations, 463
    sample size requirement, 468–470
    statistical model, 466–467
complete response rate (CRR), 356
complex innovative design (CID), 2
    adaptive trial design, 22–24
    Bayesian approach, 26–27
    master protocol, concept, 26
    n-of-1 trial design, 24–25
composite hypothesis, 182
concurrent adaptations, 23–24
conditional imputation, 159–162
conditional power, 369, 374
conditional probability, 18
confidence bound (CB) approach,
    449–453, *452, 453*
    adjustment approach, **451, 452, 453**
    reproducibility probability,
    228–230, **229**
confidence interval (CI) approach, 2, 28,
    65–66, 78–80, 391
    false negative, 71–72, 72, **72**
    false positive, 72, 74, 74, **74**
    level $1 - \alpha$ *versus* $1 - 2\alpha$, 78
    with multiple reference, 70–76,
    72, **72**
    power, 83, **84–85**
    significance level *versus* size, 79
    with single reference, 69–70
    tests sizes to, 79–80
    TOST *versus*, 76–78
confounding effects, 2–6

consistency
   index, assessment, 267–269
   rates, **279–281, 283–285**
   test, 37–38, 266–267, 332–333
consistency evaluation
   example, 286–290, **289**
   multi-regional clinical trials, issues
      in, 264–266
   simulation study, 276–286
   statistical methods, 266–276
constancy condition, 130, 354
contingency table, 159–162
continual re-assessment method (CRM),
      348, 355–356
controversial issue, 65, 94–95
   in big data analytics, 455
   practical, challenging, and, 27–43
   regulatory perspective and, 180–182
co-primary endpoints, dose selection,
      372–373, **384, 385**
   conditional power, 374
   means and standard errors, 376, **377**
   precision analysis, confidence
      interval, 374–375
   predictive probability of success, 375
   probability, best dose, 375–376
covariate adjustment, 259–260
coverage probability, 368, 423
CP (Chinese Pharmacopoeia), 333
CPMPs (Committee for Proprietary
      Medicinal Products), 180
CQAs (critical quality attributes), 27, 47,
      405, 407
Critical Path Initiative, 22, 341
Critical Path Opportunities List, 22, 341
Critical Path Report, 341
critical quality attributes (CQAs), 27, 47,
      405, 407
CRM (continual re-assessment method),
      348, 355–356
CRR (complete response rate), 356
CWE (comparison-wise error rate), 180

D

data
   compression, 455
   provenance, 455
   quality/integrity/validity, 436

safety monitoring board, 343
   visualization, 455
   wrangling, 455
data monitoring committee
      (DMC), 208
decision-making at interim, 30
decision matrix, 189, **190**, 191
delta method, 256
depression scores, 174, **175**
derived study endpoint, 95–98, 201
design selection, 343, 355
diagnostic devices, 413
disease-free survival, 112
dissolution profile comparison, 397
DLT (dose limiting toxicity), 364
DMC (data monitoring committee), 208
dose de-escalation, 355
dose escalation, 355
   trial, 355–356
dose jump, 355–356
dose limiting toxicity (DLT), 364
dose-response study, 187
dose selection, 368–369
   conditional power, 369
   co-primary endpoints, 372–377, **384,**
      384–385, **385**
   precision analysis, confidence
      interval, 370
   predictive probability, success, 370
   probability, best dose, 370–371
   single primary endpoint, 371–372,
      377, **378–379, 380–381,**
      382–383
   VMS, 376
dose titration studies, 3–4, 454–455
dose-toxicity model, 356
double-blind randomized trial, 237
dropouts, missingness mechanisms, 153
drop-the-losers design, 346–347
   MM trial, 357
drug
   discovery, 293, 298
   interchangeability, 398–399, 407
   prescribability, 390, 400
   switchability, 390
Drug Price Competition and Patent
      Term Restoration Act, 387
drug-to-drug interaction, 301
Dunnett's test, 184–185, 191

E

early efficacy stopping, 473–475
early futility stopping, 475–476
early virologic response (EVR), 358
EC (European Community), 37, 263
Edgeworth-type expansion, 36
efficacy, 174
  assessment, 336–339
  early stopping, 473–475
  endpoint, 6, 34–35, 288, **288**
  health index and measure, 336
  measure, 336
EGFR (epidermal growth factor
      receptor), 260
eight principles, 295, 325
EMA (European Medicines Agency),
      387, 409
EM algorithm, *see*
      expectation–maximization
      (EM) algorithm
EMEA (European Agency for the
      Evaluation of Medicinal
      Products), 180–181
endpoint selection, 29–30, 93–95,
      121–122
  clinical strategies, 95–96, 99–109
  therapeutic index function, 110–121
  translations, 96–99
enrichment design, 361, 413, 415–416
  alternative, 423–430
  clinical trials, *416*
  for patients, *416*, *424*
  for targeted clinical trials, *424*
enrichment process, 360, 412–413, 459
EP approach, *see* estimated power (EP)
      approach
epidermal growth factor receptor
      (EGFR), 260
equivalence, 31–32
  clinical significance and clinical,
      14–17
  criterion, 70–71
  hypotheses for, 125, *126*, 197, *197*
  limits for binary responses, **130**
  non-inferiority *versus*, 124–127
  test, 405–406
E-step, 420
estimand, 165–166

estimated power (EP) approach, 18, 222
  adjustment approach, **451**, *452*, **452**,
      453, **453**
  with equal variances, 222–225
  parallel-group designs, 227
  with unequal variances, 225–227
EU-approved reference product, 70
European Agency for the Evaluation of
      Medicinal Products (EMEA),
      180–181
European Community (EC), 37, 263
European Medicines Agency (EMA),
      387, 409
evidence-based WM, 293, **295**
EVR (early virologic response), 358
expectation–maximization (EM)
      algorithm, 33, 154, 248–249,
      419–423
experience-based TCM, 293, **295**
experiment-wise error rate (EWE), 180
extrapolation, 38–39, 241–242
  sensitivity index, assessment,
      244–253
  statistical inference, 253–258
  target patient population, shift,
      242–243

F

false negative, 71–72
false positive, 72–76
family-wise error rate (FWER), 180
FDA, *see* Food and Drug Administration
      (FDA)
FDA Modernization Act (FDAMA), 17,
      235
FDA Part 11 compliance, 437
feasibility, 152, 365
five-element theory, 295
five Zang, 325
fixed margin method, 131, 135
fixed power approach, 468
fixed sample size approach, 356–357
fixed *versus* flexible dose, 298–299
flexibility, 22–23, 352
flexible SSRE design, 346
FOB, *see* follow-on biologics (FOB)
follow-on biologics (FOB), 387–388, 397,
      399–400, 404

Food and Drug Administration (FDA),
        1, 195
    advisory committees, 40
    biosimilar regulatory submissions, 52
    critical clinical initiatives, 41–43
forward selection process, 440
frequentist approach, 237
functional characterization, 396
Fundamental Bioequivalence
        Assumption, 387, 395, 404
Fundamental Biosimilarity
        Assumption, 52, 404
FWER (family-wise error rate), 180

G

gamma distribution, 232, 247, 426
gate-keeping procedures, **190**
    multiple endpoints, 187–188
    testing, 188–192
GCP (Good Clinical Practice),
        437–438
GDMP (Good Data Management
        Practice), 437
gene expression, 429
generalizability, 17, 19–22, 38–39,
        241, 460
    model, 441–442
    probability, 238, **239**
    and reproducibility, 436
generalized estimating equations
        (GEE), 164
generic drugs, 67, **80**
    bioequivalence assessment, 392–395
    PK parameters, 289
    statistical methods, 391
    study design, 390
genomic data, 408
global clinical trial, 211, 275
global dynamic balance, 296, 335
    and harmony, 39
gold standard, 31, 124, 301–302
Good Clinical Practice (GCP),
        437–438
Good Data Management Practice
        (GDMP), 437
goodness-of-fit tests, 161–162, 441
Good Statistics Practice (GSP), 1–2, 437
group sequential design, 345

H

Hamilton depression (Ham-D) scale,
        15–16, **16**
Hatch-Waxman Act, 387
health index/efficacy measure, 336
hepatitis C virus (HCV), 358–360
HER2 over-expression, treatment
        effects, 413, **413**
Herceptin (trastuzumab), 413
    biosimilar (MYL-1401O), 52–54
    plus chemotherapy, 413
heterogeneity, 435
HGP (Human Genome Project), 360
highly similar, 49–50, 396, 405
historical data, 133, 458–459
    statistics, **133**
Hochberg's method, 187
Holm's method, 186–187
Hommel's method, 186–187
Human Genome Project (HGP), 360
human model, 333
hypothesis setting, non-inferiority
    and clinically meaningful margin,
        128–129
    regulatory requirements, 127–128
    retention ratio, 129
hypothesis(es) testing, 8–9, 66–67,
        197, 311
    interval, 68–69
    LOCF analysis, 157–158
    point, 67
    probability, inconclusiveness, 69

I

IBE (individual bioequivalence), 390,
        392–393
ICH E9 guidelines, 180
ICH (International Conference on
        Harmonization) guideline, 32,
        124, 127
identity, 323
IDMC (independent data monitoring
        committee), 364
imaging medicine (IM), 2, 43
immune response, 389
immunogenicity, 47, 50, 395–396
inconsistent, 37, 58, 71, 455

IND (Investigational New Drug Application), 40

independence, test for, 162–164

independent data monitoring committee (IDMC), 364

indication, 334

individual bioequivalence (IBE), 390, 392–393

individualized medicine approach, 40

inequality test
 reproducibility probability ratio for, 272–273
 specificity reproducibility probability for, 270–271

inflation factor, 243, 269, 441

innovative trial designs
 Bayesian approach, 466
 master protocol, 464–465
 n-of-1 trial design, 461–463
 two-stage seamless adaptive designs, 463–464, **464**

inspection, 297, 324

instability, sample size, 215–216

institutional review board (IRB), 459

intention-to-treat analysis, 6

intent-to-treat (ITT) principle, 33, 154

interaction effects, 3, 6–8

interchangeability, 398–399

interchangeable product, 398

interim analysis, 205, 357–358, 376

International Conference on Harmonization (ICH) guideline, 32, 124, 127

interrogation, 297

interval hypotheses testing, 68–69, 391

inverse gamma distribution, 247

Investigational New Drug Application (IND), 40

in vitro diagnostic multivariate index assays (IVDMIA), 361

IRB (institutional review board), 459

item non-respondents, 33, 154

IVDMIA (in vitro diagnostic multivariate index assays), 361

J

jackknife method, 159

Japanese approach, 264, 275

K

Kronecker product, 161, 172

L

label, 334

last observation carried forward (LOCF), 34, 155–156
 bias-variance trade-off, 156
 hypothesis testing, 157–158

least squares (*LS*) estimator, 168–169

likelihood-based method, 33, 154

linearization method, 33, 154

LOCF, *see* last observation carried forward (LOCF)

logistic regression model, 438, 440

log-transformation, 82, 196, 253

M

machine learning, 43

manufacturing process, 47, 388–389

MAR (missing at random), 32, 153

marginal/conditional imputation, 159–160
 goodness-of-fit test, 161–162
 simple random sampling, 160–161

margin selection strategy, 135–136
 numerical studies, 143–149
 RED, 150
 risk assessment criteria, 136–138
 risk assessment with continuous endpoints, 138–143
 RR, 150–151
 SSR, 150
 TERI, 151

masking effect, 243

master protocol, 26, 464–465

matching placebo, 303

maximum likelihood, 165, 403

maximum likelihood estimator (MLE), 33, 154

maximum tolerated dose (MTD), 347, 355–356, 363

MCAR (missing completely at random), 32, 153–154

mean/median imputation, 158–159
mean/standard deviation,
    effects, **238**
mechanism of action (MOA), 27
medical theory/mechanism and
    practice, 295–296
meta-analysis, big data, 442–444
MIDD (model-informed drug
    development), 42, 42
MIM (mobile individualized medicine),
    2, 41
minimum sample size, 36, 196, 216
missing at random (MAR), 32, 153
missing completely at random (MCAR),
    32, 153–154
missing data, 153–155, 176–177,
    437–438, 456
    estimand, 165–166
    imputation, 155–159
    marginal/conditional imputation,
        159–162
    methods for, 164–165
    statistical methods, 166–176
    testing for independence, 162–164
    treatment, 32–36
missing data imputation
    LOCF, 155–158
    mean/median, 158–159
    regression, 159
missingness mechanisms, 153
missing not at random (MNAR),
    32, 153
missing value, 32–34, 153–154; *see also*
    missing data
mixed effects model for repeated
    measures (MMRM), 165
mixed non-inferiority margin, 152
mixed null hypothesis, 152
mixture distribution, 39, 209, 242
MLE (maximum likelihood estimator),
    33, 154
MM (multiple myeloma), 357, 363
MMRM (mixed effects model for
    repeated measures), 165
MNAR (missing not at random),
    32, 153
MOA (mechanism of action), 27
mobile individualized medicine (MIM),
    2, 41

model-informed drug development
    (MIDD), 42, 42
modernization, 293–294
molecular target drug, patient
    population
    Bernoulli random variables, PPV,
        419–420
    complete-data log-likelihood
        function, 19
    confidence interval, 418
    diagnosis, 417–418
    EM algorithm, 419, 421–422
    means and variances, 421
    parametric bootstrap method,
        421–422
    unpaired two-sample t-test
        approach, 418
moment-based criteria, 59
Monte Carlo method, 232, **233**
moving target patient population, 209,
    352–353, 365
MRCT (multi-regional clinical trial), 37,
    264–266
M-step, 420
MTD (maximum tolerated dose), 347,
    355–356, 363
multi-center trials, 264–266
multiple adaptive design, 350–351
multiple comparisons, 179, 188, 192
multiple components, drug products
    with, 39–40
multiple endpoints, 187–188
multiple imputation, 164
multiple inferences, 179
multiple myeloma (MM), 357, 363
multiple-stage adaptive designs,
    205–208, 350–351, 353
    NHL trial, 356–357
multiple testing procedure, 191
multiplicity
    controversial issues, 181–182
    gate-keeping procedures, 187–192, **190**
    overview, 179–180
    regulatory perspectives, 180–181
    statistical method for adjustment,
        182–187
multi-regional clinical trial (MRCT), 37,
    264–266
multi-regional trials, 37, 263

N

NAFLD Activity Score (NAS), 29
NASH (Non-Alcoholic SteatoHepatitis)
    study, 359–360, *360*
New Drug Application (NDA), 40
new drugs *versus* generics/biosimilars,
    CI for, 28–29
Newton-Raphson algorithm, 249
NHL (Non-Hodgkin's Lymphoma) trial,
    356–357
n-of-1 trial design, 24–25
Non-Alcoholic SteatoHepatitis (NASH)
    study, 359–360, *360*
Non-Hodgkin's Lymphoma (NHL) trial,
    356–357
non-inferiority, hypotheses for, *197*
non-inferiority margin, 123–124,
    151–152
    determination, 102
    versus equivalence, 124–127
    hypothesis, 127–129
    methods, selection, 130–135
    selection, 31
    strategy, selection, 135–151
non-inferiority testing
    on absolute and relative difference,
        106, 108
    on relative difference, **105, 107, 109**
non-inferiority trials, 123–124, 133
    endpoint selection in, **96**
non-medical switch, 408
non-respondent, 33, 154, 160
normal data/superiority test
    matched-pair parallel design, case,
        276–277
    two-group parallel design, case,
        281–282, **282**
null hypotheses, 8–9, 18
numerical study, 102
    absolute *versus* relative difference,
        103, **104**
    responders' rate on, 103

O

objective *versus* subjective criteria, 297
O'Brien-Fleming method, 358
observed *p*-value, 11

ODAC (Oncologic Drugs Advisory
    Committee), 70
olfaction, 297, 324
Oncologic Drugs Advisory Committee
    (ODAC), 70
one sample problem, 37
one-sided equivalence testing, 125
one-sided test/hypotheses, 12
    proof level for, **13**
    sampling distribution, *13*
    versus two-sided, 12–14
one-size-fits-all approach, 411
one-way ANOVA, 184
    with repeated measures, 34
Orphan Drug Act of 1983, 457, 460
overall response rate (ORR),
    356–357
overall survival (OS), 112, **113**
over-expression, 413, **413**

P

pairwise comparisons method, 70,
    73, 75
pair-wise conditional technique, 164
PANSS (positive and negative symptom
    score), 20, **21**
parallel design, 196
    -group designs, 227
    matched-pair case, 225, 276–277
    two-group case, 228, 235, 281–282,
        282, 416
parametric bootstrap method, 421–422
pattern-mixture models, 165
PBE (population bioequivalence),
    390, 392
PCI (predictive confidence interval), 477
Pearson correlation coefficient, 440
performance characteristics, 81–82
personalized medicine approach, 40
    versus PM, 414, **415**
PFS (progression-free survival),
    356–357
PHA (Public Health Agency), 387
pharmaceutical development process, 2,
    339–340
Pharmaceutical Research and
    Manufacturers of America
    (PhRMA), 342, 344

pharmacokinetic and pharmacodynamic
(PK/PD), 27–28, 47
assessment, *52*
parameters, 289
similarity, 48, 51
phase I, 355, 363
phase II, 15, 260, 350, 364, 376
phase III, 350, 356–357, 364, 376
phase I/II, 350, 363, 464
phase II/III, 41, 350, 358–360, 429
phase II/III/IV, 359–360
PK/PD, *see* pharmacokinetic and
pharmacodynamic (PK/PD)
placebo-control clinical trial, 258
placebo-controlled trials, rare
diseases, 459
platform trials, rare diseases drug
development, 465, **465**
PM, *see* precision medicine (PM)
point hypotheses testing, 67
polio-vaccine trial, 3
population bioequivalence (PBE), 390, 392
population means, treatment/diagnosis,
416, **417**, **425**
positive and negative symptom score
(PANSS), 20, **21**
positive predicted value (PPV), 415, 417,
419–420
potency, 50
potential biases, big data analytics,
445–446, **446**
power analysis, 195
power and sample size determination,
99–102
power function $p(\theta)$, values, **224**
PPV (positive predicted value), 415, 417,
419–420
precision analysis, 195, 347, 367
confidence interval, 370–371, 374–375
precision medicine (PM), 412
alternative enrichment designs,
423–430
biomarker-driven clinical trials,
412–414
definition, 412
development, 431
enrichment design, 415–416, *416*
Obama's Precision Medicine
Initiative, 411, 430

versus personalized medicine, 414, **415**
population means, treatment/
diagnosis, 416, **417**
simulation results, 422–423
statistical methods, 417–422
predictive confidence interval (PCI), 477
predictive model, 43, 219, 348, 438, 441
predictive probability, success, 367, 370,
372, 375
prescribability, 297, 399
primary endpoints, 110, 350
co-, 372–377, 384–385
efficacy, *288*, **288**
single, 371–372, 377
probability
-based criteria, 59
generalizability, 460
inconclusiveness, 69
monitoring procedure, 215, 481–482
proceeding, 206
reproducibility, 477–478
statement, 195
success, 293, 370, 372, 375
process validation, 340
progression-free survival (PFS),
356–357
propensity score matching, 439
proposed bias adjustment approach
different power, 450, **451**, *452*
negative studies, **452**, *453*
number, studies, 450, **453**
prospective adaptations, 23
protocol amendments, 208–211, 216
Public Health Agency (PHA), 387
pulse taking and palpation, 297, 324
purity, 50
p-values, 10–12

Q

QC (quality control), 304
QOL (quality of life), 8, 31, 123
QR (quality range) method, 405–406
quality, 26, 130, 295, 333, 340, 396,
408, 434
quality attribute, 52, 395, 406
quality control (QC), 304
quality of life (QOL), 8, 31, 123
quality range (QR) method, 405–406

quantitative evaluation
    biosimilars, 395–400
    drug safety and efficacy, 387
    generic drugs, 389–395
quantitative instrument, validation,
    301–302

R

randomized clinical trial (RCT), 39,
    42, 335
random target patient population, 216
rare diseases drug development
    biomarkers, 459–460
    clinical trials, evaluation, 476–478
    defined, 457
    effectiveness/not ineffectiveness,
        478–480, *479*
    ethical consideration, 459
    FDA's incentives program, 457
    generalizability, 460
    historical data, 458–459
    innovative trial designs, 461–466
    probability monitoring procedure,
        sample size, 481–482
    sample size, 460–461
    statistical methods, data analysis,
        466–476
    two-stage adaptive trial design,
        480–481
RCT (randomized clinical trial), 39,
    42, 335
Real Word Evidence (RWE), **42**
real world data, 41–43
reduction, power, 176
reference product, 59–60, 396, 407
regional requirements, **287**
regression imputation, 159
regulatory process, 40
regulatory requirements, 331–332
regulatory statistics, 1
relative change, 93–94
relative difference in power (RED), 137,
    140–141, 145–146, *147,* 150
relative ratio in power/relative risk (RR),
    137–138, 142, 144, 146, 150–151
reliability, 156, 301, 328–329
repeated measures, 34
representativeness, big data, 435, 455

reproducibility, 17–19; *see also*
        reproducibility probability
    analysis, 215
    clinical results, 19
reproducibility probability, 18, **19**, 59–60
    applications, 235–239
    Bayesian approach, 230–232, **233**, 234
    concept, 220–222
    confidence bound approach,
        228–230, **229**
    with equal variances, 222–223, 225
    parallel-group designs, 227
    with unequal variances, 225–227
re-sampling method, 33–34, 155
responder analysis, 93, 121, 200
response-adaptive randomization, 345
response rate (RR), 110, 112, **113**
responsiveness, 301
restricted maximum likelihood, 32, 153
retention ratio, 129
retrospective adaptations, 24, 354
risk and sample size, 147–148
risk assessment, 136
    with continuous endpoints, 138–143
    criteria, 136–138
robustness, 36, 166
Rom's method, 187
RR, *see* relative ratio in power/relative
        risk (RR); response rate (RR)
ruggedness, 329–331

S

SABE (scaled average bioequivalence),
        392–393
sample size, 195–196
    adjustment, 208–211, 216
    allocation, 212–213, 351
    calculation, 196–200, **199**, 303–304
    on confidence interval approach,
        216–217
    instability, 215–216
    multiple-stage adaptive designs,
        205–208
    multi-regional clinical trials, 211–214
    power calculation, 214–215
    with protocol amendments,
        208–211, 216
    reproducibility probability, 236

sample size (*Continued*)
  requirement, 36–37, 126–127
  study endpoints, selection, 200–205
sample size difference (SSD), 137, *144*
sample size ratio (SSR), 136–137, 139–140,
    143, 150
sample size re-estimation (SSRE)
    design, 346, 480–481
sampling plan, 308–311, **310**
SAP (statistical analysis plan), 349
scale change, 243
scaled average bioequivalence (SABE),
    392–393
scaled criterion for drug
    interchangeability (SCDI),
    393–395
seamless adaptive design, 350
seamless adaptive trial design, 350
Second International Study of Infarct
    Survival (ISIS-2 Group), 6–7
  cumulative vascular mortality, **7**
  treatment, **7**
selection bias, 435
semi-parametric, 32, 153
sensitivity analysis, 36, 165–166
sensitivity index, 39, 243, **243**
  evaluation, 269–270
shelf-life determination, 320–321
shelf-life estimation, 333–334
shift in location parameter, 39, 242–244,
    269, 441
shift in population mean, 39, 208–209, 269
Shih's method, 285–286
signal-to-noise ratio, 20–21
significance level, 79
Simes' method, 186
similar biological drug products
    (SBDP), 387
similarity, 27–28, 242
  degree, 405
  margin selection, 31–32
simple linear regression model, 326
simple random sampling, 160–161
simulation studies, 82–83, 86
  abbreviations list, **278**
  big data analytics, 449–453
  clinical trial, 377–385
  consistency rates, **279–281, 283–285**
  global and regions, **278, 279, 282, 283**

matched-pair parallel design, 276–278
  parameter specification for, **83**
  PM, 422–423
  subgroup analysis, categories, **277**
  two-group parallel design, 281–285
simultaneous CI approach, 70–71, **73, 75**
single active ingredient *versus* multiple
    components, 298
single primary endpoint, dose
    selection, 377
  conditional power, 371–372
  precision analysis, confidence
      interval, 371
  predictive probability of success, 372
  simulation results for, **378–379,**
      380–381, 382–383
Slutsky theorem, 100, 203, 317
Spearman rank correlation
    coefficient, 440
SSD (sample size difference), 137, *144*
SSR (sample size ratio), 136–137, 139–140,
    143, 150
SSRE (sample size re-estimation) design,
    346, 480–481
stability analysis, 317–319
  discussion, 323
  models and assumptions, 319–320
  shelf-life determination, 320–321
standard error, 132, 138, 176
statistical analysis plan (SAP), 349
statistical difference, 179
statistical methods, 1
  achieving reproducibility/
      generalizability, 270–273
  adaptive trial design, 471–476
  applicability, 275–276, **276**
  biosimilars, 398
  complete *n*-of-1 trial design, 466–471
  consistency index, assessment, 267–269
  consistency test, 266–267
  estimation of Δ, 446–447
  estimation of $\mu_P - \mu_N$, 448
  generic drugs, 391
  PM, 417–422, 425–428
  sensitivity index, evaluation, 269–270
statistical methods, incomplete data
    structure, 166–168
  special case, 172–174
  2 × 3 crossover designs, 168–172

statistical principles, 1–2
statistical quality control, strategy for, 311, 314–315
statistics, 1
stepwise approach, 27, 47, 49–50, *50*, 395–396
stopping boundaries, 207, 345, 472–474
two-stage efficacy and futility designs, **475**
for two-stage efficacy designs, **474**
for two-stage futility design, **476**
stopping rule, 206, 208, 472
Stratified samplings, 162–163
strength, 12
streptokinase, effect, 6–7, **7**
strict TER (STER), 356
structural characterization, 47
study endpoints, calibration
calibration, 325–326
CDP, 323–325
reliability, 328–329
ruggedness, 329–331
validity, 326–328
subgroup analysis, categories, **277, 287**
substantial evidence, 1, 12, 48–49
totality-of-the-evidence *versus*, 57–58
success rate, 22, 341
superiority, 123–125, *126*
hypotheses for, *197*
hypothesis test, 281
reproducibility probability, 271–272
test, reproducibility probability ratio for, 273
trial, 12
supplement new drug application (SNDA), 237
survival analysis method, 32, 153
sustained virologic response (SVR), 358
switchability, 390, 392, 397, 399
synthesis method, 131

T

TAILORx (Trial Assigning Individualized Options for Treatment) trial, 360
targeted clinical trials, 360–363, *362*
target patient population, 208, 439
protocol amendments, 352
shift in, 209, 242–243

Taylor expansion, 117
TCM, *see* traditional Chinese medicines (TCM)
TCM drug development
stability analysis, 317–319
statistical QC method, 304–308, **312–313**
study endpoints, calibration, 323–331
TER, *see* traditional dose escalation rule (TER)
TERI, *see* type I error inflation (TERI)
test drug, 197
test for consistency, 266–267, 332–333
testing for independence, 162
large strata (*H*), 163–164
stratified samplings, 162–163
testing procedure, 311
closed, 185–186
gate-keeping, 188–192
multiple, 191
Tukey's multiple range, 184
test product, 68, 390, 400; *see also* biosimilar product
test statistics, 99–102
test treatment, 5, 16, 31, 48–49, 96, 123, 303, 411
therapeutic index function, 110–112
e, distribution, 115
endpoints, sample size calculation, **113**
$f_i(.)$, determination, 115
inequation and parameter settings, **119–120**
ω_i, selection, 114–115
Pr $(I_i \mid e_j)$ and Pr $(e_j \mid I_i)$, 115–118
therapeutic window, 392
time to disease progression (TTP), 112, **113**
titration trial, graphical display, 4, *4*
TOST, *see* two one-sided tests (TOST)
totality-of-the-evidence, 27–28
analytical, PK/PD and clinical similarity, 54–57
assessment, **53, 57**
biosimilarity assumptions, 51–52
biosimilar regulatory submissions, 52–53
development, 59–62
overview, 47–48
regulatory standards, 58–59

totality-of-the-evidence (*Continued*)
    stepwise approach, 49–50
    versus substantial evidence, 57–58
traditional Chinese medicines (TCM),
        39–40, 293
    clinical trial, 299–300, 303
    components, **322**
    elements, 296
    experience-based, 293
    in humans, 293
    stability data, **322**
    study design, 300–301
    Westernization, 294
    WMs *versus*, 294, **295**
traditional dose escalation rule (TER), 355
    dose de-escalation, 355
    3 + 3 TER, 355
traditional medicine, 411, **415**, 430
training data set, 441
translations, clinical endpoints, 96–99,
        200–202
treatment-by-center interaction, 8,
        265, 444
treatment imbalance, 444
TST, *see* two-sided test (TST)
TTP (time to disease progression), 112, **113**
Tukey's multiple range testing
        procedure, 184
2 × 3 crossover designs, statistical
        methods for, 168–172
two one-sided tests (TOST), 29, 58, 68,
        216, 391, 398
    versus CI approaches, 76
    procedure, 76–78
two-sample problem, 37
two-sided test (TST), 10, 29, 58
    one-sided *versus*, 12–14
    sampling distribution, *11*
two-stage adaptive design, 345, 473–476
two-stage seamless adaptive design,
        208, 358–360, 463, **464**
two-stage seamless adaptive trial
        designs, *359*, 367–368
    categories, 464
    early efficacy/futility stopping,
        474–475, **475**

early efficacy stopping,
        473–474, **474**
early futility stopping, 475–476, **476**
    HCV trial, 358–360
    phases, 463
    p-values, 470–473
    for rare disease product
        development, 480–481
    types, 464, **464**
two-way contingency tables,
        159–160
type I and type II errors, 8–9, *9*, **9**,
        82–83, **84**
    for binary responses, 86, **87**
type I error inflation (TERI), 138,
        142–143, 148–149, 151

U

unbalance, 168
uniform prior, 356
United States Food and Drug
        Administration guidance,
        127–128, 130–131
unit non-respondents, 33, 154
US-licensed drug product, 27, 63
utility-adaptive randomization, 345
utility function, 29–30

V

validation data set, 441
validity, 28, 52, 326–328, 436
valid statistics, 1
variability, 36, 59, 220–221, 397,
        406–407, 414
vasomotor symptoms (VMS), 376

W

Westernization, 293–294, 339
Western medicines (WMs), 295,
        298, 303
    approach, 299
    evidence-based, 293
    versus TCM, 294, **295**

Printed in the United States
by Baker & Taylor Publisher Services